Lecture Notes in Computer Science 16687

The series Lecture Notes in Computer Science (LNCS), including its subseries Lecture Notes in Artificial Intelligence (LNAI) and Lecture Notes in Bioinformatics (LNBI), has established itself as a medium for the publication of new developments in computer science and information technology research, teaching, and education.

LNCS enjoys close cooperation with the computer science R & D community, the series counts many renowned academics among its volume editors and paper authors, and collaborates with prestigious societies. Its mission is to serve this international community by providing an invaluable service, mainly focused on the publication of conference and workshop proceedings and postproceedings. LNCS commenced publication in 1973.

Alexander Kazakov · Michael Khachay ·
Yury Kochetov · Panos Pardalos
Editors

Mathematical Optimization Theory and Operations Research

25th International Conference, MOTOR 2026
Irkutsk, Russia, July 6–11, 2026
Proceedings

Springer

Editors

Alexander Kazakov
Matrosov Institute for System Dynamics and Control Theory
Irkutsk, Russia

Michael Khachay
Krasovsky Institute of Mathematics and Mechanics
Yekaterinburg, Russia

Yury Kochetov
Sobolev Institute of Mathematics
Novosibirsk, Russia

Panos Pardalos
University of Florida (USA), HSE University (Russia)
Nizhny Novgorod, Russia

ISSN 0302-9743 ISSN 1611-3349 (electronic)
Lecture Notes in Computer Science
ISBN 978-3-032-31389-8 ISBN 978-3-032-31390-4 (eBook)
https://doi.org/10.1007/978-3-032-31390-4

This Springer imprint is published by the registered company Springer Nature Switzerland AG
The registered company address is: Gewerbestrasse 11, 6330 Cham, Switzerland

Preface

This volume contains the refereed proceedings of the 25th International Conference on Mathematical Optimization Theory and Operations Research (MOTOR 2026) [1] held during July 6–11 in a picturesque place near Lake Baikal, Irkutsk, Russia.

The MOTOR conference is a direct successor and scientific inheritor of several prominent events on mathematical programming, combinatorial and stochastic optimization, operations research, and applications that have taken place in the Soviet Union and Russia, namely in the Urals, Siberia, and the Russian Far East over the decades. Earlier conferences in the MOTOR series were hosted in Novosibirsk, Omsk, Yekaterinburg, Irkutsk, and Petrozavodsk.

As per tradition, the conference scope included, but was not limited to, mathematical programming, discrete and combinatorial optimization, stochastic programming, scheduling, approximation algorithms, heuristics and metaheuristics, operations research, optimal control and game theory, mathematical economics, and various applications in industry and artificial intelligence.

In response to the call for papers, MOTOR 2026 received 167 submissions. Out of 56 full papers considered for review (111 abstracts and short communications were excluded for formal reasons), only 26 full papers and 2 short papers were selected by the Program Committee (PC) for publication in this volume. Each submission was single-blind reviewed by at least three PC members or invited reviewers, experts in the topic in question, in order to ensure the high quality of the results and supply helpful comments for the authors.

The conference featured five invited lectures:

- Saeid Alikhani, Yazd University, Iran
  Majority distinguishing colorings of graphs
- Mikhail Kovalev, United Institute of Informatics Problems, National Academy of Sciences, Belarus
  Routing and charging scheduling for regular electric buses
- Huynh Thi Thanh Binh, Hanoi University of Sciences and Technology, Vietnam
  Evolutionary multitasking recent advances and applications
- Mikhail Marchenko, Institute of Computational Mathematics and Mathematical Geophysics, Siberian Branch of the Russian Academy of Sciences, Russia
  Parallelization in statistical modeling methods
- Elena Konstantinova, Sobolev Institute of Mathematics, Siberian Branch of the Russian Academy of Sciences, Russia
  CayleyPy project: Artificial intelligence methods for Cayley graphs

and two Tutorial lectures:

- Alexander Strekalovsky, Matrosov Institute of System Dynamics and Control Theory, Siberian Branch of the Russian Academy of Sciences, Russia

[1] https://conference.icc.ru/event/9/page/66-about.

New principles of non-convex optimization

– Anton Eremeev, Sobolev Institute of Mathematics and Mechanics, Siberian Branch of the Russian Academy of Sciences, Russia
  Expected runtime analysis of evolutionary algorithms

A special *Round Table* session for business applications was organized to improve communication between the academic and industrial communities. The following main presentation was invited:

– Alexey Chernov, LLC "Quantum Algorithms and Optimization", Russia
  The universal solver QORETEX (Orsima). What was changed over the past year

We thank the authors for their submissions, members of the Program Committee, and all the external reviewers for their efforts in providing exhaustive reviews. We thank our sponsors and partners: Mathematical Center in Akademgorodok (Novosibirsk) and Ural Mathematical Center (Yekaterinburg), Novosibirsk State University, the Higher School of Economics (Nizhny Novgorod), Teraplan (Moscow), and BIA Technologies. We are grateful to the staff of Springer LNCS for their kind and helpful support.

July 2026

Alexander Kazakov
Michael Khachay
Yury Kochetov
Panos Pardalos

Organization

Program Committee Honorary Chair

Panos Pardalos	University of Florida, USA HSE University, Russia

Program Committee Chairs

Alexander Kazakov	Matrosov Institute for System Dynamics and Control Theory, Russia
Michael Khachay	Krasovsky Institute of Mathematics and Mechanics, Russia
Yury Kochetov	Sobolev Institute of Mathematics, Russia

Program Committee Members

Kamil Aida-zade	Institute of Control Systems, Azerbaijan National Academy of Sciences, Azerbaijan
Saeid Alikhani	Yazd University, Iran
Anatoly Antipin	FRC Computer Science and Control, Russia
Alexander Arguchintsev	Irkutsk State University, Russia
Adil Bagirov	Federation University Australia, Australia
Olga Battaïa	KEDGE Business School, France
Rene van Bevern	Novosibirsk State University, Russia
Christian Blum	Artificial Intelligence Research Institute (IIIA-CSIC), Spain
George Bolotashvili	Georgian Technical University, Georgia
Vladimir Bushenkov	University of Évora, Portugal
Maxim Buzdalov	Aberystwyth University, UK
Igor Bychkov	Matrosov Institute for System Dynamics and Control Theory, Russia
Igor Bykadorov	Sobolev Institute of Mathematics, Russia
Xujin Chen	Institute of Applied Mathematics, CAS, China
Yukun Cheng	Jiangnan University, China
Tatiana Davidović	Mathematical Institute of the Serbian Academy of Sciences and Arts, Serbia
Alexander Dolgui	IMT Atlantique, France
Anton Eremeev	Dostoevsky Omsk State University, Russia
Vladimir Erokhin	Mozhaisky Military Space Academy, Russia

Additional Reviewers

Industry Session Chair

Igor Vasilyev	Matrosov Institute for System Dynamics and Control Theory, Russia

Organizing Committee

Alexander Kazakov (Chair)	Matrosov Institute for System Dynamics and Control Theory (ISDCT SB RAS), Irkutsk
Maria Barkova	ISDCT SB RAS, Irkutsk
Irina Gribanova	ISDCT SB RAS, Irkutsk
Stepan Kochemazov	ISDCT SB RAS, Irkutsk
Victor Kondratiev	ISDCT SB RAS, Irkutsk
Alexey Kumachev	ISDCT SB RAS, Irkutsk
Alexandra Lomukhina	ISDCT SB RAS, Irkutsk
Taras Madzhara	ISDCT SB RAS, Irkutsk
Nadezhda Maltugueva	ISDCT SB RAS, Irkutsk
Timur Medvedev	Higher School of Economics, Nizhny Novgorod
Ilya Minarchenko	Melentiev Energy Systems Institute, Irkutsk
Maria Okhotina	ISDCT SB RAS, Irkutsk
Andrei Orlov	ISDCT SB RAS, Irkutsk
Ilya Otpuschennikov	ISDCT SB RAS, Irkutsk
Denis Sidorov	Sino-Russian International Cooperation Laboratory, China
Maxim Staritsyn	ISDCT SB RAS, Irkutsk
Alexander Stolbov	ISDCT SB RAS, Irkutsk
Anton Ushakov	ISDCT SB RAS, Irkutsk
Igor Vasiliev	ISDCT SB RAS, Irkutsk
Oleg Zaikin	ISDCT SB RAS, Irkutsk
Maxim Zharkov	ISDCT SB RAS, Irkutsk

Organizers

Russian Operations Research Society
Matrosov Institute for System Dynamics and Control Theory, Irkutsk
Mathematical Center in Akademgorodok, Novosibirsk
Novosibirsk State University
Sobolev Institute of Mathematics, Novosibirsk
Krasovsky Institute of Mathematics and Mechanics, Ekaterinburg
Ural Mathematical Center, Ekaterinburg
Higher School of Economics, Nizhny Novgorod

Abstracts of Invited Talks

Majority Distinguishing Colorings of Graphs

Saeid Alikhani

Yazd University, Yazd, Iran
alikhani206@gmail.com

Abstract. We study graph colorings that simultaneously satisfy two constraints: the majority condition and the distinguishing property. A majority coloring requires that, at each vertex, no color appears on more than half of the adjacent vertices or incident edges, while a distinguishing coloring breaks all nontrivial automorphisms of the graph. Motivated by recent work on majority distinguishing edge colorings, we introduce and formalize the concept of majority distinguishing vertex colorings. Accordingly, we define and study two new graph parameters, called the majority distinguishing number and the majority distinguishing index. We determine exact values of these parameters for several fundamental graph families, including paths, cycles, friendship graphs, book graphs, and certain classes of trees. In addition, we investigate how these parameters behave under the corona product of graphs and establish general bounds and relationships with classical distinguishing parameters. Our results extend and unify previous studies on symmetry breaking and majority colorings, and they suggest several open problems and directions for further research.

Keywords: graph colorings · symmetry breaking · majority condition

Routing and Charging Scheduling for Regular Electric Buses

Mikhail Y. Kovalyov

United Institute of Informatics Problems, National Academy of Sciences of Belarus, Minsk, Belarus
kovalyov_my@yahoo.co.uk

Abstract. This work is inspired by the results of the project "Planning Process and Tool for Step-by-Step Conversion of the Conventional or Mixed Bus Fleet to a 100% Electric Bus Fleet" of the Electric Mobility Europe initiative. The public transport operators involved in this project faced the challenge of effectively routing and charging electric buses operating on a fixed timetable. They have all the data to make appropriate decisions. Currently, decisions are made primarily based on previous routing schemes and the "first come, first charged" principle. The number of chargers at a charging station is determined by the maximum number of electric buses that can simultaneously reside at the station between trips. We provide an overview of research in the field of routing and charging scheduling for urban electric buses. We discuss the relationship between these problems and classic vehicle routing and machine scheduling problems.

Keywords: public transport · efficient routing · charging schemes

Evolutionary Multitasking: Recent Advances and Applications

Huỳnh Thị Thanh Bính

School of Information and Communication Technology, Hanoi University of Science and Technology, Hanoi, Vietnam
binhht@soict.hust.edu.vn

Abstract. Evolutionary multitasking optimization is a cutting-edge topic in the field of computational intelligence that merges evolutionary computation and multitasking methodologies to address multiple optimization problems concurrently. By leveraging the interactions and dependencies between problems, evolutionary multitasking allows for the sharing and transfer of valuable information between tasks, thereby enhancing the overall search performance. This talk will provide an overview of the fundamental concepts, principles, and recent advancements of evolutionary multitasking optimization. Additionally, the talk will highlight the practical significance of evolutionary multitasking algorithms and demonstrate how these techniques can effectively tackle complex real-world optimization problems. Through these problems, the talk will delve into algorithm design issues such as solution encoding and the knowledge transfer mechanism in the multitasking environment. Performance evaluations on benchmark datasets will also be presented to demonstrate the effectiveness of evolutionary multitasking algorithms.

Keywords: computational intelligence · multitasking optimization · evolutionary algorithms

Parallelization of Statistical Modeling Methods

Mikhail A. Marchenko

Institute of Computational Mathematics and Mathematical Geophysics SB RAS, Novosibirsk, Russia
marchenko@sscc.ru

Abstract. The talk will focus on methods of parallelization of statistical modeling algorithms. The main emphasis is placed on the problems of statistical modeling of kinetic processes, which is due to their practical significance. An important area touched upon in the talk is the optimization of parallel statistical modeling algorithms in order to reduce their computational complexity. The level of performance of modern supercomputers makes the use of probabilistic models of kinetic processes extremely relevant. The first reason is that such models adequately describe physical phenomena considering the influence of rare events, which is practically impossible for other approaches. The next reason is that they can be effectively implemented as parallel programs. In numerical statistical modeling applications, it is necessary to use parallel pseudorandom number generators, effectively implemented on multiprocessor computers. Then it is necessary to apply a distributed computing methodology that allows for parametric analysis of probabilistic models. The last problem is related to the study of the characteristics of probabilistic models, such as estimation error and computational labor intensity, depending on the parameters of the problem and the algorithm. To solve these problems, it is also necessary to have standard parallelized software tools.

Keywords: statistical modeling · parallel algorithms · distributed computations

CayleyPy Project: Artificial Intelligence Methods for Cayley Graphs

Elena V. Konstantinova

Sobolev Institute of Mathematics SB RAS, Novosibirsk, Russia
e_konsta@math.nsc.ru

Abstract. In this talk we review some new results obtained by using AI methods to check open problems in Cayley graph theory and to state new conjectures. The results discussed in the talk are published in the recent paper by Chernov, Fedoriaka, and Konstantinova. This is the third paper of the CayleyPy project, https://github.com/CayleyPy/CayleyPy, applying artificial intelligence methods to problems in group theory. In this paper, the public release of CayleyPy is announced. CayleyPy is an open–source Python library for computations with Cayley and Schreier graphs. Compared with state-of-the-art systems based on classical methods, such as GAP and Sage, CayleyPy handles significantly larger graphs and performs several orders of magnitude faster. For many Cayley graphs over the symmetric group Symn quasi–polynomial diameter formulas are observed: a small set of quadratic or linear polynomials on n, and it is conjectured that it is a general phenomenon. These lead to efficient diameter computation, despite the problem being NP-hard in general. A refinement of the Babai-type conjecture on diameters of Symn is proposed as follows: $1/2n^2 + 4n$ upper bounds for the diameters in the standard undirected case, as compared to prior conjectural bounds of $O(n^2)$. Explicit generator families, related to involutions in a simple "square-with-whiskers" pattern, are conjectured to maximize the diameter; extensive search confirms this for all $n \leq 15$. Some conjectures are "LLM-friendly"—they can be stated as sorting problems, which are easy to formulate for LLM, and their solutions can be given by an algorithm or by a Python code, which is easy to verify, so they can be used to test LLM's abilities to solve research problems. To benchmark various methods of path-finding on Cayley graphs more than 10 benchmark datasets were created in the form of Kaggle challenges, making benchmarking easy and public to the community. CayleyPy works with arbitrary permutation or matrix groups, and supports a pre-defined collection of more than a hundred generators including puzzle groups. The code for direct growth computation outperforms similar functions on the standard computer algebra system GAP/SAGE up to 1000 times both in speed and in maximum sizes of the graphs that it can handle.

Keywords: Cayley graphs · CayleyPy project

Abstracts of Tutorials

New Principles of Nonconvex Optimization

Alexander Strekalovskiy

Matrosov Institute for System Dynamics and Control Theory SB RAS, Irkutsk, Russia
strekal@icc.ru

Abstract. We address the non-convex optimization problem with the cost function, equality and inequality constraints given by DC functions (difference of two convex functions). With the help of Exact Penalty Theory, one reduces the problem with constraints to an unconstrained (penalized) problem with a DC cost function. A Global Search Scheme (GSS) for solving the resulting non-convex problem combines (a) Special local search methods (LSM) providing a critical (regarding an LSM) point. Within an LSM one can use classical optimization methods and modern computational software, which looks like an advantage of the developed approach; (b) Global Search procedures, suggested by the new approximate Global Optimality Conditions (GOC), and, due to the constructive (algorithmic) property of the GOC, we are able to escape critical, local and stationary points with improvement of the goal function. Hence, a convergency proof of the GSS is not only important for substantiation of the methodology, but crucial for solving a big number of non-convex problems arising in various applied fields. As is well-known, optimization problems can be divided into two main types: convex and non-convex. The first kind can be solved with classical numerical techniques such as Gradient methods, Newtonian ones, IPM, SQP, and TRM among others. Besides, modern software (CPLEX, Gurobi, etc.) is highly effective for solving these problems without the "curse of dimensionality". Meanwhile, for a huge area of non-convex problems, convex optimization methods turn out to be completely inefficient and extremely unworkable when they attempt to find a global solution to the problems in question. This makes some simplifying ideas very popular, such as B-n-B, bio-inspired methods, genetic algorithms and neural networks etc., while these approaches lack any mathematical justifications and background. Numerical testing on high-dimensional test problems certified the comparative effectiveness of the developed approach.

Keywords: non-convex optimization · DC optimization · global optimality conditions

Expected Runtime Analysis of Evolutionary Algorithms

Anton Eremeev

Omsk Department of Sobolev Institute of Mathematics SB RAS, Omsk, Russia
eremeev@ofim.oscsbras.ru

Abstract. The theory of runtime analysis of evolutionary algorithms, including genetic algorithms, has been developed during the last three decades, starting with the works of Ingo Wegener and his co-authors. This theory allows us to obtain provable mathematical results that give some guarantees on the expected computational cost till first hitting an optimal solution by the EA population (expected runtime), or to show the inefficiency of some EAs for some parameter settings. This tutorial surveys some recent techniques that prove to be useful in runtime analysis of EAs and discusses the open problems in the area.

Keywords: evolutionary algorithms · expected runtime analysis

Contents

Stochastic Optimization

Scheduling

Applications in Machine Learning and Mathematical Economics KL-means C

Mathematical Programming

Aggregation of Variables in Mixed-Integer Linear Programming

Roland Hildebrand(✉) and Alexey Chernov

Moscow Institute of Physics and Technology (National Research University), Institutskiy Per., 9, Dolgoprudny, Moscow Region 141701, Russia
khildebrand.r@mipt.ru

Abstract. In mixed-integer linear programs a subset of variables may appear in the problem formulation only in the form of a certain linear combination, either in the original problem or after previous presolving steps eliminating variables and constraints. We present a multi-stage method to aggregate such combinations into a fewer number of variables.

Keywords: Mixed-integer linear programming · Presolve reductions · Variable aggregation · Parallel columns

1 Introduction

Mixed-integer linear programs (MILP) are a widely used problem class to model difficult scheduling, planning, and optimization tasks in various branches of economy, e.g., transportation, production, finance, logistics, and energy distribution. The size and complexity of these problems often lead to MILP models with millions of variables and constraints. Even the most sophisticated branch and bound methods struggle with such problem dimensions. Automated simplifications of the MILP prior to the launch of the main algorithm are hence an indispensable part of the overall solution process. These so-called presolving procedures comprise a variety of methods, e.g., bound and coefficient strengthening, free or implied free variable elimination, detection and removal of redundant constraints, problem decomposition, non-zero cancellation, dual fixing, and probing, this list not pretending to be exhaustive. Overviews over presolving reductions in MILP and LP can be found in [1,3], a performance analysis in [2].

In this paper we consider a specific presolving technique, which may be called *merging of variables* and is applicable if the problem contains parallel columns. Informally, this means that a subset of variables appear in the problem only together and in a certain linear combination. Then this combination may be replaced by a new variable, effectively reducing both the number of variables and the number of non-zero elements in the data of the problem.

MSC 2020: 90C11, 90C26

A. Kazakov et al. (Eds.): MOTOR 2026, LNCS 16687, pp. 3–11, 2026.
https://doi.org/10.1007/978-3-032-31390-4_1

Parallel column detection is one of the less expensive presolving techniques and can be performed using normalization and hashing. In [3] it has been proposed to first detect columns with identical supports using a hash function, and then to compare columns with the same hash value pair-wise. A speed-up of this procedure was suggested in [1, Section 5.2], where in the second step the columns are normalized, first dividing each column by the maximum modulus element, and then multiplying by the sign of the non-zero element with the least index. Then a second hash function can be applied, because previously parallel columns should at this stage be identical.

Merging *pairs* of variables is nowadays a standard presolving technique implemented in MILP solvers [1, Section 6.3]. It meets limitations, however, when integer variables are present. The reason is the need of taking into account the bounds on the merged variables. As a result, the feasible set of the considered linear combination may be too complicated for the combination to be replaceable by a single new variable of continuous or integer type, or by an affine function of such a variable.

In the context of presolving MILP pairs of variables and pairs of constraints have been considered also in [5].

In the literature one can find the presolving technique of aggregating variables [4,6]. However, in that context variable elimination by replacing a variable by a combination of other variables derived from an equality constraint is considered, which is different from merging.

In this contribution we consider merging an arbitrary number of variables, which may be both integer or real. For the procedure which detects parallel columns it makes no difference whether two or more columns are to be returned, because it identifies batches of mutually parallel columns.

The remainder of the paper is structured as follows. In the next section we recall the formulation of the MILP and describe the parallel column presolving technique which merges pairs of variables. In Sect. 3 we describe our algorithm for merging of an arbitrary number of variables, and in Sect. 4 we discuss the results.

2 Preliminaries

We consider a MILP of the form

$$\min_{x\in\mathbb{R}^n} \langle c, x\rangle : \quad Ax = b,\ Cx \leq d,\ l \leq x \leq u,\ x_I \in \mathbb{Z}^{|I|},$$

where $l_i \in \mathbb{R} \cup \{-\infty\}$, $u_i \in \mathbb{R} \cup \{+\infty\}$, $l_i \leq u_i$ for all $i = 1, \ldots, n$, and $I \subset \{1, \ldots, n\}$ is the subset of indices of the *integer* variables. The remaining variables are called *real* variables.

Remark 1. For convenience we shall call a set of the form $[l_i, u_i] \setminus \{\pm\infty\}$, i.e., the feasible set of a real variable, an *interval*. This includes the case of a compact interval, a closed ray, which may be infinite in either direction, and the whole real line.

The individual variable x_i multiplies with the i-th columns A_{*i}, C_{*i} of the coefficient matrices A, C in the equality and inequality constraints, respectively, as well as with the cost coefficient c_i in the objective function. We shall assume that A_{*i}, C_{*i}, c_i are not simultaneously equal to zero, for every i, otherwise the corresponding variable can be removed from tyhe problem.

Two columns i, j are said to be *parallel* if there exists $\lambda \in \mathbb{R}$, which is necessarily non-zero, such that

$$A_{*i} = \lambda A_{*j}, \quad C_{*i} = \lambda C_{*j}, \quad c_i = \lambda c_j.$$

Obviously the relation of being parallel is an equivalence relation, and the variable indices are partitioned into disjoint subsets such that the corresponding columns are mutually parallel.

If such a subset J comprises strictly more than one variable, then the variables x_j, $j \in J$, enter the MILP only together and in the form of a specific linear combination

$$y = \sum_{j \in J} \lambda_j x_j. \tag{1}$$

More precisely, for every equality k we have that $\sum_{j\in J} A_{kj} x_j = \alpha_k y$ for some $\alpha_k \in \mathbb{R}$. Likewise, for every inequality l we have $\sum_{j\in J} C_{lj} x_j = \gamma_l y$ for some $\gamma_l \in \mathbb{R}$, and $\sum_{j\in J} c_j x_j = \zeta y$ for some $\zeta \in \mathbb{R}$.

One may then consider the possibility of replacing the variables x_j, $j \in J$, by a new variable with value equal to y. In [1, Section 6.3] the case $J = \{j, k\}$, $\lambda_j = 1$ is considered. If x_k is a real variable, and x_j is real or $|\lambda_k| \cdot (u_k - l_k) \geq 1$, then y may be declared as a real variable. If both x_j, x_k are integer, then the replacement can still be valid with y being an integer variable, if certain conditions are met by the bounds and the coefficient λ_k. Namely, the set of values the linear combination y can take should be representable as the intersection of the set of integers with an interval of the real line, in the sense of Remark 1.

The question arises whether this procedure may be generalized to linear combinations consisting of more than two variables. More precisely, given a finite index set J, coefficients λ_j, lower and upper bounds l_j, u_j, for all $j \in J$, and a subset $Z \subset J$ indicating integer variables, can we find a strictly smaller index set $\tilde{J}$, coefficients $\tilde{\lambda}_j$, lower and upper bounds $\tilde{l}_j, \tilde{u}_j$, for all $j \in \tilde{J}$, and a subset $\tilde{Z} \subset \tilde{J}$, such that $\mathcal{F}(J, \lambda) = \mathcal{F}(\tilde{J}, \tilde{\lambda}) + const$? Here

$$\mathcal{F}(J, \lambda) = \left\{ \sum_{j \in J} \lambda_j x_j \mid l_j \leq x_j \leq u_j,\ x_Z \in \mathbb{Z}^{|Z|} \right\} \tag{2}$$

is the set of values achievable by the linear combination (1), and $\mathcal{F}(\tilde{J}, \tilde{\lambda})$ is defined similarly. We may then replace the variables x_j, $j \in J$ by new variables y_j, $j \in \tilde{J}$. Technically, the initial coefficients $\alpha_k \lambda_j$ in the equations, $\gamma_l \lambda_j$ in the inequalities, and $\zeta \lambda_j$ in the objective vector, for all $j \in J$, have to be replaced by new coefficients $\alpha_k \tilde{\lambda}_j$, $\gamma_l \tilde{\lambda}_j$, $\zeta \tilde{\lambda}_j$, for $j \in \tilde{J}$. Ideally we would like to find the smallest such index set $\tilde{J}$. In the next section we shall tackle this problem.

3 Merging Several Variables of Different Types

In this section we study when a subset of variables can be merged into one. Let J, Z, λ, l, u be as above, and let (2) be the set of values achievable by the corresponding linear combination of variables. Without loss of generality we shall assume that finite bounds of integer variables are integer, and that $l_j < u_j$ for all $j \in J$.

In a MILP the feasible set of a single variable is either an interval on the real line if the variable is real, or the intersection of an interval with the set of integers $\mathbb{Z}$ if the variable is integer. For convenience we shall define the following, slightly larger class of sets.

Definition 1. *A subset $X \subset \mathbb{R}$ will be called admissible if it is either an interval in the sense of Remark 1, or the intersection of such an interval with a regularly spaced grid $\alpha\mathbb{Z} + \beta$, $\alpha \neq 0$.*

An admissible set can be represented as the set of values of an affine image of the feasible set of a single variable. Hence the subset of variables can be merged into one variable if and only if the set (2) is admissible.

Let us now highlight two trivial but important observations. Let the set J of variables to merge be represented by a disjoint union $J_1 \cup J_2$. Then

- that the variables in J can be merged into one variable does not imply that the variables in J_1 can be merged into one variable;
- if the variables in J_1 can be merged into one variable, then the resulting variable can be merged with the variables in J_2 if and only if the variables in J can be merged into one variable.

In other words, by merging subsets of variables we cannot spoil the possibility of merging all variables together. On the other hand, if some variables cannot be merged in isolation, this does not preclude the possibility that all variables still can be merged. We may hence proceed iteratively, merging variables in smaller subsets whenever this is possible.

We shall now describe several special cases of mergers of subsets of variables. These are straightforward generalizations of the merger of two variables described in [1].

Real variables: Every subset J' of real variables can be merged without restrictions. The resulting variable is also real, with value $\sum_{j \in J'} \lambda_j x_j$ and bounds

$$\tilde{l} = \sum_{j \in J'} \min(\lambda_j l_j, \lambda_j u_j), \quad \tilde{u} = \sum_{j \in J'} \max(\lambda_j l_j, \lambda_j u_j).$$

Real and Integer Variable: If $J' = \{j, k\}$, the variable x_j is real and x_k is integer, then these variables can be merged if and only if $|\lambda_j| \cdot (u_j - l_j) \geq |\lambda_k|$. The resulting variable is real, with value $\lambda_j x_j + \lambda_k x_k$ and bounds

$$\tilde{l} = \min(\lambda_j l_j, \lambda_j u_j) + \min(\lambda_k l_k, \lambda_k u_k), \quad \tilde{u} = \max(\lambda_j l_j, \lambda_j u_j) + \max(\lambda_k l_k, \lambda_k u_k).$$

Applying the first of these rules, we may merge all real variables into a single one x_r with bounds l_r, u_r. Then we may consecutively merge integer variables x_j with this real variable, starting from the one with smallest modulus $|\lambda_j|$. This is possible as long as $|\lambda_j|$ does not exceed the length $u_r - l_r$ of the feasibility interval of the real variable. Note that with each such merger, the length of the feasibility interval increases. In particular, if at any step a variable with infinite feasibility interval is involved, then all variables can be merged into a single real variable.

However, if not all integer variables have been consumed by the above procedure or if no real variables have been present at all, we still may be able to merge together integer variables. We recall a classical result on the Bezout identity.

Lemma 1 (Bezout identity). *Let p, q be non-zero co-prime integers. Then there exist integers a, b such that the Bezout identity $pa + qb = 1$ holds.*

We first consider two special cases with infinite bounds.

Two semi-infinite integer variables: We have the following result.

Theorem 1. *Let $J' = \{j, k\}$ and both variables x_j, x_k be integer, such that $\min(\lambda_j l_j, \lambda_j u_j) = -\infty$, $\max(\lambda_k l_k, \lambda_k u_k) = +\infty$, and $\frac{\lambda_j}{\lambda_k} = \frac{p}{q} \in \mathbb{Q}$ with p, q co-prime. Then these variables can be merged into a single free integer variable x_i (here free means $l_i = -\infty$, $u_i = +\infty$) with value $x_i = px_j + qx_k$ proportional to the value of the linear combination $\lambda_j x_j + \lambda_k x_k$.*

Proof. Set $\lambda_i := \frac{\lambda_k}{q} = \frac{\lambda_j}{p}$, $x_i = px_j + qx_k$. Clearly x_i takes only integer values. Moreover,

$$\lambda_j x_j + \lambda_k x_k = \lambda_i(px_j + qx_k) = \lambda_i x_i,$$

and the linear combination in question is indeed a multiple of x_i.

It remains to show that x_i can take any integer value if x_j, x_k run through the corresponding feasible sets. By assumption the combination $\lambda_j x_j + \lambda_k x_k$ can take positive and negative values with arbitrarily large modulus. This holds consequently also for the integer x_i.

Since p, q are co-prime, by Lemma 1 there exist integers a_j, a_k such that $pa_j + qa_k = 1$. Substituting x_j by $x_j + \sigma a_j$ and x_k by $x_k + \sigma a_k$ hence changes the value of x_i by σ. If there exists $\sigma \in \{-1, +1\}$ such that this substitution is always possible without violating the bounds on x_j, x_k, then $x_i = n$ is feasible implies $x_i = n + \sigma$ is feasible, and every integer value for x_i is feasible.

Assume the contrary. Suppose there exist feasible values of x_j, x_k such that the above substitution is not possible for $\sigma = 1$, i.e., that $x_j + a_j$ or $x_k + a_k$ are outside of the respective bound. This means that $\lambda_j x_j \leq \max(\lambda_j l_j, \lambda_j u_j) < \lambda_j x_j + \lambda_j a_j$, or $\lambda_k x_k + \lambda_k a_k < \min(\lambda_k l_k, \lambda_k u_k) \leq \lambda_k x_k$. This in turn implies that $\lambda_j a_j > 0$ or $\lambda_k a_k < 0$.

Likewise, setting $\sigma = -1$ leads to the conclusion that $\lambda_j a_j < 0$ or $\lambda_k a_k > 0$.

Therefore the products $\lambda_j a_j$, $\lambda_k a_k$ have the same sign. But then also the integer-valued products pa_j, qa_k have the same sign, leading to a contradiction with the identity $pa_j + qa_k = 1$.

Note that the larger the integers p, q are, the smaller will be the coefficient λ_i at the newly created integer variable x_i, and the easier it will be to merge this variable with a real variable. In the limiting case of an irrational ratio $\frac{\lambda_j}{\lambda_k}$ we have the following result.

Theorem 2. *Let $J' = \{j,k\}$ and both variables x_j, x_k be integer, such that $\min(\lambda_j l_j, \lambda_j u_j) = -\infty$, $\max(\lambda_k l_k, \lambda_k u_k) = +\infty$, and $\frac{\lambda_j}{\lambda_k} \notin \mathbb{Q}$. Then the set $\mathcal{F}$ of feasible values of the combination $\lambda_j x_j + \lambda_k x_k$ is dense in $\mathbb{R}$.*

Proof. As before, the set $\mathcal{F}$ contains both positive and negative values of arbitrarily large modulus.

For every $\epsilon > 0$, there exist integers a_j, a_k such the number $\delta = \lambda_j a_j + \lambda_k a_k$ obeys $0 < \delta \leq \epsilon$. Repeating the reasoning in the previous proof, we arrive at the conclusion that for every feasible value $x_i = \xi$, the value $\xi + \sigma\delta$ is also feasible, where σ can be set to one of the values ± 1.

The two facts combined readily imply the claim of the theorem.

If we allow for a positive tolerance in the constraints of the MILP, then a variable with a feasible set which is dense in $\mathbb{R}$ can be treated as a free real variable. Hence if the situation described in Theorem 2 occurs, all variables of the considered subset J can be merged into a single free real variable. However, the smaller the tolerance, the larger will be the optimal integer values of x_j, x_k after uncrushing the optimal solution of the presolved MILP.

Free and Bounded Integer Variable: We now consider the possibility of merging a free integer variable with a bounded one.

Theorem 3. *Let $J' = \{j,k\}$ and both variables x_j, x_k be integer, such that $l_j = -\infty$, $u_j = +\infty$. Let $\frac{\lambda_j}{\lambda_k} = \frac{p}{q} \in \mathbb{Q}$ with $p > 0$ and p, q co-prime, and suppose that $u_k - l_k \geq p-1$. Then these variables can be merged into a single free integer variable x_i with value $x_i = px_j + qx_k$ proportional to the value of the linear combination $\lambda_j x_j + \lambda_k x_k$.*

Proof. As before, we have with $\lambda_i := \frac{\lambda_k}{q} = \frac{\lambda_j}{p}$ that $\lambda_j x_j + \lambda_k x_k = \lambda_i x_i$.

Since the feasible set of the variable x_j is $\mathbb{Z}$, the summand px_j in x_i can equal every integer which is divisible by p. Hence if some value n is feasible for x_i, so are all integers which lie in the same residue class modulo p as n.

We hence need to show only that the second summand qx_k assumes values in every residue class. Since p, q are co-prime, for every p consecutive integer values of x_k the expression qx_k will fall into p pair-wise distinct residue classes, and hence cover all residue classes.

Finally, x_k can assume at least p consecutive values if and only if $u_k - l_k \geq p - 1$.

On the other hand, if the condition $u_k - l_k \geq p - 1$ is not met, then there are less than p values which x_k can assume, and hence at least one residue class will not intersect with the feasible set for x_i.

Another favorable case is the combination of integer variables which appear with coefficients of the same modulus.

Theorem 4. *Let x_j, $j \in J'$ be integer variables with coefficients λ_j having the same modulus, $\lambda_j = \sigma_j \cdot \Lambda$ with $\Lambda > 0$ and $\sigma_j \in \{-1, +1\}$, for all $j \in J'$. Then the variables x_j can be merged into an integer variable x_i with bounds*

$$l_i = \sum_{j \in J'} \min(\sigma_j l_j, \sigma_j u_j), \quad u_i = \sum_{j \in J'} \max(\sigma_j l_j, \sigma_j u_j)$$

and value $x_i = \sum_{j \in J'} \sigma_j x_j$.

Proof. Clearly $x_i = \sum_{j \in J'} \sigma_j x_j$ takes only integer values which are bounded by l_i, u_i below and above, respectively. Moreover, if l_i or u_i is finite, then it can be attained, and if it is infinite, then x_i can take values with arbitrarily large modulus and negative sign in the case $l_i = -\infty$, positive sign in the case $u_i = +\infty$.

On the other hand, every summand $\sigma_j x_j$ takes consecutive integer values in the corresponding feasible interval, hence the set of values of the sum also consists of consecutive integers.

After merging integer variables with coefficients having the same modulus, without loss of generality we may assume that $|\lambda_j|$ are pair-wise different.

It remains to consider linear combinations which are bounded to at least one side. Then the conditions allowing a merger of variables are much more restrictive. We have the following negative result.

Theorem 5. *Suppose the linear combination $y = \sum_{j \in J'} \lambda_j x_j$ of integer variables x_j is such that $\min(\lambda_j l_j, \lambda_j u_j) > -\infty$ for all $j \in J'$, and that $|\lambda_j|$ are pair-wise different. Let $\mu = \min_{j \in J'} |\lambda_j|$.*

If there exists $j \in J'$ such that $\frac{\lambda_j}{\mu} \notin \mathbb{Z}$, then the feasible set $\mathcal{F}(J', \lambda)$ of the linear combination y is not admissible.

Proof. Assume the contrary, i.e., that $\mathcal{F}(J', \lambda)$ is admissible.

Clearly the lower bound $l_i = \sum_{j \in J'} \min(\lambda_j l_j, \lambda_j u_j)$ on y can be attained, all variables x_j being at their respective bounds. Then the next largest value which x_i can attain equals $l_i = \mu$. Hence the step size in the admissible set $\mathcal{F}(J', \lambda)$ equals μ.

However, all values of the form $l_i + |\lambda_j|$, $j \in J'$ are also feasible for y. It follows that $|\lambda_j|$ must be an integer multiple of μ.

The theorem yields a necessary condition for the merger of the variables x_i if the linear combination in question is bounded from below (the case of bounded-ness from above is treated similarly). If this condition is met, then an algorithm exists which merges the variables or establishes that merging them is not possible. At each iteration the algorithm attempts to merge the two variables with the coefficients having least absolute values.

Theorem 6. *Suppose the linear combination $y = \sum_{j \in J'} \lambda_j x_j$ of integer variables x_j is such that $\min(\lambda_j l_j, \lambda_j u_j) > -\infty$ for all $j \in J'$, that $|\lambda_j|$ are pair-wise different, and that $\mu = \min_{j \in J'} |\lambda_j|$ divides all λ_j. Let further j_1, j_2 be the indices corresponding to the smallest and second-smallest value $|\lambda_j|$.*

Then there are two possibilities:

- *if* $\frac{|\lambda_{j_2}|}{\mu} - 1 > u_{j_1} - l_{j_1}$, *then merging the variables* x_j *is not possible;*
- *if* $\frac{|\lambda_{j_2}|}{\mu} - 1 \leq u_{j_1} - l_{j_1}$, *then merging the variables* x_{j_1}, x_{j_2} *is possible.*

Proof. Let $L_i = \sum_{j \in J'} \min(\lambda_j l_j, \lambda_j u_j)$. Then the value $L_i + |\lambda_{j_2}| - \mu$ is attainable for y if and only if $\frac{|\lambda_{j_2}|}{\mu} - 1 \leq u_{j_1} - l_{j_1}$, because the only way to attain this value is to leave all variables x_j except x_{j_1} at their bound and to increase or decrease x_{j_1} by a difference of $\frac{|\lambda_{j_2}|}{\mu} - 1$.

Since, however, $L_i + |\lambda_{j_2}|$ is attainable, the set $\mathcal{F}(J', \lambda)$ cannot be admissible if the first condition in the theorem is met.

If the second condition is met, however, then one can merge the variables x_{j_1}, x_{j_2} into an integer variable $x_i = \frac{\lambda_{j_1} x_{j_1} + \lambda_{j_2} x_{j_2}}{\mu}$ with bounds

$$l_i = \sum_{k=1}^{2} \min(\lambda_{j_k} l_{j_k}, \lambda_{j_k} u_{j_k}), \quad u_i = \sum_{k=1}^{2} \max(\lambda_{j_k} l_{j_k}, \lambda_{j_k} u_{j_k}).$$

By Theorem 6, either at some step the impossibility to merge all variables is detected, or the variables are merged.

4 Conclusion

In this work we established a number of results on the possibility or impossibility to merge variables entering a MILP only as a certain linear combination, into one variable. If it is impossible to do so, it may be possible to merge subsets of the variables in separate new variables, and thus still to simplify the MILP. It has been also established that merging variables facilitates the merging of additional variables with the newly created aggregated variable. The following situations have been considered:

- merging of real variables
- merging of a real variable and an integer variable
- merging two integer variables such that the corresponding summands in the linear combination are unbounded in opposite directions (Theorem 1)
- merging a free integer variable with an additional integer variable (Theorem 3)
- merging integer variables with coefficients having the same modulus (Theorem 4)
- merging integer variables such that the corresponding summands in the linear combination are bounded in the same direction (Theorem 6)

In addition, situations have been considered where the linear combination of integer variables can produce a set of feasible values which is dense in $\mathbb{R}$. In this case the variables can effectively be merged into a free real variable, although the uncrushing of the optimal solution may produce large integer values of the original variables.

Acknowledgements. Roland Hildebrand (contribution of Theorems 1, 2,3) was supported by the Ministry of Science and Higher Education of the Russian Federation, project No. FSMG-2024-0011. Alexey Chernov (contribution of Theorems 4, 5, 6) was supported by the Russian Science Foundation, project No. 21-71-30005-Π, https://rscf.ru/en/project/21-71-30005/.

References

1. Achterberg, T., Bixby, R.E., Zonghao, G., Rothberg, E., Weninger, D.: Presolve reductions in mixed integer programming. INFORMS J. Comput. **32**(2), 473–506 (2020)
2. Achterberg, T., Wunderling, R.: Mixed Integer Programming: Analyzing 12 Years of Progress, pp. 449–481. Springer, Berlin, Heidelberg (2013)
3. Andersen, E.D., Andersen, K.D.: Presolving in linear programming. Math. Program. **71**, 221–245 (1995)
4. Gamrath, G., Melchiori, A., Berthold, T., Gleixner, A.M., Salvagnin, D.: Branching on multi-aggregated variables. In: Michel, L. (ed.) Integration of AI and OR Techniques in Constraint Programming, pp. 141–156. Springer, Cham 2(015)
5. Gemander, P., Chen, W.-K., Weninger, D., Gottwald, L., Gleixner, A., Martin, A.: Two-row and two-column mixed-integer presolve using hashing-based pairing methods. EURO J. Comput. Optim. **8**, 205–240 (2020)
6. Zhang, H., Yuan, J.: A combined variable aggregation presolving technique for mixed integer programming. Oper. Res. Lett. **53**, 107074 (2024)

A Method of Scaling Variables in Large-Dimensional Problems

V. N. Krutikov[1], E. M. Tovbis[2], and L. A. Kazakovtsev[2,3(✉)]

[1] Kemerovo State University, Kemerovo, Russia
[2] Reshetnev Siberian State University of Science and Technology, Krasnoyarsk, Russia
levk@bk.ru
[3] Bauman Moscow State Technical University, Moscow, Russia

Abstract. Continuous optimization problems often require methods capable of solving large-scale problems under noise. Such methods exist today, for example, the ADAM algorithm. However, there is still no theoretical basis for its acceleration properties. The goal of this study is to formulate a minimization method that utilizes variable scaling to accelerate convergence using adjustable diagonal matrices and is suitable for solving problems with noise. In this paper, we present a prototype of a new method with finite convergence on quadratic functions and investigate its properties. The origins of this prototype lie in the field of subgradient methods. The prototype was transformed into a minimization method suitable for solving large-scale problems, robust to noise, and not requiring a one-dimensional search. A design for a method with variable scaling using diagonal metric matrices was defined. Step adaptation, which has proven effective in various minimization methods for solving minimization problems, including those with noise, was used to select the step factor of the minimization method. The algorithm was developed using ideas from modern stochastic gradient methods. The resulting algorithm has the structure of the ADAM algorithm with step adaptation, and its metric matrix transformations are largely similar to the ADAM algorithm metric. The new algorithm was numerically studied on test functions and compared with the well-known ADAM method with step adaptation. The experiment showed that the minimization method is effective for solving problems with noise and poor variable scaling.

Keywords: Relaxation Process · Minimization Algorithm · Subgradient Method

1 Introduction

Today, it seems relevant to develop optimization methods that work under ill-conditioned optimization problems, in the presence of gradient noise, and with inability to calculate function values with minimal memory requirements. Such methods (in particular, minimization methods) are widely used in neural network training.

Adaptive methods that adjust the step size have become widespread in large-scale optimization for their ability to converge robustly [1]. In [2], a method was proposed, where each dimension has its own dynamic rate: large gradients have small learning rates

A. Kazakov et al. (Eds.): MOTOR 2026, LNCS 16687, pp. 12–31, 2026.
https://doi.org/10.1007/978-3-032-31390-4_2

and small gradients have large learning rates. The AdaGrad family of methods [3–8] is based on this principle and shows good results on large-scale learning problems. For example, in [3] the author fixed the window of past gradients that are accumulated. The ADAM algorithm [9] become one of the most popular optimizers due to its simple implementation, computationally efficiency and robustness in data-intensive and parameter-intensive problems. ADAM has three main ideas: gradient smoothing, variable scaling and step selection. In [10, 11], AdaGrad-Norm with a single step size adaptation based on the gradient norm was developed. In [12], new variants of the ADAM algorithm with a long-term memory of past gradients were proposed. The authors in [13] apply the variance reduction technique to construct the adaptive step size in ADAM. Adam-based solver (QuasiGrad) exploits parallelized projection methodologies to accelerate convergence in [14]. Algorithm ADOPT [15] addresses the non-convergence issue of ADAM by removing the current gradient from the second moment estimate and changing the order of the momentum update and the normalization by the second moment estimate. The Cycle-Norm-ADAM algorithm based on the cyclic exponential decay learning rate is proposed in [16]. In [17], authors propose ADAM with adaptive step size named h_ADAM.

The use of diagonal metric matrices in minimization methods is intended to accelerate their convergence in high-dimensional problems. Diagonal metric matrices are helpful in eliminating the ill-conditioned nature of the problem when variables are poorly scaled. Therefore, studying the mechanisms and origins of algorithms for obtaining effective methods for correcting the space metric is relevant.

In this paper, a new method for correcting diagonal metric matrices is proposed. For these purposes, an efficient prototype of a method with a space metric change is used, which is transformed into a minimization method with a diagonal metric matrix. Note that the prototype method has an analogy with the CGM. It converges on quadratic functions in a finite number of iterations, while a different implementation of the method with a space metric change provides the ability to solve non-smooth problems.

The main strategy for transforming metric matrices is to increase the degree of orthogonality of the current descent vector to previous gradients. When using full-scale metric matrices, iterative methods show good results in minimizing smooth functions, comparable to the results of quasi-Newton methods. In this paper, this idea is implemented using diagonal metric matrices. It turned out that the new metric matrix transformation formula is partially similar to the ADAM algorithm formula. This partially explains the high convergence rate of the h_ADAM algorithm.

The method was developed using the h_ADAM algorithm structure, which combines modern algorithmic design techniques to overcome noise and poor scaling. The method also uses a smoothed gradient from the MOMENTUM algorithm. The result is a minimization algorithm with step adaptation and scaling diagonal matrices to generate a descent direction from a smoothed gradient.

The rest of this paper is organized as follows. In Sect. 2, principles of new minimization method are described and its prototype is given. In Sect. 3, a new minimization algorithm is proposed. In Sect. 4, numerical results are presented. Section 5 concludes the work.

2 A Prototype of Minimization Method

Let a smooth problem $f(x) \to \min_{x \in R^n}$ be solved by some relaxation method

$$x_{k+1} = x_k - \gamma_k s_k, \quad \gamma_k = \underset{\gamma}{argmin} f(x_k - \gamma s_k). \tag{1}$$

Here, k is iteration number, g_k is gradient of the function at the point of the current minimum $x_k \in R^n$, $g_k \equiv g(x_k) \equiv \nabla f(x_k)$, $s_k \in R^n$ is descent direction and n is problem dimension.

Let's define the main idea of choosing the descent direction. Denote ε-neighborhood of a point x_k by $X_\varepsilon(x_k)$. Similar to the relaxation processes of the ε-subgradient type [18], we will select the descent direction s_k from a set of directions $s_k \in S_\varepsilon(x_k)$ gradient-matched with the gradients of a certain ε-neighborhood of the point x_k, where

$$S_\varepsilon(x) = \left\{ s \in R^n | \min_{g \in G}(s, \nabla f(y)) > 0, \forall y \in X_\varepsilon(x_k) \right\}.$$

Here, g is the gradient, and G is gradient set. This direction enables one to leave the neighborhood of the point x_k and reduce the function at the same time. If the set $S_\varepsilon(x)$ is not empty, then any vector $s \in S_\varepsilon(x)$ is a solution to a set of inequalities:

$$(s, g(x)) > 0, \forall x \in X_\varepsilon(x_k) \tag{2}$$

In order to be able to use known learning algorithms, we will use, instead of (2), a system of equalities for the set of neighborhood gradients. Subgradient methods use a system of equalities that can be written for gradients of differentiable functions as $(s, g(x)) = 1$.

Under certain constraints on the function, subgradient methods find such a direction in a finite number of iterations [19], using subgradients obtained during the minimization process. The same approach can be applied to organize algorithms for minimizing smooth functions. The system of such equalities has the form [19]:

$$(s, g_i) = y_i, \mathrm{i} = 1, \ldots \mathrm{N}., \mathrm{yi} = 1. \tag{3}$$

The solution to system (3) can be obtained as a minimum point in the problem of minimizing the function [19]

$$W_N(s) = \frac{1}{2}\sum_{i=1}^{N} w_i \left[y_i - g_i^T s \right]^2 + \frac{1}{2}\sum_{i=0}^{N} s_i^2. \tag{4}$$

We will denote a symmetric positive definite matrix $A \in R^{n \times n}$ by $A > 0$. The solution to the minimization problem (4) has the form

$$s_N = A_N^{-1} Y_N, A_N = I + \sum_{i=1}^{N} w_i g_i g_i^T, Y_N = \sum_{i=1}^{N} w_i y_i g_i A_N > 0.$$

It can be found at $w_i = 1$, $i = 1, 2, \ldots, N$ by iterative least squares method (ILSM) [20]:

$$s_{i+1} = s_i + \frac{H_i g_{i+1}(y_{i+1} - (s_i, g_{i+1}))}{1 + (H_i g_{i+1}, g_{i+1})}, s_0 = 0, i = 0, 1, 2, \ldots, N-1, \tag{5}$$

$$H_{i+1} = H_i - \frac{H_i g_{i+1} g_{i+1}^T H_i^T}{1 + (g_{i+1}, H_i g_{i+1})}, H_0 = I, H_i > 0\ i = 0, 1, 2, \ldots, N-1. \quad (6)$$

In [19], it is proposed to use (5), (6) for scaled data:

$$\hat{g}_i = g_i\big[q(g_i, H_i g_i)\big]^{-0.5}, \hat{y}_i = y_i\big[q(g_i, H_i g_i)\big]^{-0.5}, \mathrm{q} > 0.$$

Then, after returning to the previous data in (5), (6), we obtain:

$$s_{i+1} = s_i + \frac{H_i g_i (y_i - (s_i, g_i))}{(1+q)(g_i, H_i g_i)} = s_i + \beta \frac{H_i g_i (y_i - (s_i, g_i))}{(g_i, H_i g_i)}, s_0 = 0, \quad (7)$$

$$H_{i+1} = H_i - \frac{H_i g_i g_i^T H_i^T}{(1+q)(g_i, H_i g_i)} = H_i - \beta \frac{H_i g_i g_i^T H_i^T}{(g_i, H_i g_i)}, H_0 = I, \quad (8)$$

where $\beta = 1/(1+q)$ and $1 > \beta \geq 0$. Further we will use the variant of process (7), (8) with constants β.

The inverse matrix for $H_i = [A_i]^{-1}$ is transformed as follows

$$A_{i+1} = A_i + \frac{\beta}{(1-\beta)} \times \frac{g_{i+1} g_{i+1}^T}{(g_{i+1}, H_i g_{i+1})}. \quad (9)$$

Let's prove this. Let the matrix $H^{1/2} > 0$ be such that $H^{1/2} H^{1/2} = H$. We transform (8)

$$H_{i+1} = H_i^{1/2} \left[I - \beta \frac{H_i^{1/2} g_i g_i^T H_i^{1/2}}{\left(H_i^{1/2} g_i, H_i^{1/2} g_i\right)} \right] H_i^{1/2} = H_i^{1/2} \left[I - \beta \frac{v v^T}{(v, v)} \right] H_i^{1/2}, \quad (10)$$

where $v = H_i^{1/2} g_i$. The vector v is an eigenvalue of the inner matrix from (10) with eigenvalue $1 - \beta$, and the remaining eigenvalues are equal to 1. When inverting it, we obtain a matrix with the same eigenvectors

$$\left[I - \beta \frac{v v^T}{(v, v)} \right]^{-1} = I + \frac{\beta}{(1+\beta)} \frac{v v^T}{(v, v)}$$

The eigenvector v now has the eigenvalue $1/(1 - \beta)$. Inverting the entire matrix (10) we obtain (9).

In [21], a method is proposed in which the descent directions are calculated by an iterative process similar to the process (7), (8). We write an analogue of the iteration of this process, where the operations of updating the metric matrices, necessary for theoretical justification and practical implementation, are excluded:

$$x_{k+1} = x_k - \gamma_k\, s_k,\ \ \gamma_k = \underset{\gamma}{\mathrm{argmin}} f(x_k - \gamma s_k),\ \ k = 0, 1, 2, \ldots, \quad (11)$$

$$s_k = \begin{cases} s_{k-1} + \frac{(1-(s_{k-1}, g_k)) H_k g_k}{(g_k, H_k g_k)}, & \textit{if } k \textit{ is not a multiple of } m \\ \frac{H_k g_k}{(g_k, H_k g_k)}, & \textit{if } k \textit{ is a multiple of } m, m \leq n. \end{cases} \quad (12)$$

$$H_{k+1} = H_k - \beta \frac{H_k g_k g_k^T H_k}{(g_k, H_k g_k)} .1 > \beta \geq 0, H_0 > 0. \tag{13}$$

Here, $x_0 \in R^n$ is an initial point, the method iterations are carried out starting from $k = 0$.

A method similar to (11)–(13) exhibits high convergence rates on smooth functions [21]. Its first property is invariance under linear coordinate transformations. This means that its form is the same in different coordinate systems related by a linear transformation. Therefore, its properties can be justified using a convenient coordinate system.

Let's define a linear transformation of variables.

$$\hat{x} = Px, \tag{14}$$

where $P \in R^n$ is a non-singular matrix. In the new coordinate system, the function to be minimized will have the following form:

$$f_P(\hat{x}) = f\left(P^{-1}\hat{x}\right) = f(x). \tag{15}$$

We will show that algorithm (11)–(13), applied to the function *f(x),* and algorithm (11)–(13), applied to the function $f_P(\hat{x})$ defined in (15), under appropriate initial conditions, construct sequences of minimization points related by the transformation (14). This will allow us to use estimates of the convergence rate of algorithm (11)–(13) in the preferred coordinate system.

Next, we consider process (11)–(13) for minimizing a quadratic function in a coordinate system with identity initial metric matrix and show that the method is identical to the standard conjugate gradient method (CGM). This means that method (11)–(13) constructs a sequence of conjugate vectors after updating for k multiples of m, without requiring an update of the metric matrix.

For quadratic functions, such use of the metric is theoretically useless, since the process of method (11)–(13) will reach a minimum after n iterations for $m \leq n$. In practice, for ill-conditioned quadratic and non-quadratic functions, the use of the metric proves useful, as confirmed in computational experiments in [19, 21].

Theorem 1. *Let the initial conditions of the algorithm (11)–(13), applied to minimize the functions f(x) and* $f_P(\hat{x})$*, defined in (15), be related by the equalities*:

$$\hat{x}_0 = Px_0, \hat{H}_0 = PH_0P^T. \tag{16}$$

Then, while $g_k \neq 0$*, characteristics of these processes are related by the relations*

$$f_P(\hat{x}_k) = f(x_k), \hat{x}_k = Px_{k,} \nabla f_P(\hat{x}_k) = P^{-T}\nabla f(x_k), \hat{H}_k = PH_kP^T, (k = 0, 1, 2, \ldots). \tag{17}$$

Proof. For the gradients of functions $f_P(\hat{x})$ and *f(x)* the following relation holds:

$$\nabla f_P(\hat{x}) = P^{-T}\nabla f(x).$$

From this and assumption (16), Eq. (17) follows for $k = 0$. Let us assume that equalities (17) are satisfied for some k. Let us show that they are satisfied for $k + 1$. The last relation from (17) follows from (13):

$$PH_{k+1}P^T = PH_kP^T - \beta\frac{PH_kP^TP^{-T}g_kg_k^TP^{-1}PH_kP^T}{\left(P^{-T}g_k, PH_kP^TP^{-T}g_k\right)} = \bar{H}_k - \beta\frac{\bar{H}_k\bar{g}_k\bar{g}_k^T\bar{H}_k}{\left(\bar{g}_k, \bar{H}_k\bar{g}_k\right)} = \bar{H}_{k+1}$$

Let's multiply (12) by P and multiply by P^T and P^{-T}:

$$Ps_k = \begin{cases} Ps_{k-1} + \frac{(1-(Ps_{k-1},P^{-T}g_k))PH_kP^TP^{-T}g_k}{(P^{-T}g_k,PH_kP^TP^{-T}g_k)}, \textit{ if } k \textit{ is not a multiple of } m \\ \frac{PH_kP^TP^{-T}g_k}{(P^{-T}g_k,PH_kP^TP^{-T}g_k)}, \textit{ if } k \textit{ is a multiple of } m, m \le n. \end{cases}$$

These expressions can be rewritten taking into account equalities (17), which are true for k

$$\bar{s}_k = \begin{cases} \bar{s}_{k-1} + \frac{(1-(\bar{s}_{k-1},\bar{g}_k))\bar{H}_k\bar{g}_k}{(\bar{g}_k,\bar{H}_k\bar{g}_k)}, \textit{ if } k \textit{ is not a multiple of } m, \\ \frac{\bar{H}_k\bar{g}_k}{\bar{g}_k,\bar{H}_k\bar{g}_k}, \textit{ if } k \textit{ is a multiple of } m, m \le n. \end{cases}$$

From (11) for $k = i$, after multiplying by P on the left we obtain

$$Px_{k+1} = Px_k - \gamma_kPs_k = \bar{x}_k - \gamma_k\bar{s}_k = \hat{x}_{k+1}, k = 0, 1, 2, \ldots$$

Taking this relation into account, as follows from the definition of the function f_P, at the new points x_{k+1} and $\hat{x}_{k+1}$ values of the functions are equal $f(x_{k+1}) = f_P(\hat{x}_{k+1})$. Due to the relation of the gradients $\nabla f_P(\hat{x}_{k+1}) = P^{-T}\nabla f(x_{k+1})$, the minimum condition for one-dimensional minimization at a new point in the new coordinate system is also satisfied

$$(\bar{s}_k, \hat{g}_{k+1}) = \left(Ps_k, P^{-T}g_{k+1}\right) = (s_k, g_{k+1}) = 0.$$

Therefore, under the condition of exact one-dimensional descent, the step sizes in the old and new coordinate systems are equal $\gamma_i = \bar{\gamma}_i$. This implies the equality $\hat{x}_{k+1} = \bar{x}_{k+1}$. Therefore, the first three conditions of (17) are satisfied for $k + 1$. Continuing the process of induction, we obtain a proof of the theorem.

□

To justify the similarity of method (11)–(13), we compare it with the CGM. The iterations of the Fletcher-Reeves conjugate gradient method are carried out according to the scheme (see, for example, [18]):

$$\bar{x}_{k+1} = \bar{x}_k - \bar{\gamma}_ks_k, \ \bar{\gamma}_k = \underset{\bar{\gamma}}{argmin} f(\bar{x}_k - \bar{\gamma}\bar{s}_k), k = 0, 1, 2, \ldots, \tag{18}$$

$$\bar{s}_k = \begin{cases} \bar{g}_k, \textit{ if } k \textit{ is not a multiple of } m, \\ \bar{g}_k + \frac{(\bar{g}_k,\bar{g}_k)}{(\bar{g}_{k-1},\bar{g}_{k-1})}\bar{s}_{k-1}, \textit{ if } k \textit{ is a multiple of } m. \end{cases} \tag{19}$$

The following theorem provides conditions for the equivalence of methods (11)–(13) and (18), (19) in minimizing quadratic functions:

$$f(x) = \frac{1}{2}(x, Ax) + (b, x) + c, A \in R^{n\times n}, A > 0, b \in R^n.$$

Theorem 2. *Let the function f(x) be quadratic and for the process (11)–(13) the matrix* $H_0 = I$. *Then, if the initial points for the processes (11), (13) and (18), (19) coincide* $x_0 = \bar{x}_0$, $m = \bar{m} \leq n$, *then for any* $k \geq 0$ *while* $g_k \neq 0$:

$$x_k = \bar{x}_k, \tag{20}$$

$$s_k = \bar{s}_k/(g_k, g_k), \tag{21}$$

i.e. both processes generate coinciding successive approximations.

Proof. Let us proceed by induction. For $k = 0$, equalities (20), (21) are satisfied by the conditions of the theorem and the construction of methods. Suppose that equalities (20), (21) are satisfied for $k = 0, 1, ..., l$, where $l > 0$. We will show that they are satisfied for $k = l + 1$.

Since $x_l = \bar{x}_l$, the vectors s_l and $\bar{s}_l$ are collinear, we will have $x_{l+1} = \bar{x}_{l+1}$, due to the exact one-dimensional descent condition. If $k = l + 1$ is a multiple of m, then equality (20) will be satisfied according to the construction of methods (18) and (19).

For the conjugate gradient method on quadratic functions, the gradient vectors calculated during the algorithm operation are mutually orthogonal. Therefore, as a result of (13), with $H_0 = I$ we will have the formula

$$H_{k+1} = H_k - \beta \frac{g_k g_k^T}{(g_k, g_k)} = I - \beta \sum_{i=0}^{k} \frac{g_i g_i^T}{(g_i, g_i)} \tag{22}$$

Therefore, taking into account the orthogonality of the gradients, the equality holds $H_{l+1} g_{l+1} = g_{l+1}$.

Due to the condition of exact one-dimensional descent, $(s_l, g_{l+1}) = 0$. Therefore, according to (12), taking into account (22), the transformation for the vector s when l is not a multiple of m will take the form:

$$s_{l+1} = s_l + \frac{H_{l+1} g_{l+1}}{(g_{l+1}, H_{l+1} g_{l+1})} = \frac{\bar{s}_l}{(g_l, g_l)} + \frac{g_{l+1}}{(g_{l+1}, g_{l+1})}.$$

Multiplying the last equality by the value (g_{l+1}, g_{l+1}), we obtain a proof of equality (21) for $k = l + 1$:

$$s_{l+1}(g_{l+1}, g_{l+1}) = \bar{s}_l \frac{(g_{l+1}, g_{l+1})}{(g_l, g_l)} + g_{l+1} = \bar{s}_{l+1}.$$

□

Note. For an arbitrary initial matrix $H_0 > 0$, we move to a coordinate system where the initial matrix is $\tilde{H}_0 = I$, and use the results of Theorem 1 in the new coordinate system. Then, due to the invariance of the method, we can again use the analogy with the CGM. Thus, the method has finite convergence on quadratic functions for any initial matrix $H_0 > 0$.

Method (11)–(13) generates directions similar to those in the conjugate gradient method. Transformation (11) suppresses the components of previous gradients in the gradient during multiplication $H_k g_k$. With high computational accuracy in simple minimization problems, the new gradient is already orthogonal to previous gradients. Therefore, the metric transformation plays no special role here. However, in ill-conditioned problems, such suppression of components of previous gradients plays a significant role in generating vectors with properties closer to conjugate ones. The metric matrix preserves the accumulated experience of the orthogonality of the current gradient involved in forming the descent direction to previous gradients. This property is useful even for quadratic functions in the case of their high conditionality [21, 22]. Method (11)–(13) is similar to process (7), (8). Its properties allow us to assume that the minimization methods with the descent direction (7), (8) have accelerating abilities. In what follows, we will use (7), (8) to obtain a minimization process with diagonal metric matrices for solving large-scale problems.

3 Algorithm with Step Adaptation

Algorithms with diagonal metric matrices include RMSProp [23] and ADAM [9]. These algorithms were improved by using a special step adaptation method (algorithm h_RMSProp was proposed in [24] and h_ADAM in [17]). The structure of the h_ADAM algorithm has three important elements. These are the step adjustment method, the descent direction selection method, and the diagonal metric matrix adjustment method. We will fill in these elements of the algorithm structure based on the iterative process (7), (8).

Let's consider the basic idea of step size selection based solely on the function gradient values. Let's consider the problem of minimizing the function $f(x)$, $x \in R^n$. In relaxation minimization methods, successive approximations are constructed using the formulas:

$$x_{k+1} = x_k - h_k s_k,\ s_k \in R^n \tag{23}$$

Here, s_k is descent direction, h_k is one-dimensional search step, $f(x_{k+1}) \le f(x_k)$. In exact one-dimensional descent $h_k = \underset{h \in \boldsymbol{R}}{\operatorname{argmin}} f(x_k - h s_k)$. Let us consider the situation of one-dimensional descent in Fig. 1. The iteration starts from the point x.

The principle for adjusting the search step during an iteration is as follows. Depending on the search step size in Fig. 1, there are three gradient values: g_1, g_2, and g_3. Gradient g_2 forms a 90° angle with the direction s. If the step is insufficient, gradient g_1 is calculated. Its angle with the direction s is acute. In this case, when adjusting the step, it should be increased. If the step is excessive, gradient g_3 is calculated. Its angle with the direction s is obtuse. In this case, the step should be reduced. Thus, the principle of the step adaptation scheme is based on reproducing the state in which the descent direction and the new gradient are found during exact one-dimensional descent. During exact one-dimensional descent, the angle between these directions is right angle. During imprecise descent, the step is adjusted according to the following rule: if the angle between the descent direction and the new gradient is obtuse, then the step is too large and should be reduced; if the angle between the descent direction and the new gradient is acute, then the step should be increased. In this case, the characteristics of the gradient magnitudes are not used, but

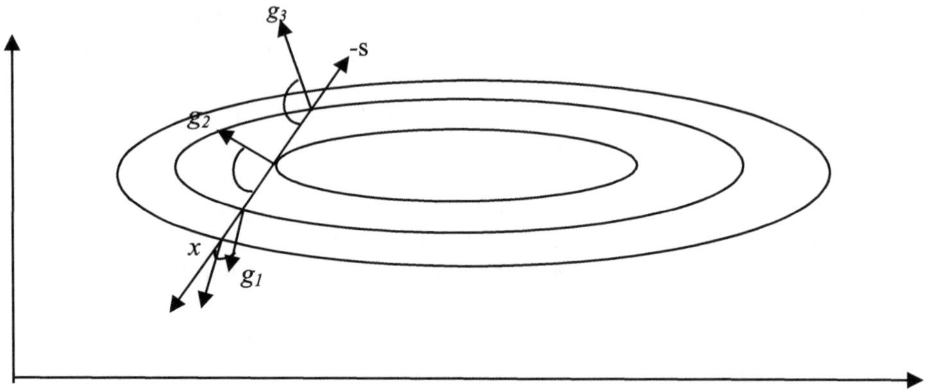

Fig. 1. Scheme of one-dimensional descent.

only the directions of the normals of the tangent hyperplanes to the isosurface. Unlike known methods of step adjustment in variations of the stochastic gradient method, the step adaptation method used here uses only the structural characteristics of the isosurface topology.

Several step adaptation algorithms for the gradient method, which do not involve a one-dimensional search, are proposed in [25]. In this section, we will extend the application of one of them to a new algorithm. We use the technique of decreasing or increasing the step in accordance with the idea described earlier. The simplest way to organize an algorithm of this kind is from the standpoint of machine learning.

Denote an optimal minimization step in (23) as h_k^*, then form a quality functional:

$$\Phi(h) = \left(h_k - h_k^*\right)^2/2. \tag{24}$$

In [25], several ways of this transformation are used. The most suitable for solving minimization problems under noise conditions is the following method:

$$h_{k+1} = h_k - q_k \text{sign}\left(h_k - h_k^*\right).$$

Here, possible oscillations due to large changes in $h_k - h_k^*$ values when estimating h_k^* are excluded. To organize the process here, we only need information about whether the step h_k is greater or less than the step h_k^*. Denote $g_k = \nabla f(x_k)$. According to Fig. 1, the step is small for $(s, g_{k+1}) > 0$, and is large for $(s, g_{k+1}) < 0$. Therefore, the step transformation formula can also be represented as follows

$$h_{k+1} = h_k + q_k \text{sign}(s, g_{k+1}).$$

Due to significant changes in h_k values, a more suitable method is the step adaptation method in which the transformation is consistent with the step size

$$h_{k+1} = h_k (1+\alpha)^{\text{sign}(s, g_{k+1})}, \alpha \in [0.3 - 0.001]. \tag{25}$$

An algorithm with step adaptation will be used to minimize functions with gradient noise. Therefore, for a random variable $sign(s, g_{k+1})$ with zero mean on a segment of iterations, we must have equality $h_{k+m} = h_k$ for the same number of positive and negative values of $sign(s, g_{k+1})$. Transformation (25) ensures the required equality, i.e., the absence of a trend component, for the same number of positive and negative values of $sign(s, g_{k+1})$ on a segment of m transformations.

In respect that the algorithms described below can be applied to essentially non-quadratic and noisy minimization problems, we will use normalization of the descent vector:

$$x_{k+1} = x_k - h_k s_k / \|s_k\|.$$

Here, the influence of the gradient magnitude on the bias obtained at the iteration is excluded.

The following steps concern the selection of the descent direction based on formulas (7) and (8), similar to the scheme corresponding to the ADAM method with step adaptation (h_ADAM) [17].

The h_ADAM algorithm uses a diagonal metric matrix. We denote it by A_k. Here, k is the iteration number. A smoothed gradient is used to generate the descent direction s_k:

$$p_k = \beta_1 \cdot p_{k-1} + (1 - \beta_1) g_{k-1}, p_0 = 0. \tag{26}$$

The descent direction is calculated using the diagonal metric matrices

$$s_k = \left(A_k^{1/2} + \varepsilon I\right)^{-1} p_k \tag{27}$$

Here, I is identity matrix, $A_k^{1/2} > 0$ is diagonal matrix such that $A_k^{1/2} A_k^{1/2} = A_k > 0$. Matrix A_k is transformed by the law $A_k = A(A_{k-1}, g_{k-1}, \beta_2)$, where the operator of diagonal elements transformation has the form

$$A(A, y, \beta)_{i,i} = \beta_2 A_{i,i} + (1 - \beta_2) g_i^2, i = 0, 1, 2, \ldots, n. \tag{28}$$

The new search step is calculated according to the step adaptation rule (25) as

$$h_k = \begin{cases} h_{k-1}, \quad if \mid (g_k, s_k) \mid \leq 10^{-15}, \\ h_{k-1}(1 + \alpha), \quad if \ |g_k, s_k| > 10^{-15} and \ (g_k, s_k) \geq 0, \\ \frac{h_{k-1}}{1+\alpha}, \quad if \ |g_k, s_k| > 10^{-15} \ and \ (g_k, s_k) < 0. \end{cases} \tag{29}$$

Here, the (g_k, s_k) value is checked to eliminate the influence of small or zero gradient values. Thus, we have outlined the elements that must be present in the new algorithm. Since the new algorithm will be used in problems where the gradient is calculated with noise and one-dimensional search is not available, the method for calculating the descent direction in (7) should be modified. The new direction will be:

$$s_k = H_k p_k, p_k = \beta_1 \cdot p_{k-1} + (1 - \beta_1) g_{k-1}, p_0 = 0. \tag{30}$$

Instead of transforming the matrices in (8), we will use the inverse matrix transformation (9) to organize the process of transforming diagonal matrices, and then use the inverse of the diagonal matrix.

The diagonal matrix transformation operator $A = H^{-1}$ corresponds to the transformation of only the diagonal elements in (9)

$$A(A, y, \beta)_{i,i} = A_{i,i} + \frac{\beta g_i^2}{(1-\beta)(g_i, H_i g_i)} = A_{i,i}\left(1 + \frac{\frac{\beta g_i^2}{A_{i,i}}}{(1-\beta)\sum_{i=1}^{n}\frac{g_i^2}{A_{i,i}}}\right), i = 0, 1, 2, \ldots, n. \quad (31)$$

In the new algorithm, we will use $H = A^{-1}$. Note the similarity between the transformations of the matrices A in the h_ADAM algorithm and the new algorithm (31).

According to (31), the matrix A is increasing. Therefore, it is sometimes necessary to normalize the main diagonal. Here, ill-conditioning may also arise. Therefore, it is necessary to increase the minor elements of the main diagonal by some small amount. Here, we present a version of the modified algorithm (7), (8) in accordance with formulas (30), (31).

Algorithm 1(h_AG).

1. Set the initial point x_0*, gradient* $g_0 = \nabla f(x_0)$*, the initial search step* h_0*, the initial diagonal matrix* $A_0 > 0$, $p_{k-1} = 0$, $p_{k-1} \in R^n$*, and parameters* $\varepsilon \in [10^2 - 10^9]$, $\beta_1 \in [0 - 0.999]$, $\beta_2 \in [0.001 - 0.999]$*, step adaptation parameter* $1 > \alpha > 0$.

2. For $k = 0,1,2,\ldots$ *do*

2.1 Set

$$p_{k+1} = \beta_1 \cdot p_k + (1-\beta_1) g_k.$$

2.2 Find a new minimum approximation

$$x_{k+1} = x_k - h_k s_k / \|s_k\|, \; s_k = A_k^{-1} p_{k+1}.$$

2.3 Perform a metric matrix transformation

2.3.1 Normalize the elements

$$A_{i,i,k} = A_{i,i,k} / \max_{1 \le j \le n} A_{j,j,k}, i = 0,1,2,\ldots,n.$$

2.3.2 Limit the minimum element to the ε *value*

$$A_{i,i,k} = \max\{A_{i,i,k}, \varepsilon\}, i = 0,1,2,\ldots,n.$$

2.3.3 Transform the diagonal elements

$$A_{k+1} = A(A_k, g_k, \beta_2),$$

where the transformation operator of matrix elements is

$$A(A,y,\beta)_{i,i} = A_{i,i} + \frac{\beta}{1-\beta} \times \frac{g_i^2}{(g, A^{-1} g)}, i = 0,1,2,\ldots,n.$$

2.4 Compute $g_{k+1} = \nabla f(x_{k+1})$.

2.5 Compute a new search step

2.5.1 *If* $|(g_{k+1}, s_k)| \le 10^{-15}$ *then* $h_{k+1} = h_k$ *else*

2.5.2 *If* $|(g_{k+1}, s_k)| > 10^{-15}$ *and* $(g_{k+1}, s_k) \ge 0$ *then* $h_{k+1} = h_k(1+\alpha)$ *else*

2.5.3 *If* $|(g_{k+1}, s_k)| > 10^{-15}$ *and* $(g_{k+1}, s_k) < 0$ *then* $h_{k+1} = h_k/(1+\alpha)$.

2.6 If the stopping criterion is met, then stop.

In (32), the smoothed gradient is calculated, step 2.1 corresponds to the calculations of the h_ADAM algorithm. In step 2.2, a search step is made along the direction $s_k = H_k p_{k+1} = A_k^{-1} p_{k+1}$. Here, the inversion of matrices A is used. Matrix A corresponds to the matrix from (9) of algorithm (7), (8). While the matrix transformations of the h_ADAM algorithm (28) are similar to those of the new algorithm (36), their use differs in the degree of the matrix. In the h_ADAM and ADAM algorithms, according to (27), $s_k = \left(A_k^{1/2} + \varepsilon I\right)^{-1} p_k$. In step 2.3, a metric transformation is performed. First, matrix (33) is normalized. According to (31), multiplying a matrix by a number does not affect the relation between the elements. With further matrix transformations, normalization does not affect the factors of the matrix elements, as can be seen in (31). Changing elements that are too small in (34) reduces the conditionality of the matrix, if present. In (36), the elements are transformed in accordance with (9). In step 2.5, the step adaptation is performed based on information about the new gradient.

4 Numerical Results

For the methods under study, we will adopt the following notation:

h_AG – minimization algorithm with scaling metric matrices and step adaptation;

h_ADAM – ADAM minimization algorithm with step adaptation;

HS – Hestenes–Stiefel conjugate gradient method [26].

All methods involved in the study used the step adaptation algorithm [24–26]. The listed methods lack full-scale metric matrices, which allows them to be helpful in solving high-dimensional problems. One of the most successful methods for solving large-scale problems with a high degree of conditionality is the conjugate gradient method. Therefore, when testing the proposed methods, we use the Hestenes–Stiefel conjugate gradient method as a benchmark, which uses a one-dimensional search [22]. To perform a one-dimensional search in CGM, additional gradient and function value calculations are used. Due to the presence of a one-dimensional search, the iteration costs increase. For the CGM, the tables indicate the number of iterations and the number of gradient and function calculations required to obtain a solution.

For test functions we use the notation $f(x, [parameters])$. In computations we will use the gradient of the function with noise $\nabla f_r = \nabla f + r \times \| \nabla f \| \times V$, where V is a random vector uniformly distributed in the unit sphere, and the value of r varies.

All the methods except HS use only gradient values. The function values are calculated for the purpose of assessing the method's progress in minimizing the function and as a stopping criterion. The algorithm stops under the condition $\boldsymbol{f - f* \leq 10^{-10}}$, where f^* is a function value at a minimum point. For all test functions $f^* = 0$. The initial point x_0 and the value ε are given in the description of the corresponding function.

Tables 1, 2, 3, 4, 5, 6 and 7 show the results of minimization. The tables indicate the number of algorithm iterations (gradient computations) required to achieve the specified accuracy. For the HS method a pair of values is given: number of iterations / number of function computations. A comparison with HS enables to determine the acceleration properties of the new h_AG algorithm. A comparison with h_ADAM provides insight into the efficiency of the new algorithm compared to a well-known efficient method using scaling matrices.

4.1 Quadratic Function

The following quadratic function is taken:

$$f_Q(x, [amax]) = \frac{1}{2}\sum\nolimits_{i=1}^{n} a_i x_i^2,\ a_i = a\max^{\frac{i-1}{n-1}}.$$

Eigenvalues a_i of this function has the limits $\lambda_{min} = 1$ and $\lambda_{max} = a_{max}$. The choice of the starting point affects the convergence of the methods, so we performed calculations for various starting points. In Table 1, the first starting point is located outside the ravines; the main initial residual in variables is in the region of large eigenvalues of the matrix of second derivatives, which significantly simplifies the operation of the gradient method. For the second starting point, the residual in variables is the same for all eigenvalues. For all tables $i = 1, 2, 3, \ldots, n$.

Table 1. Function f_Q $(x, [a_{max} = 10^4])$ minimization results.

n	$x_i^0 = i/n$			$x_i^0 = 1$		
	h_AG $\beta_1 = 0.9$ $\beta_2 = 0.1$	h_ADAM $\beta_1 = 0.9$ $\beta_2 = 0.999$	HS	h_AG $\beta_1 = 0.9$ $\beta_2 = 0.1$	h_ADAM $\beta_1 = 0.9$ $\beta_2 = 0.999$	HS
100	350	322	244 / 490	384	467	290 / 582
1000	327	464	567/1140	343	357	701 / 1409
10000	379	405	663/1339	397	403	809 / 1638

At the next starting point, the components of the variables corresponding to small eigenvalues predominate. The point is located near the bottom of a multidimensional ravine. The following results are obtained on a quadratic function with a higher degree of conditioning.

Table 2. Function f_Q $(x, [a_{max} = 10^6])$ minimization results.

n	$x_i^0 = 100(n - i + 1)$		$x_i^0 = i/n$		
	h_AG $\beta_1 = 0.9$ $\beta_2 = 0.1$	h_ADAM $\beta_1 = 0.9$ $\beta_2 = 0.999$	h_AG $\beta_1 = 0.9$ $\beta_2 = 0.1$	h_ADAM $\beta_1 = 0.9$ $\beta_2 = 0.999$	HS
100	2910	8659	380	356	1011 / 2310
1000	3141	30518	391	553	4911 / 11499
10000	4117	8613	437	447	6281 / 13306

Based on the Tables 1, 2, 3 and 4 results, the following conclusions can be drawn.

1. H_AG and h_ADAM methods are comparable in efficiency.
2. The convergence of gradient method is significantly affected by roundoff errors and the precision of one-dimensional descent. As a result, its convergence rate decreases, especially as the degree of problem conditioning increases.
3. The efficiency of the methods depends on the choice of the starting point. At a starting point close to the bottom of the ravine (Table 2), the convergence of h_ADAM method significantly lower compared to h_AG method.
4. On this function, the problem's conditioning can be removed by changing the coordinate system using diagonal matrices. The results for h_AG and h_ADAM algorithms compared to the CGM confirm the assumption that h_AG algorithm has acceleration properties, inheriting the acceleration properties of algorithms (11)–(13), as stated in Theorem 2.

Function with Multidimensional Ellipsoidal Ravine

The following function has a multidimensional ellipsoidal ravine. Minimization occurs when moving along a curvilinear ravine to a minimum point following function has a multidimensional ellipsoidal ravine. Minimization occurs when moving along a curvilinear ravine to a minimum point

$$f_{EX}(x, [amax, bmax]) = (1 - x_1)^2 + \text{amax}\left(1 - \sum\nolimits_{i=1}^{n} x_i^2/b_i\right)^2 + \frac{1}{2}\sum\nolimits_{i=2}^{n} x_i^2/b_i, .b_i = \text{bmax}^{\frac{i-1}{n-1}}$$

The initial point is $x^0{}_1 = -1, x^0{}_i = i, i = 2,3,\ldots n, \varepsilon = 10^{-10}$.

Table 3. Function f_{EX} $(x, [a_{max} = 100, b_{max} = 1000])$ minimization results.

n	$x_i^0 = -1/(nb_i)^{1/2}$		
	h_AG $\beta_1 = 0.9$ $\beta_2 = 0.9$	h_ADAM $\beta_1 = 0.9$ $\beta_2 = 0.999$	HS
100	1224	1501	15490 / 38053
1000	934	911	29296 / 72678
10000	1921	2712	11544 / 28213

1. The results of h_AG and h_ADAM step-adaptive methods are practically equivalent.
2. The acceleration properties arising from the use of diagonal metric matrices are evident on this function. The results for h_AG and h_ADAM methods are significantly better than those of HS method.

4.2 Non-Quadratic Function

The following function has topological similarity of level surfaces to a quadratic function.

$$f_{Q^{\wedge}2}(x, [amax]) = \left(\sum\nolimits_{i=1}^{n} a_i x_i^2\right)^2, a_i = \text{amax}^{\frac{i-1}{n-1}}$$

1. The results of h_AG method are significantly better than those of h_ADAM method.
2. The acceleration properties of h_AG and h_ADAM algorithms using scaling matrices are confirmed by comparison with HS method.

4.3 Function with Variables Scaling

The third function used additional variables c_i to change the scales a_i for each variable.

$$f_{abc}(x, [amax, bmax]) = \frac{1}{2}\sum\nolimits_{i=1}^{n} a_i c_i x_i^2, a_i = \text{amax}^{\frac{i-1}{n-1}},$$

Table 4. Function $f_{Q\wedge 2}(x, [a_{max} = 10^6])$ minimization results.

n	$x_i^0 = 100(n - i + 1)^2/n$		
	h_AG $\beta_1 = 0.9$ $\beta_2 = 0.2$	h_ADAM $\beta_1 = 0.9$ $\beta_2 = 0.999$	HS
100	998	1717	9749 / 23532
1000	1321	3164	10436 / 24335
10000	3746	4105	12122 / 28707

$$c_i = \frac{bmax}{b_i}\left(\frac{x_i^2}{1 + x_i^2}\right) + b_i\left(1 - \frac{x_i^2}{1 + x_i^2}\right), b_i = \text{bmax}^{\frac{i-1}{n-1}},$$

This function near the extremum will have the form

$$f_{abc}(x, [amax, bmax]) \approx \frac{1}{2}\sum\nolimits_{i=1}^{n} a_i b_i x_i^2.$$

Far from the extremum, we obtain a function in which the coefficients b_i are used in reverse order

$$f_{abc}(x, [amax, bmax]) \approx \frac{1}{2}\sum\nolimits_{i=1}^{n} a_i \frac{\text{bmax}}{b_i} x_i^2.$$

Changes in the scales of the coefficients c_i according to the presented limiting variants of the function can be represented by the transition: $a_i bmax/b_i \rightarrow a_i b_i$. Therefore, the first coefficient decreases by a factor of *bmax*, and the last coefficient increases by a factor of *bmax*.

Table 5. Function f_{abc} $(x, [a_{max} = 10^4, b_{max} = 10^3])$ minimization results.

n	$x_0 = (100{,}100, \ldots, 100)$		
	h_AG $\beta_1 = 0.9$ $\beta_2 = 0.2$	h_ADAM $\beta_1 = 0.9$ $\beta_2 = 0.999$	HS
100	3408	43971	24005 / 58476
1000	34038	39495	30602 / 71454
10000	35841	42319	48022 / 105163

1. The efficiency of the h_AG algorithm is significantly higher than that of the h_ADAM and HS methods.
2. The acceleration properties of using scaling matrices are confirmed by comparison with the conjugate gradient method.

4.4 Functions with Noise

The following test is performed on a functions with gradient noise. Noise level is set by parameter r.

Table 6. Minimization results with gradient noise, $n = 100$.

r	$f_Q(x, [amax = 10^4])$	$f_{EX}(x, [amax = 10^2, bmax = 10^3])$		
	$x_i^0 = 100(n - i + 1)$	$x_i^0 = -1/(nb_i)^{1/2}$		
	h_AG $\beta_1 = 0.9\ \beta_2 = 0.1$	h_ADAM $\beta_1 = 0.9\ \beta_2 = 0.999$	h_AG $\beta_1 = 0.9\ \beta_2 = 0.1$	h_ADAM $\beta_1 = 0.9\ \beta_2 = 0.999$
1	3159	10366	7654	29635
2	4193	41410	31165	73560
3	77101	105045	75318	140862
4	130008	185586	185294	251508
5	203569	287944	348088	403907
6	289803	413094	501755	591178
7	399299	567439	644012	790679
8	538112	743540	845341	1093874
9	692707	946125	1046815	1336462
10	898585	1183209	1655108	1807074

1. In the case of quadratic function, the results for the h_AG and h_ADAM methods do not reveal significant differences between the algorithms. Due to the presence of noise during optimization, there are no situations where the convergence rate of the methods changes abruptly depending on the starting point. The methods demonstrate the feasibility of solving the problem with a noise level significantly exceeding the gradient norm.
2. The results of the h_ADAM algorithm on function with curvilinear ravine f_{EX} are somewhat worse than h_AG algorithm.

4.5 Regression Problem

One of the potential problems in estimating the parameters of a linear model is poor scaling. A similar problem also occurs in neural network approximation problems. Unlike neural network models, which face many other challenges, the solution to a linear model is unambiguous. This makes it possible to purely model situations with a given degree of complexity and explore the capabilities of algorithms for cases of poorly scaled variables. The main function of the optimization (learning) method is to effectively overcome the problems of ill-conditioning in a minimization problem. In this regard, estimating the parameters of a linear model enables us to model the conditionality of the problem and evaluate the performance of the learning algorithm.

Denote $x^*\in R^n$, $x_i^* = 1, i = 1,..,n\text{-}1, x_n^* = 0$, matrix $m\in R^{N\times n}$, matrix rows $m_i\in R^n$, model values at observation points $y\in R^N$ and

$m_{i,j} = q \in [-a_i, a_i], j = 1, 2, \ldots, n-1, m_{i,n} = 1, y_i = (m_i, x^*), i = 1, 2, \ldots, N,$

where $q\in[-a_i, a_i]$ is a random uniformly distributed number, $a_i = amax^{\frac{i-1}{n-1}}$. The goal is to find unknown parameters $x^*\in R^n$, where m, y are given. The solution can be obtained by minimizing the function

$$f_r(x, [amax]) = \frac{1}{2}\sum_{i=1}^{N}\left(\left(m_{k(i)}, x\right) - y_{k(i)}\right)^2.$$

We will search for a solution using stochastic gradient methods, based on the gradients of a function formed on a randomly selected segment of data

$$f_{rb}(x, Nc, Nb, [amax]) = \frac{1}{2}\sum_{i=Nc+1}^{Nc+Nb}\left(\left(m_{k(i)}, x\right) - y_{k(i)}\right)^2,$$

where Nc is a random number on a segment $Nc\in[0,N\text{-}1]$, $k(i) = i \bmod N$ is remainder of division. The first stopping criterion is $\varepsilon = 10^{-10}$. If the algorithm fails to complete the task within the given number of iterations $n\_max = 500000$, then a dash is indicated in the table cell. Here, Nb is a mini batch size. Starting point is $x_i^0 = 0, i = 1, 2, 3, \ldots, n$.

Table 7. Linear regression without noise. Function f_r $(x, [a_{max}])$ minimization using gradient information f_{rb} $(x, Nc, Nb, [a_{max}])$, $n = 100$, $N = 1000$.

Nb	$a_{max} = 10$		$a_{max} = 100$		$a_{max} = 1000$		$a_{max} = 10000$	
	h_ADAM $\beta_1 = 0.9$ $\beta_2 = 0.999$	h_AG $\beta_1 = 0.9$ $\beta_2 = 0.1$	h_ADAM $\beta_1 = 0.9$ $\beta_2 = 0.999$	h_AG $\beta_1 = 0.9$ $\beta_2 = 0.1$	h_ADAM $\beta_1 = 0.9$ $\beta_2 = 0.999$	h_AG $\beta_1 = 0.9$ $\beta_2 = 0.1$	h_ADAM $\beta_1 = 0.9$ $\beta_2 = 0.999$	h_AG $\beta_1 = 0.9$ $\beta_2 = 0.1$
1	15330	13790	71989	13193	–	13533	–	23912
10	2406	3323	9195	3194	–	3257	–	3618
100	592	602	2272	756	2354	766	–	1187
200	511	495	1860	552	2043	1061	–	1971
300	406	440	1789	567	2007	540	2206	1044
1000	161	372	617	451	483	514	1561	657

Thus, as mismatch of data scaling increases, the efficiency of the h_ADAM algorithm decreases. Some problems can only be solved by increasing the mini batch size. The efficiency of the h_AG algorithm in these conditions decreases in lesser extent. However, it becomes acceptable to use very small mini batches for gradient calculation. For these problems, the advantages of the h_AG algorithm are obvious.

5 Conclusion

The algorithm presented in this paper combines the ideas of using diagonal metric matrices to solve ill-conditioned problems, step adaptation and smoothed gradient to eliminate problems caused by noise. The result is h_AG minimization algorithm with step adaptation and scaling diagonal matrices to generate a descent direction from the smoothed gradient. The presented algorithm is designed to solve ill-conditioned, large-scale minimization problems that arise in cases of poor scaling of the variables of the function being minimized, both with and without gradient noise.

A numerical study of the proposed method was performed using a series of tests and a comparison with the ADAM method with step adaptation (h_ADAM) and the Hestenes–Stiefel conjugate gradient method. The h_AG algorithm is effective when solving problems with poor variable scaling. It exhibits acceleration properties on such problems, as demonstrated by comparisons with the conjugate gradient method.

The h_AG algorithm is effective when solving conditioned problems with noise. For functions with linear ravines, the efficiency of the h_AG and h_ADAM algorithms is almost identical, but when problem conditionality increase and a starting point locates near the ravine, h_AG algorithm dominates. For functions with curved ravines, the efficiency of the h_AG algorithm is higher than that of the h_ADAM method.

In the regression problem, as the degree of data scaling deteriorates, the efficiency of the h_ADAM algorithm decreases. Some problems can only be solved by increasing the batch size for gradient calculation. As the degree of data scaling deteriorates, the efficiency of the h_AG algorithm decreases to a lesser extent. This makes it possible to use very small data batches for gradient calculation. A numerical experiment showed that when minimizing functions with noise, the h_AG and h_ADAM algorithms withstand tenfold noise on the gradient, however, the h_AG algorithm is more efficient with large noise levels. Thus, the use of the proposed modifications in situations with significant noise and a high degree of problem conditionality seems reasonable.

References

1. Duchi, J., Hazan, E., Singer, Y.: Adaptive subgradient methods for online learning and stochastic optimization. J. Mach. Learn. Res. **12**, 2121–2159 (2011)
2. Zeiler, M.D.: ADADELTA: An Adaptive Learning Rate Method. arXiv. (2012). https://doi.org/10.48550/arXiv.1212.5701.
3. Orabona, F., Pál, D.: Scale-free algorithms for online linear optimization. In: Chaudhuri, K., Gentile, C., Zilles, S. (eds.) Algorithmic Learning Theory ALT 2015, Lecture Notes in Computer Science, vol. 9355. Springer, Cham, Switzerland (2015). https://doi.org/10.1007/978-3-319-24486-0_19
4. Vaswani, S., Laradji, I., Kunstner, F., Meng, S.Y., Schmidt, M., Lacoste-Julien, S.: Adaptive Gradient Methods Converge Faster with Over-Parameterization (But You Should Do a Line-Search). arXiv. (2020). https://arxiv.org/abs/2006.06835.
5. Vaswani, S., Mishkin, A., Laradji, I., Schmidt, M., Gidel, G., Lacoste-Julien, S.: Painless Stochastic Gradient: Interpolation, Line-Search, and Convergence Rates. arXiv. (2019). https://arxiv.org/abs/1905.09997.

6. Reddi, S.J., Kale, S., Kumar, S.: On the convergence of Adam and beyond. In: Proceedings of the 6th International Conference on Learning Representations (ICLR), Vancouver, BC, Canada (2018)
7. Tan, C., Ma, S., Dai, Y.-H., Qian, Y.: Barzilai-Borwein step size for stochastic gradient descent. In: Proceedings of the Advances in Neural Information Processing Systems, pp. 685–693, Barcelona, Spain (2016)
8. Chen, J., Zhou, D., Tang, Y., Yang, Z., Cao, Y., Gu, Q.: Closing the Generalization Gap of Adaptive Gradient Methods in Training Deep Neural Networks. arXiv. (2018). https://arxiv.org/abs/1806.06763.
9. Kingma, D.P., Ba, J.: Adam: A Method for Stochastic Optimization. arXiv. (2014). https://arxiv.org/abs/1412.6980.
10. Ward, R., Wu, X., Bottou, L.: AdaGrad stepsizes: sharp convergence over nonconvex landscapes. In: Proceedings of the 36th International Conference on Machine Learning, Long Beach, **97**, pp. 6677–6686. PMLR, CA, USA (2019)
11. Xie, Y., Wu, X., Ward, R.: Linear convergence of adaptive stochastic gradient descent. In: Proceedings of the 33th International Conference on Artificial Intelligence and Statistics, **108**, pp. 1475–1485. PMLR (2020)
12. Cheng, W., Pu, R., Wang, B.: AMC: adaptive learning rate adjustment based on model complexity. Mathematics **13**, 650 (2025)
13. Liu, L., et al.: On the Variance of the Adaptive Learning Rate and Beyond. arXiv. (2019), doi:https://arxiv.org/abs/1908.03265.
14. Chevalier, S.: A parallelized, Adam-based solver for reserve and security constrained AC unit commitment. Electr. Power Syst. Res. **235**, 110685 (2024). https://doi.org/10.1016/j.epsr.2024.110685
15. Taniguchi, S., et al.: ADOPT: modified Adam can converge with any β2 with the optimal rate. In: Proceedings of the 38th Conference on Neural Information Processing Systems, Vancouver, BC, Canada (2024)
16. Shao, Y., et al.: An improvement of Adam based on a cyclic exponential decay learning rate and gradient norm constraints. Electronics **13**, 1778 (2024)
17. Krutikov, V., Tovbis, E., Kazakovtsev, L.: Adam algorithm with step adaptation. Algorithms **18**, 268 (2025). https://doi.org/10.3390/a18050268
18. Polyak, B. T.: Introduction to optimization. Optimization Software, New York (1987).
19. Krutikov, V.N., Petrova, T.: Relaxation method of minimization with space extension in the subgradient direction. Ekon. Mat.Met. **39**, 106–119 (2003)
20. Zhukovskii, E.L., Liptser, R.S.: On a recurrence method for computing normal solutions of linear algebraic equations. USSR Comput. Math. Math. Phys. **12**(4), 1–18 (1972)
21. Krutikov, V.: A new relaxation subgradient method with metric change. Vestnik KemGU. **4**, 16–22 (2001) In Russian
22. Shor, N.: Minimization Methods for Nondifferentiable Functions. Springer, Berlin/Heidelberg, Germany (1985)
23. Tieleman, T., Hinton, G.: Lecture 6.5-rmsprop: divide the gradient by a running average of its recent magnitude. COURSERA neural Netw. Mach. Learn. (2012)
24. Krutikov, V., Tovbis, E., Kazakovtsev, L.: Stochastic gradient descent methods with step adaptation. In: Y. Kochetov et al. (Eds.): MOTOR 2025, Lecture Notes in Computer Science (LNCS) 15681, pp. 62–79 (2025). https://doi.org/10.1007/978-3-031-97077-1_5
25. Krutikov, V., Tovbis, E., Gutova, S., Rozhnov, I., Kazakovtsev, L.: Gradient Method with Step Adaptation. Mathematics **13**(1), 61 (2025). https://doi.org/10.3390/math13010061
26. Hestenes, M.R., Stiefel, E.: Method of conjugate gradients for solving linear systems. J. Res. Natl. Bur. Stand. **49**, 409–436 (1952)

Decentralized Gradient Descent with an Increasing Consensus Procedure Accuracy

Anastasiia Murzina[1,2(✉)], Ilya Kuruzov[3,4], and Alexander Gasnikov[3,4,5]

[1] Sirius University of Science and Technology, Sirius Federal Territory 354340, Russia
murzinanastasiia@gmail.com
[2] ITMO University, St. Petersburg 197101, Russia
[3] Moscow Institute of Physics and Technology, Moscow 117303, Russia
[4] Innopolis University, Innopolis 420500, Russia
[5] IITP RAS, Moscow 127051, Russia

Abstract. This research explores the sum of the L-smooth functions minimization problem in the decentralized setting. A wide class of iterative algorithms for such problems alternates between local optimization steps and consensus procedures. Fixing the number of communication rounds per gradient iteration in advance can be problematic when the function characteristics are unknown. To overcome this, we propose a decentralized gradient descent method in which the number of communications grows polynomially or logarithmically with the iteration count. We establish convergence guarantees and analyze how the resulting sequence of consensus errors depends on the chosen parameters, both for general L-smooth functions and for functions satisfying the PL condition. Numerical experiments validate the theoretical findings and identify optimal parameter choices for achieving a prescribed accuracy.

Keywords: Decentralized optimization · Decentralized Gradient Descent · Increasing consensus accuracy

1 Introduction

1.1 An Optimization Problem

In this paper, we focus on the following smooth sum-optimization problem:

$$\min_{x \in \mathbb{R}^d} \{f(x) := \frac{1}{n} \sum_{i=1}^{n} f_i(x)\}, \tag{1}$$

where each function $f_i : \mathbb{R}^d \to \mathbb{R}$ is held privately by node i in some network. Note, that such problem statement allows significantly accelerate computations of large models [5,7,33] that is especially important for modern models in DL [3,12]. Besides, it is important for problems involving signal processing [19],

A. Kazakov et al. (Eds.): MOTOR 2026, LNCS 16687, pp. 32–47, 2026.
https://doi.org/10.1007/978-3-032-31390-4_3

power systems control [11,26], statistics [21,23], classic machine learning [10], spectrum sensing [1] etc.

We assume that this network contains the constant set of vertices V and time-varying edges E^t. In this paper, we consider on undirected networks $G^t = (V, E^t)$ where each node can communicate by any information with immediate neighbors. Changes in real-world network is ubiquitous phenomena [4,22]. Therefore, methods for such networks play important role in modern optimization.

One of the first such methods is the decentralized consensus framework proposed in [31], which laid the theoretical groundwork for distributed agreement protocols in multi-agent systems. It is one of the early exact first-order methods for decentralized consensus optimization, introducing fixed-step schemes that inspired later Adapt-Then-Combine (ATC) developments [18]. The main disadvantage of such approach is convergence for only some neighborhood of solution. There are different ways to solve this problem. Especially, increase of communications per iteration allows to improve accuracy of solution. Such approaches base on so-called consensus procedure [29]. It allows to effectively construct decentralized algorithms based on their classic version for different class of problems: (strongly) convex deterministic [28] and stochastic [27] optimization, minimization problems with PL-condition and saddle-point problems with two-sided PL-condition [16]. These approaches based on idea of inexact oracle and multiple communications per iteration. Nevertheless, the most of them use the same number of communications per each iteration that leads to excess communications for early iterations.

Other common approaches to working with time-varying networks are gradient tracking procedures. Such method was proposed in [32,34] and developed for case with non-constant network in [20] to achieve geometric convergence. Further, this method was developed for different class of problems with constant or time-varying networks in works [15,17,30]. Despite the inspired performance of such algorithms, the main complexity is non-trivial tuning for network parameters.

In this paper we develop idea of consensus procedure [29] in the following way. It is clear that when method is far enough from solution it requires less communication. In this stage it tries to optimize object function. At the same time, when it moves to optimal point one need to increase number of communication to decrease consensus error. It gives us natural idea to present consensus procedure with increasing number of communication per iteration.

We provide theoretical analysis for general non-convex function demonstrating that our method can achieve arbitrary accuracy without the need to tune network parameters. This is a key difference compared to standard consensus-based methods [16,29] that requires knowledge about network parameters and different hard calculable parameters of function. Moreover, even gradient tracking [8,25] requires knowledge about network parameters.

Further, we generalize our analysis for well-known PL-condition proposed in [24]. This condition meets in many real-world problems [9,13]. Special simple case is solution of over-parametrized systems that is very relevant for modern large models (see details in [2]).

1.2 Assumptions and Notation

We denote by $\|\cdot\|$ the Euclidean norm for vectors and the Frobenius norm for matrices. The original problem can be equivalently reformulated as follows:

$$\min_{X \in \mathbb{R}^{n\times d}, \frac{1}{n}\mathbf{1}\mathbf{1}^\top X = X} \left[F(X) = \sum_{i=1}^{n} f_i(x_i) \right],$$

where x_i is a local copy of variable in node i and $X = (x_1 \dots x_n)^\top$ is row-stacked variable. Let $\mathbf{1}$ denote the vector of all ones. The projection of a matrix $X \in \mathbb{R}^{n\times d}$ onto the consensus subspace is defined as $\overline{X} = \frac{1}{n}\mathbf{1}\mathbf{1}^\top X$. The quantity $\|X - \overline{X}\|$ thus provides a measure of distance from consensus. Additionally, we define the aggregate mean variable $\overline{x} = \sum_{i=1}^{n} x_i$.

During the consensus procedure, the goal is to make the local copies x_i become approximately equal across all nodes. It could be done by a sequence of communication rounds, each consisting of the exchange procedure between neighboring nodes and the averaging of the obtained vector values [6].

Using notation above, let us formalize our main assumptions. We consider a sequence of non-directed communication graphs $\{G^t = (V, E^t)\}_{k=0}^{\infty}$ and a sequence of corresponding mixing matrices $\{W^t\}_{t=0}^{\infty} \subseteq \mathbb{R}^{n\times n}$ associated with it. We assume the following assumption.

Assumption 1. Mixing matrix sequence $\{W^t\}_{t=0}^{\infty}$ satisfies the following properties:

(i) (Decentralized property) If $(i, j) \notin E^t$, then the corresponding elements of the mixing matrix are zero $W_{ij}^t = 0$,
(ii) (Double stochasticity) $W^t\mathbf{1}_n = \mathbf{1}_n$, and $\mathbf{1}_n^T W^t = \mathbf{1}_n^T$,
(iii) (Contraction property) There exist $\tau \in \mathbb{Z}_{++}$ and $\lambda \in (0, 1)$ such that for every $t \geq \tau - 1$, it holds the following inequality

$$\|W_\tau^t X - \overline{X}\| \leq (1 - \lambda)\|X - \overline{X}\|, \tag{2}$$

where $W_\tau^t = W^t \dots W^{t-\tau+1}$.

In this work, we focus on a class of objective functions that are smooth in the sense of having a Lipschitz-continuous gradient. Specifically, throughout the paper we assume that all functions under consideration are L-smooth,

Assumption 2. (L-smoothness) A differentiable function $f : \mathbb{R}^d \to \mathbb{R}$ is L-smooth if

$$f(y) \leq f(x) + \nabla f(x)^T (y - x) + \frac{L}{2}\|y - x\|^2.$$

Relying solely on this L-smoothness property, we are able to prove convergence to a stationary point for the general non-convex setting without any further convexity or curvature conditions. In other words, we establish that the iterates generated by our algorithm converge to a point where the gradient norm vanishes, which is the best guarantee one can hope for in the worst-case non-convex scenario.

We describe L from the Lipschitz-continuous gradient inequality for all f_i in the set of functions from problem (1) with $L_l = \max_{1 \le i \le n} L_i$ and $L_g = \frac{1}{n}\sum_{i=1}^{n} L_i$.

At the same time, we also examine a more structured subclass of non-convex problems that satisfy the PŁ-condition.

Assumption 3. (PŁ-condition) The function f satisfies the Polyak-Łojasiewicz-condition, i.e., it holds the following inequality

$$f(x) - f^* \le \frac{1}{2\mu}\|\nabla f(x)\|^2.$$

While being much weaker than strong convexity, this condition still enables linear convergence rates for many first-order methods and covers a variety of important applications. Especially, an overparametrized non-linear system satisfies such condition (see [2,14]).

A useful consequence of combining L-smoothness with the PŁcondition is that the function automatically satisfies the quadratic-growth condition. More precisely, under Assumptions 2 and 3 one can show that

$$\|x - x^*\|^2 \le \frac{2}{\mu}(f(x) - f^*); \quad \forall x \in \mathbb{R}^d, \tag{3}$$

This equation is useful for estimation of gradient method trajectory.

Further we will consider two optimization tasks briefly named L-smooth and PL tasks are satisfying Assumptions 2 and 2 together with (3) correspondingly. The PL task property of the gradient of an objective function, guarantees the linear convergence [14] of the function value sequence generated by the gradient descent method in the centralized case.

1.3 Proposed Algorithm

We now present the decentralized gradient method with a an increasing number of communication rounds from iteration to iteration.

Having established the contraction Property (2) of the consensus procedure over τ rounds, we now describe the decentralized optimization Algorithm 1.1 with a fixed communication schedule. At each outer iteration k, every node first performs a local gradient descent step, then the algorithm executes exactly T^k communication rounds. The values T^k form a predetermined sequence, which may be chosen, for example, as a constant [14] or as a function of the iteration index.

Algorithm 1.1: Decentralized gradient descent

Input: Starting point $X^0 = \overline{X}^0$, step size $\gamma > 0$, number of steps N, the sequence of number for communication rounds $T^k = \lceil k^\beta \rceil$ in consensus with $\beta > 0$.

```
for k = 0, ..., N − 1 do
    Z^{k+1} = X^k − γ∇F(X^k);
    X^{k+1} = Consesus(Z^{k+1}, T^k);
end
return X^N;
```

The paper is organized as follows. In Sect. 2, we derive an upper bound on the maximum inexactness $|X - \overline{X}|$ and analyze the convergence of gradient descent in the presence of a sequence of iteration-wise inexactness. Section 3 establishes sufficient conditions on the growth of the number of communication rounds (Theorems 1 and 2) that guarantee the convergence rate of gradient descent is preserved. In Sect. 4, we present numerical experiments that confirm the theoretical findings.

2 Preliminary Convergence Bounds

If the consensus errors diminish over time, convergence guarantees akin to those of exact gradient descent can often be recovered. This motivates the study of gradient descent with increasing communication rounds (or equivalently, decaying inexactness), where the consensus error is initially larger than the desired tolerance δ of averaging at the first decent iterations but decrease as optimization progress.

We introduce a consensus error constraint for the argument $\|X^k - \overline{X}^k\|^2 \le \delta^k$ and the gradient deviation rate relative to the gradient of the averaged argument $\|\nabla f(x) - \nabla f(\overline{x})\|^2 \le \frac{L_l^2 \delta^k}{n}$.

The given below lemmas provide the limitations of the indicated sequence $\{\delta^k\}_{k=0}^{\infty}$ produced Algorithm 1.1 with the increasing consensus communication numbers $\{T^k\}_{k=0}^{\infty}$, that corresponding a non-increasing δ^k-inexact consensus result after the N_0-th gradient descent iteration. Let us start from lemma with estimation of gradient norm over a trajectory of algorithm.

Lemma 1. *As the number of iterations increases and the gradient norm decreases on the iterations, we have a sequence $\{\delta^k\}$ of consensus errors $\|X^k - \overline{X}^k\|^2 \le \delta^k$. The minimum and maximum norm of $\|\nabla f(\overline{x}^k)\|$ during first N steps are limited if the cumulative sequence error $s_{N-1} = \sum_{i=0}^{N-1} \delta^i$ is limited:*

$$\min_{0 \le k \le N-1} \|\nabla f(\overline{x}^k)\|^2 \le \frac{L_l^2/n \sum_{k=0}^{N-1} \delta^k + 2L_g\left(f\left(\overline{x}^0\right) - f\left(\overline{x}^*\right)\right)}{N}, \qquad (4)$$

$$\max_{0 \le k \le N-1} \|\nabla f(\overline{x}^k)\|^2 \le \frac{L_l^2}{n} \sum_{k=0}^{N-1} \delta^k + 2L_g \left(f\left(\overline{x}^0\right) - f\left(\overline{x}^*\right)\right). \tag{5}$$

Proof. If we are going to increase the communication number T^k with iterations from a few times which obviously cannot satisfy the arbitrary δ in the consensus accuracy. Let's consider the deviations from the average in iterations $X^{k+1} = W_\tau^t(X^k - \gamma \nabla F(X^k)) = W_\tau^t Z^{k+1}$ of Algorithm 1.1 with the following limited inexactness $\|X^k - \overline{X}^k\| \le (1-\lambda)^{\lfloor \frac{T^k}{\tau} \rfloor} \|Z^k - \overline{X}^k\| \le \sqrt{\delta^k}$.

Then we consider a process like iterations performed on a single node with an inaccurate gradient $\overline{x}^{k+1} = \overline{x}^k - \frac{\gamma}{n} \sum_{i=1}^n \nabla f_i(x_i^k) = \overline{x}^k - \gamma g^k$ and the deviation norm of the displacement vector during the iteration is no more than

$$\|g^k - \nabla f(\overline{x}^k)\| = \|1/n \sum_{i=1}^n \nabla f_i(x_i^k) - \nabla f_i(\overline{x}^k)\| \le 1/n \sum_{i=1}^n \|\nabla f_i(x_i^k) - \nabla f_i(\overline{x}^k)\| \le$$

$$1/n \sum_{i=1}^n L_i \|x_i^k - \overline{x}^k\| \le 1/n \sqrt{\sum_{i=1}^n L_i^2} \sqrt{\sum_{i=1}^n \|x_i^k - \overline{x}^k\|^2} \le L_l \sqrt{\delta^k / n}.$$

Using Assumption 2, we get

$$f(\overline{x}^{k+1}) \le f(\overline{x}^k) + \nabla f(\overline{x}^k)^T (\overline{x}^{k+1} - \overline{x}^k) + \frac{L_g}{2} \|\overline{x}^{k+1} - \overline{x}^k\|^2$$
$$= f(\overline{x}^k) - \frac{1}{L_g} \nabla f(\overline{x}^k)^T g^k + \frac{1}{2L_g} \|g^k\|^2,$$

add and subtract terms to group them relative to the bias $g^k - \nabla f(\overline{x}^k)$ due to consensus error:

$$f(\overline{x}^{k+1}) \le f(\overline{x}^k) - \frac{1}{L_g} \left(\|f(\overline{x}^k)\|^2 + \nabla f(\overline{x}^k)^T (g^k - \nabla f(\overline{x}^k)) \right)$$
$$+ \frac{1}{2L_g} (\|g^k - \nabla f(\overline{x}^k)\|^2 + \|\nabla f(\overline{x}^k)\|^2 + 2 \nabla f(\overline{x}^k)^T (g^k - \nabla f(\overline{x}^k)))$$
$$= f(\overline{x}^k) - \frac{1}{L_g} \|\nabla f(\overline{x}^k)\|^2 + \frac{1}{2L_g} \|\nabla f(\overline{x}^k) - g^k\|^2 \le f(\overline{x}^k) - \frac{1}{L_g} \|\nabla f(\overline{x}^k)\|^2 + \frac{L_l^2 \delta^k}{2nL_g}.$$

As a result of iterating gradient descent with an inexactness, we obtain a decrease in the function by at least

$$f(\overline{x}^{k+1}) - f(\overline{x}^k) \le \frac{L_l^2 \delta^k}{2nL_g} - \frac{1}{2L_g} \|\nabla f(\overline{x}^k)\|^2. \tag{6}$$

From the last inequality we notice that $\frac{L_l^2 \delta^k}{n} < \|\nabla f(\overline{x}^k)\|^2$ leads to a decrease in the value of the function and with Assumption 3 and inequality (3) to the solution $\overline{x}^*$. After summarization the iterations from 0 to $N-1$ for (6):

$$f(\overline{x}^*) - f(\overline{x}^0) \le f(\overline{x}^N) - f(\overline{x}^0) \le \frac{L_l^2}{2nL_g} \sum_{k=0}^{N-1} \delta^k - \frac{1}{2L_g} \sum_{k=0}^{N-1} \|\nabla f(\overline{x}^k)\|^2.$$

The minimum value $\|\nabla f(\overline{x}^k)\|^2$ is no more than the average and the maximum is no more than the total sum.

Further, using result above and aggressively enough increase of communications round, we can guarantee that starting from some iteration consensus error will be smaller then any desired value.

Lemma 2. *For all δ the number $N = N(\delta)$ of an iteration of Algorithm 1.1 exists such that $\forall k \geq N \Rightarrow \|X^k - \overline{X}^k\|^2 \leq \delta^k \leq \delta$ and before the inexactness reaches the required accuracy, it will not exceed $\delta_{\max} = 2\|X^0 - X^*\|$.*

Proof. It is worth noting that during the first steps $t < K$, where $\|X^t - \overline{X}^t\|^2 \leq \delta$ is not guarantee for any δ. The inaccuracy vector in the next iteration equals $Z^{k+1} - \overline{Z}^{k+1} = X^k - \overline{X}^k - \gamma(\nabla F(X^k) - \overline{\nabla F}(X^k))$. Let's apply the triangle inequality several times to estimate its norm:

$$
\begin{aligned}
&\|Z^{k+1} - \overline{Z}^{k+1}\| \leq \|X^k - \overline{X}^k\| + \gamma\|\nabla F(X^k) - \overline{\nabla F}(X^k)\| \\
&\leq \sqrt{\delta^k} + \gamma(\|\nabla F(X^k) - \nabla F(\overline{X}^k)\| + \|\nabla F(\overline{X}^k) - \overline{\nabla F}(\overline{X}^k)\| + \|\overline{\nabla F}(\overline{X}^k) - \overline{\nabla F}(X^k)\|).
\end{aligned}
$$

The norm of the gradient difference of the function at different points is limited by the distance between the points and the smoothness constant $\|Z^{k+1} - \overline{Z}^{k+1}\| \leq \sqrt{\delta^k}(1 + \gamma(L_l + L_g)) + \gamma\|\nabla F(\overline{X}^k) - \overline{\nabla F}(\overline{X}^k)\|$. If we denote the problem heterogeneity σ such $\forall k \geq 0, i : \|\nabla f_i(\overline{x}^k) - \nabla f(\overline{x}^k)\| \leq \sigma$ then

$$
\|Z^{k+1} - \overline{Z}^{k+1}\| \leq \sqrt{\delta^k}\left(2 + \frac{L_l}{L_g}\right) + \frac{\sigma\sqrt{n}}{L_g} \tag{7}
$$

We propose any increasing sequence $\{T^k\}_{k\geq 0}$ and accordingly decreasing one $\{a_k\}_{0\leq k\leq N} : (1-\lambda)^{\lfloor \frac{T^k}{\tau} \rfloor} = a_k < a_{k-1}$, the sequence δ^k is produced $\sqrt{\delta^{k+1}} \leq a_k(C\sqrt{\delta^k} + D)$ after revealing the recursion

$$
\sqrt{\delta^{k+1}} \leq C^k\sqrt{\delta^0}\prod_{i=0}^{k} a_i + D\sum_{i=0}^{k} C^{k-i}\prod_{j=i}^{k} a_j. \tag{8}
$$

From $k = 0, \delta^0 = 0$ and before getting $k = N_0$ the inexact averaging accumulates steps $\sqrt{\delta^{K+1}} \leq D\sum_{i=0}^{N_0} C^{N_0-1-i}\prod_{j=i}^{N_0} a_j = V_k/C = a_{N_0}(1/C + V_{k-1})$. We are looking for $V_{N_0} < V_{N_0-1} \Rightarrow a_N(1 + CV_{N_0-1}) < V_{N_0-1}$, it is guaranteed to be true under the condition

$$
a_{N_0} < \frac{V_{N_0-1}}{1 + CV_{N_0-1}} = \frac{1}{C} - \frac{1}{C(1 + CV_{N_0-1})} < \frac{1}{C}.
$$

After reaching $a_K \leq 1/C$, the sequence δ^k will decrease with increasing T^k. For example, there are two ways to set increasing sequences $T^k = k^\beta \Rightarrow N_0 < \lceil \frac{\tau}{\lambda}\log(2 + \frac{L_l}{L_g})\rceil^{1/\beta}$.

On the other hand, the maximum possible inexactness is limited by

$$
\|X^k - \overline{X}^k\| \leq \|X^k - X^*\| + \|\overline{X}^k - X^*\| \leq 2\|X^k - X^*\| \leq 2\|X^0 - X^*\|.
$$

The last inequality is valid as a consequence of the following two inequalities: the matrix W has eigenvalues less than one

$$\|W(X^k - \gamma\nabla F(X^k)) - X^*\| = \|W(X^k - \gamma\nabla F(X^k) - X^*)\| \leq \|X^k - \gamma\nabla F(X^k) - X^*\|$$

and for $\gamma \leq 1/L_g$ it holds

$$\|X^k - \gamma\nabla F(X^k) - X^*\| \leq \|X^k - X^*\|.$$

As a result, we got $\delta_{\max} \leq 4R$.

Lemma 2 states that, in the worst case, the performance of our algorithm will not be worse than that of an algorithm with a sufficiently large constant number of communication rounds. As it will be shown later, the number of iterations of the gradient descent is proportional to the maximum of the consensus inexactness. Without knowing the constants from the Assumption 2, it is possible to limit the maximum inaccuracy by choosing δ^k and modifying the consensus procedure, with communication terminated after achieving a given accuracy.

3 Optimal Communication Schedule with Consensus Accuracy Growth

3.1 Main Results

We consider a T^k monotonic increase strategy for Algorithm 2 [16], at each iteration of it we have access only to X^k, where $X = (x_1^T \ldots x_n^T)^T \in \mathbb{R}^{n\times d}$.

We formulate the main results of the paper. In the initial phase of the decentralized gradient descent, we observe a limited consensus error due to an insufficient number of the communication rounds among nodes. However, as the number of communications exceeds N_0, the system achieves the desired consensus accuracy, ultimately attaining the target optimization performance ε.

Theorem 1. *Let f be a function meets Assumptions 2 and 3 and let the local gradients ∇f_i be L_i-Lipschitz. Consider Algorithm 1.1 with step size $\gamma \leq 1/L_g$ and a communication schedule $\{T^k\}_{k\geq 0}$. Assume the consensus errors satisfy $\delta^k \leq C\left(1 - \frac{\mu}{L_g}\right)^k / k^{1+\alpha}$ with some constants $C, \alpha > 0$. Then after*

$$N = \mathcal{O}\left(\frac{L_g}{\mu}\log\frac{\max\left(f(\overline{x}^0) - f^*, \frac{L_l^2\delta_{\max}}{2n\alpha\mu}\right)}{\varepsilon}\right)$$

iterations of the gradient computation at each node for $f(\overline{x}^N) - f^* \leq \varepsilon$. To achieve such a geometric decay, the communication rounds must satisfy

$$T^k = \frac{\tau}{\lambda}\left(\frac{k}{2}\log\frac{1}{1-\mu/L_g} + \frac{1+\alpha}{2}\log k\right) \tag{9}$$

Consequently, to reach an ε-optimal solution, the total number of communication rounds is $T_{tot} = \mathcal{O}\left(\log^2\frac{1}{\varepsilon}\right)$.

We propose a communication strategy that dynamically increases the consensus accuracy per gradient step, achieving consensus convergence while maintaining the same asymptotic total communication complexity as established in prior work [16, Theorem 3.1]. The number of communication rounds in such case is linear in the iteration number from the presented above Theorem.

The increases communication rounds based on local consensus progress, matching the optimal $\|\nabla f(\overline{x}^N)\|^2 = \mathcal{O}(1/N)$ convergence rate in gradient norm under Assumption 2 alone. So for the smooth case, we consider the convergence estimate based on the gradient norm $\|\nabla f(\overline{x}^k)\|^2 < \varepsilon_\nabla$, where the gradient norm at the required point $\|\nabla f(\overline{x}^*)\| = 0$.

Theorem 2. *Let f be a function meets Assumption 2 and the local gradients ∇f_i be L_i-Lipschitz. Consider Algorithm 1.1 with step size $\gamma = 1/L_g$ and a communication schedule $\{T^k\}_{k\geq 0}$. Assume the partial sums of the consensus errors are limited $\sum_{k=0}^{N} \delta^k \leq \frac{C\delta_{\max}\varepsilon_\nabla}{2}$ with some constants $C, \alpha > 0$. Then the squared gradient norm satisfies $\|\nabla f(\overline{x}^N)\|^2 < \varepsilon_\nabla$ after N iterations*

$$N = \mathcal{O}\left(\frac{2}{\varepsilon_\nabla} \max\left(\frac{L_l^2 C\delta_{\max}}{n}, 2L_g(f(\overline{x}^0) - f^*)\right)\right).$$

In the process of following the strategy of increasing the number of communications with a lack of the inaccuracy control can lead to the first iterations without limiting the maximum deviation from the average argument during the optimization iteration.

Corollary 1. *The communication rounds per iteration must be $T^k = \frac{\tau(1+\alpha)}{2\lambda} \log k$ with the total amount is $T_{tot} = \mathcal{O}(\frac{1}{\varepsilon} \log \frac{1}{\varepsilon})$ is a sufficient strategy, starting from some N_0, for choosing the number of communications to maintain $\frac{1}{N} \sum_{k=N_0}^{N_0+N} \delta^k \leq \frac{\delta_{\max}\varepsilon_\nabla(1+\alpha)}{2\alpha}$ after N iterations of the gradient descent.*

Remark 1. The result above is obtained to the expression for T^k to simplify the result. At the same time, one can use $\tilde{T}^k = cT^k$ with a constant $c < 1$. This will deteriorate the convergence rate by a fixed factor depending on c, thereby allowing us to avoid exact knowledge of the parameters.

We propose a communication strategy for decentralized gradient descent that eliminates the need for prior knowledge of the problem parameters (network topology, algebraic connectivity etc.). Unlike existing optimization methods for time-varying network requiring hyperparameter tuning or worst-case assumptions, our approach achieves suboptimal rates without parameter dependence.

3.2 Proof of Main Results

This chapter provides proofs of the main theorems and corollaries for the special case of increasing the number of communications for smooth functions.

Proof (Theorem 1). Let us assume that f is μ-Polyak-Lojasiewicz and L-smooth (Assumptions 3 and 2), for some $L_g \geq \mu > 0$. Consider $(x^k)_{k\in\mathbb{N}}$ a sequence generated by the GD algorithm, with a stepsize satisfying $\gamma = \frac{1}{L_g}$, then with (6):

$$f(\overline{x}^{k+1}) - f^* \leq \frac{L_l^2\delta^k}{2nL_g} - \frac{2\mu\left(f\left(\overline{x}^k\right) - f^*\right)}{2L_g}. \tag{10}$$

Recursively insert the error of the function into the expression

$$\begin{aligned} f(\overline{x}^{k+1}) - f^* &\leq \left(1 - \frac{\mu}{L_g}\right)\left(f\left(\overline{x}^k\right) - f^*\right) + \frac{L_l^2\delta^k}{2nL_g} \\ &\leq \left(1 - \frac{\mu}{L_g}\right)^{k+1}\left(f\left(\overline{x}^0\right) - f^*\right) + \frac{L_l^2}{2nL_g}\sum_{i=0}^{k}\delta^i\left(1 - \frac{\mu}{L_g}\right)^{k-i} \end{aligned}$$

After replacing δ^i with chosen sequence $\left(1 - \frac{\mu}{L_g}\right)^i/i^{1+\alpha}$ we got

$$\begin{aligned} f(\overline{x}^{k+1}) - f^* &\leq \left(1 - \frac{\mu}{L_g}\right)^{k+1}\left(f\left(\overline{x}^0\right) - f^*\right) + \frac{L_l^2\delta_{\max}\left(1 - \frac{\mu}{L_g}\right)^k}{2nL_g}\sum_{i=0}^{k}\frac{1}{i^{1+\alpha}} \\ &\leq \left(1 - \frac{\mu}{L_g}\right)^{k+1}\left(f\left(\overline{x}^0\right) - f^*\right) + \frac{L_l^2\delta_{\max}\left(1 - \frac{\mu}{L_g}\right)^k}{2n\alpha L_g} \end{aligned}$$

and $(\log(1/\varepsilon)$ Complexity) for exact gradient descent was preserved in an inexact case

$$k \geq \frac{L_g}{\mu}\log\left(\frac{2\max\left[\frac{L_l^2\delta_{\max}}{2n\alpha L_g}, f(x^0) - f^*\right]}{\varepsilon}\right) \Rightarrow f(x^k) - f^* \leq \varepsilon. \tag{11}$$

ε-convergence may require an additional number of iterations performed without the required accuracy due to insufficient communication number T^k during the first iterations of descent. Next, we will consider a suitable strategy for selecting the number of communications T^k to ensure the required accuracy δ^k.

The inaccuracy δ^k and the number of iterations T^k are related through inequalities (2) and (7) by the compression value

$$a_k \leq \frac{\sqrt{\delta^{k+1}}}{2\max\left(\left(2 + \frac{L_l}{L_g}\right)\sqrt{\delta^k}, \frac{\sigma\sqrt{n}}{L_g}\right)}, \tag{12}$$

and the amount of communications $T^k = \frac{\tau}{\lambda}\log\frac{1}{a_k}$.

We find the bound between the before mentioned $\delta^i = C(1 - \frac{\mu}{L_g})^{i-1}/i^{1+\alpha}$ and T^k through the compression value to the average

$$a_k = \frac{L_g k^{-(1+\alpha)/2}}{2\max\left((2L_g + L_l)\left(1 - \frac{\mu}{L_g}\right)^{\alpha(k-1)/2}, \sigma\sqrt{n}\right)}\left(1 - \frac{\mu}{L_g}\right)^{k/2},$$

neglecting the constant of the problem $\sigma\sqrt{n}$, we obtain the asymptotic growth of the number of communications per k-th iteration

$$T^k = \mathcal{O}\left(\frac{\tau}{\lambda}\left(\frac{k}{2}\log\frac{L_g}{L_g-\mu} + \frac{1+\alpha}{2}\log k\right)\right).$$

Despite the known dependence of the number of communications on the iteration number, it is difficult to estimate $T^i, i < N_0$. Therefore, it is proposed to estimate and control the inaccuracy of the δ^1 consensus using special estimations which can be measured by communicating with neighboring nodes.

Proof of Theorem 2. We are going to consider $\{\delta^k\}$ such $\sum_k \delta^k$ is limited in according to (4).

$$\min_{0\leq k\leq N-1} \|\nabla f(\overline{x}^k)\|^2 \leq \frac{2}{N}\max\left[L_l^2/n\sum_{k=0}^{N-1}\delta^k, \left(f\left(\overline{x}^0\right) - f\left(\overline{x}^*\right)\right)\right] \leq \varepsilon_\nabla$$

We know that after some iteration number N_0 the sequence δ^k is decreasing. The specified ε_∇-accuracy will be achieved after N iterations provided $\sum_{k=0}^{N}\delta^k \leq N_0\delta_{\max} + \sum_{k=N_0}^{\infty}\delta^k \leq \frac{nN\varepsilon_\nabla}{2L_l^2}$. Determining the sequence δ^k that meets these requirements will allow us to develop a strategy for the number of communications in iterations.

The acceptable sequences by Theorem 2 may include, but is not limited to one of the $\delta^k = 1/k^{1+\alpha}$ sequence with positive valued α. We will find the appropriate strategy of the number of communications T^k for this.

Proof of Corollary 1. The infinite sums are limited with $\sum_{k=0}^{\infty}\delta^k = \frac{1}{\alpha}$, then the number of iterations is $N = \frac{2}{\varepsilon_\nabla}\max\left[\frac{L_l^2}{n\alpha}, 2L_g(f(\overline{x}^0) - f(\overline{x}^*))\right]$

Let's take into account the first N_0 steps, where there may be an increase in inaccuracy due to following not the δ-strategy, but the number of communications T^k. Then the inaccuracies $\delta^k, k > N_0$ will be limited as follows $\delta^k \leq \delta_{\max}\left(N_0/k\right)^{1+\alpha}$. After finding the amount of compression and logarithm, we get $T^k = \mathcal{O}\left(\frac{\tau(1+\alpha)}{2\lambda}\log k\right)$ and the number of gradient descent iterations

$$N = \mathcal{O}\left(\frac{2}{\varepsilon_\nabla}\max\left[\frac{4L_l^2RN_0^{1+\alpha}}{n}\left(1+\frac{1}{\alpha}\right), 2L_g(f(\overline{x}^0) - f(\overline{x}^*))\right]\right)$$

In this section, we have established two key results regarding the communication strategy for decentralized optimization. First, for functions satisfying the PL condition, we proved that a linearly increasing number of consensus steps per iteration, is sufficient to guarantee linear convergence to the exact solution. Second, for general L-smooth (but not necessarily strongly convex) functions, we derived a condition on the consensus accuracy sequence $\{\delta^k\}$ that must be driven to zero at a prescribed rate. Importantly, this condition does not directly require T^k to grow to achieve a sufficiently accurate averaging at each iteration. Thus, while the PL case allows a simple increasing schedule, the smooth case demands a more refined control of consensus accuracy, which can be realized either by increasing T^k.

4 Numerical Experiments

To show the practical performance of the proposed Algorithm 1.1 we performed a series of numerical experiments. Let us consider the following least squares minimization problem

$$\min_{x \in \mathbb{R}^d} \sum_{i=1}^{n} \|A_i x - b_i\|^2, \tag{13}$$

for given matrices $A_i \in \mathbb{R}^{m \times d}$, vectors $b_i \in \mathbb{R}^m$, when A_i, b_i placed in different nodes connected according to the time-varying network Erdős-Rényi graph $G(n,p)$ with mixing matrices W_k satisfied Assumption 1. We run Algorithm 1.1 for the problem (13), with dimension parameters $d = 100$, count of nodes $n = 25$, probability of the existence of an edge between vertices $p = 0.3$. We can choose the number of communications policy $T^k = \text{const}$ or $T^k = \alpha k$ between gradient descent iterations.

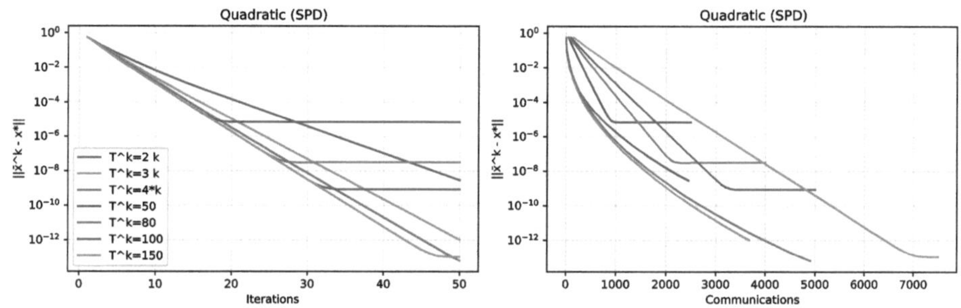

Fig. 1. Comparison between a constant and a growing number of communications. Left: relative error $\|\overline{x}^k - x^*\|$ versus the number of gradient descent iterations. Right: the same error versus the total number of exchanged messages.

Let's compare the parameters in both rules for the number of messages based on gradient descent steps and the number of messages with the metric $\|\overline{x}^k - x^*\|$ in Fig. 1. With some ratios of the growth parameter $\alpha = 3$ and a constant number of $T^k = 50$, the increasing-communication strategy achieves consistently smaller error at every stage in terms of total messages. Moreover, the growing policy drives the error to arbitrarily small values, while the constant policy eventually stagnates. The choice of the growth parameter also matters: $\alpha = 3$ yields slightly faster convergence than $\alpha = 2$ or $\alpha = 4$ for this problem.

The logistic regression problem was considered as a minimization of a smooth function:

$$\min_{x \in \mathbb{R}^d} \sum_{i=1}^{n} \sum_{j=1}^{m_i} \log\bigl(1 + \exp(-y_{ij} a_{ij}^\top x)\bigr) + \frac{\lambda}{2}|x|^2, \tag{14}$$

the setting of the problem $d = 100; n = 25; m_i = 25; p = 0.3$ is comparable to the previous experiment.

In addition to the logistic regression problem, we test our algorithm on a more challenging non-convex benchmark – the Rosenbrock function. The Rosenbrock function is a classic testbed for optimization: it is non-convex, has a curved, narrow valley, yet possesses a known unique global minimum. Moreover, its additive structure lends itself naturally to a distributed partitioning, making it a good test case for evaluating decentralized algorithms under substantial heterogeneity and non-convexity.

The global objective is the d-dimensional Rosenbrock function

$$F(x) = \sum_{j=1}^{d-1} \Big[(1 - x_j)^2 + 100\,(x_{j+1} - x_j^2)^2\Big], \tag{15}$$

with unique minimizer $x^* = (1, 1, \ldots, 1)^\top$ and $F(x^*) = 0$. We distribute the $d - 1 = 50$ pairwise terms among the $n = 25$ agents.

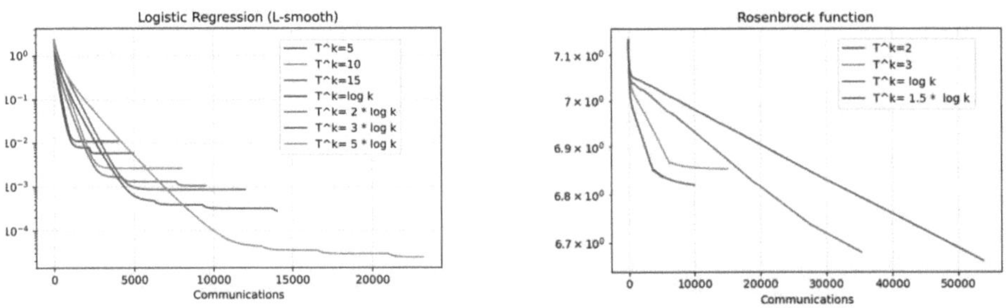

Fig. 2. Comparison of constant and logarithmic increasing communication policies for logistic regression and Rosenbrock function on Erdős-Rényi graph in terms of total messages.

In the Fig. 2, we do not observe a dominant selection strategy $T^k = \alpha \log k$ for any accuracy. Different accuracy values are better achieved by different parameter α values. As the algorithm approaches the minimum, the stagnation in consensus accuracy turns into a growing disadvantage: the distance to the true solution stops decreasing, or does so at an unacceptably slow rate. The right part of the Fig. 2 shows the evolution of the distance to the true solution $\|\bar{x}^k - x^*\|^2$ for the distributed Rosenbrock problem (15). This experiment highlights that even for a non-convex objective with tight variable coupling, a logarithmic increase in communication rounds suffices to avoid stagnation and keep the average iterate on a descending path.

In addition to the Erdős-Rényi random graphs, we also evaluate our method on a deterministic cycle topology. The cycle graph connects the $n = 25$ agents in a ring, where each agent communicates only with its two immediate neighbors. Despite having the smallest possible number of edges for a connected network, the cycle is 2-regular and has a diameter of $\lfloor n/2 \rfloor$, making consensus spread

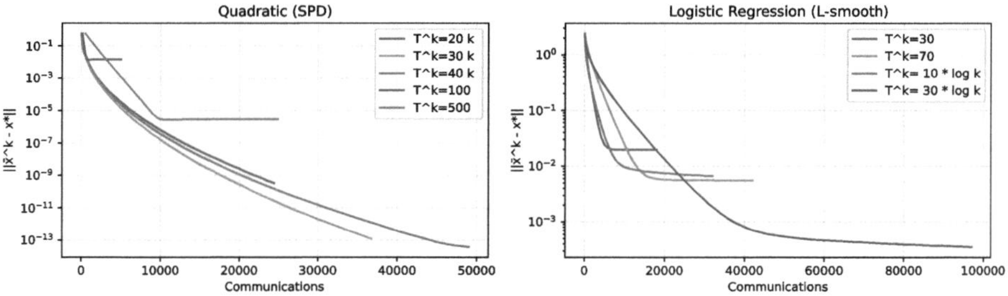

Fig. 3. Comparison of constant and increasing communication round policies for quadratic and logistic regression problem for the cycle graph.

slowly across the network. This extreme sparsity provides a challenging test for communication efficiency.

The quadratic (13) and logistic regression (14) tasks on the cycle graph $n = 100$ with worse τ and λ values relative to the Erdős-Rényi graph require an order of magnitude more iterations, but linear and logarithmic growth of communications are still the best strategies (Fig. 3) for the tasks correspondingly.

A strategy that would bypass these curves from the bottom left would definitely be more optimal than the suggested ones. Perhaps adding the adaptability of consensus regarding inaccuracy in iterations will allow us to build such a method.

Acknowledgments. The research was supported by Russian Science Foundation (project No. 21-71-30005-P), https://rscf.ru/en/project/21-71-30005/.

Disclosure of Interests. The authors have no competing interests to declare that are relevant to the content of this article.

References

1. Bazerque, J.A., Giannakis, G.B.: Distributed spectrum sensing for cognitive radio networks by exploiting sparsity. IEEE Trans. Signal Process. **58**(3), 1847–1862 (2009)
2. Belkin, M.: Fit without fear: remarkable mathematical phenomena of deep learning through the prism of interpolation. Acta Numer **30**, 203–248 (2021)
3. Ben-Nun, T., Hoefler, T.: Demystifying parallel and distributed deep learning: an in-depth concurrency analysis. ACM Comput. Surv. (CSUR) **52**(4), 1–43 (2019)
4. Boccaletti, S., et al.: The structure and dynamics of networks with higher order interactions. Phys. Rep. **1018**, 1–64 (2023)
5. Borzunov, A., et al.: Distributed inference and fine-tuning of large language models over the internet. Adv. Neural. Inf. Process. Syst. **36**, 12312–12331 (2023)
6. Boyd, S., Ghosh, A., Prabhakar, B., Shah, D.: Randomized gossip algorithms. IEEE Trans. Inf. Theory **52**(6), 2508–2530 (2006)
7. Dehghani, M., Yazdanparast, Z.: From distributed machine to distributed deep learning: a comprehensive survey. J. Big Data **10**(1), 158 (2023)

8. Di Lorenzo, P., Scutari, G.: Next: in-network nonconvex optimization. IEEE Trans. Signal Inf. Process. Netw. **2**(2), 120–136 (2016)
9. Du, S., Lee, J., Li, H., Wang, L., Zhai, X.: Gradient descent finds global minima of deep neural networks. In: International Conference on Machine Learning, pp. 1675–1685. PMLR (2019)
10. Forero, P.A., Cano, A., Giannakis, G.B.: Consensus-based distributed linear support vector machines. In: Proceedings of the 9th ACM/IEEE International Conference on Information Processing in Sensor Networks, pp. 35–46 (2010)
11. Gan, L., Topcu, U., Low, S.H.: Optimal decentralized protocol for electric vehicle charging. IEEE Trans. Power Syst. **28**(2), 940–951 (2012)
12. Goodfellow, I., Bengio, Y., Courville, A., Bengio, Y.: Deep Learning, vol. 1. MIT Press Cambridge (2016)
13. Jacot, A., Gabriel, F., Hongler, C.: Neural tangent kernel: convergence and generalization in neural networks. In: Advances in Neural Information Processing Systems, p. 31 (2018)
14. Karimi, H., Nutini, J., Schmidt, M.: Linear convergence of gradient and proximal-gradient methods under the Polyak-Łojasiewicz condition. In: Frasconi, P., Landwehr, N., Manco, G., Vreeken, J. (eds.) ECML PKDD 2016, Part I. LNCS (LNAI), vol. 9851, pp. 795–811. Springer, Cham (2016). https://doi.org/10.1007/978-3-319-46128-1_50
15. Koloskova, A., Lin, T., Stich, S.U.: An improved analysis of gradient tracking for decentralized machine learning. Adv. Neural. Inf. Process. Syst. **34**, 11422–11435 (2021)
16. Kuruzov, I., Alkousa, M., Stonyakin, F., Gasnikov, A.: Gradient-type methods for decentralized optimization problems with Polyak-łojasiewicz condition over time-varying networks. In: Optimization Methods and Software, pp. 1–28 (2025)
17. Liu, Y., Lin, T., Koloskova, A., Stich, S.U.: Decentralized gradient tracking with local steps. Optim. Methods Softw. **40**(5), 1153–1180 (2025)
18. Lopes, C.G., Sayed, A.H.: Diffusion least-mean squares over adaptive networks: formulation and performance analysis. IEEE Trans. Signal Process. **56**(7), 3122–3136 (2008)
19. Neal, P., Eric, C., Borja, P., Jonathan, E.: Distributed optimization and statistical learning via the alternating direction method of multipliers. Found. Trends® Mach. Learn. **3**(1), 1–122 (2011)
20. Nedic, A., Olshevsky, A., Shi, W.: Achieving geometric convergence for distributed optimization over time-varying graphs. SIAM J. Optim. **27**(4), 2597–2633 (2017)
21. Nedić, A., Olshevsky, A., Uribe, C.A.: Fast convergence rates for distributed non-bayesian learning. IEEE Trans. Autom. Control **62**(11), 5538–5553 (2017)
22. Olfati-Saber, R., Murray, R.: Consensus problems in networks of agents with switching topology and time-delays. IEEE Trans. Autom. Control **49**(9), 1520–1533 (2004)
23. Parikh, N., Boyd, S.: Proximal algorithms. Found. Trends Optim. **1**(3), 127–239 (2014)
24. Polyak, B.T.: Gradient methods for solving equations and inequalities. USSR Comput. Math. Math. Phys. **4**(6), 17–32 (1964)
25. Pu, S., Nedić, A.: Distributed stochastic gradient tracking methods. Math. Program. **187**(1), 409–457 (2021)
26. Ram, S.S., Veeravalli, V.V., Nedic, A.: Distributed non-autonomous power control through distributed convex optimization. In: IEEE INFOCOM, pp. 3001–3005. IEEE (2009)

27. Rogozin, A., Bochko, M., Dvurechensky, P., Gasnikov, A., Lukoshkin, V.: An accelerated method for decentralized distributed stochastic optimization over time-varying graphs. In: 2021 60th IEEE Conference on Decision and Control (CDC), pp. 3367–3373. IEEE (2021)
28. Rogozin, A., Lukoshkin, V., Gasnikov, A., Kovalev, D., Shulgin, E.: Towards accelerated rates for distributed optimization over time-varying networks. In: Olenev, N.N., Evtushenko, Y.G., Jaćimović, M., Khachay, M., Malkova, V. (eds.) OPTIMA 2021. LNCS, vol. 13078, pp. 258–272. Springer, Cham (2021). https://doi.org/10.1007/978-3-030-91059-4_19
29. Scaman, K., Bach, F., Bubeck, S., Lee, Y.T., Massoulié, L.: Optimal algorithms for smooth and strongly convex distributed optimization in networks. In: International Conference on Machine Learning, pp. 3027–3036. PMLR (2017)
30. Song, Z., Shi, L., Pu, S., Yan, M.: Optimal gradient tracking for decentralized optimization. Math. Program. **207**(1), 1–53 (2024)
31. Tsitsiklis, J., Bertsekas, D., Athans, M.: Distributed asynchronous deterministic and stochastic gradient optimization algorithms. IEEE Trans. Autom. Control **31**(9), 803–812 (1986)
32. Xu, J., Zhu, S., Soh, Y.C., Xie, L.: Augmented distributed gradient methods for multi-agent optimization under uncoordinated constant stepsizes. In: 54th IEEE Conference on Decision and Control (CDC), pp. 2055–2060. IEEE (2015)
33. Zeng, F., Gan, W., Wang, Y., Yu, P.S.: Distributed training of large language models: a survey. Nat. Lang. Process. J., 100174 (2025)
34. Zhu, M., Martínez, S.: Discrete-time dynamic average consensus. Automatica **46**(2), 322–329 (2010)

Decentralized gradient descent with an increasing consensus procedure accuracy

Anton Novitskii[1,4(✉)], Alexander Rogozin[1], and Alexander Gasnikov[1,2,3]

[1] Moscow Institute of Physics and Technology, Moscow, Russia
novitskii.ag@phystech.edu
[2] Institute for Information Transportation Problems, Moscow, Russia
[3] Caucasus Mathematic Center, Adygh State University, Moscow, Russia
[4] Trusted AI Research Center, Ivannikov Institute for System Programming, Moscow, Russia

Abstract. Decentralized optimization is well studied for smooth unconstrained problems. However, constrained problems or problems with composite terms are an open direction for research. We study structured (or composite) optimization problems, where the functional is a sum of a convex smooth function and a proper convex proximal-friendly term. Our method builds upon an accelerated proximal gradient descent and makes several consensus iterations between computations.

Keywords: convex optimization · distributed optimization · proximal method

1 Introduction

Distributed optimization has a wide range of applications. The fields where distributed optimization problems arise include power system control [9,20], formation control [10,17,22], distributed statistical inference and machine learning [8,15,16,19] distributed coordination and control [23], distributed averaging [5,18,31], distributed spectrum sensing [2]. Distributed optimization takes place when the data is separated between several computational entities due to a large amount of datasets, privacy constraints or a split structure of the data itself.

In this paper we focus on decentralized systems. Several computational nodes, or agents, locally hold objective functions and can communicate to each other. A centralized aggregator is not present in the network, so the agents directly communicate to each other. The structure of the network is represented as an undirected graph, and the agents exchange information with their immediate neighbors.

The communication network may also change with time, which typically happens due to technical instabilities [25]. A time-varying network corresponds to a changing communication graph.

The research was supported by Russian Science Foundation (project No. 21-71-30005-P), https://rscf.ru/en/project/21-71-30005/.

A. Kazakov et al. (Eds.): MOTOR 2026, LNCS 16687, pp. 48–63, 2026.
https://doi.org/10.1007/978-3-032-31390-4_4

Informally speaking, the complexity of a decentralized method depends on objectives condition number κ and on graph condition number χ. For static networks, optimal methods require $O(\sqrt{\chi}\sqrt{\kappa}\log(1/\varepsilon))$ communication rounds to reach ε-accuracy. For optimization over time-varying networks, optimal algorithms require $O(\chi\sqrt{\kappa}\log(1/\varepsilon))$ communications.

Related Work. Previously in the literature the classes or time-varying and time-static graphs have been studied. For functions with smooth gradients, lower communication and local computation complexity bounds were proposed in [27]. Optimal dual [27] and primal [13] algorithms were developed, as well. Lower bounds for problems with non-smooth Lipschitz objectives were proposed in [28] along with methods optimal up to a factor dependent on space dimension. Paper [7] proposed an optimal primal scheme based on gradient sliding. A method for problems with composite terms was developed in [32].

Distributed proximal methods typically arise in sparse signal recovery [12,14, 21] and are used in classical machine learning for feature selection [1,33]. In signal recovery, a linear regression problem with ℓ1-regularizer is considered. Papers [14,21] give decentralized proximal methods with asymptotic convergence rates and [12] proposes a linearly convergent algorithm. For general convex objectives, linear convergence rates are given by [1,33]. See Fig. 1 for details.

Loss	Comm. complexity
[14] lin. regr.	asymptotic
[21] lin. regr.	asymptotic
[12] lin. regr.	$O(\chi\kappa \cdot dm\ln(m)\ln(1/\varepsilon))$
[1] convex	linear
[33] convex	$O(\sqrt{\chi}\sqrt{\kappa}\ln(\widetilde{\kappa})\ln(1/\varepsilon))$
Ours convex	$O(\tau\chi\sqrt{\kappa}\ln^2(1/\varepsilon))$

Fig. 1. Overview of decentralized proximal methods. For static graphs, $\tau = 1$. Ambient dimension is denoted d, and $\widetilde{\kappa}$ is a quantity defined in [33].

There is a group of algorithms that use a consensus subroutine technique. Initially this approach has been proposed in [11] for time-static graphs. After that, it was applied to time-varying graphs. Deterministic setup was studied in [26] and stochastic setup with mini-batching was studied in [24]. The technique has also been applied to saddle-point problems [3,4]. Paper [4] supports composite min-max problems, but the analysis requires a bounded constraint set. Our assumptions do not require bounded gradient norm or bounded constraint set.

Our Contribution. We continue a series of works on consensus subroutine, proposing a decentralized proximal method. Our analysis is relatively easy and we do not require bounded gradients or bounded constraint set. Our approach only requires Lipschitz gradient and strong convexity. To the best of our knowledge, previously decentralized proximal methods for objectives of general type were studied in [1,33]. In comparison with these works, our contribution is summarized as follows.

- Our method is capable to work over time-varying graphs.
- Our approach enables to work with stochastic gradients.
- In comparison with [33], we have an additional logarithmic factor $\ln(1/\varepsilon)$ instead of $\ln(\widetilde{\kappa})$.

2 Problem Statement

In this paper, we study a sum-type minimization problem

$$\min_{x\in Q}\left[f(x)+g(x):=\frac{1}{m}\sum_{i=1}^{m}f_i(x)+g(x)\right],\quad f_i(x):=\mathbb{E}_{\xi_i\sim\mathcal{D}_i}\mathbf{f}_i(x,\xi_i). \tag{1}$$

Here ξ_i's are random variables with distributions $\mathcal{D}_i$, set Q is a closed convex subset of $\mathbb{R}^d$, and function $g(x)$ is a proper convex closed function, possibly non-smooth. Functions f_i are locally stored at computational agents. Each node has access to the stochastic gradient $\nabla\mathbf{f}_i(x,\xi_i)$. The agents also communicate through a time-varying computational network represented by a sequence of undirected graphs. At each time step, each of the nodes can only exchange information with its immediate neighbors.

2.1 Notation

Let $\otimes$ denote the Kronecker product. Let $\mathbf{x}=\mathrm{col}[x_1,\ldots,x_m]=[x_1^\top,\ldots,x_m^\top]^\top\in\mathbb{R}^{md}$ denote a column vector. Let $\lambda_{\max}(\cdot)$ and $\lambda_{\min}^{+}(\cdot)$ denote maximum and minimum nonzero eigenvalues of a matrix. Also let $\mathcal{L}=\{x_1=\ldots=x_m\}$ denote the consensus constraint set. We denote $\mathbf{1}$ a vector of all ones and introduce a projection operator $\mathbf{P}=(1/m)\mathbf{1}\mathbf{1}^\top\otimes\mathbf{I}$ (the dimension is known from the context).

We introduce $\mathbf{x}=\mathrm{col}[x_1,\ldots,x_m]$ and denote

$$F(\mathbf{x})=\sum_{i=1}^{m}f_i(x_i),\;\; G(\mathbf{x})=\sum_{i=1}^{m}g_i(x_i). \tag{2}$$

We use notation $\nabla^r F(\mathbf{x},\xi)$ for batched gradient of F, where r is batch size and $\xi=\left\{\xi_i^j\right\}_{i=1,j=1}^{m,r}$ is the set of stochastic realizations, i.e.

$$\nabla^r F(\mathbf{x},\xi)=\left[\frac{1}{r}\sum_{i=1}^{r}\nabla\mathbf{f}_1(x_1,\xi_1^i)\ldots\frac{1}{r}\sum_{i=1}^{r}\nabla\mathbf{f}_m(x_m,\xi_m^i)\right].$$

For brevity we will write just $\nabla^r F$. We also introduce a prox-operator for g w.r.t. set Q:

$$\text{prox}_g^\gamma(x) = \arg\min_{y \in Q} \left(g(y) + \frac{1}{2\gamma} \|y - x\|_2^2 \right).$$

Analogously, a prox-operator for G w.r.t. $Q^m = \{\mathbf{x} \in \mathbb{R}^{md} : x_i \in Q, i = 1, \ldots, m\}$ writes as

$$\text{prox}_G^\gamma(\mathbf{x}) = \arg\min_{\mathbf{y} \in Q^m} \left(G(\mathbf{x}) + \frac{1}{2\gamma} \|\mathbf{y} - \mathbf{x}\|_2^2 \right) = \text{col}[\text{prox}_g^\gamma(x_1) \ldots \text{prox}_g^\gamma(x_m)].$$

Note that $\text{prox}_G^\gamma(\mathbf{x})$ is a separable operator, i.e. it can be computed separately for $x_1, \ldots, x_m$.

2.2 Objective Functions

Our paper focuses on smooth strongly convex functions.

Assumption 1. *For each $i = 1, \ldots, m$ function f_i is μ_i-strongly convex, i.e. for any $x, y \in \mathbb{R}^d$ it holds*

$$f_i(y) \geq f_i(x) + \langle \nabla f_i(x), y - x \rangle + \frac{\mu_i}{2} \|y - x\|_2^2.$$

Assumption 2. *For each $i = 1, \ldots, m$ function $\mathbf{f}_i$ has gradient $\nabla \mathbf{f}_i(x, \xi_i)$ which is $L_i(\xi_i)$-Lipschitz continuous, i.e. for any $x, y \in \mathbb{R}^d$ it holds*

$$\mathbf{f}_i(y, \xi_i) \leq \mathbf{f}_i(x, \xi_i) + \langle \nabla \mathbf{f}_i(x, \xi_i), y - x \rangle + \frac{L_i(\xi_i)}{2} \|y - x\|_2^2.$$

Also there is a constant $L_i \geq 0$ such that $\sqrt{\mathbb{E}_{\xi_i} L_i(\xi_i)} \leq L_i \leq +\infty$.

Note that under Assumption 2, $\mathbb{E}_{\xi_i} \nabla \mathbf{f}_i(x, \xi_i) = \nabla f_i(x)$ and f is L_i-smooth

Assumption 3. *For each $i = 1, \ldots, m$ and $x \in \mathbb{R}^d$ it holds:*

$$\mathbb{E}_{\xi_i} [\|\nabla \mathbf{f}_i(x, \xi_i) - \nabla f_i(x)\|_2^2] \leq \sigma_i^2. \tag{3}$$

Also introduce local and global constants characterizing problem optimization parameters.

$$L_l = \max_{i=1,\ldots,m} L_i, \ \mu_l = \min_{i=1,\ldots,m} \mu_i, \tag{4a}$$

$$L_g = \frac{1}{m} \sum_{i=1}^m L_i, \ \mu_g = \frac{1}{m} \sum_{i=1}^m \mu_i, \tag{4b}$$

$$\sigma^2 = \frac{1}{m} \sum_{i=1}^m \sigma_i^2. \tag{4c}$$

It is known that global and local constants may significantly differ [25,27]. Moreover, we introduce worst-case smoothness constant over stochastic realizations L_ξ and a maximum gradient norm M_ξ.

$$L_\xi = \max_{i=1,\dots,m} \max_\xi L_i(\xi), \quad M_\xi = \max_{i=1,\dots,m} \max_\xi \|\nabla f_i(x^*, \xi)\|.$$

Function $g(x)$ is a proper convex closed function, possibly non-smooth, and Q is a closed convex set. We assume that $g(x)$ is proximal-friendly, i.e. its proximal operator can be easily computed.

We rewrite problem (1) as

$$\min_{\mathbf{x}\in\mathbb{R}^{md}} F(\mathbf{x}) + G(\mathbf{x}) \quad \text{s.t.} \quad x_1 = \ldots = x_m \tag{5}$$

2.3 Communication Network

Assumption 4. *We assume that nodes are connected via a time-varying network represented by a sequence of graphs $\{\mathcal{G}^k = (\mathcal{V}, \mathcal{E}^k)\}_{k=0}^\infty$. The graphs have a common set of vertices $\mathcal{V}$ but may have different edge sets $\mathcal{E}^k$. With each of the graphs, we associate a mixing matrix W^k.*

Mixing matrix sequence $\{W^k\}_{k=1}^\infty$ satisfies the following properties:

1. *(Network compatibility) For each $k = 1, 2, \ldots$ it holds $[W^k]_{ij} = 0$ if $(i,j) \notin \mathcal{E}^k$.*
2. *(Double stochasticity) For each $k = 1, 2, \ldots$ it holds $W^k\mathbf{1} = \mathbf{1}$, $\mathbf{1}^\top W^k = \mathbf{1}^\top$.*
3. *(Spectral property) There exist an integer $\tau \geq 1$ and $\chi > 1$ such that for all $k \geq \tau - 1$ and for all x it holds*
   $\left(\left\|(W_\tau^k - \frac{1}{m}\mathbf{1}\mathbf{1}^\top)x\right\|_2^2\right) \leq (1 - \chi^{-1}) \|x\|_2^2,$

   where $W_\tau^k = W^k \ldots W^{k-\tau+1}$. We also introduce $\mathbf{W}^k = W^k \otimes \mathbf{I}$.

3 Inexact Oracle Framework

Let us construct an inexact model for function $h(x) = \dfrac{1}{m}\sum_{i=1}^m f_i(x) + g(x)$.

Lemma 1. *Consider* $y \in \mathbb{R}^d,\ z \in \mathbb{R}^d,\ \mathbf{x} = \mathrm{col}[x_1, \ldots, x_m] \in \mathbb{R}^{md}$. *Define*

$$\eta = \frac{1}{2m}\left(\frac{L_l^2}{L_g} + \frac{2L_l^2}{\mu_g} + L_l - \mu_l\right), \tag{6}$$

$$\delta = \eta \sum_{i=1}^{m} \|x_i - y\|_2^2, \tag{7}$$

$$f_\delta(y, \mathbf{x}) = \frac{1}{m}\sum_{i=1}^{m}\left[f_i(x_i) + \langle \nabla f_i(x_i), y - x_i\rangle + \frac{1}{2}\left(\mu_l - \frac{2L_l^2}{\mu_g}\right)\|y - x_i\|^2\right],$$

$$\psi_\delta(z, y, \mathbf{x}) = \frac{1}{m}\sum_{i=1}^{m}\left[\langle \nabla f_i(x_i), z - y\rangle + g(z) - g_i(x_i)\right],$$

$$\widetilde{\psi}_\delta(z, y, \mathbf{x}, \xi) = \frac{1}{m}\sum_{i=1}^{m}\frac{1}{r}\sum_{j=1}^{r}\left[\left\langle \nabla \mathbf{f}_i(x_i, \xi_i^j), z - y\right\rangle + g(z) - g_i(x_i)\right].$$

Then $(f_\delta(y, \mathbf{x}), \psi_\delta(z, y, \mathbf{x}))$ *is a* $(\delta, 2L_g, \mu_g/2)$*-model of* f *at point* $\overline{x}$, *i.e.*

$$\frac{\mu_g}{4}\|z - y\|^2 \le f(z) - f_\delta(y, \mathbf{x}) - \psi_\delta(z, y, \mathbf{x}) \le L_g\|z - y\|^2 + \delta.$$

Moreover, we have

$$\mathbb{E}\widetilde{\psi}_\delta(z, y, \mathbf{x}, \xi) = \psi_\delta(z, y, \mathbf{x}), \tag{8a}$$

$$\mathbb{E}\left\|\widetilde{\psi}_\delta(z, y, \mathbf{x}, \xi) - \psi_\delta(z, y, \mathbf{x})\right\|^2 \le \frac{\sigma^2}{mr}. \tag{8b}$$

Proof. Relations (8) directly follow from the definition of $\widehat{\psi}_\delta$ and ψ_δ. Denote $\mathbf{y} = \mathbf{1} \otimes y,\ \mathbf{z} = \mathbf{1} \otimes z$. It is convenient to use the notation of $F(\mathbf{x}), G(\mathbf{x})$. First, write the following lower bound.

$$\begin{aligned}
F(\mathbf{z}) + G(\mathbf{z}) &\overset{①}{\ge} F(\mathbf{x}) + \left[\langle \nabla F(\mathbf{x}), \mathbf{y} - \mathbf{x}\rangle + \tfrac{\mu_l}{2}\|\mathbf{x} - \mathbf{y}\|^2\right] \\
&\quad + \left[\langle \mathbf{P}\nabla F(\mathbf{y}), \mathbf{z} - \mathbf{y}\rangle + \tfrac{\mu_g}{2}\|\mathbf{z} - \mathbf{y}\|^2\right] + G(\mathbf{z}) \\
&= \left[F(\mathbf{x}) + \langle \nabla F(\mathbf{x}), \mathbf{y} - \mathbf{x}\rangle + \tfrac{\mu_l}{2}\|\mathbf{x} - \mathbf{y}\|^2\right] + \langle \mathbf{P}\nabla F(\mathbf{x}), \mathbf{z} - \mathbf{y}\rangle \\
&\quad + \langle \mathbf{P}(\nabla F(\mathbf{y}) - \nabla F(\mathbf{x})), \mathbf{z} - \mathbf{y}\rangle + \tfrac{\mu_g}{2}\|\mathbf{y} - \mathbf{z}\|^2 + G(\mathbf{z}) \\
&\overset{②}{\ge} \left[F(\mathbf{x}) + \langle \nabla F(\mathbf{x}), \mathbf{y} - \mathbf{x}\rangle + \tfrac{\mu_l}{2}\|\mathbf{x} - \mathbf{y}\|_2^2\right] + \langle \mathbf{P}\nabla F(\mathbf{x}), \mathbf{z} - \mathbf{y}\rangle \\
&\quad - \tfrac{1}{\mu_g}\|\mathbf{P}(\nabla F(\mathbf{y}) - \nabla F(\mathbf{x}))\|_2^2 - \tfrac{\mu_g}{4}\|\mathbf{z} - \mathbf{y}\|_2^2 + \tfrac{\mu_g}{2}\|\mathbf{z} - \mathbf{y}\|_2^2 + G(\mathbf{z}) \\
&\overset{③}{\ge} \left[F(\mathbf{x}) + G(\mathbf{x}) + \langle \nabla F(\mathbf{x}), \mathbf{y} - \mathbf{x}\rangle + \left(\tfrac{\mu_l}{2} - \tfrac{L_l^2}{\mu_g}\right)\|\mathbf{x} - \mathbf{y}\|_2^2\right] \\
&\quad + \left[\langle \mathbf{P}\nabla F(\mathbf{x}), \mathbf{z} - \mathbf{y}\rangle + G(\mathbf{z}) - G(\mathbf{x})\right] + \tfrac{\mu_g}{4}\|\mathbf{z} - \mathbf{y}\|_2^2,
\end{aligned} \tag{9}$$

where ① holds since F is μ_l-strongly convex over R^{md} and μ_g-strongly convex over $\mathcal{L}$, ② holds by Young inequality and ③ holds by L_l-smoothness of F.

Second, we get an upper estimate on $F(\mathbf{z}) + G(\mathbf{z})$.

$$
\begin{aligned}
F(\mathbf{z}) + G(\mathbf{z}) &\overset{①}{\leq} F(\mathbf{x}) + \left[\langle \nabla F(\mathbf{x}), \mathbf{y} - \mathbf{x}\rangle + \tfrac{L_l}{2}\|\mathbf{y}-\mathbf{x}\|_2^2\right] \\
&\quad + \left[\langle \mathbf{P}\nabla F(\mathbf{y}), \mathbf{z} - \mathbf{y}\rangle + \tfrac{L_g}{2}\|\mathbf{z}-\mathbf{y}\|_2^2\right] + G(\mathbf{z}) \\
&= \left[F(\mathbf{x}) + \langle \nabla F(\mathbf{x}), \mathbf{y} - \mathbf{x}\rangle + \tfrac{L_l}{2}\|\mathbf{y}-\mathbf{x}\|_2^2\right] \\
&\quad + \langle \mathbf{P}\nabla F(\mathbf{x}), \mathbf{z} - \mathbf{y}\rangle + \tfrac{L_g}{2}\|\mathbf{z}-\mathbf{y}\|_2^2 \\
&\quad + \langle \mathbf{P}(\nabla F(\mathbf{y}) - \nabla F(\mathbf{x})), \mathbf{z} - \mathbf{y}\rangle + G(\mathbf{z}) \\
&\overset{②}{\leq} \left[F(\mathbf{x}) + \langle \nabla F(\mathbf{x}), \mathbf{y} - \mathbf{x}\rangle + \tfrac{L_l}{2}\|\mathbf{y}-\mathbf{x}\|_2^2\right] \\
&\quad + \langle \mathbf{P}\nabla F(\mathbf{x}), \mathbf{z} - \mathbf{y}\rangle + \tfrac{L_g}{2}\|\mathbf{z}-\mathbf{y}\|_2^2 \\
&\quad + \tfrac{1}{2L_g}\|\mathbf{P}(\nabla F(\mathbf{y}) - \nabla F(\mathbf{x}))\|_2^2 + \tfrac{L_g}{2}\|\mathbf{z}-\mathbf{y}\|_2^2 + G(\mathbf{z}) \\
&\overset{③}{\leq} \left[F(\mathbf{x}) + G(\mathbf{x}) + \langle \nabla F(\mathbf{x}), \mathbf{y} - \mathbf{x}\rangle + \left(\tfrac{\mu_l}{2} - \tfrac{L_l^2}{\mu_g}\right)\|\mathbf{y}-\mathbf{x}\|_2^2\right] \\
&\quad + \left[\langle \mathbf{P}\nabla F(\mathbf{x}), \mathbf{z} - \mathbf{y}\rangle + G(\mathbf{z}) - G(\mathbf{x})\right] + L_g\|\mathbf{z}-\mathbf{y}\|_2^2 \\
&\quad + \left(\tfrac{L_l^2}{2L_g} + \tfrac{L_l^2}{\mu_g} - \tfrac{\mu_l}{2} + \tfrac{L_l}{2}\right)\|\mathbf{y}-\mathbf{x}\|_2^2,
\end{aligned}
\tag{10}
$$

where ① holds since F is L_l-smooth over $\mathbb{R}^{md}$ and L_g-smooth over $\mathcal{L}$, ② holds by Young inequality and ③ holds by L_l-smoothness of F.

It remains to recall the definitions of $F(\mathbf{x}), G(\mathbf{x})$ from (2) and combine (9) and (10) to get the desired inequality.

4 Accelerated Algorithm and Convergence

Our algorithm is based on the Similar Triangles Method with inexactness [30]. Our method is decentralized and uses a multi-step consensus procedure. This procedure is defined as a multiplication by communication matrix several times.

Introduce $\mathbf{W}_T^k = \mathbf{W}^k \ldots \mathbf{W}^{k-T+1}$ for $k \geq T-1$. Assuming that the number of consensus steps T is a multiple of τ, we can group the matrices into T/τ blocks. By Assumption 4, each block strictly contracts the space:

$$
\left\|\mathbf{W}_T^k - \mathbf{P}\right\|_2 \leq \left(\sqrt{1-\chi^{-1}}\right)^{T/\tau} = (1-\chi^{-1})^{\frac{T}{2\tau}}.
$$

Also introduce $\lambda = (1-\chi^{-1})^{\frac{T}{2\tau}}$.

Define $\gamma^k = \frac{\alpha_{k+1}}{1+A_{k+1}\mu_g/2}$, $\widehat{\mathbf{x}}^0 = \mathbf{x}^0$, $\widehat{\mathbf{y}} = \mathbf{y}^0$, $\widehat{\mathbf{u}}^0 = \mathbf{u}^0$ and consider a method which trajectory lies in $\mathcal{L}$:

$$
\begin{aligned}
\widehat{\mathbf{y}}^{k+1} &= \frac{\alpha_{k+1}\widehat{\mathbf{u}}^k + A_k\widehat{\mathbf{x}}^k}{A_{k+1}} \\
\widehat{\mathbf{u}}^{k+1} &= \mathrm{prox}_G^{\gamma_k}\left[\mu\gamma_k\widehat{\mathbf{y}}^{k+1} + (1-\mu\gamma_k)\widehat{\mathbf{u}}^k - \gamma_k\mathbf{P}\nabla^r F(\mathbf{y}^{k+1})\right] \\
\widehat{\mathbf{x}}^{k+1} &= \frac{\alpha^{k+1}\widehat{\mathbf{u}}^{k+1} + A_k\widehat{\mathbf{x}}^k}{A_{k+1}}.
\end{aligned}
$$

Algorithm 1. Accelerated decentralized proximal method with consensus subroutine

Require: Initial guess $\mathbf{x}^0 \in \mathcal{L}$, constants $L, \mu > 0$, $\mathbf{u}^0 = \mathbf{x}^0$, $\alpha^0 = A^0 = 0$
1: **for** $k = 0, 1, 2, \ldots$ **do**
2: Find α^{k+1} as the greater root of $(A^k + \alpha^{k+1})(1 + A^k \mu_g/2) = 2L_g(\alpha^{k+1})^2$
3: $A^{k+1} = A^k + \alpha^{k+1}$
4: $\mathbf{y}^{k+1} = \dfrac{\alpha^{k+1}\mathbf{u}^k + A^k \mathbf{x}^k}{A^{k+1}}$
5: $\mathbf{v}^{k+1} = \dfrac{\alpha^{k+1}(\mu_g/2)\mathbf{y}^{k+1} + (1 + A^k\mu_g/2)\mathbf{u}^k}{1 + A^{k+1}\mu_g/2} - \dfrac{\alpha^{k+1}\nabla^r F(\mathbf{y}^{k+1})}{1 + A^{k+1}\mu_g/2}$
6: $\mathbf{u}^{k+1} = \mathrm{prox}_G^{\gamma_k}\left(\mathbf{W}_T^{kT-1}\mathbf{v}^{k+1}\right)$
7: $\mathbf{x}^{k+1} = \dfrac{\alpha^{k+1}\mathbf{u}^{k+1} + A^k \mathbf{x}^k}{A^{k+1}}$
8: **end for**

To bound coefficients A^N, let us recall a corresponding result from [26,29].

Lemma 2. *The following relations hold.*

$$A^N \geq \frac{1}{2L_g}\left(1 + \frac{1}{4}\sqrt{\frac{\mu_g}{L_g}}\right)^{2(N-1)}, \quad \frac{\sum_{k=0}^{N-1} A^{k+1}}{A^N} \leq 1 + 2\sqrt{\frac{L_g}{\mu_g}}.$$

Lemma 3. *For $k \geq 0$ define $\beta_k = \max(\mathbb{E}\left\|\mathbf{y}^k - \widehat{\mathbf{y}}^k\right\|_2, \mathbb{E}\left\|\mathbf{u}^k - \widehat{\mathbf{u}}^k\right\|_2, \mathbb{E}\left\|\mathbf{x}^k - \widehat{\mathbf{x}}^k\right\|_2)$. We have*

$$\beta_{k+1} \leq (1 + \lambda)\beta_k + \lambda\gamma_k \mathbb{E}\left\|\nabla^r F(\mathbf{y}^{k+1})\right\|.$$

Proof. First, we have

$$\begin{aligned}\mathbb{E}\left\|\mathbf{y}^{k+1} - \widehat{\mathbf{y}}^{k+1}\right\|_2 &\leq \frac{\alpha_{k+1}}{A_{k+1}}\mathbb{E}\left\|\mathbf{u}^k - \widehat{\mathbf{u}}^k\right\|_2 + \frac{A_k}{A_{k+1}}\mathbb{E}\left\|\mathbf{x}^k - \widehat{\mathbf{x}}^k\right\|_2 \leq \beta_k \\ &\leq (1 + \lambda)\beta_k + \lambda\gamma_k\mathbb{E}\left\|\nabla^r F(\mathbf{y}^{k+1})\right\|_2.\end{aligned}$$

After that, consider an update rule for $\mathbf{u}^{k+1}$. We have:

$$
\begin{aligned}
&\mathbb{E}\|\mathbf{u}^{k+1} - \widehat{\mathbf{u}}^{k+1}\|_2 \\
&\quad \leq \mathbb{E}\Big\| \operatorname{prox}_G^{\gamma_k} \left[\mathbf{W}_T^{(k+1)T-1} \left(\mu\gamma_k \mathbf{y}^{k+1} + (1-\mu\gamma_k)\mathbf{u}^k - \gamma_k \nabla^r F(\mathbf{y}^{k+1})\right)\right] \\
&\qquad - \operatorname{prox}_G^{\gamma_k} \left[\mu\gamma_k \widehat{\mathbf{y}}^{k+1} + (1-\mu\gamma_k)\widehat{\mathbf{u}}^k - \gamma_k \mathbf{P}\nabla^r F(\mathbf{y}^{k+1})\right] \Big\|_2 \\
&\quad \overset{①}{\leq} \mathbb{E}\Big\| \left[\mathbf{W}_T^{(k+1)T-1} \left(\mu\gamma_k \mathbf{y}^{k+1} + (1-\mu\gamma_k)\mathbf{u}^k - \gamma_k \nabla^r F(\mathbf{y}^{k+1})\right)\right] \\
&\qquad - \left[\mu\gamma_k \widehat{\mathbf{y}}^{k+1} + (1-\mu\gamma_k)\widehat{\mathbf{u}}^k - \gamma_k \mathbf{P}\nabla^r F(\mathbf{y}^{k+1})\right] \Big\|_2 \\
&\quad \leq \mathbb{E}\|\mu\gamma_k(\mathbf{W}_T^{(k+1)T-1}\mathbf{y}^{k+1} - \widehat{\mathbf{y}}^{k+1}) + (1-\mu\gamma_k)(\mathbf{W}_T^{(k+1)T-1}\mathbf{u}^k - \widehat{\mathbf{u}}^k) \\
&\qquad - \gamma_k(\mathbf{W}_T^{(k+1)T-1}\nabla^r F(\mathbf{y}^{k+1}) - \mathbf{P}\nabla^r F(\mathbf{y}^{k+1}))\|_2 \\
&\quad \leq \mu\gamma_k \mathbb{E}\left\|(\mathbf{W}_T^{(k+1)T-1} - \mathbf{P})(\mathbf{y}^{k+1} - \mathbf{P}\mathbf{y}^{k+1}) + (\mathbf{P}\mathbf{y}^{k+1} - \mathbf{P}\widehat{\mathbf{y}}^{k+1})\right\|_2 \\
&\qquad + (1-\mu\gamma_k)\mathbb{E}\left\|(\mathbf{W}_T^{(k+1)T-1} - \mathbf{P})(\mathbf{u}^k - \mathbf{P}\mathbf{u}^k) + (\mathbf{P}\mathbf{u}^k - \mathbf{P}\widehat{\mathbf{u}}^k)\right\|_2 \\
&\qquad + \gamma_k \mathbb{E}\left\|(\mathbf{W}_T^{(k+1)T-1} - \mathbf{P})\nabla^r F(\mathbf{y}^{k+1})\right\|_2 \\
&\quad \leq \mu\gamma_k \left(\lambda \mathbb{E}\left\|\mathbf{y}^{k+1} - \mathbf{P}\mathbf{y}^{k+1}\right\|_2 + \mathbb{E}\left\|\mathbf{P}\mathbf{y}^{k+1} - \mathbf{P}\widehat{\mathbf{y}}^{k+1}\right\|_2\right) \\
&\qquad + (1-\mu\gamma_k)\left(\lambda \mathbb{E}\left\|\mathbf{u}^k - \mathbf{P}\mathbf{u}^k\right\|_2 + \mathbb{E}\left\|\mathbf{u}^k - \widehat{\mathbf{u}}^k\right\|_2\right) + \lambda\gamma_k \mathbb{E}\left\|\nabla^r F(\mathbf{y}^{k+1})\right\|_2 \\
&\quad \leq \mu\gamma_k(\lambda\beta_k + \beta_k) + (1-\mu\gamma_k)(\lambda\beta_k + \beta_k) + \lambda\gamma_k \mathbb{E}\left\|\nabla^r F(\mathbf{y}^{k+1})\right\|_2 \\
&\quad = (1+\lambda)\beta_k + \lambda\gamma_k \mathbb{E}\left\|\nabla^r F(\mathbf{y}^{k+1})\right\|_2,
\end{aligned}
$$

where ① holds by the non-expansiveness property of the prox-operator. Finally, for $\mathbf{x}^{k+1}$ we obtain

$$
\begin{aligned}
\mathbb{E}\left\|\mathbf{x}^{k+1} - \widehat{\mathbf{x}}^{k+1}\right\|_2 &\leq \frac{\alpha_{k+1}}{A_{k+1}}\mathbb{E}\left\|\mathbf{u}^{k+1} - \widehat{\mathbf{u}}^{k+1}\right\|_2 + \frac{A_k}{A_{k+1}}\mathbb{E}\left\|\mathbf{x}^k - \widehat{\mathbf{x}}^k\right\|_2 \\
&\leq (1+\lambda)\beta_k + \lambda\gamma_k \mathbb{E}\left\|\nabla^r F(\mathbf{y}^{k+1})\right\|_2.
\end{aligned}
$$

As a result, we have

$$
\begin{aligned}
\beta_{k+1} &= \max\left(\mathbb{E}\left\|\mathbf{y}^{k+1} - \widehat{\mathbf{y}}^{k+1}\right\|_2, \mathbb{E}\left\|\mathbf{u}^{k+1} - \widehat{\mathbf{u}}^{k+1}\right\|_2, \mathbb{E}\left\|\mathbf{x}^{k+1} - \widehat{\mathbf{x}}^{k+1}\right\|_2\right) \\
&\leq (1+\lambda)\beta_k + \lambda\gamma_k \mathbb{E}\left\|\nabla^r F(\mathbf{y}^{k+1})\right\|_2.
\end{aligned}
$$

Theorem 1. *Let Assumptions 1, 2, 3 and 4 hold. Then Algorithm 1 requires* $N_{comp} = O\left(\frac{\sigma^2}{\mu_g \varepsilon}\log\left(\frac{1}{\varepsilon}\right)\right)$ *oracle calls per node and* $N_{comm} = O\left(\tau\chi\sqrt{\frac{L_g}{\mu_g}}\log^2\left(\frac{1}{\varepsilon}\right)\right)$ *communication rounds to reach* ε*-accuracy.*

Proof. Let $\delta = \eta \sum_{k=0}^{N-1} \beta_k^2$. According to Theorem 4.2 in [24] we have

$$\mathbb{E}\left\|\widehat{\mathbf{u}}^N - \mathbf{x}^*\right\|_2^2 \le \frac{\left\|\widehat{\mathbf{u}}^0 - \mathbf{x}^*\right\|_2^2}{1 + A_N\mu_g/2} + \frac{4\sum\limits_{k=0}^{N-1} A_{k+1}\left(\frac{\sigma^2}{L_g r} + \delta\right)}{1 + A_N\mu_g/2}$$
$$\le \frac{2\left\|\widehat{\mathbf{u}}^0 - \mathbf{x}^*\right\|_2^2}{A_N\mu_g} + \frac{8\sum\limits_{k=0}^{N-1} A_{k+1}\left(\frac{\sigma^2}{L_g r} + \delta\right)}{A_N\mu_g},$$
$$\mathbb{E}\left\|\widehat{\mathbf{x}}^N - \mathbf{x}^*\right\|_2^2 \le \frac{2}{\mu_g/2}\left(\frac{\left\|\widehat{\mathbf{u}}^0 - \mathbf{x}^*\right\|_2^2}{2A_N} + \frac{2\sum\limits_{k=0}^{N-1} A_{k+1}\left(\frac{\sigma^2}{L_g r} + \delta\right)}{A_N}\right)$$
$$= \frac{2\left\|\widehat{\mathbf{u}}^0 - \mathbf{x}^*\right\|_2^2}{A_N\mu_g} + \frac{8\sum\limits_{k=0}^{N-1} A_{k+1}\left(\frac{\sigma^2}{L_g r} + \delta\right)}{A_N\mu_g}.$$

First, by Lemma 2 we have $A^N \ge \frac{1}{2L_g}\left(1 + \frac{1}{4}\sqrt{\frac{\mu_g}{L_g}}\right)^{2(N-1)}$. To ensure condition $\frac{2\left\|\widehat{\mathbf{u}}^0 - \mathbf{x}^*\right\|_2^2}{A_N\mu_g} \le \varepsilon/3$ it is sufficient to set $N = O\left(\sqrt{L_g/\mu_g}\log(1/\varepsilon)\right)$.

Let us bound $\mathbb{E}\left\|\nabla^r F(\mathbf{y}^{k+1})\right\|_2$:

$$\begin{aligned}
&\mathbb{E}\left\|\nabla^r F(\mathbf{y}^{k+1})\right\|_2\\
&\le \mathbb{E}\left\|\nabla^r F(\mathbf{y}^{k+1}) - \nabla^r F(\widehat{\mathbf{y}}^{k+1})\right\| + \left\|\nabla^r F(\widehat{\mathbf{y}}^{k+1}) - \nabla^r F(\mathbf{x}^*)\right\|_2 + \mathbb{E}\left\|\nabla^r F(\mathbf{x}^*)\right\|_2\\
&\le L_\xi \mathbb{E}\left\|\mathbf{y}^{k+1} - \widehat{\mathbf{y}}^{k+1}\right\|_2 + L_\xi \mathbb{E}\left\|\widehat{\mathbf{y}}^{k+1} - \mathbf{x}^*\right\|_2 + \mathbb{E}\left\|\nabla^r F(\mathbf{x}^*)\right\|_2\\
&\le L_\xi \beta_{k+1} + L_\xi \mathbb{E}\left\|\widehat{\mathbf{y}}^{k+1} - \mathbf{x}^*\right\|_2 + \mathbb{E}\left\|\nabla^r F(\mathbf{x}^*)\right\|_2.
\end{aligned}$$

For $\mathbb{E}\left\|\widehat{\mathbf{y}}^{k+1} - \mathbf{x}^*\right\|$, we have:

$$\mathbb{E}\left\|\widehat{\mathbf{y}}^{k+1} - \mathbf{x}^*\right\|_2^2 = \mathbb{E}\left\|\frac{\alpha^{k+1}\widehat{\mathbf{u}}^k}{A^{k+1}} + \frac{A^k\widehat{\mathbf{x}}^k}{A^{k+1}} - \mathbf{x}^*\right\|_2^2 = \mathbb{E}\left\|\frac{\alpha^{k+1}}{A^{k+1}}\left(\widehat{\mathbf{u}}^k - \mathbf{x}^*\right) + \frac{A^k}{A^{k+1}}\left(\widehat{\mathbf{x}}^k - \mathbf{x}^*\right)\right\|_2^2$$
$$\le \frac{\alpha^{k+1}}{A^{k+1}}\mathbb{E}\left\|\widehat{\mathbf{u}}^k - \mathbf{x}^*\right\|_2^2 + \frac{A^k}{A^{k+1}}\mathbb{E}\left\|\widehat{\mathbf{x}}^k - \mathbf{x}^*\right\|_2^2 \le \frac{2\left\|\widehat{\mathbf{u}}^0 - \mathbf{x}^*\right\|_2^2}{A_{k+1}\mu_g} + \frac{8\sum\limits_{t=0}^{k} A_{t+1}\left(\frac{\sigma^2}{L_g r} + \delta\right)}{A_{k+1}\mu_g}$$
$$\le \frac{4L_g}{\mu_g}\left\|\widehat{\mathbf{u}}^0 - \mathbf{x}^*\right\|_2^2 + \frac{8\left(\frac{\sigma^2}{L_g r} + \delta\right)}{\mu_g}\left(1 + 2\sqrt{\frac{L_g}{\mu_g}}\right).$$

Define $R_0 = \left\|\widehat{\mathbf{u}}^0 - \mathbf{x}^*\right\|_2$. For brevity introduce $a = \frac{4L_\xi^2 L_g}{\mu_g}$, $b = \frac{8L_\xi^2}{\mu_g}\left(1 + 2\sqrt{\frac{L_g}{\mu_g}}\right)$.

Using this, for $\mathbb{E}\left\|\nabla^r F(\mathbf{y}^{k+1})\right\|_2$ we have:

$$\begin{aligned}
\mathbb{E}\left\|\nabla^r F(\mathbf{y}^{k+1})\right\|_2 &\le L_\xi \mathbb{E}\left\|\mathbf{y}^{k+1} - \widehat{\mathbf{y}}^{k+1}\right\|_2 + L_\xi\sqrt{\mathbb{E}\left\|\widehat{\mathbf{y}}^{k+1} - \mathbf{x}^*\right\|_2^2} + \mathbb{E}\left\|\nabla^r F(\mathbf{x}^*)\right\|_2\\
&\le L_\xi \beta_{k+1} + \sqrt{aR_0^2 + b\delta + \frac{b\sigma^2}{L_g r}} + \mathbb{E}\left\|\nabla^r F(\mathbf{x}^*)\right\|_2.
\end{aligned}$$

Notice that $\gamma_k = \frac{\alpha_{k+1}}{1+A_{k+1}\mu_g/2} \le \frac{2\alpha_{k+1}}{A_{k+1}\mu_g} \le \frac{2}{\mu_g} = \gamma$ and $\mathbb{E}\left\|\nabla^r F(\mathbf{x}^*)\right\|_2 \le \sqrt{m}M_\xi$. Then, we can write:

$$\begin{aligned}\beta_{k+1} &\le (1+\lambda)\beta_k + \lambda\gamma\mathbb{E}\left\|\nabla^r F(\mathbf{y}^{k+1}\right\|_2 \\ &\le (1+\lambda)\beta_k + \lambda\gamma L_\xi \beta_{k+1} + \lambda\gamma \cdot \sqrt{aR_0^2 + b\delta + \tfrac{b\sigma^2}{L_g r}} + \lambda\gamma\sqrt{m}M_\xi,\end{aligned}$$

Replacing the terms and unfolding the recurrence:

$$\begin{aligned}\beta_{k+1} &\le \tfrac{1}{1-\lambda\gamma L_\xi}\left((1+\lambda)\beta_k + \lambda\gamma \cdot \sqrt{aR_0^2 + b\delta + \tfrac{b\sigma^2}{L_g r}} + \lambda\gamma\sqrt{m}M_\xi\right) \\ &\le \frac{\lambda\gamma\cdot\sqrt{aR_0^2+b\delta+\frac{b\sigma^2}{L_g r}}+\lambda\gamma\sqrt{m}M_\xi}{(1-\lambda\gamma L_\xi)} \sum_{t=0}^{k} \left(\tfrac{1+\lambda}{1-\lambda\gamma L_\xi}\right)^k \\ &\le \frac{\gamma\cdot\sqrt{aR_0^2+b\delta+\frac{b\sigma^2}{L_g r}}+\gamma\sqrt{m}M_\xi}{(1+\gamma L_\xi)} \left(\left(\tfrac{1+\lambda}{1-\lambda\gamma L_\xi}\right)^{k+1} - 1\right).\end{aligned}$$

Rename $\frac{\gamma}{1+\gamma L_\xi} = \gamma_0$. Then, summing over k, we obtain

$$\begin{aligned}\delta &= \eta \sum_{k=0}^{N-1} \beta_k^2 \le \eta N \beta_{N-1}^2 \\ &\le \eta\gamma_0^2 \left(\sqrt{aR_0^2 + b\delta + \tfrac{b\sigma^2}{L_g r}} + \sqrt{m}M_\xi\right)^2 \cdot N\left(\left(\tfrac{1+\lambda}{1-\lambda\gamma L_\xi}\right)^{N-1} - 1\right)^2.\end{aligned}$$

Denote $c(N) = N\left(\left(\frac{1+\lambda}{1-\lambda\gamma L_\xi}\right)^{N-1} - 1\right)^2$.

$$\begin{aligned}\frac{\delta}{c(N)} &\le \eta\gamma_0^2\left(\sqrt{aR_0^2 + b\delta + \tfrac{b\sigma^2}{L_g r}} + \gamma\sqrt{m}M_\xi\right)^2 \\ &\le 2\eta\gamma_0^2\left(aR_0^2 + b\delta + \tfrac{b\sigma^2}{L_g r} + \mathbb{E}\left\|\nabla^r F(\mathbf{x}^*)\right\|_2^2\right) \\ \delta\left(\frac{1}{c(N)} - 2\eta\gamma_0^2 b\right) &\le 2\gamma_0^2\left(aR_0^2 + \tfrac{b\sigma^2}{L_g r} + mM_\xi^2\right), \\ \delta &\le 2\gamma_0^2\left(aR_0^2 + \tfrac{b\sigma^2}{L_g r} + mM_\xi^2\right)\left(\frac{1}{c(N)} - 2\eta\gamma_0^2 b\right)^{-1}, \\ &= \left(aR_0^2 + \tfrac{b\sigma^2}{L_g r} + mM_\xi^2\right)\frac{1}{\eta b}\left(\frac{1}{1-2\gamma_0^2 bc(N)} - 1\right).\end{aligned}$$

We would like to choose number of consensus iterations T such that $\delta \leq \varepsilon/3$. Setting

$$\begin{aligned}
T &\geq 2\tau\chi \log\left[\frac{192N^3}{\eta\mu_g L_g \varepsilon}\left(aR_0^2 + \tfrac{b\sigma^2}{L_g r} + \mathbb{E}\left\|\nabla^r F(\mathbf{x}^*)\right\|_2^2\right)\right] \\
&\geq 2\tau\chi \log\left[\frac{192N^3}{\eta\mu_g L_g \varepsilon}\left(\frac{4L_\xi^2 L_g}{\mu_g}\left\|\mathbf{u}^0 - \mathbf{x}^*\right\|_2^2 + \tfrac{b\sigma^2}{L_g r} + \mathbb{E}\left\|\nabla^r F(\mathbf{x}^*)\right\|_2^2\right)\right] \\
&= O\left(\tau\chi \log\left(\frac{N^3}{\varepsilon}\right)\right) = O\left(\tau\chi \log\left(\frac{1}{\varepsilon}\right)\right),
\end{aligned}$$

we obtain

$$\begin{aligned}
\lambda &= (1 - \chi^{-1})^{\frac{T}{2\tau}} \leq \exp\left(-\tfrac{T}{2\tau\chi}\right) \\
&\leq \frac{\sqrt{\eta\mu_g L_g}}{8N}\sqrt{\frac{\varepsilon}{3N}}\left(\frac{4L_\xi^2 L_g}{\mu_g}\left\|\mathbf{u}^0 - \mathbf{x}^*\right\|_2^2 + \tfrac{b\sigma^2}{L_g r} + \mathbb{E}\left\|\nabla^r F(\mathbf{x}^*)\right\|_2^2\right)^{-1/2} \\
&= \frac{1}{8N}\left(\frac{\eta\varepsilon}{3N\gamma_0^2\left(aR_0^2 + \tfrac{b\sigma^2}{L_g r} + mM_\xi^2\right)}\right)^{1/2}.
\end{aligned}$$

We define $\lambda_0 = \frac{\lambda(1+L_\xi)}{1-\lambda\gamma L_\xi}$. Therefore, assuming that ε is sufficiently small, we have $\lambda_0 \leq 2\lambda(1 + \gamma L_\xi)$ and, using the fact that for any $0 \leq x \leq 1$ the following inequality holds: $1 - \frac{1}{1+x} \geq 1 - \left(1 - \frac{x}{2}\right) = \frac{x}{2}$, we obtain

$$\begin{aligned}
\lambda_0 &\leq \frac{1}{N}\left[1 - \left(1 + \left(\frac{\eta\varepsilon}{12N\gamma_0^2\left(aR_0^2 + \tfrac{b\sigma^2}{L_g r} + mM_\xi^2\right)}\right)^{1/2}\right)^{-1}\right], \\
1 - N\lambda_0 &\geq \left[1 + \left(\frac{\eta\varepsilon}{12N\gamma_0^2\left(aR_0^2 + \tfrac{b\sigma^2}{L_g r} + mM_\xi^2\right)}\right)^{1/2}\right]^{-1}, \\
\frac{1}{1 - N\lambda_0} - 1 &\leq \left(\frac{\eta\varepsilon}{12N\gamma_0^2\left(aR_0^2 + \tfrac{b\sigma^2}{L_g r} + mM_\xi^2\right)}\right)^{1/2}.
\end{aligned}$$

After that, let us note that

$$\left(\tfrac{(1+\lambda)}{1-\lambda\gamma L_\xi}\right)^{N-1} - 1 = \sum_{k=1}^{N-1}\binom{N-1}{k}\lambda_0^k \leq \sum_{k=1}^{N-1} N^k \lambda_0^k \leq \frac{N\lambda_0}{1 - N\lambda_0} = \frac{1}{1 - N\lambda_0} - 1.$$

Consequently,

$$c(N) = N\left(\left(\frac{1+\lambda}{1-\lambda\gamma L_\xi}\right)^{N-1} - 1\right)^2 \le \frac{\eta\varepsilon}{12\gamma_0^2\left(aR_0^2 + \frac{b\sigma^2}{L_g r} + mM_\xi^2\right)},$$

$$1 - 2\gamma_0^2 bc(N) \ge 1 - \frac{\eta b\varepsilon}{6\left(aR_0^2 + \frac{b\sigma^2}{L_g r} + mM_\xi^2\right)}.$$

We once again use that $\frac{1}{1+x} \le 1 - \frac{x}{2}$ for $0 \le x \le 1$.

$$1 - 2\gamma_0^2 bc(N) \ge \left(1 + \frac{\eta b\varepsilon}{3\left(aR_0^2 + \frac{b\sigma^2}{L_g r} + mM_\xi^2\right)}\right)^{-1}$$

$$\frac{1}{1 - 2\gamma_0^2 bc(N)} \le 1 + \frac{\eta b\varepsilon}{3\left(aR_0^2 + \frac{b\sigma^2}{L_g r} + mM_\xi^2\right)}.$$

Finally, we get

$$\delta \le \frac{aR_0^2 + \frac{b\sigma^2}{L_g r} + mM_\xi^2}{\eta b}\left(\frac{1}{1 - 2\gamma_0^2 bc(N)} - 1\right) \le \frac{\varepsilon}{3}.$$

In order to satisfy $\sigma^2/(\mu_g r) \le \varepsilon/3$ it is sufficient to choose $r = \frac{3\sigma^2}{\varepsilon\sqrt{L_g \mu_g}}$.

Summing up, to reach ε-accuracy Algorithm 1 requires $N_{comm} = NT = O\left(\tau\chi\sqrt{\frac{L_g}{\mu_g}}\log^2\left(\frac{1}{\varepsilon}\right)\right)$ and $N_{comp} = Nr = O\left(\frac{\sigma^2}{\mu_g \varepsilon}\log\left(\frac{1}{\varepsilon}\right)\right)$

5 Numerical Experiments

We run Algorithm 1 on the sparse logistic regression problem:

$$F(x) = \frac{1}{m}\sum_{i=1}^{m} f_i(x) + \sigma_1\|x\|_1 + \frac{\sigma_2}{2}\|x\|^2;\ f_i(x) = \frac{1}{n}\sum_{j=1}^{n}\log[1 + \exp(-b_j\langle a_j, x\rangle)],$$

where $a_j \in \mathbb{R}^d$ is the j-th input vector, and $b_j \in \{-1, 1\}$ is the corresponding label.

Setup. We simulate a network of $m = 100$ agents connected via an Erdős-Rényi random graph ($p = 0.1$). The evaluation is performed on the a9a dataset from the LIBSVM library [6], with regularization parameters $\sigma_1 = \sigma_2 = 10^{-4}$. Algorithms DPA [1] and DAPG [33] are well-tuned for optimal practical performance. We use different settings for batch-size r and number of consensus iterations T, these parameters did not significantly alter the convergence trends.

Results. As shown in Fig. 2, Algorithm 1 achieves a competitive initial convergence rate before plateauing due to stochastic mini-batch variance and the

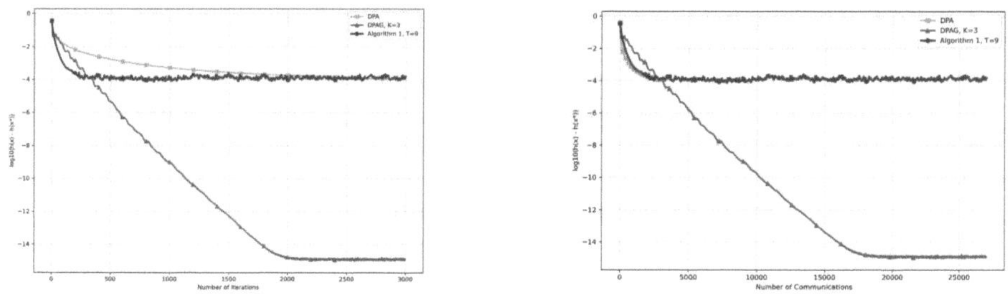

Fig. 2. Comparison of the proposed Algorithm 1 against DPA and DAPG. Here $h(x) = F(x) + G(x)$.

lack of gradient tracking (unlike the exact DAPG baseline). However, its primary advantage is structural simplicity. By requiring only one consensus inner loop for a single combined variable (compared to DAPG's three) Algorithm 1 is highly practical for real-world distributed systems constrained by memory, computation, and communication limits.

6 Conclusion

We propose a consensus subroutine approach that works for composite optimization. The algorithm is an accelerated proximal method that performs several communication rounds after each proximal step. The novelty of the approach is that we do not bound the gradient norm. We apply a novel proof technique and thus get an algorithm that recovers lower communication complexity bounds up to a logarithmic factor.

References

1. Alghunaim, S., Yuan, K., Sayed, A.H.: A linearly convergent proximal gradient algorithm for decentralized optimization. Adv. Neural. Inf. Process. Syst. **32** (2019)
2. Bazerque, J.A., Giannakis, G.B.: Distributed spectrum sensing for cognitive radio networks by exploiting sparsity. IEEE Trans. Signal Process. **58**(3), 1847–1862 (2009)
3. Beznosikov, A., Rogozin, A., Kovalev, D., Gasnikov, A.: Near-optimal decentralized algorithms for saddle point problems over time-varying networks. In: Olenev, N.N., Evtushenko, Y.G., Jaćimović, M., Khachay, M., Malkova, V. (eds.) OPTIMA 2021. LNCS, vol. 13078, pp. 246–257. Springer, Cham (2021). https://doi.org/10.1007/978-3-030-91059-4_18
4. Beznosikov, A., Samokhin, V., Gasnikov, A.: Distributed saddle point problems: lower bounds, near-optimal and robust algorithms. Optim. Methods Softw. **40**(5), 1249–1266 (2025)
5. Cai, K., Ishii, H.: Average consensus on arbitrary strongly connected digraphs with time-varying topologies. IEEE Trans. Autom. Control **59**(4), 1066–1071 (2014)

6. Chang, C.-C.: A library for support vector machines (2001). http://www.csie.ntu.edu.tw/~cjlin/libsvm
7. Dvinskikh, D., Gasnikov, A.: Decentralized and parallel primal and dual accelerated methods for stochastic convex programming problems. J. Inverse Ill-Posed Probl. **29**(3), 385–405 (2021)
8. Forero, P.A., Cano, A., Giannakis, G.B.: Consensus-based distributed support vector machines. J. Mach. Learn. Res. **11**(5) (2010)
9. Gan, L., Topcu, U., Low, S.H.: Optimal decentralized protocol for electric vehicle charging. IEEE Trans. Power Syst. **28**(2), 940–951 (2012)
10. Jadbabaie, A., Lin, J., Morse, A.S.: Coordination of groups of mobile autonomous agents using nearest neighbor rules. IEEE Trans. Autom. Control **48**(6), 988–1001 (2003)
11. Jakovetić, D., Xavier, J., Moura, J.M.: Fast distributed gradient methods. IEEE Trans. Autom. Control **59**(5), 1131–1146 (2014)
12. Ji, Y., Scutari, G., Sun, Y., Honnappa, H.: Distributed sparse regression via penalization. J. Mach. Learn. Res. **24**(272), 1–62 (2023)
13. Kovalev, D., Salim, A., Richtárik, P.: Optimal and practical algorithms for smooth and strongly convex decentralized optimization. Adv. Neural. Inf. Process. Syst. **33** (2020)
14. Mateos, G., Giannakis, G.B.: Distributed sparse linear regression. IEEE Trans. Signal Process. **58**(10), 5262–5276 (2010)
15. Nedic, A.: Distributed gradient methods for convex machine learning problems in networks: distributed optimization. IEEE Signal Process. Mag. **37**(3), 92–101 (2020)
16. Nedić, A., Olshevsky, A., Uribe, C.A.: Fast convergence rates for distributed non-Bayesian learning. IEEE Trans. Autom. Control **62**(11), 5538–5553 (2017)
17. Olshevsky, A.: Efficient information aggregation strategies for distributed control and signal processing, arXiv preprint arXiv:1009.6036 (2010)
18. Olshevsky, A.: Linear time average consensus on fixed graphs and implications for decentralized optimization. SIAM J. Control. Optim. **53**(6), 3199–3223 (2015)
19. Rabbat, M., Nowak, R.: Distributed optimization in sensor networks. In: Proceedings of the 3rd International Symposium on Information Processing in Sensor Networks, pp. 20–27 (2004)
20. Ram, S.S., Veeravalli, V.V., Nedic, A.: Distributed non-autonomous power control through distributed convex optimization. In: IEEE INFOCOM 2009, pp. 3001–3005. IEEE (2009)
21. Ravazzi, C., Fosson, S.M., Magli, E.: Distributed soft thresholding for sparse signal recovery. In: 2013 IEEE Global Communications Conference (GLOBECOM), pp. 3429–3434. IEEE (2013)
22. Ren, W.: Consensus based formation control strategies for multi-vehicle systems. In: 2006 American Control Conference, pages 6–pp. IEEE (2006)
23. Ren, W., Beard, R.W.: Distributed Consensus in Multi-vehicle Cooperative Control, vol. 27. Springer, London (2008)
24. Rogozin, A., Bochko, M., Dvurechensky, P., Gasnikov, A., Lukoshkin, V.: An accelerated method for decentralized distributed stochastic optimization over time-varying graphs. In: Conference on Decision and Control (2021)
25. Rogozin, A., Gasnikov, A., Beznosikov, A., Kovalev, D.: Decentralized convex optimization over time-varying graphs. In: Encyclopedia of Optimization, pp. 1–17. Springer (2023)

26. Rogozin, A., Lukoshkin, V., Gasnikov, A., Kovalev, D., Shulgin, E.: Towards accelerated rates for distributed optimization over time-varying networks. In: Olenev, N.N., Evtushenko, Y.G., Jaćimović, M., Khachay, M., Malkova, V. (eds.) OPTIMA 2021. LNCS, vol. 13078, pp. 258–272. Springer, Cham (2021). https://doi.org/10.1007/978-3-030-91059-4_19
27. Scaman, K., Bach, F., Bubeck, S., Lee, Y.T., Massoulié, L.: Optimal algorithms for smooth and strongly convex distributed optimization in networks. In: Proceedings of the 34th International Conference on Machine Learning-Volume 70, pp. 3027–3036. JMLR.org (2017)
28. Scaman, K., Bach, F., Bubeck, S., Massoulié, L., Lee, Y.T.: Optimal algorithms for non-smooth distributed optimization in networks. In: Advances in Neural Information Processing Systems, pp. 2740–2749 (2018)
29. Stonyakin, F., et al.: Inexact model: a framework for optimization and variational inequalities. Optim. Methods Softw., 1–47 (2021)
30. Stonyakin, F.S., et al.: Gradient methods for problems with inexact model of the objective. In: Khachay, M., Kochetov, Y., Pardalos, P. (eds.) MOTOR 2019. LNCS, vol. 11548, pp. 97–114. Springer, Cham (2019). https://doi.org/10.1007/978-3-030-22629-9_8
31. Xiao, L., Boyd, S., Kim, S.-J.: Distributed average consensus with least-mean-square deviation. J. Parallel Distrib. Comput. **67**(1), 33–46 (2007)
32. Ye, H., Luo, L., Zhou, Z., Zhang, T.: Multi-consensus decentralized accelerated gradient descent, arXiv preprint arXiv:2005.00797 (2020)
33. Ye, H., Zhou, Z., Luo, L., Zhang, T.: Decentralized accelerated proximal gradient descent. Adv. Neural. Inf. Process. Syst. **33** (2020)

Relaxation Variants of the Proximal Method and the Regularized Proximal Method Using Auxiliary Functions Approximation

Igor Zabotin, Oksana Shulgina, and Rashid Yarullin(✉)

Kazan (Volga Region) Federal University, Kazan, Russia
yarullinrs@gmail.com

Abstract. We propose two methods for solving the convex programming problem. One of them is a variant of the classical proximal method, and the second is a modification of the also well-known regularized proximal method. A key feature of these methods is that the sequence of approximations is constructed using a relaxation condition. In addition, the methods use the operation of embedding the epigraphs of auxiliary functions into some polyhedral sets in order to simplify the solving problems of constructing iteration points. We discuss the implementation of these methods and prove their convergence.

Keywords: constrained minimization · iteration point · relaxation · optimal value · approximating set · epigraph of the function · cutting plane · convergence of the method

1 Introduction

Regularization methods, which are well-known in the literature (e.g., [1–3]), comprise one of the classes of constrained minimization methods. Like many other optimization methods, they are based on reducing the original problem to a sequential solution of problems of minimizing various auxiliary functions on an admissible set or sets of simpler structure. In regularization methods, auxiliary functions are constructed by adding to the objective function some strongly convex functions that change from step to step. These additions make the problems of constructing iteration points with these auxiliary objective functions more convenient from a practical view point and give them better properties than the original problem.

The methods of this class include, in particular, the classical proximal method and its various variants (e.g., [1,2,4–8]). This method determines each subsequent approximation by solving a minimization problem for the above-mentioned auxiliary function on the feasible set, where the second term depends on the current iteration point and is defined by the Euclidean norm. The regularized

A. Kazakov et al. (Eds.): MOTOR 2026, LNCS 16687, pp. 64–77, 2026.
https://doi.org/10.1007/978-3-032-31390-4_5

proximal method developed on its basis belongs to the same class [1]. It uses a similar auxiliary function with the first term, i.e. the objective function, replaced by the Tikhonov function.

The proposed convex programming methods can naturally be considered as variants of the proximal method and, accordingly, the regularized proximal method, since the same strongly convex auxiliary functions are used to construct the iteration points. The developed versions of known methods are characterized by the following. Firstly, the iterative process involves approximating the epigraphs of auxiliary functions using polyhedral sets. Secondly, the sequence of approximations is generated using relaxation-based techniques, which enhances the practical efficiency of these methods. Thirdly, new convergent methods can be easily constructed based on the proposed methods. These methods are obtained by incorporating any suitable algorithms for solving the original problem into the construction of the main iterative sequence. Note that these and some other features of the methods will be discussed in detail after the formal description of the methods.

2 The Problem Settings

We solve the following problem:

$$\min\{f(x) : x \in D\}, \tag{1}$$

where $f(x)$ is a convex function defined in the n-dimensional Euclidean space $\mathrm{R_n}$ that takes a finite value at each point $x \in \mathrm{R_n}$, $D \subseteq \mathrm{R_n}$ is a convex closed set.

Suppose $X^* = \operatorname{Arg\,min}\{f(x) : x \in D\} \neq \emptyset$, $x^* \in X^*$. Let $f^* = f(x^*)$, $\operatorname{epi}(f, \mathrm{R_n}) = \{(x, \gamma) \in \mathrm{R_{n+1}} : x \in \mathrm{R_n}, \gamma \geq f(x)\}$, $W(z, Q) = \{a \in \mathrm{R_{n+1}} : \|a\| = 1, \langle a, v - z\rangle \leq 0 \ \forall v \in Q\}$ be a set of normalized generalized support vectors for the set $Q \subset \mathrm{R_{n+1}}$ at the point $z \in \mathrm{R_{n+1}}$, $\operatorname{int} Q$ be an interior of the set Q, $K = \{0, 1, \ldots\}$.

3 Variants of the Proximal Method and the Regularized Proximal Method

The proposed methods for solving the problem (1) are conceptually similar and differ primarily in the type of auxiliary functions that are used in constructing the iteration points x_k. Consequently, both methods follow a unified description scheme. We first present the relaxation version of the proximal method, which then serves as the basis for describing the second method. It should be noted that these methods are applicable to problem (1) even in the case where $D = \mathrm{R_n}$.

The first method produces a sequence of approximations $\{x_k\}$, $k \in K$, as follows. A closed convex set $G \subset \mathrm{R_n}$ is constructed such that $x^* \in G$ and the intersection $D \cap G$ is bounded. The starting point $x_0 \in D \cap G$ is chosen. The numbers α_k, β_k, Δ_k, ε_k, $k \in K$, are specified such that $0 < \alpha' \le \alpha_k \le \alpha'' < +\infty$, $0 < \beta' \le \beta_k \le \beta'' < +\infty$, $0 < \Delta' \le \Delta_k \le \Delta'' < +\infty$, $\varepsilon_k > 0$, $k \in K$,

$$\varepsilon_k \to 0, \quad k \to \infty. \tag{2}$$

Let the point $x_k \in D \cap G$, $k \ge 0$, already be constructed.

1. Define

$$\Phi_k(x) = \alpha_k f(x) + \frac{1}{2}\beta_k \|x - x_k\|^2, \tag{3}$$

$$v_k = (x_k, \Phi_k(x_k) + \Delta_k),$$

$$M_k^0 = \{v \in \mathrm{R_{n+1}} : \langle a_k, v - (x_k, \Phi_k(x_k))\rangle \le 0\},$$

where $a_k \in W((x_k, \Phi_k(x_k)), \mathrm{epi}(\Phi_k, \mathrm{R_n}))$. Assign $i = 0$.

2. The point (y_k^i, γ_k^i) is found, where $y_k^i \in \mathrm{R_n}$, $\gamma_k^i \in \mathrm{R_1}$, as a solution of the problem

$$\min\{\gamma : (x, \gamma) \in M_k^i, x \in D \cap G\}. \tag{4}$$

3. If

$$\Phi_k(y_k^i) - \gamma_k^i \le \varepsilon_k, \tag{5}$$

then proceed to Step 6. Otherwise, go to Step 4.

4. The point $z_k^i \in \mathrm{R_{n+1}}$ is chosen on the segment $(v_k, (y_k^i, \gamma_k^i))$ such that $z_k^i \notin \mathrm{int\,epi}\,(\Phi_k, \mathrm{R_n})$ and the point

$$\bar{v}_k^i = (y_k^i, \gamma_k^i) + q_k^i(z_k^i - (y_k^i, \gamma_k^i))$$

satisfies the inclusion

$$\bar{v}_k^i \in \mathrm{epi}\,(\Phi_k, \mathrm{R_n}).$$

for some $q_k^i \in [1, q]$, where $q < +\infty$.

5. The finite set $A_k^i \subset W(z_k^i, \mathrm{epi}\,(\Phi_k, \mathrm{R_n}))$ is chosen, the set

$$M_k^{i+1} = M_k^i \cap \{v \in \mathrm{R_{n+1}} : \langle a, v - z_k^i\rangle \le 0 \ \forall a \in A_k^i\}, \tag{6}$$

is constructed, the index i is incremented by one, and go to Step 2.

6. Define $i_k = i$, $u_k = y_k^{i_k}$, choose the point $x_{k+1} \in D \cap G$ in accordance with

$$f(x_{k+1}) \le \min\{f(x_k), f(u_k)\}. \tag{7}$$

The method is thus established. Next, we present the second proposed algorithm: a modification of the regularized proximal method (e.g., [1, p. 874]) intended for solving (1). It generates a sequence of iteration points $\{x_k\}$, $k \in K$, as follows.

According to the conditions stated in the preliminary step of the first method, the set G, the point x_0 and the numbers α_k, β_k, Δ_k, ε_k, $k \in K$ are selected. In addition, numbers δ_k, $k \in K$, are given such that

$$\delta_k \to 0, \quad k \to \infty. \tag{8}$$

Let the point $x_k \in D \cap G$ be constructed, $k \ge 0$.

1. Put

$$\Phi_k(x) = \alpha_k T_k(x) + \frac{1}{2}\beta_k \|x - x_k\|^2, \tag{9}$$

where

$$T_k(x) = f(x) + \delta_k \|x\|^2$$

is a Tikhonov regularization function. The point v_k is fixed and the set M_k^0 is constructed according to Step 1 of the first method, put $i = 0$.

Steps 2–5 are identical to Steps 2–5 of the first method, respectively.

4 Discussion of the Methods

We now provide several remarks concerning the proposed methods, justify the solvability of problem (4), and establish an optimality criterion.

Remark 1. The construction of the set M_k^0 for each $k \in K$ in Step 1, together with equality (6), ensures that all M_k^i, $i \geq 0$, are polyhedral used to approximate the epigraph of the function $\Phi_k(x)$. Therefore, it is advantageous to define the set G using linear constraints during the initialization phase, regardless of whether $D = \mathrm{R_n}$ or $D \neq \mathrm{R_n}$. If $D = \mathrm{R_n}$ or the constraints in (1) are linear, then problem (4) reduces to a linear programming problem for all k, $i \in K$.

Remark 2. The condition defining the set G ensures the existence of a solution to problem (4) and guarantees the boundedness of the sequences $\{u_k\}$, $\{x_k\}$ generated by the methods. If the set D is bounded and defined by linear constraints, the special construction of the set G becomes unnecessary, in this case, it suffices to take $G = D$.

Let us now show that problem (4) for each pair of numbers k, $i \in K$ is indeed solvable.

Lemma 1. *Let $k \in K$, and the function $\Phi_k(x)$ is given by (3) or (9). Then for all $i \in K$ the inclusion*

$$\mathrm{epi}\,(\Phi_k, \mathrm{R_n}) \subset M_k^i \tag{10}$$

holds.

Proof. For $i = 0$, inclusion (10) holds by the definition of the set M_k^0 in Step 1 of the methods. Suppose that (10) holds for $i = l \in K$. Let us show that

$$\mathrm{epi}\,(\Phi_k, \mathrm{R_n}) \subset M_k^{l+1}, \tag{11}$$

then the statement of the lemma will be proved. Indeed, $\mathrm{epi}\,(\Phi_k, \mathrm{R_n}) \subset M_k^l$ holds by the induction hypothesis. According to the choice of the point $z_k^l \notin \mathrm{int\,epi}\,(\Phi_k, \mathrm{R_n})$ and the set A_k^l of generalized support vectors for $\mathrm{epi}\,(\Phi_k, \mathrm{R_n})$ at the point z_k^l, we have $\mathrm{epi}\,(\Phi_k, \mathrm{R_n}) \subset T_k^l$, where $T_k^l = \{v \in \mathrm{R_{n+1}} : \langle a, v - z_k^l \rangle \leq 0 \;\forall a \in A_k^l\}$. Since $M_k^{l+1} = M_k^l \cap T_k^l$, inclusion (11) holds. The lemma is proved.

Henceforth, x_k^* denotes the solution to the problem

$$\min\{\Phi_k(x) : x \in D \cap G\}, \tag{12}$$

$k \in K$, where $\Phi_k(x)$ is given by (3) or (9). Under the conditions of problem (1), functions (3) and (9) are strongly convex. Specifically, the strong convexity parameter $\varkappa_k$ for function (3) is given by $\varkappa_k = \beta_k$ (see, e.g., [1, pp. 206, 207]), and for the function (9) it can be readily shown, taking into account the remark [1, p. 876], that $\varkappa_k = \beta_k + 2\delta_k\alpha_k$. We immediately note that the solution x_k^* of problem (12) is unique (see, e.g., [1, p. 207]), and we put

$$\Phi_k^* = \Phi_k(x_k^*).$$

Since $(x_k^*, \Phi_k^*) \in \text{epi}\,(\Phi_k, \mathrm{R_n})$, $k \in K$, it follows from Lemma 1 that $(x_k^*, \Phi_k^*) \in M_k^i$ for all i, $k \in K$. Moreover, $x_k^* \in D \cap G$, $k \in K$. This means that the feasible set of problem (4) is non-empty for all i, $k \in K$ and both types of the function $\Phi_k(x)$. Since the set $D \cap G$ is closed and bounded, and the sets M_k^i are closed, it follows that $\gamma \geq \gamma_k$ under some $\gamma_k > -\infty$ for all $(x, \gamma) \in M_k^i$ with $x \in D \cap G$ according to boundedness of $D \cap G$. Consequently, problem (4) is solvable for all i, $k \in K$.

We establish and further substantiate an optimality criterion for problem (12) based on the solution to problem (4), assuming $k \in K$ to be fixed.

Lemma 2. *Let $k \in K$, the function $\Phi_k(x)$ has form (3) or (9) in (12). Then the inequality*

$$\gamma_k^i \leq \Phi_k^* \tag{13}$$

holds for the solution (y_k^i, γ_k^i) of problem (4) under any $i \in K$.

Proof. Since (y_k^i, γ_k^i) is a solution of problem (4), then the inequality

$$\gamma_k^i \leq \gamma, \quad i \in K \tag{14}$$

is fulfilled for any point (x, γ) satisfying the constraints of this problem. Note that $x_k^* \in D \cap G$ according to (12), and the inclusion $(x_k^*, \Phi_k(x_k^*)) \in M_k^i$, $i \in K$, holds according to Lemma 1. This implies that (x_k^*, Φ_k^*), $i \in K$, is a feasible point for problem (4). Then inequality (14) also holds for this point, i.e. (13) is satisfied. This completes the proof of the lemma.

Lemma 3. *Suppose the assumptions of Lemma 3 hold, and the equality*

$$\Phi_k(y_k^i) = \gamma_k^i. \tag{15}$$

is satisfied for some $i \in K$ at the point (y_k^i, γ_k^i). Then y_k^i is a solution to problem (12).

Proof. Since $y_k^i \in D \cap G$, then $\Phi_k(y_k^i) \geq \Phi_k^*$. On the other hand, taking into account condition (15) and inequality (13) we have $\Phi_k(y_k^i) \leq \Phi_k^*$. Thus, $\Phi_k(y_k^i) = \Phi_k^*$, and the statement of the lemma is proved.

Lemma 3 is used to justify the optimality criterion for an iterate point, which directly relates to original problem (1). We now proceed to discuss the optimality criterion, which can be incorporated into Step 2 of the first proposed method.

Theorem 1. *Suppose that* $\bar{x} \in D \cap G$, $\alpha > 0$, $\beta > 0$,

$$\Phi(x) = \alpha f(x) + \frac{1}{2}\beta\|x - \bar{x}\|^2.$$

The inclusion

$$\bar{x} \in X^* \tag{16}$$

is satisfied if and only if

$$\bar{x} = \arg\min\{\Phi(x) : x \in D \cap G\} \tag{17}$$

holds.

Proof. Assume that inclusion (16) holds for the point $\bar{x}$, i.e. $f(\bar{x}) \le f(x)$ for all $x \in D \cap G$. Thus, $\alpha f(\bar{x}) \le \alpha f(x)$ for all $x \in D \cap G$. Since $\Phi(\bar{x}) = \alpha f(\bar{x})$, then $\Phi(\bar{x}) \le \alpha f(x)$ for all $x \in D \cap G$. It follows that

$$\Phi(\bar{x}) \le \alpha f(x) + \frac{1}{2}\beta\|x - \bar{x}\|^2 = \Phi(x) \quad \forall x \in D \cap G,$$

i.e. (17) is proved.

Let (17) now be satisfied. Since the function $\Phi(x)$ is strongly convex on D with the constant of strong convexity β, then the inequality

$$\frac{1}{2}\beta\|x - \bar{x}\|^2 \le \Phi(x) - \Phi(\bar{x}),$$

holds according to a well-known theorem (e.g., [1, p. 207]) for all $x \in D \cap G$, and the inequality

$$\frac{1}{2}\beta\|x - \bar{x}\|^2 \le \alpha f(x) + \frac{1}{2}\beta\|x - \bar{x}\|^2 - \alpha f(\bar{x})$$

is fulfilled too. Thus, $\alpha(f(x) - f(\bar{x})) \ge 0$ for all $x \in D \cap G$. Hence, taking into account that $\alpha > 0$, the inclusion (16) follows. The theorem is proved.

By specifying $\Phi(x) = \Phi_k(x)$ Theorem 1, where $\Phi_k(x)$ is given by (3), and considering $\bar{x} = x_k$, $\alpha = \alpha_k$, $\beta = \beta_k$, we obtain

Consequence 1. For the point x_k constructed by the first proposed method, the inclusion $x_k \in X^*$ holds if and only if $x_k = x_k^*$.

Theorem 2. *Suppose that the point* x_k, $k \in K$, *be constructed using the first of the proposed methods. Let the solution* (y_k^i, γ_k^i) *to problem (4) satisfy equality (15) for some* $i \in K$, *and, in addition,*

$$y_k^i = x_k. \tag{18}$$

Then x_k *is a solution to problem (1).*

Proof. By Lemma 1, the point y_k^i is a solution to problem (12), where the function $\Phi_k(x)$ has form (3). Given the uniqueness of the solution to this problem, it follows that $y_k^i = x_k^*$. In view of condition (18), the equality $x_k = x_k^*$ also holds. Then, by Consequence 1, the inclusion $x_k \in X^*$ holds. The theorem is proved.

Remark 3. According to Theorem 2 at Step 2 of the variant of the proximal method (with the function $\Phi_k(x)$ of form (3)) the following stopping criterion can be added. If the equalities

$$\Phi_k(y_k^i) = \gamma_k^i, \quad y_k^i = x_k,$$

are simultaneously satisfied, then x_k is a solution to the original problem, and the process terminates.

Remark 4. Suppose the function $\Phi_k(x)$ takes form (3) or (9). If equality (15) is satisfied for some $i \in K$, then condition (5) is also met, and the point u_k is thus found. Now, let $i \in K$ be such that $\gamma_k^i < \Phi_k(y_k^i)$, i.e. $(y_k^i, \gamma_k^i) \notin \text{epi}\,(\Phi_k, \mathrm{R_n})$. In this case, the point z_k^i must be determined according to Step 6 to construct the cutting plane. If we put $q_k^i = 1$, then $z_k^i = \bar{v}_k^i$, and z_k^i is the intersection of the interval $(v_k, (y_k^i, \gamma_k^i))$ with the boundary of $\text{epi}\,(\Phi_k, \mathrm{R_n})$. The flexibility to choose q_k^i such that $1 < q_k^i \leq q$ allows for an approximate determination of this intersection point, which is advantageous for practical implementation.

Remark 5. When solving problem (1) using classical cutting-plane methods based on the approximation of the set $\text{epi}\,(f, \mathrm{R_n})$ by polyhedral sets (e.g., [9,10]), an accumulation of cuts occurs as these sets are sequentially constructed. Since the construction of approximating sets M_k^i, $i \geq 1$, begins with the updated set M_k^0 for each $k \in K$, then there is no infinite accumulation of cutting sets.

Remark 6. Let $\alpha_k = 1$ for all $k \in K$ in (3), (9), which means that the inclusions $\text{epi}\,(\Phi_k, \mathrm{R_n}) \subset \text{epi}\,(f, \mathrm{R_n})$ hold. In this case, we can construct a convex closed polyhedral set $M \subset \mathrm{R_{n+1}}$ with the condition $\text{epi}\,(f, \mathrm{R_n}) \subset M$ and put $M_k^0 = M$ for all $k \in K$. Then $(y_k^0, \gamma_k^0) = (y_0^0, \gamma_0^0)$ for all $k \geq 1$.

Let us discuss condition (7) for selecting the points of the main sequence constructed by any of the described methods. According to (7) for each $k \in K$ it is permissible to put $x_{k+1} = x_k$ if $f(x_k) \leq f(u_k)$, or $x_{k+1} = u_k$ if $f(u_k) \leq f(x_k)$. However, (7) provides the possibility of constructing $\{x_k\}$ with the strict relaxation condition, choosing $x_{k+1} \in D \cap G$ such that

$$f(x_{k+1}) < \min\{f(x_k), f(u_k)\}. \tag{19}$$

Additionally, taking into account conditions of problem (1) any known or newly developed algorithms can be used to construct iterative points from the feasible set. Let any suitable algorithm be chosen for this purpose. The point x_k or

u_k with the best objective value is then retained. Starting from this point, the algorithm performs a sufficient number of steps to ensure that the approximation found within $D \cap G$ can be designated as x_{k+1} satisfying condition (19). This observation proves the possibility of constructing new mixed algorithms for solving the problem (1) based on each of the two methods. Moreover, the convergence of such algorithms will already be justified by the convergence of the methods proposed here.

5 Validity of the Methods

To justify the validity of the methods, we will show that the methods will construct an auxiliary point u_k for each $k \in K$, which means that it will be possible to select a point x_{k+1} of the main sequence. To this end, we prove the existence the number $i = i_k$ for each $k \in K$, for which inequality (5) holds for the point (y_k^i, γ_k^i).

The proof of the existence of such a number $i = i_k$ will be carried out by contradiction, i.e. assume that the inequality

$$\Phi_k(y_k^i) - \gamma_k^i > \varepsilon_k \quad \forall i \in K \tag{20}$$

holds for some fixed $k \in K$. Then the process of constructing points (y_k^i, γ_k^i), $i \in K$, using methods will be infinite. In this regard, we will first substantiate auxiliary statements, where the infinity of such a process will be assumed.

Lemma 4. *Let the number $k \in K$ be fixed, the function $\Phi_k(x)$ has form (3) or (9), and the sequence $\{(y_k^i, \gamma_k^i)\}$, $i \in K$, be constructed according to Steps 2–5 of the methods. Then any limit point of this sequence belongs to the set* $\mathrm{epi}\,(\Phi_k, \mathrm{R_n})$.

Proof. Since the set $D \cap G$ is bounded, then the sequence $\{(y_k^i, \gamma_k^i)\}$, $i \in K$, is also bounded. Let us select a convergent subsequence $\{(y_k^i, \gamma_k^i)\}$, $i \in K' \subset K$, and let $(\tilde{y}_k, \tilde{\gamma}_k)$ be its limit point. Let us show that

$$(\tilde{y}_k, \tilde{\gamma}_k) \in \mathrm{epi}\,(\Phi_k, \mathrm{R_n}), \tag{21}$$

then the lemma will be proved.

The sequence of points z_k^i, $i \in K'$, is also constructed in addition to the $\{(y_k^i, \gamma_k^i)\}$, $i \in K'$, and are given by

$$z_k^i = (y_k^i, \gamma_k^i) + \lambda_k^i (v_k - (y_k^i, \gamma_k^i)) \tag{22}$$

according to Step 4 of the methods, where $\lambda_k^i \in [0, 1)$. We establish the equality

$$\lim_{k \in K'} \|z_k^i - (y_k^i, \gamma_k^i)\| = 0 \tag{23}$$

following the approach [11, p. 99]. Let l, $p \in K'$ be such that $l < p$. The inclusions $(y_k^p, \gamma_k^p) \in M_k^p$ and $M_k^p \subset M_k^l$ hold by virtue of (4) and (6), respectively. Moreover, any vector from the set A_k^l is a generalized support vector for the set

M_k^p at the point z_k^l. This implies that $\langle a, (y_k^p, \gamma_k^p) - z_k^l \rangle \leq 0$ for all $a \in A_k^l$, and taking into account (22) we obtain

$$\langle a, (y_k^l, \gamma_k^l) - (y_k^p, \gamma_k^p) \rangle \geq \lambda_k^l \langle a, (y_k^l, \gamma_k^l) - v_k \rangle \quad \forall a \in A_k^l.$$

According to Lemma 1 of [12], there exists $\beta > 0$ such that $\langle a, (y_k^l, \gamma_k^l) - v_k \rangle \geq \beta$ for all $a \in A_k^l$. Therefore, $\langle a, (y_k^l, \gamma_k^l) - (y_k^p, \gamma_k^p) \rangle \geq \lambda_k^l \beta$ for all vectors $a \in A_k^l$, and since these vectors have unit norm, it follows that

$$\|(y_k^l, \gamma_k^l) - (y_k^p, \gamma_k^p)\| \geq \lambda_k^l \beta.$$

From this and from the convergence of the sequence $\{(y_k^i, \gamma_k^i)\}$, $i \in K'$, it follows that $\lambda_k^i \to 0$, $k \in K'$. Then from (22), taking into account the boundedness of the sequence $\{\|v_k - (y_k^i, \gamma_k^i)\|\}$, $i \in K'$, we obtain assertion (23).

Next, we select a convergent subsequence $\{\bar{v}_k^i\}$, $i \in K'' \subset K'$, from the sequence $\{\bar{v}_k^i\}$, $i \in K'$, constructed at Step 4 of the methods. Let $\tilde{v}_k$ be its limit point. Note that

$$\tilde{v}_k \in \text{epi}\,(\Phi_k, \mathrm{R_n}) \tag{24}$$

due to the closedness of the set epi $(\Phi_k, \mathrm{R_n})$. Since $\bar{v}_k^i = (y_k^i, \gamma_k^i) + q_k^i (z_k^i - (y_k^i, \gamma_k^i))$ for numbers $i \in K''$, where $1 \leq q_k^i \leq q$, it follows from (23) and the boundedness of the sequence $\{q_k^i\}$, $i \in K''$ that $\tilde{v}_k = (\tilde{y}_k, \tilde{\gamma}_k)$. Then, according to (24), inclusion (21) holds. The lemma is proved.

Lemma 5. *Suppose the conditions of Lemma 4 hold. Then*

$$\lim_{i \in K} (\Phi_k(y_k^i) - \gamma_k^i) = 0. \tag{25}$$

Proof. Since $y_k^i \in D \cap G$, $i \in K$, then $\Phi_k(y_k^i) \geq \Phi_k^*$, $i \in K$, and taking into account (13) the inequality $\Phi_k(y_k^i) - \gamma_k^i \geq 0$ holds for all $i \in K$. Assume that the assertion of the lemma is false. Then there exists an infinite subset K_1 of the set K such that

$$\Phi_k(y_k^i) - \gamma_k^i \geq \alpha > 0 \quad \forall i \in K_1.$$

Select a convergence subsequence $\{(y_k^i, \gamma_k^i)\}$, $i \in K' \subset K_1$, from $\{(y_k^i, \gamma_k^i)\}$, $i \in K_1$. Let $(\tilde{y}_k, \tilde{\gamma}_k)$ be its limit point. Then

$$\Phi_k(\tilde{y}_k) - \tilde{\gamma}_k \geq \alpha.$$

The resulting inequality is contradictory, since inclusion (21) holds by Lemma 4, and, therefore, $\Phi_k(\tilde{y}_k) - \tilde{\gamma}_k \leq 0$. The lemma is proved.

Theorem 3. *There is a number $i = i_k$ such that inequality (5) holds for each $k \in K$ while constructing points (y_k^i, γ_k^i), $i \in K$, by either of the two methods described above.*

Proof. Let us assume that the statement of the theorem is false, i.e., (20) holds for the points (y_k^i, γ_k^i) constructed by the methods for some $k \in K$. Then the sequence $\{(y_k^i, \gamma_k^i)\}$, $i \in K$, is constructed for this $k \in K$ according to Steps 2–5. By Lemma 5 equality (25) holds for this sequence, which contradicts assumption (20). The theorem is proved.

6 Convergence Analysis of the Methods

We now establish the convergence of the proposed variants of the proximal method and the regularized proximal method.

Lemma 6. *The following equality*

$$\lim_{k\in K} \|u_k - x_k^*\| = 0 \tag{26}$$

holds for the sequence $\{u_k\}$, $k \in K$, *constructed by the first method, where* x_k^* *is a solution to problem (12) with the function* $\Phi_k(x)$ *given by (3) for all* $k \in K$.

Proof. By Theorem [1, p. 207] mentioned above, the following inequality

$$\frac{1}{2}\varkappa_k \|x - x_k^*\|^2 \le \Phi_k(x) - \Phi_k(x_k^*) \quad \forall k \in K, x \in D \cap G, \tag{27}$$

holds, where $\varkappa_k = \beta_k$ is the strong convexity constant of the function $\Phi_k(x)$, which the constant of strong convexity of the function $\Phi_k(x)$, which takes form (3) according to the conditions of the lemma. Since $u_k \in D \cap G$, $k \in K$, then from (27) is follows that

$$\frac{1}{2}\varkappa_k \|u_k - x_k^*\|^2 \le \Phi_k(u_k) - \Phi_k(x_k^*), \quad k \in K. \tag{28}$$

According to (5) we have $\Phi_k(u_k) = \Phi_k(y_k^{i_k}) \le \gamma_k^{i_k} + \varepsilon_k$, $k \in K$, but $\gamma_k^{i_k} \le \Phi_k(x_k^*)$, $k \in K$, by (13), and, consequently,

$$\Phi_k(u_k) \le \Phi_k(x_k^*) + \varepsilon_k, \quad k \in K. \tag{29}$$

Then in view of (28), (29) we obtain

$$\frac{1}{2}\varkappa_k \|u_k - x_k^*\|^2 \le \varepsilon_k, \quad k \in K, \tag{30}$$

where $\varkappa_k = \beta_k$. Hence and from (2), the condition for specifying the numbers β_k, $k \in K$, it follows (26). The lemma is proved.

Lemma 7. *Let the sequence* $\{u_k\}$ *be constructed using the second of the proposed methods. Then equality (26) holds, where* x_k^* *is a solution to problem (12) with the function* $\Phi_k(x)$ *given by form (9) for all* $k \in K$.

Proof. By the same theorem [1, p. 207], assertion (27) holds, implying inequality (28), where $\varkappa_k$ is a strong convexity constant of the function $\Phi_k(x)$ given by (9). From (28), taking into account (29), we obtain (30), where $\varkappa_k = \beta_k + 2\delta_k\alpha_k$. Note that the sequence $\{\beta_k + 2\delta_k\alpha_k\}$, $k \in K$, is bounded from above and, moreover, $(\beta_k + 2\delta_k\alpha_k) \not\to 0$, $k \in K$, according to the choice of the numbers β_k, α_k, δ_k. Then from (30) taking into account (2) it follows (26). The lemma is proved.

Lemma 8. *Let the sequences* $\{u_k\}$, $\{x_k\}$, $k \in K$, *be constructed by any of the two proposed methods. Then*

$$\lim_{k\in K} \|u_k - x_k\| = 0. \tag{31}$$

Proof. We will establish assertion (31) by considering each method separately.

1) First, let the function $\Phi_k(x)$ be given in form (3) for all $k \in K$. From (29) we have

$$\Phi_k(u_k) = \alpha_k f(u_k) + \frac{1}{2}\beta_k\|u_k - x_k\|^2 \leq \Phi_k^* + \varepsilon_k$$

for each $k \in K$. But

$$\Phi_k^* \leq \Phi_k(x_k), \tag{32}$$

and $\Phi_k(x_k) = \alpha_k f(x_k)$. It means that

$$\frac{1}{2}\beta_k\|u_k - x_k\|^2 \leq \alpha_k(f(x_k) - f(u_k)) + \varepsilon_k,$$

and since $f(x_{k+1}) \leq f(u_k)$, then

$$\frac{1}{2}\beta_k\|u_k - x_k\|^2 \leq \alpha_k(f(x_k) - f(x_{k+1})) + \varepsilon_k, \quad k \in K. \tag{33}$$

Since the sequence $\{f(x_k)\}$, $k \in K$, is monotonically decreasing and bounded, it converges and

$$\lim_{k\in K}(f(x_k) - f(x_{k+1})) = 0. \tag{34}$$

Then from (33), (34), taking into account (2), as well as the conditions for choosing the numbers α_k, β_k, $k \in K$, it follows (31).

2) Now let the function $\Phi_k(x)$, $k \in K$, has form (9). Then from (29), (32), (9) it follows that

$$\alpha_k(f(u_k) + \delta_k\|u_k\|^2) + \frac{1}{2}\beta_k\|u_k - x_k\|^2 \leq \alpha_k(f(x_k) + \delta_k\|x_k\|^2) + \varepsilon_k, \quad k \in K.$$

From here, taking into account (7) and the positivity of the numbers α_k, δ_k, the following inequality is easily obtained:

$$\frac{1}{2}\beta_k\|u_k - x_k\|^2 \leq \alpha_k(f(x_k) - f(x_{k+1})) + \alpha_k\delta_k\|x_k\|^2,$$

$k \in K$. From this inequality, taking into account equality (34), conditions (2), (8) of the choice of the numbers ε_k, δ_k, and the boundedness of the sequences $\{\|x_k\|\}$, $\{\alpha_k\}$, $\{\beta_k\}$, it follows (31). The lemma is proved.

Put

$$\bar{X}^* = \operatorname{Arg\,min}\{f(x) : x \in D \cap G\}.$$

Lemma 9. *Let* $\bar{x} \in \bar{X}^*$. *Then* $\bar{x} \in X^*$.

Proof. Since $\bar{x} \in D \cap G$, and therefore $\bar{x} \in D$, then $f(\bar{x}) \geq f^* = f(x^*)$. On the other hand, $f(\bar{x}) \leq f(x^*)$, since the inclusion $x^* \in D \cap G$ holds according to the condition of choosing the set G. Thus, $f(\bar{x}) = f^*$, and the lemma is proved.

Using Lemmas 6–9, we will finally prove the convergence of the sequence constructed by each of the proposed methods to the set X^*.

Theorem 4. *Let the sequence* $\{x_k\}$, $k \in K$, *be constructed by any of the two methods. Then any limit point of this sequence belongs to the solution set to problem (1).*

Proof. Let $\{x_k\}$, $k \in K' \subset K$, be a convergent subsequence of $\{x_k\}$, $k \in K$, and let $\bar{x}$ be its limit point. Lets show that

$$\bar{x} \in \bar{X}^*, \tag{35}$$

then according to Lemma 9 the statement of the theorem will be proved. We immediately note that $\bar{x} \in D \cap G$, since $x_k \in D \cap G$, $k \in K$, by construction, and the sets D and G are closed. In addition, we will assume that the point x_k^* used below relates to problem (12) with functions $\Phi_k(x)$ of form (3) or (9) depending on the method by which the sequence $\{x_k\}$ was constructed.

Let us select from the set K' an infinite subset of numbers K'' such that the sequences $\{u_k\}$, $\{x_k\}$, $\{x_k^*\}$, $\{\alpha_k\}$, $\{\beta_k\}$, $k \in K''$, are convergent, and let $\bar{\alpha} = \lim_{k \in K''} \alpha_k$, $\bar{\beta} = \lim_{k \in K''} \beta_k$. By Lemma 8, the sequence $\{u_k\}$, $k \in K''$, converges to the point $\bar{x}$, and therefore, the sequence $\{x_k^*\}$, $k \in K''$, also converges to the same point $\bar{x}$ according to Lemmas 6, 7. Let us prove inclusion (35) for the point $\bar{x}$ in case of constructing $\{x_k\}$, $k \in K$, by the first method, and then by the second method.

1) Let the function $\Phi_k(x)$ be given in form (3) for all $k \in K$, when constructing $\{x_k\}$. By (27) under $\varkappa_k = \beta_k$ the equality

$$\frac{1}{2}\beta_k \|x - x_k^*\|^2 \leq \alpha_k(f(x) - f(x_k^*)) + \frac{1}{2}\beta_k \|x - x_k\|^2 - \frac{1}{2}\|x_k^* - x_k\|^2$$

holds for all $x \in D \cap G$, $k \in K$. Consequently,

$$\frac{1}{2}\beta_k \|x - x_k^*\|^2 \leq \alpha_k(f(x) - f(x_k^*)) + \frac{1}{2}\beta_k \|x - x_k\|^2 \quad \forall k \in K'', x \in D \cap G.$$

Let us pass in the last inequality to the limit with respect to $k \to \infty$, $k \in K''$, for any fixed $x \in D \cap G$, taking into account that the sequences $\{x_k\}$, $\{x_k^*\}$, $k \in K''$, converge to the point $\bar{x}$. Then $\bar{\alpha}(f(x) - f(\bar{x})) \geq 0$ for all $x \in D \cap G$. From $\bar{\alpha} > 0$, it follows (35).

2) We will now assume that the sequence $\{x_k\}$ is constructed using the function $\Phi_k(x)$ given by (9). According to (27), where $\varkappa_k = \beta_k + 2\delta_k \alpha_k$, for all $x \in D \cap G$, $k \in K$, the inequality

$$(\frac{1}{2}\beta_k + \delta_k \alpha_k)\|x - x_k^*\|^2 \leq \alpha_k(f(x) + \delta_k \|x\|^2) + \frac{1}{2}\beta_k \|x - x_k\|^2 - \alpha_k(f(x_k^*) + \delta_k \|x_k^*\|^2)$$

holds. Thus,

$$(\frac{1}{2}\beta_k + \delta_k\alpha_k)\|x - x_k^*\|^2 \leq \alpha_k(f(x) - f(x_k^*)) \\ + \alpha_k\delta_k\|x\|^2 + \frac{1}{2}\beta_k\|x - x_k\|^2 \quad \forall k \in K'', x \in D \cap G.$$

Let us pass in this inequality for a fixed $x \in D \cap G$ to the limit with respect to $k \in K''$, taking into account condition (8). Then

$$\frac{1}{2}\bar{\beta}\|x - \bar{x}\|^2 \leq \bar{\alpha}(f(x) - f(\bar{x})) + \frac{1}{2}\bar{\beta}\|x - \bar{x}\|^2 \quad \forall x \in D \cap G.$$

From here, in view of the positivity of $\bar{\alpha}$, it follows (35). The theorem is proved.

Consequence 2. The sequence $\{x_k\}$, $k \in K$, constructed by any of the two described methods, converges to the set X^*.

Proof. Let $\{x_k\}$, $k \in K' \subset K$, be a convergent subsequence of $\{x_k\}$, $k \in K$, and $\bar{x}$ be its limit point. As already noted, the sequence $\{f(x_k)\}$, $k \in K$, converges. Then, by Theorem 4, the following equalities hold:

$$\lim_{k \in K} f(x_k) = \lim_{k \in K'} f(x_k) = f(\bar{x}) = f^*.$$

This, together with a well-known theorem (e.g., [1, p. 62]), implies the assertion being proved

Remark 7. The parameter α_k is used in the function $\Phi_k(x)$ given by (3) according to [1, p. 327], and the parameter β_k is introduced following [2, p. 206]. As noted in [2, p. 207], an appropriate choice of β_k ensures that the function $\Phi_k(x)$ is strongly convex, even if the objective function in (1) is only weakly convex. Namely, if $f(x)$ is weakly convex with the weak convexity constant $\tau > 0$ [2, p. 48], then, the function $\Phi_k(x)$ becomes strongly convex by choosing $\alpha_k = 1$, $\beta_k > \tau$, and problem (12) has a unique solution x_k^*.

References

1. Vasiliev, F.P.: Optimization Methods: vol. 1, vol. 2. MCCME, Moscow (2011). (in Russian)
2. Konnov, I.V.: Nonlinear Optimization and Variational Inequalities. Kazan University, Kazan (2013). (in Russian)
3. Tikhonov, A.N., Arsenin, V.A.: Methods for Solving Ill-Posed Problems. Nauka, Moscow (1979). (in Russian)
4. Antipin, A.S.: Continuous and Iterative Processes with the Projection Operator. Collection "Problems of Cybernetics. Computational Issues in the Analysis of Large Systems", pp. 5–43. Publishing House of the USSR Academy of Sciences, Moscow (1989). (in Russian)

5. Antipin, A.S.: Equilibrium programming: proximal methods. J. Comput. Math. Math. Phys. **37**(11), 1327–1339 (1997). (in Russian)
6. Kaplan, A., Tichatschke, R.: Multi-step-prox-regularization method for solving convex variational problems. Optimization **33**, 287–319 (1995)
7. Lemaire, B.: The proximal algorithm. Intern. Ser. Numer. Math. **87** (1989). Birkhauser Verlag. Basel
8. Rockafellar, R.T.: Monotone operators and the proximal point algorithm. SIAM J. Control Optim., 877–898 (1976)
9. Bulatov, V.P.: Embedding Methods in Optimization Problems. Nauka, Novosibirsk (1977). (in Russian)
10. Polyak, B.T.: Introduction to Optimization. Nauka, Moscow (1983). (in Russian)
11. Zabotin, I.Ya., Yarullin, R.S.: Cutting-plane methods without nested approximating sets. arXiv preprint (2023). https://arxiv.org/abs/2311.12520 (in Russian)
12. Zabotin, I.Ya.: On the several algorithms of immersion-severances for the problem of mathematical programming. Bull. Irkutsk State Univ. Ser. Math. **4**(2), 91–101 (2011). (in Russian)

Discrete and Combinatorial Optimization

Solving Branch Location Problem Using a Special Variant of VNS

Maxim Khabarov[1], Stepan Kochemazov[2(✉)], Ilya Otpuschennikov[2], and Alexander Semenov[2]

[1] ITMO University, Saint Petersburg, Russia
[2] ISDCT SB RAS, Irkutsk, Russia
veinamond@gmail.com

Abstract. We study a class of location problems formulated for a bipartite graph: the vertices from one part of the graph interpret clients while the vertices from the second part correspond to facilities that serve the clients. The goal is to close as many facilities as possible so that the number of dissatisfied clients does not exceed the pre-specified threshold. This formulation corresponds to the so-called Branch Location Problem (BLP). In the paper we reduce BLP to Integer Linear Programming (ILP) and Maximum Satisfiability (MaxSAT). To solve the instances of these combinatorial problems we employ complete 0-1-ILP and MaxSAT solvers, as well as one metaheuristic (MVP) which can be viewed as a variant of the well known Variable Neighborhood Search (VNS) algorithm. We construct families of random BLP instances of varying dimensions and demonstrate in the computational experiments that the MVP-based algorithms show good performance on such instances: in most cases they find the global optimum and spend significantly less time on solving compared to complete solvers.

Keywords: Location problems · Branch location problem (BLP) · 0-1-Integer linear programming (0-1-ILP) · Maximum satisfiability (MaxSAT) · Variable neighborhood search (VNS) · Merging variables principle (MVP)

1 Introduction

In the paper we study the subclass of location problems known as Branch Location Problem (BLP) [17]. It can be reduced to the variants of the well known Set Cover problem (SCP) the decision variant of which was among the first 21 NP-complete problems compiled by R.M. Karp [13]. Both SCP and BLP have a wide spectrum of practical applications and thus it is relevant to construct algorithms for their solving. The feature of SCP is that to solve it one can employ greedy approximate algorithms for which the approximation factor grows slowly with the power of the covered set (the so-called ground set): [7,9,14,20,28] etc. Since BLP generalizes SCP, these algorithms can be adapted for it. To construct

A. Kazakov et al. (Eds.): MOTOR 2026, LNCS 16687, pp. 81–97, 2026.
https://doi.org/10.1007/978-3-032-31390-4_6

exact BLP solutions one can use combinatorial algorithms, such as Integer Linear Programming (ILP) solvers and Maximum Satisfiability (MaxSAT) solvers. It is quite interesting that randomly generated BLP instances of large dimension turn out to be relatively simple for 0-1-ILP solvers while being extremely hard for complete MaxSAT solvers.

In the present paper, the main focus of our study is on the solving of BLP using the so-called *Merging Variable Principle* (MVP) [27] which is a special variant of the Variable Neighborhood Search (VNS) concept [22] adapted for pseudo-Boolean optimization. More specifically, we consider BLP as a MaxSAT instance and solve it using MVP. We compare the effectiveness of MVP-based algorithms to that of other approaches to BLP solving including those based on the use of 0-1-ILP solvers, complete MaxSAT solvers and greedy algorithms. The main contributions of the paper are as follows:

1. We analyze and compare the effectiveness of several combinatorial algorithms on randomly generated BLP instances. Specifically, we reduce BLP to ILP and MaxSAT and apply to corresponding instances the following solvers: SCIP (ILP), RC2, Loandra (MaxSAT) and MVP metaheuristic (MaxSAT).
2. We generate random BLP instances with different parameters and show that even for relatively small dimensions such instances remain extremely hard for modern MaxSAT solvers.
3. We also show that on the generated test instances the MVP-based algorithms show good performance: specifically, in most cases they yield the global solution of the problem and they manage to construct it significantly faster than the complete solvers.

2 Preliminaries

This section contains notations and notions which lie in the basis of the main results of presented paper.

2.1 Set Cover Problem and Branch Location Problem

The basic formulation of the Set Cover problem looks as follows. Consider an n-element set $I = \{i_1, \ldots, i_n\}$, and the set $J = \{j_1, \ldots, j_m\}$ such that for each $q \in \{1, \ldots, m\}$ it holds that $j_q \in 2^I$, $\bigcup_{j \in J} j = I$, and some natural number d. The question is whether there exists such $J' \subseteq J : |J'| \leq d$, that $\bigcup_{j \in J'} j = I$. The set J' is called the *cover* of I. In this formulation, SCP is NP-complete [13]. In the optimization variant one needs to find a cover of minimum cardinality, and this problem is NP-hard.

We can interpret sets I and J as parts of a bipartite graph so that any pair $(i, j) \in I \times J$ is an edge of this graph only if $i \in j$. This formulation is a convenient base for many *location problems* [17], which have a wide spectrum of practical applications. The latter include the so called *Branch Location Problem*(BLP), which in its basic formulation can be stated as follows. Consider a bipartite graph $G = (I, J, E)$, in which I is the set of vertices called *clients*, and J is the set of vertices called *facilities* or *branches* (here, it is implied that they are the branches of some single company that serve the clients). Denote by E the set of edges: $E, E \subseteq I \times J$. If $(i, j) \in E$ then we say that the facility j serves the client i. With each $i \in I$ we can associate its *neighborhood* in G: the set of facilities that serve this client $J^i = \{j \in J : (i, j) \in E\}$. We say that the client $i \in I$ is *satisfied* in G, if $J^i \neq \emptyset$. Let us delete from J some set of vertices J^* and remove the edges incident to them from E. Denote the resulting graph as $G' = (I, J', E')$. Let K be an arbitrary natural number such that $1 \leq K \leq |I|$. Hereinafter, we consider the following variant of BLP: to maximize the value of $|J^*|$ while the number of satisfied clients in G' remains at least K. Let us refer to the set $J' = J \setminus J^*$ that satisfies this requirement as to K-cover of I. It is clear that for $K = |I|$ this BLP formulation is equivalent to the optimization variant of SCP.

BLP makes perfect sense from the industry's point of view. Indeed, it arises naturally, e.g., when a company needs to optimize the number of service points (bank branches, transit stops, pickup locations) based on client coverage.

2.2 ILP, SAT and MaxSAT

Integer Linear Programming (ILP) (see e.g. [25]) is one of the classical combinatorial problems with the widest spectrum of practical applications. In the original formulation it consists in the following. Given a linear inequality system of the kind

$$x \cdot A \geq b \tag{1}$$

which defines a set of *feasible solutions*, one needs to maximize or minimize the following linear form $\sum_{i=1}^{n} x_i \cdot c_i$. In the general case, A in (1) is an $n \times m$-matrix with components from some field (e.g. from $\mathbb{R}$ or $\mathbb{Q}$); $x = (x_1, \ldots, x_n)$ is the so-called vector of variables, and $b = (b_1, \ldots, b_m)$ and $c = (c_1, \ldots, c_n)$ are vectors the coordinates of which are the elements of the considered field. If in addition to that one needs the variables $x_i, i \in \{1, \ldots, n\}$ to take values from $\{0, 1\} \cap \mathbb{Z}$, then the corresponding problem is called 0-1-ILP. This problem in its optimization formulation is NP-hard.

Recall that a Boolean formula can be considered as a word over a finite alphabet comprising Boolean variables, brackets, and special symbols called *Boolean connectives*. Usually, the Boolean connectives form the so-called *complete basis* using which one can represent by formula any total Boolean function of arbitrary arity [31]. If X is a set of Boolean variables from formula F then an arbitrary set of values from $\{0, 1\}$ of variables from X is referred to as an *assignment* over X. For an arbitrary assignment α over X it is possible to define the rules of interpretation of the corresponding formula F on α by substituting values from α into corresponding variables in F [6]. By interpreting a Boolean formula F on each assignment of variables X from F we specify a Boolean function of the kind $f_F : \{0, 1\}^n \to \{0, 1\}$, $n = |X|$, where by $\{0, 1\}^n$ we denote an *n-dimensional Boolean hypercube* i.e. the set of all possible binary words of length n. A Boolean formula is called *satisfiable* if there exists an assignment α s.t. $f_F(\alpha) = 1$, called a *satisfying assignment*. If formula F does not have any satisfying assignments then it is called *unsatisfiable*. Deciding, whether an arbitrary Boolean formula is satisfiable constitutes the *Boolean satisfiability problem* (SAT). SAT is a classical NP-complete problem.

The simplest Boolean formulas are Boolean variables or their negations are called *literals*. A pair of literals $(x, \neg x)$ where x is some Boolean variable is called *complementary*. A disjunction of literals in which no pair is a complementary pair is called *clause*. A conjunction of different clauses is called a *Conjunctive Normal Form*(CNF). Using Tseitin transformations [30] one can reduce SAT for an arbitrary Boolean formula F to SAT for some CNF in polynomial time in the length of binary description of formula F. Thus, in practice SAT is usually considered for formulas in CNF.

Consider an arbitrary CNF C over a set of variables $X : |X| = n$, which contains m clauses. Hereinafter, the notation $\{0,1\}^{|B|}$ means the set of assignments of Boolean variables comprising set B. Let us associate with C and $\{0,1\}^n$ the following function:

$$\phi_C : \{0,1\}^{|X|} \to \{0,1,\dots,m\} \tag{2}$$

The value of (2) on an arbitrary assignment α of variables from X is defined as the number of clauses in CNF C which take the value of 1 on α. The Maximum Satisfiability (MaxSAT) problem is to maximize the value of function (2) over the hypercube $\{0,1\}^{|X|}$. It is clear, that $\phi_C(\alpha) = m$ if and only if C is satisfiable. Thus MaxSAT is NP-hard problem.

During the recent 25 years there have been developed the algorithms for solving 0-1-ILP, SAT and MaxSAT which show impressive effectiveness in practice. For example, MILP (Mixed Integer Linear Programming) solvers, such as Gurobi [10] and CPLEX [11] are well-known commercial tools for optimizing business-related processes. The SCIP solver [1] provides an academic alternative for them. SAT solvers and SMT solvers based on them serve as the main tools for the areas of symbolic verification and testing programs (see e.g. [16,23], etc.). The MaxSAT solvers are also used more and more in practice each year.

2.3 Algorithms for Solving BLP

In this subsection, we describe some algorithms which are used to solve BLP. As we mentioned above, greedy algorithms work well on SCP and BLP. In our experiments we used the simplest greedy algorithm presented in pseudocode as Algorithm 1.

Algorithm 1: Greedy algorithm

Input: $J; \phi; K$
Output: J'_{k^*-1}

$J'_0 \leftarrow J$
for $k \leftarrow 1$ **to** $|J|$ **do**
  $x \leftarrow \operatorname{argmax}_{x \in J'_{k-1}} \phi(J'_{k-1} \setminus \{x\})$
  $J'_k \leftarrow J'_{k-1} \setminus \{x\}$
  if $\phi(J'_k) < K$ **then**
    return J'_{k-1}
return $\emptyset$

This algorithm performs a number of iterations. By J'_k it denotes the set of facilities that remained open after iteration number k. The algorithm starts from the set $J'_0 = J$ and at each iteration number $k \geq 1$ it reduces the number of open facilities by 1. The value of $\phi(J'_k)$ is equal to the number of clients satisfied by the set J'_k. The algorithm terminates after iteration number k^*, at which $\phi(J'_{k^*}) < K$. The solution it returns is the set J'_{k^*-1}, i.e. the set containing $|J| - k^* + 1$ facilities. The described algorithm in its naive implementation works in time $O(|G| \cdot |J|^2)$, however, it can be optimized to $O(|J| \cdot |G|)$ if we do not recompute the number of satisfied clients after deleting each facility.

In paper [33] they used a MaxSAT-based approach to solve one variant of BLP. This paper initially featured a complete bipartite graph, in which the edges had weights, and the number of satisfied clients was evaluated based on the change in the minimal edge weight of a satisfied client after the transition from G to G'. However, as it follows from [33] one can effectively transition from this formulation to the one described above.

From the above one can conclude that in practical applications J can be of size up to tens or hundreds of elements, while the size of I is in the range of hundreds of thousands or even millions. Therefore, below we consider the BLP variants with $|J| \ll |I|$.

There is a clear and natural reduction from BLP to 0-1-ILP. Indeed, consider an arbitrary BLP instance for a graph $G = (I, J, E)$. With every $i \in I$ associate a Boolean variable x_i, and denote the resulting set of variables as X. Let us assume that $x_i = 1$ if and only if client i is satisfied, otherwise $x_i = 0$. With every $j \in J$ associate a Boolean variable y_j and assume that $y_j = 1$ if $j \in J \setminus J^*$, and $y_j = 0$, if $j \in J^*$. Denote by Y the set of variables y_j. It is obvious that the client is satisfied if and only if $\sum_{j \in J^i} y_j \geq 1$. Taking this into account we can write the following 0-1-ILP formulation of BLP.

Maximize the linear form $\sum_{j \in J}(1 - y_j)$ under the following constraints:

$$\begin{cases} \sum_{j \in J^i} y_j \geq x_i, \text{for each } i \in I; \\ \sum_{i \in I} x_i \geq K. \end{cases} \tag{3}$$

Using this formulation one can use any complete MILP solver to construct an exact solution for BLP.

Other than 0-1-ILP, it is easy to reduce BLP to SAT. Indeed, it is easy to see that the constraint $\sum_{j \in J^i} y_j \geq x_i$ holds if and only if the following formula is true:

$$\Phi_i : x_i \equiv \bigvee_{j \in J^i} y_j \tag{4}$$

Thus, similar to (3) one needs to maximize $\sum_{j \in J}(1 - y_j)$ while the following CNF is satisfied:

$$C(\Phi_1) \wedge \ldots \wedge C(\Phi_{|I|}) \wedge C(\sum_{i \in I} x_i \geq K) \tag{5}$$

In (5) by $C(\Phi_i)$ we denote the CNF representations of formulas (4), $C(\sum_{i \in I} x_i) \geq K$ – CNFs which take the value of true only on those assignments α_i of variables x_i, $i \in I$, for which $\sum_{i \in I} \alpha_i \geq K$. Encoding the constraints of the kind $\sum_{i \in I} x_i \geq K$ and $\sum_{j \in J}(1 - y_j) \geq l$ to SAT can be done effectively using the techniques described in e.g. [8]. In this formulation one can solve BLP in $O(\log |J|)$ invocations of a SAT oracle.

In [33] the variant of BLP considered in that paper was studied using the MaxSAT solvers. Following [33] we have the following MaxSAT formulation with hard and soft clauses [18] for the BLP variant studied in the present paper:

- Hard clauses: $H_1 \triangleq C(\sum_{i \in I} x_i \geq K), H_2 \triangleq \wedge_{i \in I} C(\Phi_i)$;
- Soft clauses: $S \triangleq \wedge_{j \in J} \neg y_j$.

The goal is to maximize the number of satisfied soft clauses.

Let us remark on one negative issue with the latter reduction. The problem is that the SAT encodings of the constraints of the kind $\sum_{i \in I} x_i \geq p$, while formally effective, are quite large in practice. For example, for $|I| \approx 10^6$ the SAT encoding of a single constraint of the kind $\sum_{i \in I} x_i \geq K$ takes several gigabytes of memory.

In this situation we can make use of the feature of practical variants of BLP mentioned above, which consists in $|J| \ll |I|$. In particular, let l be an upper bound for the number of such $j \in J$ that $j \notin J^*$. This fact can be encoded as the requirement for a CNF $C(\sum_{j \in J} y_j \leq l)$ to be satisfiable. Thus, the hard clauses for such value of l look as follows:

$$H \triangleq C\left(\sum_{j \in J} y_j \leq l\right) \tag{6}$$

It is easy to see, that in the role of soft clauses, one can use the following CNF:

$$\bigwedge_{i\in I}\left(\bigvee_{j\in J^i} y_j\right) \tag{7}$$

The goal is to maximize the number of satisfied soft clauses. An interesting observation is that for large $|I|$, among the clauses in (7) there may be many duplicates. In this case, each unique disjunct is assigned a positive integer weight equal to the number of its occurrences in (7). The resulting problem is an instance of Weighted MaxSAT [2]. Modern MaxSAT solvers support such a weighted format of MaxSAT. Note, that exactly this MaxSAT encoding was used in [33].

The maximization of the number of satisfied clauses in (7) can then be performed by invoking an exact MaxSAT oracle for decreasing values of l. Let β be an arbitrary assignment of variables y_j, $j \in J$. The solution to the original BLP is an assignment β_* found by the MaxSAT oracle for the minimum feasible value l_*, such that:

1. β_* is a solution to the MaxSAT problem formulated above for $l = l_*$;
2. the total weight of clauses in (7) satisfied by β_* is at least K.

It is clear that the BLP solution in this formulation can be found in $O(\log|J|)$ invocations of a MaxSAT oracle. Since $|J|$ is small relative to $|I|$, the described approach appears promising.

A rather surprising fact is that the SAT and MaxSAT formulations of BLP turn out to be extremely hard for modern complete solvers for these problems (e.g. for $|I| \approx 10^5$ and $|J| \approx 40$). In contrast, 0-1-ILP solvers successfully handle BLP variants of substantial dimensions ($|I| > 10^5$, $|J| > 40$). These are the BLP variants generated by the algorithm described at the beginning of the experimental section.

3 Applying Merging Variables Principle to Solving BLP

The main goal of the present work is to find approximate solutions to BLP for instances of large dimension. To this end, we employ a metaheuristic based on the Merging Variable Principle (MVP) [27].

MVP is a special case of the well-known technique known as Variable Neighborhood Search (VNS). VNS was first described by N. Mladenovic and P. Hansen in their seminal paper [22]. The core idea of VNS is to consider a set of different *neighborhood structures* [5] in the search space. If some base heuristic converges to a local extremum within one neighborhood structure, this does not imply that the same point is a local extremum with respect to other neighborhood structures. Consequently, by changing the structure, it is possible to escape the local extremum and further improve the value of the objective function.

MVP exploits a very similar idea. The main difference is that within the MVP framework, a transition is made from the original search space, which is the Boolean hypercube with a neighborhood structure defined by Hamming balls of fixed radius, to another metric space with a non-Boolean structure. At the same time, there exists a one-to-one correspondence between the points of these spaces. We refer to the new space as *conjugated space.* A Hamming metric, and accordingly various neighborhood structures, can be defined on the conjugated space.

Consider the optimization problem for a *pseudo-Boolean function* [4], i.e., a function of the following kind:

$$f : \{0,1\}^n \rightarrow \mathbb{R} \tag{8}$$

We assume that on $\{0,1\}^n$ a neighborhood structure formed by Hamming balls of radius 1 is defined. The basic algorithm used to optimize (8) is the Hill Climbing algorithm [26].

In [27], the coordinates of points in $\{0,1\}^n$ were associated with Boolean variables forming a set $U = \{u_1, u_2, \dots, u_n\}$. Then a natural number $r : 1 < r < n$ was chosen and the set U was partitioned into r disjoint subsets, each associated with a separate variable. The set of such variables is denoted by $V = \{\nu_1, \dots, \nu_r\}$. This defines a mapping $\mu : U \rightarrow V$, called the *merging mapping.* Consider an arbitrary variable $\nu_j \in V$, $j \in \{1, \dots, r\}$ and let U_j be the set of preimages of ν_j under μ. Each Boolean vector from $\{0,1\}^{|U_j|}$ is associated with a distinct symbol, and the set of these symbols forms the domain of variable ν_j. Thus, ν_j is a variable taking $2^{|U_j|}$ values. Variables from V are called *merged.* The Cartesian product of the domains of all merged variables forms a space on which a Hamming metric can be specified, and accordingly various neighborhood structures can be defined. This metric space is denoted as D^μ. As shown in [27], there exists a bijection $\tau_\mu : D^\mu \rightarrow \{0,1\}^n$. On D^μ a special function $F_{f,\mu} : D^\mu \rightarrow \mathbb{R}$, which we refer as μ-conjugated with function (8), can be defined. For this function it holds that $\text{extr}_{\{0,1\}^n} f = \text{extr}_{D^\mu} F_{f,\mu}$, where extr denotes the minimum or maximum value of the corresponding function.

On the other hand, using τ_μ one can construct on $\{0,1\}^n$ a neighborhood structure, and in this sense different μ-mappings specify different such structures. Thus, MVP can be considered as a special case of VNS. It was shown in [27] that for fixed r and n there exist $r! \cdot S(r,n)$ different merging mappings, where $S(\cdot,\cdot)$ denotes the Stirling numbers of the second kind [29]. Therefore, the described MVP-based technique allows one to effectively generate a large number of distinct neighborhood structures in $\{0,1\}^n$.

MVP is most beneficial in situations where the search can be organized over a Boolean hypercube of relatively small dimension. In this case, the transition to the space D^μ can be advantageous, since its dimension $r < n$, and the traversal of Hamming neighborhoods in D^μ can be performed in parallel in a straightforward manner. In [24] exactly this feature was exploited in application to some special cases of the MaxSAT problem. Specifically, this applies to the situation where a Boolean formula in CNF possesses a non-trivial Strong Backdoor Set (SBS) with respect to some polynomial sub-solver in the sense of [32].

Assume that we have an arbitrary CNF C over Boolean variables X, and a polynomial algorithm A^p which when invoked on C can answer SAT (if C is satisfiable), UNSAT (if C is unsatisfiable) and INDET (if it failed to determine the satisfiability of C within the given time). Consider an arbitrary set $B : B \subseteq X$ and for any assignment $\beta \in \{0,1\}^{|B|}$ denote by $C[\beta/B]$ the CNF obtained as result of substitution of β into C (substitution is understood in the standard sense [6]). The set $B : B \subseteq X$ is SBS [32] for C w.r.t. A^p if for any assignment $\beta \in \{0,1\}^{|B|}$ it holds that: $A^p(C[\beta/B]) \in \{SAT, UNSAT\}$. In other words, if B is SBS then the satisfiability of any CNF of the kind $C[\beta/B]$ is decidable by algorithm A^p.

In such cases, solving the original MaxSAT problem can be effectively reduced to the maximization of a pseudo-Boolean function of the form (8) over a hypercube whose dimension equals the cardinality of some SBS.

It is not hard to note that X is a trivial SBS w.r.t. a simple algorithm known as Unit Propagation rule (UP) [21]. Any SBS w.r.t. UP is called *Strong Unit Propagation Backdoor Set (SUPBS)*.

In the modern SAT solvers, a role of formula $C[\beta/B]$ for specific $\beta \in \{0,1\}^{|B|}$, $B = \{z_1, \ldots, z_s\}$, is usually performed by a formula of the following kind:

$$z_1^{\beta_1} \wedge \ldots \wedge z_s^{\beta_s} \wedge C \tag{9}$$

If B is SUPBS, then applying to (9) UP as a polynomial sub-solver A^p we either derive the values of all variables in X or derive for some variable $x \in X$ its complementary pair: $(x, \neg x)$. The latter means that $C[\beta/B]$ is unsatisfiable.

It can be shown that in the case of BLP, for both SAT encodings with hard and soft clauses described above, the set of variables Y representing facilities can be used as SUPBS. We do not present a full proof of this fact here, as this is not the primary goal of the present paper. Note, however, that this fact follows from the possibility of encoding the constraints of the form (6) into a CNF H in such a way that any assignment of variables from Y either induces, via the UP rule, the values of all remaining variables in H, or leads to the derivation of a conflict. Thus, for a specific BLP instance $G_K = (I, J, E, K)$ we can define the following function of the form (8):

$$f_{G_K} : \{0,1\}^{|Y|} \rightarrow \mathbb{N} \tag{10}$$

The function (10) is defined as follows: for an arbitrary assignment $\alpha = (\alpha_1, \ldots, \alpha_n)$ of variables from $Y = \{y_1, \ldots, y_n\}$, $n = |J|$, the formula $H[\alpha/Y] = y_1^{\alpha_1} \wedge \ldots \wedge y_n^{\alpha_n} \wedge H$ is constructed, where H is the conjunction of hard clauses from (6); UP is then applied to $H[\alpha/Y]$. If as the result it derives a conflict, then the function (10) is considered undefined at point α. Otherwise, the value of f_{G_K} equals the number of clauses in (7) satisfied by the assignment $y_1 = \alpha_1, \ldots, y_n = \alpha_n$. To maximize (10), MVP is applied, in particular, the MVP-variant of the Hill Climbing strategy, which we refer as to Merging Variables Hill Climbing (MVHC).

The implemented variant of MVHC operates according to the following scheme: first it chooses a starting point $\alpha \in \{0,1\}^n$, generates a random merging mapping μ, and applies the Hill Climbing algorithm in D^μ, starting from the point $\gamma \in D^\mu$ which is the image of α under the mapping τ_μ^{-1} [27]. Then it searches for a local extremum of $F_{f,\mu}$ in D^{μ}. Once such a local extremum $\tilde{\gamma} \in D^{\mu}$ is found, a new merging mapping μ' is constructed, and HC starts from the point $\gamma' \in D^{\mu'}$ uniquely determined from $\tilde{\gamma}$. The domain sizes for variables in spaces of the form D^{μ} and the number of distinct random merging mappings are bounded by the input parameters of the algorithm. Specifically, in the experiments presented below, the configuration used was $r = \lfloor \frac{n}{4} \rfloor$, and the set of merged variables was formed according to the uniform scheme from [27]. Thus, the majority of the merged variables are formed from 4 Boolean variables; accordingly, such merged variables have domains of cardinality 16.

During the search, 250 random merging mappings are generated, after which the point achieving the best known value (BKV) of the objective function (10) is returned as the answer. The algorithm starts from a random Boolean vector α of Hamming weight l; thus, such a vector specifies an initial set of l open facilities.

A parallel version PMVHC was also implemented, based on the parallel neighborhood processing strategy in the space D^{μ} described in [27].

4 Experimental Results

In this section, we describe the settings we used to run the experiments and discuss their results.

4.1 Test Generation

To generate test instances we employed a simple model in which the facilities are represented as circles and clients as points on a plane. The points and circles are contained in a rectangular area of size $W \times H$. Each client is a point inside the rectangle. Each facility is a circle inside the rectangle. A client is served by a facility if it falls within the circle. A simple example is presented in Fig. 1.

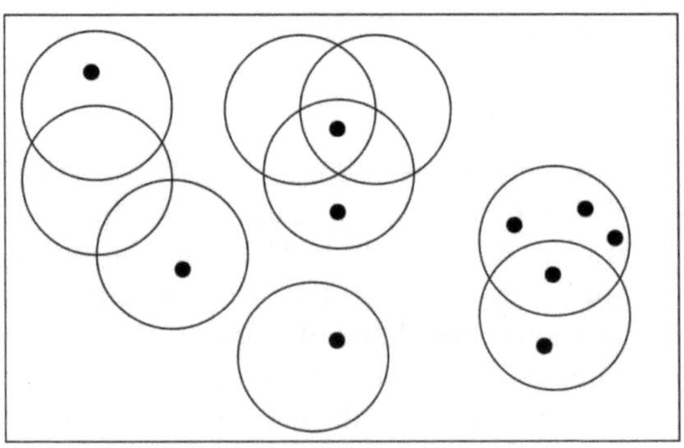

Fig. 1. An example of a BLP instance.

The input parameters of Algorithm 2 are: the number of clients $|I|$, the number of facilities $|J|$, the size of the area $W \times H$, and the bounds on the radii of the circles $r_1, r_2 : r_1 < r_2$. The output of the algorithm is a bipartite graph on the basis of which a specific BLP instance is formulated.

Algorithm 2: Generator

```
Input: W, H, r1, r2, |I|, |J|
Output: G – bipartite graph
X ~ U[0, W] — random number generator
Y ~ U[0, H] — random number generator
R ~ U[r1, r2] — random number generator
clients ← ∅
facilities ← ∅
for i ← 1 to |I| do
    clients.add({x = X(), y = Y()}) — random point
for i ← 1 to |J| do
    facilities.add({x = X(), y = Y(), r = R()}) — random circle
G ← (clients, facilities, ∅) — empty bipartite graph with clients in the left part
    and facilities in the right part
for c ∈ clients do
    for b ∈ facilities do
        if distance((b.x, b.y), (c.x, c.y)) < b.r then
            G.add_edge(c, b)
return G
```

4.2 Additional Technical Details

Section 2 described the general scheme of iteratively reducing the number of facilities required to satisfy $\geq K$ clients. Let us call any such number l *feasible*. It is easy to see, that if some l is feasible, then any value of the form $l + m$ for $m : 1 \leq m \leq |J| - l$ is also feasible. Conversely, if some l is not feasible, then no value less than l is feasible either. We refer to this property as *monotonicity in* l. This monotonicity property is precisely what enables the use of binary search to find the smallest feasible value of l. However, in this case only a complete algorithm can serve as the oracle, since monotonicity in l cannot be guaranteed for incomplete algorithms. Taking this into account, in the main part of the experiments we used a complete algorithm to find the exact smallest feasible l, and then compared the values produced by the approximate algorithms against it. The 0-1-ILP solver SCIP proved to be the most reliable complete algorithm, and the values of l found by it were used as the reference.

4.3 Experimental Evaluation

For NP-hard problems such as BLP, the runtime of exact algorithms can grow rapidly with input size. Therefore, for complete algorithms, we established the limit of 5 h on their runtime. Once the runtime exceeded this limit, the solving process was terminated. The MVP-based algorithms stopped their work after performing 250 merging mappings. This time in all considered cases turned out to be less than the time limit for complete algorithms. The greedy algorithm also shows relatively small runtime on all considered test instances. For incomplete algorithms, the point achieving the best known value (BKV) of the objective function was returned as the solution.

All runs were conducted on the ITMO university computing cluster nodes, equipped with Intel Xeon Gold 6248R CPU and 256 GB RAM. Each algorithm, except PMVHC, was run in single-threaded mode, while PMVHC utilized all available cores. It is worth noting that for the BLP instances from the family under consideration, the greedy algorithm often finds the optimal solution within less than 1 s. Therefore, in the majority of the experiments we focused on instances for which the greedy algorithm fails to find the optimal solution.

Instances of Small Dimension. The first group of benchmarks comprised instances of relatively small dimension. The goal was to use them to compare the effectiveness of the approaches based on the use of 0-1-ILP and MaxSAT solvers. The well-known RC2 solver [12], which has won at MaxSAT Evaluations on multiple occasions, was used as the complete MaxSAT solver. The MaxSAT solver Loandra [3], which in the general case yields an approximate solution to MaxSAT, was also applied to these and subsequent tests.

For all instances in this group, $|I| = 2000$ and $|J| = 20$. In Table 1, the notation $[r_1; r_2]$ indicates the range of radii of the circles generated by Algorithm 2. All algorithms produced exact solutions for all tests in this series. From the results presented in Table 1 it is clear that SCIP, MVHC, and Loandra spent very little time solving the instances in this series, whereas for RC2 they turned out to be extremely challenging.

Instances of Medium Dimension with Known Exact Solution. The second group consists of instances with $|J| = 40$ and $|I| = 100{,}000$. For this group, we also selected BLP instances for which the greedy algorithm fails to find the optimal solution. Since the costs of the found solutions start to differ here, the columns with results are split into two segments, one reporting the runtime, and the other showing the minimum number of facilities left open as reported by the corresponding algorithm (SCIP is guaranteed to find the exact solution). Dashes in the table indicate that the algorithm produced no answer within the allotted time (5 h). Since Loandra is an anytime solver, we include in the table the value of BKV it managed to find before terminating. Note that, despite being incomplete, MVHC converged to the global optimum on all considered test instances faster than the remaining solvers. Interestingly, using the greedy algorithm's solution as the starting point gives no advantage for the MVP strategy: the convergence time is approximately the same, yet for one of the tests a suboptimal solution was obtained (the corresponding value is highlighted in

Table 1. Results of experiments for BLP instances of small dimension ($|I| = 2000$, $|J| = 20$). All algorithms found the optimal solution; accordingly, only the running times are reported. Time is reported in seconds. Columns 'Fac.' show the number of active facilities in each solution. The MVHC algorithm was launched with 5 different random seeds and the corresponding running time interval provided for each problem.

	SCIP		MVHC		Loandra		RC2	
Problem	Sec.	Fac.	Sec.	Fac.	Sec.	Fac.	Sec.	Fac.
$R = [15; 20]$ $K = 1800$	0.22	9	[2; 3]	9	12	9	2049	9
$R = [10; 15]$ $K = 1600$	0.11	10	[3; 3]	10	12	10	7529	10
$R = [20; 20]$ $K = 1800$	0.27	8	[2; 2]	8	9	8	3033	8
$R = [20; 25]$ $K = 1600$	0.34	5	[2; 2]	5	12	5	3300	5
$R = [20; 25]$ $K = 1400$	0.32	4	[3; 3]	4	13	4	3739	4
$R = [20; 20]$ $K = 1600$	0.39	6	[2; 2]	6	10	6	3999	6
$R = [20; 20]$ $K = 1400$	0.38	5	[3; 4]	5	11	5	16262	5
$R = [15; 15]$ $K = 1600$	1.17	9	[3; 3]	9	12	9	17905	9
$R = [10; 15]$ $K = 1800$	0.07	12	[2; 3]	12	8	12	2725	12
$R = [25; 35]$ $K = 1800$	0.34	4	[3; 3]	4	10	4	273	4

bold). We also note the high speed at which the parallel variant PMVHC finds solutions (Table 2).

Instances of Large Dimension with Unknown Exact Solution. The last series comprised instances of relatively large dimension: with $|J| = 100$ and $|I| = 1{,}000{,}000$. Neither SCIP nor RC2 managed to find the exact solution for any of the benchmarks presented in Table 3 within 5 h. The Loandra solver also timed out on all instances. Thus, Table 3 presents a comparison of the solutions found by the Greedy algorithm (Algorithm 1) and the MVP-based algorithms. In all cases, the MVP-based algorithms found solutions that are no worse in cost than those found by the Greedy algorithm.

It is worth noting, that MVP-based algorithms have very good potential for parallelization and we are going to exploit this feature in future.

The MVHC algorithm was implemented in C++. The code is available in the GitHub repository at: https://github.com/maximkhabarov/covering-blp. The repository also contains scripts for running the SCIP, Loandra, and RC2 solvers on BLP instances. The instance generator is implemented in Python and is likewise available in the same repository.

5 Conclusion and Future Works

The main purpose of this paper is a computational study of a particular case of the Facility Location Problem known as the Branch Location Problem (BLP) [17]. In the paper, we compare, in the context of this problem, the

Table 2. Results of experiments for BLP instances of medium dimension ($|I| = 100,000$ and $|J| = 40$). The columns report the running time and the cost of the solutions found. The "-" symbol means that the algorithm was interrupted after exceeding the 5 hour time limit. The MVHC and PMVHC algorithms were launched with 5 different random seeds and the corresponding running time interval provided for each problem.

	SCIP		Loandra		MVHC		RC2		PMVHC	
Problem	Sec.	Fac.	Sec.	Fac.	Time, s.	Fac.	Sec.	Fac.	Sec.	Fac.
$R = [15;15]$ $K = 70000$	12203	10	-	20	[23; 26]	10	-	40	[4; 5]	10
$R = [20;20]$ $K = 50000$	3064	14	3146	14	[22; 28]	14	-	20	[4; 5]	14
$R = [20;30]$ $K = 90000$	1874	15	106	15	[25; 27]	15	-	20	[5; 5]	15
$R = [15;20]$ $K = 70000$	942	7	-	10	[20; 21]	7	-	20	[4; 4]	7
$R = [25;30]$ $K = 90000$	858	5	118	5	[25; 28]	5	-	10	[4; 5]	5
$R = [10;25]$ $K = 70000$	659	5	457	5	[21; 23]	5	-	20	[4; 6]	5
$R = [10;15]$ $K = 30000$	351	8	-	20	[22; 26]	8	-	40	[4; 5]	8
$R = [10;15]$ $K = 70000$	273	11	-	20	[19; 21]	11	-	40	[4; 4]	11
$R = [20;30]$ $K = 30000$	7313	13	869	13	[24; 27]	13	-	20	[5; 5]	13
$R = [5;10]$, $K = 30000$	409	9	52	9	[21; 23]	9	-	20	[4; 5]	9

effectiveness of both complete algorithms and algorithms that yield approximate solutions. The main result of the paper is what we consider a convincing demonstration of the effectiveness of a metaheuristic algorithm applied to BLP, in particular the one that can be viewed as a special case of the general methodology known as Variable Neighborhood Search (VNS) [22]. We mean the so-called Merging Variables Principle (MVP), proposed in [27]. We show that MVP-based strategies applied to BLP almost always yield an exact or near-optimal solution and exhibit short running times. These strategies demonstrate the applicability to BLP instances of significantly larger dimension compared to complete algorithms such as SCIP. It is important that MVP admits very natural parallelization procedures, and thus the construction of corresponding software implementations combining MVP with massive parallelism paradigms is of practical interest. We regard this direction as a promising area for future research.

Table 3. Results of experiments for BLP instances of large dimension ($|I| = 1,000,000$ and $|J| = 100$). The columns present the runtime and the cost of the found solutions. The MVHC and PMVHC algorithm were launched with 5 different random seeds and the corresponding running time interval provided for each problem.

	Greedy		MVHC		PMVHC	
Problem	Sec.	Fac.	Sec.	Fac.	Sec.	Fac.
$R = [10; 15]$ $K = 990000$	0.09	34	[179; 243]	29	[13; 15]	29
$R = [20; 25]$ $K = 990000$	0.2	14	[303; 327]	9	[15; 19]	9
$R = [5; 20]$ $K = 990000$	0.08	22	[317; 336]	18	[20; 22]	18
$R = [10; 15]$ $K = 950000$	0.12	26	[240; 268]	22	[18; 24]	22
$R = [10; 15]$ $K = 999000$	0.1	42	[204; 210]	38	[10; 17]	38
$R = [5; 20]$ $K = 999000$	0.08	27	[299; 308]	22	[17; 20]	22
$R = [10; 20]$ $K = 700000$	0.15	12	[203; 224]	9	[17; 18]	9
$R = [10; 20]$ $K = 999000$	0.38	26	[578; 590]	22	[35; 37]	22
$R = [5; 20]$ $K = 999000$	0.12	28	[650; 683]	22	[44; 46]	22
$R = [10; 10]$ $K = 500000$	0.15	17	[442; 467]	15	[33; 36]	15

In the nearest future, we plan to thoroughly extend the comparison of MVP-based algorithms with other widely used metaheuristics such as simulated annealing [15], evolutionary and genetic algorithms [19], etc. Another important direction for future research is to apply the method to solving real-world industrial BLP instances.

Acknowledgments. We express our deep gratitude to Viktor Kondratiev for his preliminary version of BLP instances generator used in our experiments. This research was financially supported by the Ministry of Education and Science of the Russian Federation (State Registration No 126021217232-3, code FWEW-2026-0008).

References

1. Achterberg, T.: SCIP: solving constraint integer programs. Math. Program. Comput. **1**(1), 1–41 (2009)
2. Ansótegui, C., Bonet, M.L., Levy, J.: Solving (weighted) partial MaxSAT through satisfiability testing. In: SAT. LNCS, vol. 5584, pp. 427–440. Springer (2009)
3. Berg, J., Demirovic, E., Stuckey, P.J.: Core-boosted linear search for incomplete MaxSAT. In: CPAIOR. LNCS, vol. 11494, pp. 39–56. Springer (2019)
4. Boros, E., Hammer, P.L.: Pseudo-Boolean optimization. Discret. Appl. Math. **123**(1–3), 155–225 (2002)
5. Burke, E.K.: Search Methodologies: Introductory Tutorials in Optimization and Decision Support Techniques, 2 edn. Springer (2014)
6. Chang, C.L., Lee, R.C.T.: Symbolic Logic and Mechanical Theorem Proving. Computer Science Classics. Academic Press (1973)

7. Chvatal, V.: A greedy heuristic for the set-covering problem. Math. Oper. Res. **4**(3), 233–235 (1979)
8. Eén, N., Sörensson, N.: Translating Pseudo-Boolean constraints into SAT. J. Satisf. Boolean Model. Comput. **2**(1–4), 1–26 (2006)
9. Feige, U.: A threshold of ln n for approximating set cover (preliminary version). In: STOC, pp. 314–318. ACM (1996)
10. Gurobi Optimization, LLC: Gurobi Optimizer Reference Manual (2023). https://www.gurobi.com
11. IBM: IBM ILOG CPLEX Optimizer (2023). https://www.ibm.com/products/ilog-cplex-optimization-studio/cplex-optimizer
12. Ignatiev, A., Morgado, A., Marques-Silva, J.: RC2: an efficient MaxSAT solver. J. Satisf. Boolean Model. Comput. **11**(1), 53–64 (2019)
13. Karp, R.M.: Reducibility among combinatorial problems. In: Complexity of Computer Computations, pp. 85–103. Plenum Press (1972)
14. Karpinski, M., Zelikovsky, A.: Approximating dense cases of covering problems. Electron. Colloquium Comput. Complex. **TR97-004** (1997)
15. Kirkpatrick, S., Gelatt, C.D., Vecchi, M.P.: Optimization by simmulated annealing. Science **220**(4598), 671–680 (1983)
16. Kroening, D.: Software verification. In: Handbook of Satisfiability, FAIA, vol. 336, pp. 791–818. IOS Press (2021). https://doi.org/10.3233/FAIA201004
17. Laporte, G., Nickel, S., da Gama, F.S.: Introduction to location science. In: Location Science, pp. 1–18. Springer (2015)
18. Li, C.M., Manyà, F.: MaxSAT, hard and soft constraints. In: Handbook of Satisfiability, FAIA, vol. 336, pp. 903–927. IOS Press (2021)
19. Luke, S.: Essentials of Metaheuristics. Lulu, 2nd edn. (2013)
20. Lund, C., Yannakakis, M.: On the hardness of approximating minimization problems. J. ACM **41**(5), 960–981 (1994)
21. Marques-Silva, J., Lynce, I., Malik, S.: Conflict-driven clause learning SAT solvers. In: Handbook of Satisfiability, FAIA, vol. 336, pp. 133–182. IOS Press (2021)
22. Mladenovic, N., Hansen, P.: Variable neighborhood search. Comput. Oper. Res. **24**(11), 1097–1100 (1997)
23. de Moura, L., Bjørner, N.: Z3: an efficient SMT solver. In: TACAS. LNCS, vol. 4963, pp. 337–340. Springer (2008)
24. Otpuschennikov, I.V., Semenov, A.A.: Using merging variables-based local search to solve special variants of MaxSAT problem. In: MOTOR. CCIS **1275**, 363–378. Springer (2020)
25. Papadimitriou, C.H., Steiglitz, K.: Combinatorial Optimization: Algorithms and Complexity. Prentice-Hall (1982)
26. Russell, S.J., Norvig, P.: Artificial Intelligence: A Modern Approach, 3 edn. Pearson Education (2010)
27. Semenov, A.: Merging variables: one technique of search in pseudo-Boolean optimization. In: Mathematical Optimization Theory and Operations Research, CCIS, vol. 1090, pp. 86–102. Springer (2019)
28. Slavik, P.: A tight analysis of the greedy algorithm for set cover. J. Algorithms **25**(2), 237–254 (1997). https://doi.org/10.1006/jagm.1997.0887
29. Stanley, R.P.: Enumerative Combinatorics, Cambridge Studies in Advanced Mathematics, vol. 1, 2 edn. Cambridge University Press (2011)
30. Tseitin, G.S.: On the complexity of derivation in propositional calculus. In: Studies in constructive mathematics and mathematical logic. part II, Seminars in mathematics, vol. 8, pp. 115–125. V. A. Steklov Mathematical Institute, Leningrad (1970)

31. Wegener, I.: The Complexity of Boolean Functions. Wiley-Teubner (1987)
32. Williams, R., Gomes, C.P., Selman, B.: Backdoors to Typical Case Complexity. In: IJCAI, pp. 1173–1178. Morgan Kaufmann (2003)
33. Zaikin, O., Ignatiev, A., Marques-Silva, J.: Branch location problems with maximum satisfiability. In: ECAI 2020. FAIA, vol. 325, pp. 379–386. IOS Press (2020)

A Spanning Graphs Method for Assessing Priorities Based on a Pairwise Comparison Matrix

Irik Z. Mukhametzyanov(✉)

Institute of Socio-Economic Research–subdivision of the Ufa Federal Research Centre of the Russian Academy of Sciences, Ufa, Russia
izmukhametzyanov@gmail.com

Abstract. Pairwise comparisons are used to evaluate the importance of objects (alternatives) in optimization problems with discrete variables: the number of objects and their attributes is finite. For a multi-criteria decision making problem, a new approach to object prioritization based on processing a pairwise comparison matrix (PCM) is proposed—the Mean of Spanning Graphs (MSG) method. The weights obtained by MSG are comparable to estimates obtained by minimizing the approximation error of the PC matrix. The MSG approach takes into account all possible connections and relationships in the PCM, increasing the reliability of priority assessments. Unlike other methods, MSG allows for the processing of incomplete pairwise comparison matrices. In MSG, PC inconsistency is estimated on the basis of the standard deviation in the distribution of weights obtained across the entire set of graphs. Using MSG, a procedure is proposed for identifying and sorting inconsistent judgments in the PCM by the degree of their deviation from the agreed-upon judgment. This allows for consistent re-evaluation of inconsistent judgments and bringing the PCM to an acceptable level of consistency.

Keywords: MCDM · pairwise comparisons · weights estimation · spanning graph · consistency

1 Introduction

Pairwise comparisons (PCs) are among the most common tools for multi-criteria decision making [1–3]. Pairwise comparisons are used to evaluate the importance of alternatives (objects, criteria) in problems with discrete decision variables and a limited number of alternatives and attributes. In addition to the Analytic Hierarchy Process (AHP), pairwise comparisons are an integral part of methods to identify components of the cause-and-effect chain of a complex system DEcision MAking Trial and Evaluation Laboratory (DEMATEL) [4], criteria significance assessment methods such as the Best-Worst Method (BWM) [5], in the Step-wise Weight Assessment Ratio Analysis (SWARA) [6], and other methods.

A. Kazakov et al. (Eds.): MOTOR 2026, LNCS 16687, pp. 98–115, 2026.
https://doi.org/10.1007/978-3-032-31390-4_7

The decision maker first performs linguistic pairwise comparisons, then derives numerical pairwise comparisons by selecting a specific numerical scale for their quantitative evaluation, and finally derives a priority vector from the numerical pairwise comparisons. At each stage, various approaches and methods exist with varying degrees of validity, which determines the design of the model.

Let n objects $A_1, \ldots, A_n$ be defined, and each object be assigned a positive integer w_j—the weight (significance) of object A_j $(j = 1, \ldots, n)$. We form a matrix of relative weights:

$$A = (a_{ij}), \quad [n \times n], \quad a_{ij} = w_i/w_j, \tag{1}$$

which is called the Pairwise Comparison Matrix (PCM). In general, between n objects, $n(n\text{-}1)/2$ different independent pairwise comparisons can be defined, forming a PCM.

To assess the significance (weights) of n objects, it is sufficient to have a set of $(n - 1)$ PC and use the normalization condition $\sum_j w_j = 1$ [7]. Therefore, different sets of the PC matrix produce different object priorities. However, not all sets are suitable for determining weights. This study demonstrates the structure of such sets, termed by T. Saaty as the spanning graph of the PCM [3], and presents an algorithm for identifying them for subsequent weight calculation. The number of spanning graphs is greater than the number of independent elements in the PCM. This means that the PCM actually contains more information than just its independent elements.

This paper proposes a new prioritization method based on processing the entire set of spanning graphs—the Mean of Spanning Graphs (MSG) method. The basic idea of the method is to generate all available spanning graphs of the PCM. Next, determine priorities for each spanning graph, and then determine the integral weights of the objects. This approach allows for the use of all the information contained in the PCM. The method can handle incomplete matrices. Based on this, we explain how well each well-known prioritization method utilizes this information. For example, why does the Best-Worst method yield comparable results to AHP.

In MSG, PCM inconsistency is estimated on the basis of the standard deviation in the distribution of weights obtained across the entire set of graphs. Using MSG, a procedure is proposed for identifying and sorting inconsistent judgments in the PCM. This allows for consistent re-evaluation of inconsistent judgments and bringing the PCM to an acceptable level of consistency.

The paper is structured as follows. The first section formulates the problem of developing a prioritization method that utilizes all the information contained in the PCM. The second section presents the general framework for constructing pairwise comparisons. The third section presents a spanning graph prioritization method based on PCM. The correctness of the approach used and some obvious advantages are demonstrated. Section four describes a methodology for identifying inconsistent judgments in a PCM and provides a step-by-step algorithm to improve consistency. Section five presents a numerical example in which, based on the described methodology, the consistency of a PCM was improved to an acceptable level through targeted step-by-step adjustments to 6 of the 15 pairwise comparison elements.

2 Pairwise Comparisons: A General Framework for Prioritization

2.1 Pairwise Comparison Matrix

According to the construction method, all elements of matrix A are positive, all elements on the main diagonal are equal to 1, the matrix is inversely symmetric, and the number of independent elements in the PCM is $n(n\text{-}1)/2$:

$$a_{ij} > 0, a_{ii} = 1, a_{ji} = 1/a_{ij}. \tag{2}$$

The consistency of the expert's judgments reflected in the PCM is the primary criterion for the adequacy of the results. The PCM is consistent if and only if:

$$a_{ij} \cdot a_{jk} = a_{ik}, \forall i, j, k. \tag{3}$$

However, cardinal consistency will not be observed across the entire matrix due to the following objective and subjective factors [8]:
(i) the numerical scale of paired comparisons is discrete and limited,
(ii) transitivity of judgments according to Eq. (3) is not always observed,
(iii) people's feelings and preferences remain contradictory.

2.2 Estimating the Weight of Objects Based on Pairwise Comparisons

Formally, prioritization methods based on processing pairwise comparison matrices can be divided into two groups. The first group uses elements of the PCM in the form of equations or formulas. These include methods such as the Additive Normalization Method (ANM), the Eigenvector Method (EM) [1,8], the Geometric Mean Method (GMM) [9], and others.

The other group of methods is based on the definition priorities w_j that approximate elements of the PCM: $a_{ij} \approx w_i/w_j$. The priority vector w is the solution to the problem of finding the minimum of the deviation function defined in the chosen metric:

$$w_j : \quad \|A - A^*\| \rightarrow min \tag{4}$$

$$s.t. \quad \sum w_j = 1. \tag{5}$$

The choice of distance metric in (4) is not formalized and introduces another degree of freedom. These methods include the Least Squares Method (LSM) [10], Logarithmic Least Squares Method (LLSM) [11], Minimax method [12], Absolute Deviation Method [12], Cosine Maximization Method (CMM) [13], etc.

2.3 Consistency Indices of PCM

To assess the consistency of a pairwise comparison matrix, numerical consistency measures in the form of consistency indices are used. In this study, we define a statistical consistency index based on spanning graphs and use the consistency index defined by Saaty [1,8] for comparison and analysis:

$$CI = (\lambda_{max} - n)/(n - 1). \tag{6}$$

When forming the acceptability criterion for the consistency of a PCM, the Consistency Ratio is used:

$$CR = CI/RI. \tag{7}$$

Where, $RI(n)$ (random index) is the consistency index that would be obtained if the assigned judgment values were totally random. Specifically, Saaty defined the acceptability threshold so that the CI is no more than 10 % of the RI,

If the consistency of the pairwise comparison matrix is low, judgments need to be revised. Typically, only one pair of elements is revised during the consistency improvement process, which aims to preserve as much of the original information in the pairwise comparison matrix as possible, as determined by the decision maker.

3 Spanning Graph Method for Assessing Object Priorities

3.1 The Spanning Graph of PCM

The dimension of the minimum set of elements of the PCM that can be used to calculate the weights of all objects is $(n - 1)$. Such a set must include pairwise comparisons between all objects, i.e., the set's indices include all numbers from 1 to n. Additionally, the graph generated by such a set must be connected. Following T. Saaty [1], we call such a set a Spanning Graph (SG).

Definition: A spanning graph of PCM is any unordered set of $(n - 1)$ elements located above the main diagonal, representing at least once each of the n compared objects and not containing disconnected fragments in the indices.

Remark 1: Elements below the main diagonal, due to the property of inverse symmetry of the elements of the PCM, do not introduce new information and are therefore not included in the consideration. For example, any row of the PCM, without a diagonal element, actually represents a spanning graph. But due to inverse symmetry, this spanning graph is uniquely determined by the elements of the graph built only on the upper diagonal elements.

Remark 2: A spanning graph is undirected graph by definition.

Remark 3: To simplify notation, we will identify a spanning graph and the corresponding graph formed by the numbers of rows and columns of elements of this graph.

For example, the graph $(a_{12}\ \ a_{13}\ \ a_{24}\ \ a_{25}\ \ a_{16}) \equiv (12\ \ 13\ \ 24\ \ 25\ \ 16)$.
Figure 1 shows an illustration of 3 spanning graphs for the dimension n=6. Only

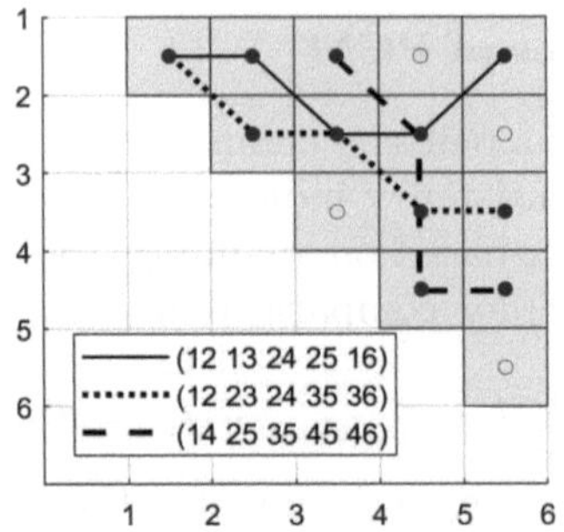

Fig. 1. Illustration of 3 spanning graphs of the PCM (n=6).

three different spanning graphs out of 1282 satisfying the definition are highlighted in this figure. All graphs are represented by their indices (rows and columns), which contain all numbers from 1 to 6 and form one connected graph.

For example, the graph with indices (12 13 14 24 25) is not a spanning graph, since it does not contain a relation for the 6th object. The graph (12 13 23 45 46) is also not a spanning graph by definition, since it is not connected and contains dependent elements $a_{12} \cdot a_{23} = a_{13}$. This graph contains two connected cycles in its structure (12 13 23) and (45 46), connecting the graph vertices.

The initial set for forming the spanning graph represents all upper-diagonal elements of the PCM. From this set of volume $n(n-1)/2$, we select without replacement all combinations consisting of $(n-1)$ elements. From this set of graphs, it is necessary to remove all graphs containing disconnected cycles, which are identified using the single-connectivity algorithm. Thus, the number of spanning graphs is determined using the combinatorial formula:

$$N_{SG} = C_{n(n-1)/2}^{n-1} - r, \tag{8}$$

Where, r is the number of graphs containing two or more disconnected cycles in their structure.

Table 1. Number of spanning graphs NSG for different dimension n of the PCM.

n	4	5	6	7	8	9
$n(n-1)/2$	6	10	15	21	28	36
$C_{n(n-1)/2}^{n-1}$	20	210	303	54264	1'184040	30'260340
r	4	85	1721	3803	940167	24'836241
N_{SG}	16	125	1282	16234	243873	5'424099

The results of calculating the number SG for different dimensions of the matrix of pairwise comparisons are shown in Table 1 below. If the expert focuses on only $n(n-1)/2$ pairwise comparisons, then the entire set of spanning graphs reflects the entire N_{SG} set of hidden relationships. If for n=5 it is necessary to perform 10 pairwise comparisons, then the number of graphs that reflect the hidden relationships of these 10 coded comparisons is 125, i.e. an order of magnitude greater. For n=8, the number of graphs already exceeds the number of comparisons by 5 orders of magnitude. The completeness of the information contained in the graphs, as a consequence, should provide an additional opportunity for consistency analysis and reliability of the results.

3.2 Calculating Priorities Using a Spanning Graph

Considering that $a_{ij} = w_i/w_j$ and the fact that the spanning graph contains the entire list of objects (1, ..., n), we have (n-1) independent equations, which are closed by the normalization condition $\sum w_i = 1$. From this it follows that each $SG^{(k)}$ allows us to calculate the priorities of the objects under study within the framework of the preferences reflected in the elements of this kth spanning graph: $w^{(k)} = (w_1, w_2, \ldots, w_n)$.

3.3 Weights of the Inconsistent PCM. Example

For a fully consistent PCM, all spanning graphs produce identical weight. If the PCM contains inconsistent elements, some spanning graphs produce different object priorities. Let consider an example of a $[4 \times 4]$ PCM presented in [7]:

$$A^* = \begin{pmatrix} 1 & 3 & 1/2 & 2 \\ 1/3 & 1 & 1/6 & 2/3 \\ 2 & 6 & 1 & 4 \\ 1/2 & 3/2 & 1/4 & 1 \end{pmatrix}$$

This matrix is fully consistent, $\lambda_{max} = 4$, and the priority vector (not normalized) is w=(1 1/3 2 1/2). The normalized vector is w=(0.261 0.087 0.522 0.130). For n=4, the number of spanning graphs is 16, which can be easily counted. These are samples of 3 elements from matrix A^*, with the following indices: (12 13 14), (12 13 24), (12 13 34), (12 14 23), (12 14 34), (12 23 24), (12 23 34), (12 24 34), (13 14 23), (13 14 24), (13 23 24), (13 23 34), (13 24 34), (14 23 24), (14 23 34), (14 24 34). It is easy to verify, that all of the listed graphs for a fully consistent PCM produce the same priority vector.

The element $a_{42} = 3/2$ of the matrix A^* does not belong to the integer scale. In fact, there is no prohibition on choosing an arbitrary numerical scale (in this case, it could be the scale: 1 1.5 2 3 4 5 6), but this entails voluntarism and uncertainty in solving problems. If we use an integer numerical scale, for example, the Saaty scale, there are two possible options for adjusting to the nearest integer: 1) $a_{24} = 1/2, a_{42} = 2$, or 2) $a_{24} = 1, a_{42} = 1$. In both cases, the matrix becomes partially inconsistent and equality (3) is violated For the first option of adjusting the elements of the PCM $a_{24} = 1/2, a_{42} = 2$, the following results were obtained: $\lambda_{max} = 4.01$, CI=0.0034. w=(0.260 0.081 0.520 0.140)—EV-method. Of the 16

Table 2. Spanning graphs and corresponding weight coefficients (1st option).

	Spanning graph (base set)	Weight coefficients
1	(12 13 14), (12 13 34), (12 14 23), (12 14 34), (12 23 34), (13 14 23), (13 23 34), (14 23 34)	$w^{(2)}$= (0.261 0.087 0.522 0.130)
2	(13 14 24), (13 24 34), (14 24 34)	$w^{(2)}$= (0.267 0.067 0.533 0.133)
3	(12 13 24), (12 23 24), (13 23 24)	$w^{(3)}$= (0.250 0.083 0.500 0.167)
4	(12 24 34)	$w^{(4)}$= (0.214 0.071 0.571 0.143)
5	(14 23 24)	$w^{(5)}$= (0.308 0.077 0.462 0.154)
	Arithmetic mean	$w^{(m)}$= (0.243 0.073 0.565 0.119)
	Standard deviation	(0.036 0.016 0.056 0.029)

basis sets for the partially consistent PCM, only 5 groups produce different priority vectors (Table 2). The arithmetic mean roughly coincides with the weights determined using the EV-method. It is obvious that the frequency determines the closeness to the mean, as we can see in Table 2 for the first group of spanning graphs. It is also noteworthy that the first group of spanning graphs does not contain the element a_{24}, which is an inconsistent judgment. Only 8 basis sets of the first group contain completely consistent judgments. This corresponds to the property that spanning graphs containing only consistent elements produce the same weights. The remaining spanning graphs contain the element a_{24}, which determines the deviation of the result from the mean. Inconsistent judgments in the spanning graphs are highlighted in yellow. Low frequency for basis sets 4 and 5 corresponds to the greatest deviation from the mean. For the second variant of adjustment $a_{24} = 1, a_{42} = 1$ the following results were obtained: $\lambda_{max} = 4.02$, CI=0.0068, w=(0.261 0.097 0.523 0.119)—EV-method. And in this case, 16 basis sets for a partially matched PCM produce only 5 groups of different vectors of weight coefficients. The difference from the first variant is only in the weights obtained using spanning graphs containing the inconsistent element a_{24}. The average values of the vector of weight coefficients change (0.26 0.10 0.52 0.11). Since the second adjustment option is worse in terms of the consistency of the A^* matrix according to the CI consistency index, this is reflected in the standard deviation for all components of the weighting coefficient vector. In the second option, the standard deviation is (0.026 0.015 0.032 0.016) and is higher than in the first option (Table 2).

Let's perform two adjustments simultaneously. Together with the first, $a_{24} = 1/2, a_{42} = 2$, we change element 34: $a_{34} = 6, a_{43} = 1/6$ and worsen the consistency of the A^* matrix. In this case, the following results were obtained: $\lambda_{max} = 4.046$, CI=0.0153, w=(0.247 0.077 0.554 0.122)—EV-method. Now 10 basis sets out of 16 for a partially consistent PCM produce different priority vectors (Table 3).

The first group of spanning graphs (only 3) does not contain elements a_{24} and a_{34}, which are inconsistent judgments. The remaining spanning graphs contain

Table 3. Spanning graphs and their corresponding weight coefficients (3th option)

	Spanning graph (base set)	Weight coefficients
1	(12 13 14), (12 14 23) (13 14 23)	$w^{(1)}$= (0.261 0.087 0.522 0.130)
2	(13 14 24)	$w^{(2)}$= (0.267 0.067 0.533 0.133)
3	(12 13 24), (12 23 24), (13 23 24)	$w^{(3)}$= (0.250 0.083 0.500 0.167)
4	(12 13 34), (12 23 34), (13 23 34)	$w^{(4)}$= (0.273 0.091 0.545 0.091)
5	(12 24 34)	$w^{(5)}$= (0.167 0.056 0.667 0.111)
6	(12 14 34)	$w^{(6)}$= (0.207 0.069 0.621 0.103)
7	(13 24 34)	$w^{(7)}$= (0.286 0.048 0.571 0.095)
8	(14 23 24)	$w^{(8)}$= (0.308 0.077 0.462 0.154)
9	(14 23 34)	$w^{(9)}$= (0.200 0.100 0.600 0.100)
10	(14 24 34)	$w^{(10)}$= (0.211 0.053 0.632 0.105)
	Arithmetic mean	$w^{(m)}$= (0.243 0.073 0.565 0.119)
	Standard deviation	(0.036 0.016 0.056 0.029)

inconsistent judgments, which determines the deviation of the result from the mean. Inconsistent judgments are highlighted in yellow. The standard deviation in this variant has increased twofold compared to the first variant of the partially inconsistent matrix.

3.4 Calculation of Criteria Weights for All SGs. MSG Method

Each spanning graph reflects one of the possible interactions of a set of n objects, and each spanning graph allows calculating priorities:

$$w^{(k)} = f(SG^{(k)}), \ \forall\, i, j \ \ (j > i), \ \ k = 1, ..., N_{SG}. \tag{9}$$

As a result, we have a spectrum of weight values N_{SG} for each object:

$$w^{(k)} = (w_1^{(k)}, w_2^{(k)}, \ldots, w_n^{(k)}), \quad k = 1, \ldots, N_{SG}.$$

The degree of deviation of the results from the mean characterizes the degree of inconsistency of judgments in the matrix of paired comparisons, and the mean itself determines the optimal value according to the criterion of minimizing deviations under the condition of weak asymmetry of the data, for example, in the case of a normal distribution of data.

Hypothesis: The distribution of the weight of each criterion in the SG set is close to the normal distribution law.

This hypothesis is based on numerical experiments for various PCs described in the literature. The normality hypothesis (Kolmogorov-Smirnov, Jarke-Bera, and Lilliefors criteria) is accepted in approximately 90% of cases at a significance level >0.9. Deviations from normality manifest themselves in the presence of "fat tails" in the distribution. When there is a deviation from normality in the data distribution, a weak asymmetry is a sufficient condition to minimize the deviations, and in this case it is better to use the median value of the data as the optimal solution. The distribution of the weight of the alternatives for the case n=6 is shown in Fig. 2. The corresponding PCM is taken from the Saaty study [1] and reflects the comparison of characteristics with respect to overall satisfaction with school:

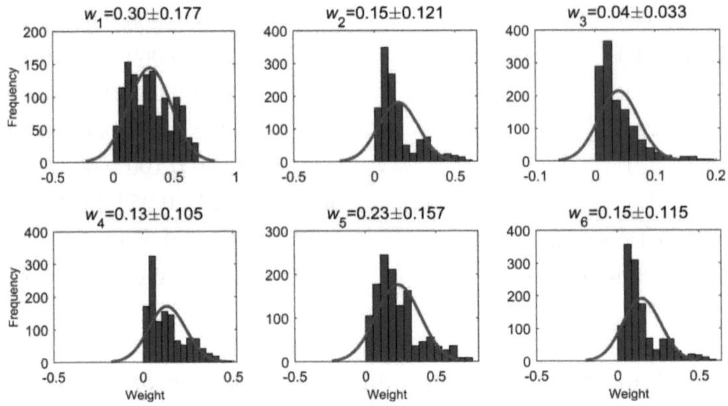

Fig. 2. Distribution of weight coefficients of objects obtained on a set of spanning graphs (n=6, $N_{SG} = 1282$).

$$A = \begin{pmatrix} 1 & 4 & 3 & 1 & 3 & 4 \\ 1/4 & 1 & 7 & 3 & 1/5 & 1 \\ 1/3 & 1/7 & 1 & 1/5 & 1/5 & 1/6 \\ 1 & 1/3 & 5 & 1 & 1 & 1/3 \\ 1/3 & 5 & 5 & 1 & 1 & 3 \\ 1/4 & 1 & 6 & 3 & 1/3 & 1 \end{pmatrix}.$$

The number of the spanning graphs for n=6 is 1282 (Table 1). Below are (for illustration) some spanning graphs and their corresponding various object weights:

...

$SG_{12} = (16\ 26\ 34\ 36\ 56) \rightarrow w^{(12)} = (0.400\ 0.100\ 0.017\ 0.083\ 0.30)$
$SG_{32} = (16\ 25\ 34\ 35\ 36) \rightarrow w^{(32)} = (0.571\ 0.024\ 0.024\ 0.119\ 0.119)$
$SG_{72} = (16\ 24\ 25\ 34\ 36) \rightarrow w^{(72)} = (0.190\ 0.119\ 0.008\ 0.040\ 0.595)$

...

$SG_{367} = (14\ 24\ 25\ 35\ 36) \rightarrow w^{(367)} = (0.024\ 0.073\ 0.073\ 0.024\ 0.366)$
$SG_{407} = (14\ 16\ 26\ 35\ 45) \rightarrow w^{(407)} = (0.270\ 0.068\ 0.054\ 0.270\ 0.270)$
$SG_{427} = (14\ 16\ 23\ 35\ 36) \rightarrow w^{(427)} = (0.358\ 0.104\ 0.015\ 0.358\ 0.075)$

...

The spread of weight values by components (objects) is significant. But these are not object weights, since the PCM contains not the data on priorities, but their encodings. It is better to interpret these values as the weights of the information matrix, reflecting all possible relationships of the preference matrix. In such situations, averaging of results is used to identify the "true" priorities. Average values represent the optimal values of object weights in the case of a normal distribution and are comparable with estimates based on minimizing the approximation error of the PCM in the selected metric:

$$q_j = \frac{1}{N}\sum_{k=1}^{N} w_j^{(k)} = w_j^*, \quad where\ w_j^* :\ \|w_j^* - w_j^{(k)}\| \to min\,. \tag{10}$$

After finding the average values for each component, normalization is required:

$$w_j = q_j / \sum q_j\,. \tag{11}$$

Let us denote the proposed approach as MSG (Mean of Spanning Graphs). The MSG method processes all available information of the PCM, reflecting all possible relationships of preferences, including those hidden to the expert (!)

3.5 Comparison of the MSG Method and Other Prioritization Methods

In this section, we will compare the MSG method and other well-known prioritization methods (see Sect. 2.4, as well as reviews by Gao et al. [12] and Zhang et al. [15] and using numerical analysis and a specific example of a PCM according to expression (10). In accordance with Sect. 2.2, prioritization methods can be represented as two groups of methods:

1) methods that directly use elements of the PCM: EigenVector method (EVM), Geometric Mean Method (GMM), Sum Method (Sum),

2) optimization methods: Least Squares Method (LSM), Minimax method (Minimax), Absolute Deviation Method (ADM), Cosine maximization method (CMM), the Logarithmic Least Squares Method (LLSM).

For a partially consistent PCM, the results differ because various known prioritization methods based on processing a PCM use different parts of the information contained in the matrix. The results of a numerical example (Fig. 3, Table 4) are presented below, showing good consistency between prioritization using the MSG method and the group of methods identified above. The consistency is based on two criteria:

$$w_j^{MSG} \in [w_j^{min}, w_j^{max}], \;\; j = 1, \ldots, n$$

$$\Delta^{MSG} \in [\Delta_{min}, \Delta_{max}],$$

Where, Δ is the approximation quality, expressing the degree of deviation between the PCMs $A = (a_{ij})$ and $A^* = (w_i/w_j)$.

The weights calculated based on the MSG method and the quality of approximation are in good agreement with similar results obtained using known analogues. This confirms the correctness of the proposed method.

In this study, the degree of consistency Δ of deviations $d_{ij} = a_{ij} - w_i/w_j$ are calculated for using 4 different matrix norms: Δ_1 is the "maximum" of the sum by columns, Δ_F is the Frobenius norm, Δ_∞ is the "maximum" of the sum by rows, Δ_{GCI} is the norm for logarithmic values.

The weights, calculated using the MSG method, are in high agreement with the arithmetic mean of the weight (A.mean in Fig. 4), calculated using a set of 7 prioritization methods based on processing the matrix of paired comparisons.

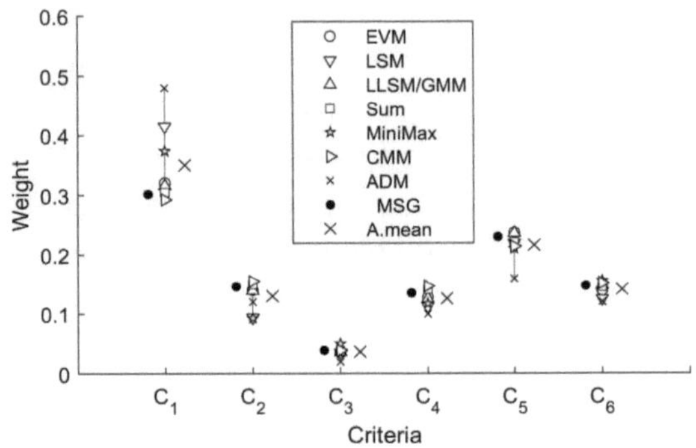

Fig. 3. Weight coefficients of objects obtained on the basis of processing the matrix of pairwise comparisons using various methods.

This is explained by the fact that the MSG method uses all the information contained in the PCM, while other methods use only partial information. For example, the trivial Sum method uses information only on the rows of the PCM.

Table 4. Weight coefficients of objects calculated by different methods based on the PC matrix by Eq. (10), and the approximation quality Δ.

Method	w_1	w_2	w_3	w_4	w_5	w_6	Δ_1	Δ_F	Δ_∞	Δ_{GCI}
EVM (CR=0.23)	0.32	0.14	0.03	0.13	0.24	0.14	2.87	0.32	2.55	1.59
LSM	0.42	0.09	0.03	0.11	0.22	0.13	3.74	0.39	2.77	1.86
Mimimax	0.37	0.09	0.05	0.12	0.21	0.16	3.19	0.32	1.90	1.96
ADM	0.48	0.12	0.02	0.10	0.16	0.12	5.00	0.74	4.96	2.27
CMM	0.29	0.16	0.04	0.15	0.21	0.15	2.27	0.29	2.25	1.65
LLSM/GMM	0.32	0.14	0.04	0.13	0.24	0.15	2.78	0.31	2.51	1.58
Sum	0.30	0.15	0.04	0.14	0.22	0.15	2.47	0.29	2.33	1.61
MSG ($N_{SG} = 1282$)	0.30	0.15	0.04	0.13	0.23	0.15	2.53	0.29	2.33	1.60

In total, the author performed verification calculations for more than 20 pairwise comparison matrices of different dimensions using the data available in numerous publications.

Note: As n increases by one, the number of spanning graphs increases by approximately one order of magnitude. This may create a problem with the calculation time. Note that for $n = 7$, 8, and 9, the calculation time in the MatLab system for the Intel CORE i5 processor is approximately 3 sec, 6 min, and 5 h, respectively. The author believes that it is possible to speed up the calculation by improving the algorithms for processing a large number of spanning graphs through parallelization, since all objects are identical in structure and processing methods. Note also that the entire set of spanning graphs does not change for each fixed n. Therefore, all $SG^{(k)}$ are calculated only once and saved in the file. And this leads to a sharp reduction in calculation time by 1-2 orders of magnitude.

3.6 Consistency of Pairwise Comparison Matrix. Spanning Graph Approach

The MSG method determines the optimal solution as the arithmetic mean of the priorities defined for each individual spanning graph. For a partially consistent matrix, part of the SG contains only consistent elements. This set produces the same vector of object priorities. The remaining spanning graphs contain inconsistent elements, which determines the deviation of the result from the mean for each component of the priority vector. The higher the degree of inconsistency in the pairwise comparison matrix, the greater the number of SGs containing inconsistent elements and the higher the standard deviation of the result from the mean. Thus, the consistency of the pairwise comparison matrix can be determined by the standard deviation from the mean over the set of spanning graphs, which corresponds to the statistical approach. The degree of inconsistency of judgments has a natural interpretation as the magnitude of dispersion in the distribution of weight coefficients—the standard deviation $std(w_j)$. To justify the proposed approach, we demonstrate the correlation between the known Saaty consistency index CR (formula (6)) and $std(w_j)$ in a numerical experiment. The numerical experiment setup is as follows:

Step 1: For the original PCM $A^{(0)}$, define CR (EV-method) and $std(w_j)$ (MSG-method).

Step 2: Randomly select a number from the pairwise comparison scale and replace the randomly selected upper-diagonal element of the PCM $A^{(0)}$ with a new value. The consistency of the new matrix $A^{(i)}$ will change.

Step 3: For the new PCM A_i, define CR_i and $std_i(w_j)$. and form an array of points (CR_i, std_i).

Step 4: Continue the procedure from step 2 a specified number of times.

Figure 4 shows a dot plot and a trend line array of dots (CR_i, std_i) for the well-known example of comparing beverage consumption from the study Saaty et al. [16]. The obtained data show a linear relationship between these indicators. The correlation coefficient is 0.85. Some fluctuations and jumps in the data are due to randomness in the generation of the matrix of paired comparisons. Thus, the consistency of judgments can be determined through the distribution of

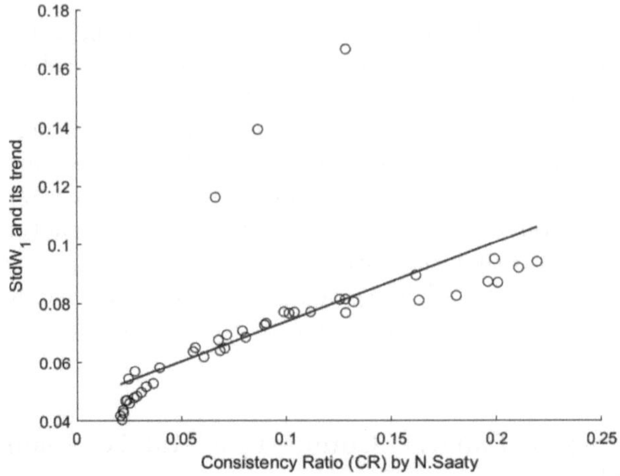

Fig. 4. Scatter plot of points (CR_i, std_i) and trend line.

weight coefficients calculated on a set of spanning graphs, as the value of the standard deviation.

3.7 Identification of Inconsistencies Elements in PCM

Let us show how to identify all judgments in the matrix of pairwise comparisons that differ significantly from the general trend. In Sect. 3.2 above, the rationale for the normal weight distribution for each criterion calculated on a set of spanning graphs was given. The hypothesis is not rejected in approximately 90% of cases at a significance level >0.9. Deviations from normality are manifested as the presence of "fat tails" in the distribution or a slight asymmetry.

We will associate the mean value of the distribution as a trend. For a normal distribution, the number of values differing from the mean by three standard deviations is 99.7%. It should be assumed that all weight values outside the 3σ interval were obtained using inconsistent SGs. A simple technique follows:

1) isolate all graphs on the SG_k set for which the weight of at least one alternative is outside the 3σ interval (or p-level). Let us denote this set as LowSG,

2) let us define the distribution (frequency) of the elements of the PCM a_{ij} on this set,

3) provided that the PCM is not consistent, the elements of the PCM with high frequency must be re-evaluated sequentially, starting with the element with the highest frequency.

Figure 5 shows an example of the distribution of a_{ij} on the LowSG set. The PCM is borrowed from [16]. The distribution of the weight coefficients of various alternatives obtained using spanning graphs for this PCM is shown above in Fig. 2. About 7% of SGs (91 out of 1282) are out of the general trend. The "dark zone" in Fig. 5 and the corresponding value show how many times the comparison a_{ij} is included in the LowSG set. For this example, the most frequent ones in the

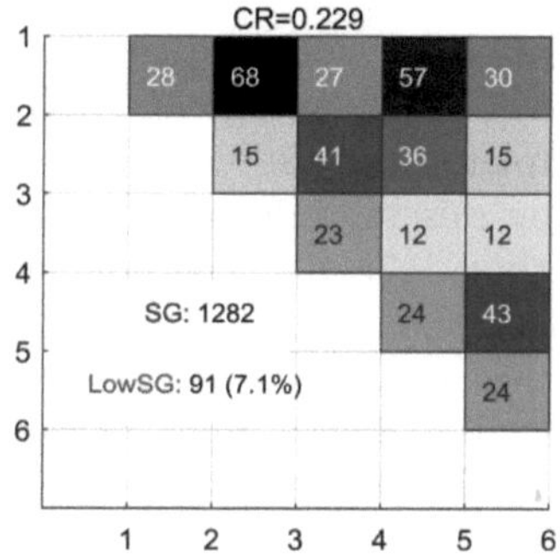

Fig. 5. Distribution of a_{ij} on the LowSG set, n=6.

"dark zone" are $a_{13}, a_{15}, a_{46}, a_{24}$. These elements (pairwise comparisons) need to be reviewed.

3.8 The Best-Worst Method as a Variant of an Incomplete PCM. MSG Solution

The BWM method is explained in detail in the study by Rezaei [5]. The transformation of BWM into MSG will be explained using an example. In accordance with the results of Sect. 3.4, we select the pairwise comparisons of the first row of the matrix A according to formula (5) as the Best-to-Other (BO), and the 3rd column as the Other-to-Worst (OW). We perform the optimization calculation of the weight coefficients using the BWM method:

w_{BWM} = (0.27 0.11 0.04 0.34 0.14 0.11).

Based on the data from these two graphs (BO and OW), we restore the elements of the PCM using the property of inverse symmetry to the following form:

$$A_{BWM} = \begin{pmatrix} 1 & 4 & 3 & 1 & 3 & 4 \\ 1/4 & 1 & 7 & 0 & 0 & 0 \\ 1/3 & 1/7 & 1 & 1/5 & 1/5 & 1/6 \\ 1 & 0 & 5 & 1 & 0 & 0 \\ 1/3 & 0 & 5 & 0 & 1 & 0 \\ 1/4 & 0 & 6 & 0 & 0 & 1 \end{pmatrix}$$

Now we can apply the MSG method and $w_{MSG/BWM}$ = (0.3 0.15 0.04 0.23 0.14 0.14), which is quite consistent with the results of Table 4. The number of spanning graphs for the A_{BWM} matrix of dimension 6 is $N_{SG} = 96$. Supplementing Table 1, the number of spanning graphs for the BWM method with $n = 7$, $N_{SG} = 112$, with $n = 8$, $N_{SG} = 256$. Thus, the BWM method is defined as a variant of MSG with an incomplete matrix of pairwise comparisons.

3.9 What Is the Advantage of the MSG Method?

Let us note the advantages of determining priorities using the MSP method. The first advantage of MSG over all other methods is that it utilizes all possible pairwise comparison relation chains reflected in the PCM, which increases the method's robustness. The second advantage of MSG over all other methods is its ability to handle incomplete pairwise comparison matrices. In such cases, various options for correcting missing information are possible. For example, in the well-known Harker method [17], if a component is missing, it is replaced with an artificial value. With the MSG method, it is sufficient to omit from consideration all SGs that are missing at least one element. In this case, the number of spanning graphs will decrease, but the result will remain stable. The third advantage is determined by consistency correction. The statistical metric of inconsistency in MSG is defined based on the standard deviation of each spanning graph from the mean. This allows for determining the sequence of judgment revisions, starting with the most problematic ones.

4 Numerical Example: Improving PCM Consistency

According to the results of the previous section, it is necessary to re-evaluate pairwise comparisons $a_{13} = 3, a_{15} = 3$, which with a frequency of 68 and 57 $\in$ LowSG (Fig. 6). The total number of graphs belonging to LowSG is 91, which is 7.1 % of the total number of SGs (1282). The degree of consistency CR (Saaty) is CR=0.23>0.1. The matrix of pairwise comparisons is not consistent. The step-by-step process of re-evaluating the elements of the matrix of pairwise comparisons is presented in Table 5, corresponding to the matrix $A^{(3)}$ and in Fig. 6. The matrix $A^{(3)}$ displays step-by-step changes in elements (from left to right) with a note of the step number in the form of a superscript. $<1>, <2>, <3>, <4>$ is adjustment numbers (or steps) of a_{ij}.

Table 5. Step-by-step process for re-evaluating elements of a PCM.

	elements to be replaced	CR	LowSG
Step 1	$a_{13} = 6, a_{15} = 2$	0.183	5.2
Step 2	$a_{14} = 1$	0.151	4.0
Step 3	$a_{13} = 8, a_{24} = 2, a_{25} = 1/3$	0.117	3.3
Step 4	$a_{46} = 1/2$	0.081	2.6

$$A^{(3)} = \begin{pmatrix} 1 & 4 & 3^{<1>}6^{<3>}8 & 1^{<2>}2 & 3^{<1>}2 & 4 \\ 1/4 & 1 & 7 & 3^{<3>}2 & 1/5^{<3>}1/3 & 1 \\ 1/3 & 1/7 & 1 & 1/5 & 1/5 & 1/6 \\ 1 & 1/2 & 5 & 1 & 1 & 1/3^{<4>}1/2 \\ 1/2 & 3 & 5 & 1 & 1 & 3 \\ 1/4 & 1 & 6 & 2 & 1/3 & 1 \end{pmatrix}$$

The optimal value of pairwise comparisons of objects at each step is obtained by simple enumeration. At each step of re-evaluation, the consistency of the PCM

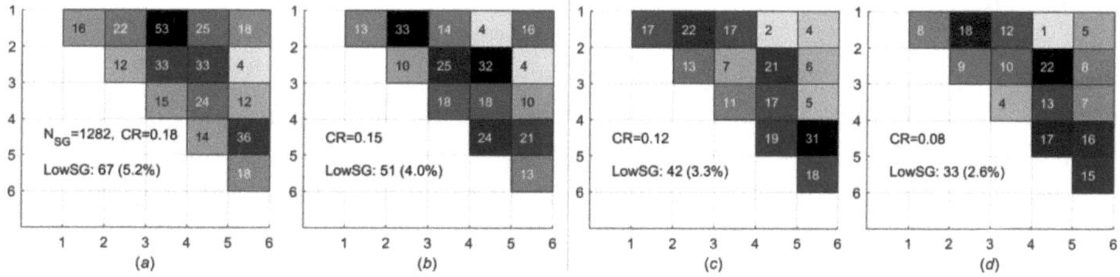

Fig. 6. Distribution of a_{ij} on the LowSG set after PCM revision. (a)-(d) successive steps 1)-4) from Sect. 3.8 .

improves (CR). Accordingly, the proportion of spanning graphs in the LowSG set also decreases, indicating improved consistency among estimates based on the spanning graph methodology. The standard deviation for all weighting coefficients also decreased.

The PCM after step 4 is consistent. The process of adjusting the elements of the PCM can be stopped. Thus, 6 pairwise comparisons out of 15 have been adjusted. Changes in the elements of the PCM (see matrix $A^{(3)}$) are performed with the priority in the relations between the alternatives preserved. Only the degree of superiority changes. In this example, the most significant change was made to element a_{13} from 3 to 8. The subject content of this relation is learning/school life, and the value 8, which is close to "absolute importance", is quite justified. Probably, many chains of relations (spanning graphs) including this pair were not consistent. The result of the weights estimation after re-estimation of the PCM is presented in Table 5 (Table 6).

Table 6. The result of the weights of alternatives before and after re-evaluation of the PCM: EVM and MSG methods.

	$A^{(1)}$, (CR=0.23)			$A^{(2)}$, (CR=0.081)			Increase, %		
w	EVM	MSG	$\pm s$	EVM	MSG	$\pm s$	EVM	MSG	$\pm s$
w_1	0.321	0.32	0.177	0.360	0.353	0.115	12.1	16.9	−35.0
w_2	0.140	0.146	0.121	0.134	0.138	0.072	−4.3	−5.5	−40.5
w_3	0.035	0.039	0.033	0.028	0.03	0.015	−20.0	−23.1	−54.5
w_4	0.128	0.135	0.105	0.123	0.126	0.063	−3.9	−6.7	−40.0
w_5	0.237	0.23	0.157	0.225	0.221	0.101	−5.1	−3.9	−35.7
w_6	0.139	0.148	0.115	0.13	0.133	0.069	−6.5	−10.1	−40.0

It should be noted that the weighting coefficients after improving the consistency of the PCM underwent changes from 4 to 23%.

5 Conclusion

The MSG approach reveals the causal relationship between the PCM and the decision through the distribution of weights calculated on a set of spanning graphs.

The MSG approach takes into account all possible connections and relations of the PCM, i.e. the analysis includes the fine structure of relations between objects contained in the already formed PCM. Now the degree of inconsistency of judgments has a natural interpretation in the form of the magnitude of dispersion in the distribution of weights.

Prioritization based on spanning graphs corresponds to AHP/EVM and analogues. One of the obvious advantages of MSG is the processing of incomplete pairwise comparison matrices. In particular, the Best-Worst method is a variant of MSG with an incomplete PCM. Despite the loss of information content, the reliability of the BWM method remains high, since the number of spanning graphs based on the Best-Wors matrix is still large. Another advantage of the SG approach is the ability to identify contradictory judgments in the matrix of pairwise comparisons. In SG, inconsistency is estimated based on the analysis of outliers in the distribution of weights obtained on the entire set of graphs. One of the confirmations of the rationality of MSG is the complete agreement in identifying contradictory judgments with the triad-based method.

More information is better only when it is possible to extract the main thing, or its essence, from this information. This is possible if we cut off the "noise" in a large amount of information, which is an integral and constituent part of the object and information about this object. MSG needs a high-quality theoretical justification, which can be the direction of further research. According to the authors, the solution to the problem of comparing weighting methods lies in the field of information theory.

Acknowledgments. The study was carried out within the framework of the state assignment of the Ufa Federal Research Center of the Russian Academy of Sciences (No. 075-00576-26-00 for 2026 and for the planning period of 2027 and 2028).

Disclosure of Interests. The authors have no competing interests to declare that are relevant to the content of this article.

References

1. David, H.A.: The Method of Paired Comparisons. Oxford University Press, New York (1988)
2. Belton, V., Stewart, T.: Multiple Criteria Decision Analysis: An Integrated Approach, Kluwer (2002)
3. Saaty, T.L.: A scaling method for priorities in hierarchical structures. J. Math. Psychol. **15**, 234–281 (1977)
4. Gabus, A., Fontela, E.: World Problems, an Invitation to Further Thought Within the Framework of DEMATEL. Battelle Geneva Research Centre, Geneva, Switzerland (1972)

5. Rezaei, J.: Best-worst multi-criteria decision-making method. Omega **53**, 49–57 (2015). https://doi.org/10.1016/j.omega.2014.11.009
6. Kersuliene, V., Zavadskas, E. K., Turskis, Z.: Selection of rational dispute resolution method by applying new Step-Wise Weight Assessment Ratio Analysis (SWARA). J. Bus. Econom. Manage. **11**, 243—258 (2010). https://doi.org/10.3846/jbem.2010.12
7. Nogin, V.D.: A simplified variant of the analytic hierarchy process based on a non-linear scalarizing function. Comput. Math. Math. Phys. **44**(7), 1194–1202 (2004)
8. Saaty, T.L.: The Analytic Hierarchy Process. McGraw-Hill, New York (1980)
9. Barzilai, J.: Deriving weights from pairwise comparison matrices. J. Oper. Res. Society **48**, 1226–1232 (1997)
10. Saaty, T.L., Vargas, L.G.: Inconsistency and rank preservation. J. Math. Psychol. **28**, 205–214 (1984)
11. Crawford, G., Williams, C.: A note on the analysis of subjective judgment matrices. J. Math. Psychol. **29**(4), 387–405 (1985)
12. Gao, S., Zhang, Z., Gao C.: Calculating weights methods in complete matrices and incomplete matrices. J. Softw. **5**(3), 304–311 (2010). https://doi.org/10.4304/jsw.5.3.304-311
13. Kou, G., Lin, C.: A cosine maximization method for the priority vector derivation in AHP. Europ. J. Oper. Res. **235**(1), 225–232 (2014). https://doi.org/10.1016/j.ejor.2013.10.019
14. Pant, S., Kumar, A., Ram, M., Klochkov, Y., Sharma, H.K.: Consistency indices in analytic hierarchy process: a review. Math. **10**, 1206 (2022). https://doi.org/10.3390/math10081206
15. Zhang, J., Kou, G., Peng, Yi., Zhang, Yu.: Estimating priorities from relative deviations in pairwise comparison matrices. Inform. Sci. **552**, 310–327 (2021). https://doi.org/10.1016/j.ins.2020.12.008
16. Saaty, T.L., Shang, J.S.: Group decision-making: Head-count versus intensity of preference. Socio-Economic Plann. Sci. **41**, 22–37 (2007). https://doi.org/10.1016/j.seps.2005.10.001
17. Harker, P.T.: Incomplete pairwise comparisons in the analytic hierarchy process. Math. Modell. **9**(11), 837–848 (1987)

An Integer Linear Programming Model for the Evolomino Puzzle

Andrei V. Nikolaev(✉) and Yuri A. Myasnikov

P.G. Demidov Yaroslavl State University, Yaroslavl, Russia
andrei.v.nikolaev@gmail.com, ya_myasnikov@vk.com

Abstract. Evolomino is a pencil-and-paper logic puzzle published by the Japanese company Nikoli, renowned for culture-independent puzzles such as Sudoku, Kakuro, and Slitherlink. Its name reflects the core mechanic: the polyomino-like blocks drawn by the player must gradually "evolve" according to the directions indicated by arrows pre-printed on a rectangular grid.

In this paper, we formalize the rules of Evolomino as an integer linear programming (ILP) model, encoding block evolution, connectivity, and consistency requirements through linear constraints. Furthermore, we introduce an algorithm for generating random Evolomino instances, utilizing this ILP framework to ensure solution uniqueness. Computational experiments on a custom benchmark dataset demonstrate that a state-of-the-art CP-SAT solver successfully handles puzzle instances of up to 10×10 within one second and up to 18×18 within one minute.

Keywords: Evolomino · Pencil-and-paper logic puzzle · Polyominoes · Integer linear programming · Puzzle generation · Solution uniqueness

1 Introduction

Nikoli is a renowned Japanese publisher specializing in board games and, in particular, logic puzzles. Founded in 1980 by Maki Kaji, often referred to as "the father of Sudoku", the company rose to worldwide prominence with the global spread of Sudoku.

Nikoli is especially known for its extensive library of culture-independent puzzles that do not rely on any specific language or alphabet and are often purely logical or numerical in nature. The company's namesake magazine was the first puzzle magazine in Japan, and over the years it has introduced numerous new puzzle genres and brought several new games to the Japanese audience.

In this paper, we study the relatively new pencil-and-paper logic puzzle Evolomino, which first appeared in *Puzzle Communication Nikoli*, Vol. 182, in March 2023 [1].

Here we present the rules of the Evolomino puzzle as they appear on Nikoli's official website [24]:

A. Kazakov et al. (Eds.): MOTOR 2026, LNCS 16687, pp. 116–133, 2026.
https://doi.org/10.1007/978-3-032-31390-4_8

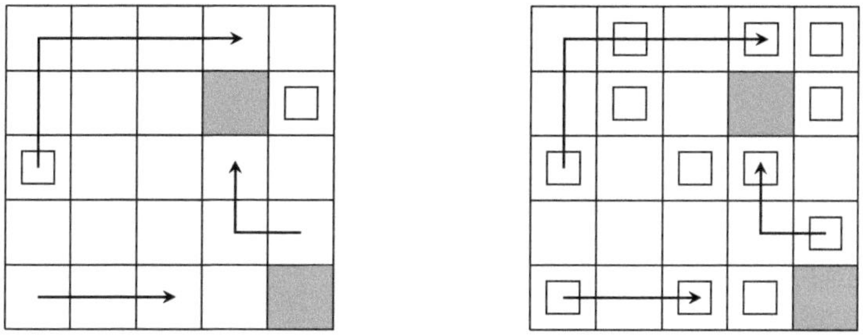

Sample puzzle Solution

Fig. 1. Example of an Evolomino puzzle.

- Evolomino is played on a rectangular board with white and shaded cells. Some white cells contain pre-drawn squares and arrows.
- The player must draw squares (□) in some of the white cells.
- A polyomino-like group of squares connected vertically and horizontally is called a *block* (including only one square). Each block must contain exactly one square placed on a pre-drawn arrow.
- Each arrow must pass through at least two blocks.
- The second and later blocks on the route of an arrow from start to finish must progress by adding one square to the previous block without rotating or flipping.

An example of the Evolomino puzzle and its solution from Nikoli's official website [24] is shown in Fig. 1.

As a relatively new puzzle, Evolomino has not yet been extensively studied from a theoretical computer science perspective. In [22], a reduction from 3-SAT was used to prove that deciding whether an Evolomino puzzle has at least one valid solution is NP-complete. Furthermore, since the reduction is parsimonious and preserves the number of distinct solutions, it follows that the associated counting problem is #P-complete.

In this paper, we formulate the rules of the Evolomino puzzle as an integer linear programming (ILP) model. ILP models are among the most established and versatile methods for solving not only combinatorial optimization problems but also a wide range of puzzles and logic challenges. Examples include MILP formulations for Battleship [20], the Eternity puzzle [5], Nonograms (both black-and-white [4] and colored [21]), Kakuro [28], Shakashaka [8], Smashed Sums [14], the Survo puzzle [29], Tantrix [15], and even letter-substitution cipher attacks [27]. ILP-based approaches have also been proposed for several well-known Nikoli puzzles, including Sudoku [2], Kakuro [28], and Futoshiki [3]. The book by Tony Hürlimann [12] offers a comprehensive overview of mathematical modeling, featuring more than 140 puzzles as illustrative case studies.

Although integer programming is one of Karp's original 21 NP-complete problems [13], many algorithms have proven remarkably effective in practice.

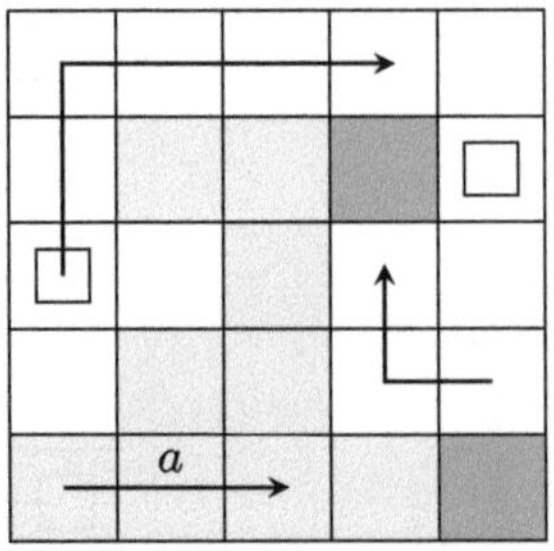

Fig. 2. Example of region $\mathcal{C}_a$ for arrow a, highlighted in green. (Color figure online)

For example, Koch et al. [18] demonstrated that between 2001 and 2020 computer hardware became roughly 20 times faster and algorithms improved by a factor of about 50, resulting in an overall speed-up of approximately 1 000 for MILP models. Moreover, many instances considered unsolvable twenty years ago can now be solved within seconds. Therefore, once a puzzle can be formulated as an integer programming problem, we may hope to use modern solvers to handle it efficiently in practice.

The rest of the paper is organized as follows. Section 2 presents the integer linear programming constraints that model the rules of the Evolomino puzzle. Section 3 outlines the algorithm for generating random puzzle instances. Section 4 reports the results of computational experiments conducted on a custom benchmark dataset, produced by the generator from Sect. 3, with CP-SAT [25] from the OR-Tools [26] library serving as the main solver.

2 ILP-Model

2.1 Preliminary Notation

- $\mathcal{C}$—the set of all cells of a rectangular $m \times n$ board.
  For simplicity, the cells of the board are numbered by a single linear index row by row, starting from the top-left corner.
- $\mathcal{A}$—the set of all arrows.
- $\mathcal{P}_a \subset \mathcal{C}$—the set of cells along arrow $a \in \mathcal{A}$.
- $\text{next}_a(i)$—the cell that follows $i \in \mathcal{P}_a$ in the direction of arrow $a \in \mathcal{A}$.
- $\mathcal{P}_a^{\prec i}$—the set of cells preceding cell $i \in \mathcal{P}_a$ in the direction of arrow $a \in \mathcal{A}$.
- K_a—the maximum possible number of blocks along arrow $a \in \mathcal{A}$, computed for each arrow as $K_a = \lceil \frac{|\mathcal{P}_a|}{2} \rceil$.
- $\mathcal{C}_a$—the set of cells that may belong to a polyomino block along arrow $a \in \mathcal{A}$. The example of region $\mathcal{C}_a$ is shown in Fig. 2 (highlighted in green). Starting from an arbitrary cell i on arrow a, we perform a breadth-first search to collect all cells j that are reachable from i by a path that does not pass through shaded cells, other arrows, or cells belonging to blocks of other arrows; all such cells are included in $\mathcal{C}_a$.

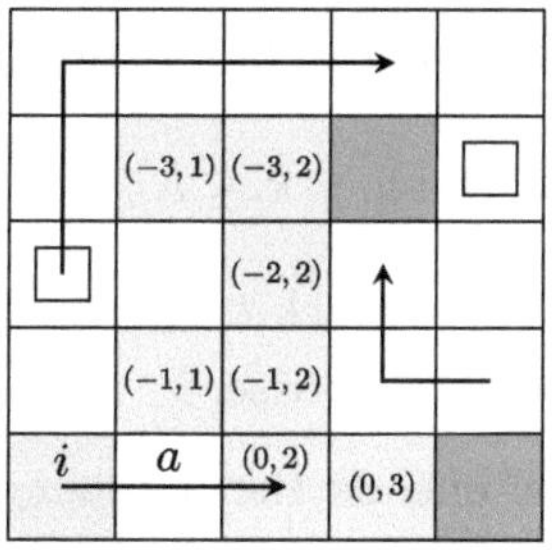

Fig. 3. Example of set $\mathcal{T}_i^a$ of admissible translations (p, q) from a cell i highlighted in orange along arrow a; admissible target cells are highlighted in green. (Color figure online)

- row(i) and col(i)—functions that return the row and column indices of cell i, respectively.
- M—a sufficiently large constant that exceeds the size of any polyomino block, for example $M = |\mathcal{C}|$.
- $\mathcal{T}$—the set of all possible two-dimensional translations (shifts) applicable to two consecutive blocks, where the second block contains a copy of a first block as a sub-shape displaced by p rows and q columns. It consists of all integer pairs (p, q) with $p \in \{-(m-1), \dots, m-1\}$ and $q \in \{-(n-1), \dots, n-1\}$ where $|p| + |q| > 1$ (no zero translations or translations to adjacent cells).
- $\text{shift}(i, p, q) = i + p \cdot n + q$—the function that maps linear index i to the index of the cell shifted by p rows and q columns.
- $\mathcal{T}_i^a \subseteq \mathcal{T}$—the set of admissible translations (p, q) from a cell $i \in \mathcal{C}_a$ along arrow $a \in \mathcal{A}$ such that the shift remains within the board: $1 \leq \text{row}(i) + p \leq m$, $1 \leq \text{col}(i) + q \leq n$, and $\text{shift}(i, p, q) \in \mathcal{C}_a$ (valid translation inside region $\mathcal{C}_a$). An example of the set $\mathcal{T}_i^a$ is shown in Fig. 3.
- $\mathcal{T}_a = \bigcup_{i \in \mathcal{P}_a} \mathcal{T}_i^a$—the set of all admissible two-dimensional translations (p, q) for the blocks associated with arrow $a \in \mathcal{A}$.

2.2 Basic Variables

- For each cell $i \in \mathcal{C}$ we introduce the *cell-variable*:

$$x_i = \begin{cases} 1, & \text{if cell } i \text{ contains a square } \square, \\ 0, & \text{otherwise.} \end{cases}$$

  The variables x_i are used to encode the initial puzzle grid: pre-placed squares ($x_i = 1$) and shaded cells ($x_i = 0$), and they will also store the final solution of the puzzle.
- For each arrow $a \in \mathcal{A}$, each block $k \leq K_a$, and each cell $i \in \mathcal{C}_a$ we introduce the *block-variable*:

$$y_i^{ak} = \begin{cases} 1, & \text{if cell } i \text{ belongs to block } k \text{ on arrow } a, \\ 0, & \text{otherwise.} \end{cases}$$

Using the variables y_i^{ak} we encode all blocks by associating them with specific cells of the board.

- For each arrow $a \in \mathcal{A}$ and each block $k \leq K_a$ we introduce the *block activation variable*:

$$b_{ak} = \begin{cases} 1, & \text{if block } k \text{ on arrow } a \text{ exists,} \\ 0, & \text{otherwise.} \end{cases}$$

Using the variables b_{ak} we encode the number of blocks along each arrow. Since any arrow passes through at least two blocks, we set $b_{a1} = b_{a2} = 1$ for every arrow $a \in \mathcal{A}$.

2.3 Basic Constraints

- Each square must be assigned to exactly one block:

$$\sum_{\substack{a \in \mathcal{A} \\ i \in \mathcal{C}_a}} \sum_{k=1}^{K_a} y_i^{ak} = x_i \qquad \forall i \in \mathcal{C}. \tag{1}$$

If a $\square$ is placed in cell i ($x_i = 1$), then it belongs to exactly one block.

- No two consecutive squares may be placed along an arrow:

$$x_i + x_{\text{next}_a(i)} \leq 1 \qquad \forall a \in \mathcal{A},\ \forall i \in \mathcal{P}_a. \tag{2}$$

- If a block is not activated, then it must be empty:

$$\sum_{i \in \mathcal{C}_a} y_i^{ak} \leq M \cdot b_{ak} \qquad \forall a \in \mathcal{A},\ k = 1, \ldots, K_a. \tag{3}$$

- If a block is activated, then it contains exactly one square along the arrow:

$$\sum_{i \in \mathcal{P}_a} y_i^{ak} = b_{ak} \qquad \forall a \in \mathcal{A},\ k = 1, \ldots, K_a. \tag{4}$$

- A subsequent block cannot be activated before the previous one:

$$b_{ak} \leq b_{a,k-1} \qquad \forall a \in \mathcal{A},\ k = 2, \ldots, K_a. \tag{5}$$

- The order of blocks must follow the direction of the arrow:

$$\sum_{j \in \mathcal{P}_a^{\prec i}} y_j^{ak} \leq 1 - y_i^{a,k-1} \qquad \forall a \in \mathcal{A},\ \forall i \in \mathcal{P}_a,\ k = 2, \ldots, K_a. \tag{6}$$

Block k cannot appear earlier than block $k-1$ along arrow a. If cell i contains the square of block $k-1$ (i.e., $y_i^{a,k-1} = 1$), then block k (and, by induction, all subsequent blocks) cannot be placed in any position from $\mathcal{P}_a^{\prec i}$.

– Two adjacent cells cannot belong to different blocks:

$$y_i^{ak} + y_{i+1}^{a'k'} \leq 1 \quad \forall a, a' \in \mathcal{A},\ \forall i \in \mathcal{C}_a,\ i+1 \in \mathcal{C}_{a'},\ k, k' = 1, \ldots, K_a, \tag{7}$$
$$y_i^{ak} + y_{i+n}^{a'k'} \leq 1 \quad \forall a, a' \in \mathcal{A},\ \forall i \in \mathcal{C}_a,\ i+n \in \mathcal{C}_{a'},\ k, k' = 1, \ldots, K_a, \tag{8}$$

where $a \neq a'$ or $k \neq k'$, $\mathrm{row}(i) < m$, and $\mathrm{col}(i) < n$.
This restriction is imposed for every pair of blocks and every pair of adjacent cells on the board, excluding cells in the last row and the last column.
– For convenience, we introduce the *block size variable* N_{ak} for block k on arrow a:

$$N_{ak} = \sum_{i \in \mathcal{C}_a} y_i^{ak} \qquad \forall a \in \mathcal{A},\ k = 1, \ldots, K_a. \tag{9}$$

This variable is not strictly necessary, but it makes the subsequent constraints slightly easier to write.
– The size of each subsequent block along arrow a must exceed the previous one by exactly one square:

$$N_{ak} \geq N_{a,k-1} + 1 - M(1 - b_{ak}) \qquad \forall a \in \mathcal{A},\ k = 2, \ldots, K_a, \tag{10}$$
$$N_{ak} \leq N_{a,k-1} + 1 + M(1 - b_{ak}) \qquad \forall a \in \mathcal{A},\ k = 2, \ldots, K_a. \tag{11}$$

Here, if block k is activated ($b_{ak} = 1$), then constraints (10)–(11) reduce to

$$N_{ak} = N_{a,k-1} + 1.$$

Otherwise, if the block is not activated, then (10)–(11) reduce to trivial bounds.

2.4 Connectivity Variables and Constraints

The following group of variables and constraints ensures that each block forms a connected polyomino. To achieve this, we adapt a classic flow-based model by Gavish and Graves [10].

For each block k on each arrow $a \in \mathcal{A}$ and for every pair of adjacent cells i and j from $\mathcal{C}_a$ (each cell i has up to four neighboring cells, which we denote by $\mathcal{N}(i)$), we introduce two integer *flow variables* f_{ij}^{ak} and f_{ji}^{ak} representing the amount of flow between the cells in the directions from i to j and from j to i, respectively.

The idea of the flow-based model is that a single *source cell* generates a flow equal to the size of the polyomino block, and this flow must be distributed across all cells of the block. If the block consists of several connected components, some cells will be unreachable by the flow, and the corresponding constraints will not be satisfied.

An example of such a flow is shown in Fig. 4. For the source cell s, the outgoing flow equals the size of the polyomino minus one. For every other cell, the difference between incoming and outgoing flow is equal to one.

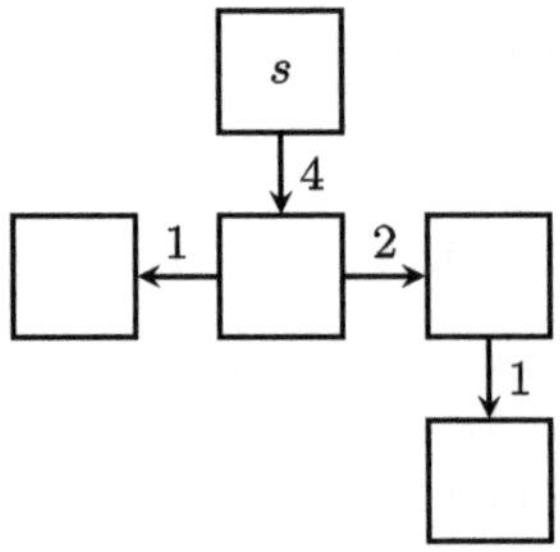

Fig. 4. An example of a flow used to verify polyomino connectivity.

We do not know in which cell of the grid the source is located, since the exact shape of the block is not known in advance. However, the key observation is that exactly one square of each block lies on the arrow, and we place the source in that square. Thus, all cells of the polyomino block are divided into *consumers* (those not lying on the arrow) and a single *source* (the one lying on the arrow).

For each block k on each arrow a, and for every cell $i \in \mathcal{P}_a$, we introduce a *flow-supply variable*

$$F_i^{ak} = \begin{cases} N_{ak}, & \text{if cell } i \text{ is the source of block } k \text{ on arrow } a, \\ 0, & \text{otherwise.} \end{cases}$$

Connectivity of the polyomino blocks is enforced by the following constraints.

- Total flow generated by the cells on the arrow equals the size of the polyomino block:

$$\sum_{i \in \mathcal{P}_a} F_i^{ak} = N_{ak} \qquad \forall a \in \mathcal{A}, k = 1, \ldots, K_a. \tag{12}$$

We sum over all cells on the arrow because the exact location of the source is not known in advance.

- Flow may be generated only by a source that belongs to the block:

$$0 \le F_i^{ak} \le M \cdot y_i^{ak} \qquad \forall a \in \mathcal{A}, \forall i \in \mathcal{P}_a, k = 1, \ldots, K_a. \tag{13}$$

- Flow can leave a cell only if it belongs to the block:

$$f_{ij}^{ak} \le M \cdot y_i^{ak} \qquad \forall a \in \mathcal{A}, \forall i \in \mathcal{C}_a, \forall j \in \mathcal{N}(i), k = 1, \ldots, K_a. \tag{14}$$

- Flow may enter only those cells that belong to the block:

$$f_{ij}^{ak} \le M \cdot y_j^{ak} \qquad \forall a \in \mathcal{A}, \forall i \in \mathcal{C}_a, \forall j \in \mathcal{N}(i), k = 1, \ldots, K_a. \tag{15}$$

- Flow cannot be negative:

$$f_{ij}^{ak} \ge 0 \qquad \forall a \in \mathcal{A},\ \forall i \in \mathcal{C}_a,\ \forall j \in \mathcal{N}(i),\ k = 1, \ldots, K_a. \tag{16}$$

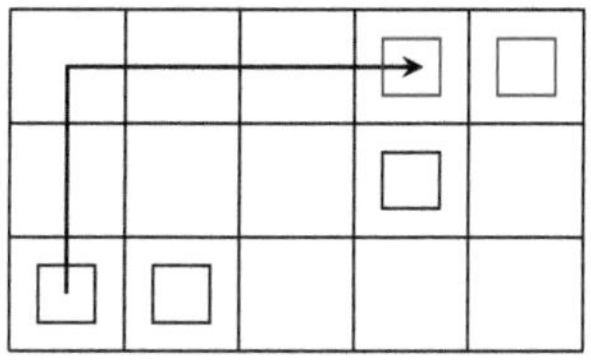

Fig. 5. Example of a translation with $t^{a2}_{-2,3} = 1$, i.e., a shift of 2 rows upward and 3 columns to the right; the copy of the first block inside the second along the arrow is highlighted in blue.

- Flow conservation for consumer cells $i \notin \mathcal{P}_a$:

$$\sum_{j \in \mathcal{N}(i)} f^{ak}_{ji} - \sum_{j \in \mathcal{N}(i)} f^{ak}_{ij} = y^{ak}_i \quad \forall a \in \mathcal{A},\ \forall i \in \mathcal{C}_a \setminus \mathcal{P}_a,\ k = 1, \ldots, K_a. \tag{17}$$

For a consumer cell, one unit of flow remains in the cell.
- Flow conservation for a cell $i \in \mathcal{P}_a$ on arrow a, which may serve as the source:

$$\sum_{j \in \mathcal{N}(i) \setminus \mathcal{P}_a} f^{ak}_{ji} - \sum_{j \in \mathcal{N}(i) \setminus \mathcal{P}_a} f^{ak}_{ij} = y^{ak}_i - F^{ak}_i \quad \forall a \in \mathcal{A},\ \forall i \in \mathcal{P}_a,\ k = 1, \ldots, K_a. \tag{18}$$

For the source cell, the outgoing flow equals $F^{ak}_i - 1$.

2.5 Block Evolution Variables and Constraints

The final group of constraints ensures that each subsequent block along the arrow adds exactly one square to the previous block, without rotation or reflection. The sizes of the blocks have already been enforced by the basic constraints; what remains is to preserve the shape of the blocks.

For each block $2 \leq k \leq K_a$ on arrow $a \in \mathcal{A}$, and for each admissible two-dimensional translation $(p, q) \in \mathcal{T}_a$, we introduce a *translation variable*

$$t^{ak}_{pq} = \begin{cases} 1, & \text{if block } k \text{ on arrow } a \text{ is translated by } p \text{ rows} \\ & \text{and } q \text{ columns relative to block } k-1, \\ 0, & \text{otherwise.} \end{cases}$$

An example of a translation is shown in Fig. 5. The second block along the arrow contains a copy of the first block (highlighted in blue) that is shifted 2 rows upward and 3 columns to the right.

The following constraints guarantee the correct evolution of the blocks.

- Only one translation can be assigned to each block:

$$\sum_{(p,q) \in \mathcal{T}_a} t^{ak}_{pq} = b_{ak} \qquad \forall a \in \mathcal{A}, k = 2, \ldots, K_a. \tag{19}$$

- Each square of the block must be translated:

$$\sum_{(p,q)\in\mathcal{T}_i^a} t_{pq}^{ak} \geq y_i^{a,k-1} + b_{ak} - 1 \qquad \forall a \in \mathcal{A}, \forall i \in \mathcal{C}_a, k = 2, \ldots, K_a, \qquad (20)$$

  where $\mathcal{T}_i^a$ is the subset of all admissible translations from cell i in the direction of arrow a.
  Here, if cell i contains a square of block $k-1$ ($y_i^{a,k-1} = 1$) and block k exists ($b_{ak} = 1$), then that square must have at least one valid translation into block k.
- Correspondence constraints between squares of block $k-1$ and block k:

$$y_{\text{shift}(i,p,q)}^{ak} \geq y_i^{a,k-1} - (1 - t_{pq}^{ak}) \quad \forall a \in \mathcal{A}, \forall i \in \mathcal{C}_a, \forall (p,q) \in \mathcal{T}_i^a, k = 2, \ldots, K_a. \qquad (21)$$

  This constraint preserves the geometric shape of the polyomino blocks across consecutive steps without rotation or reflection. If translation (p,q) is chosen for block k ($t_{pq}^{ak} = 1$), the inequality (21) enforces

$$y_{\text{shift}(i,p,q)}^{ak} \geq y_i^{a,k-1},$$

  so a square of block $k-1$ located at cell i must reappear in block k at the translated position $\text{shift}(i,p,q) = i + p \cdot n + q$. Otherwise, if $t_{pq}^{ak} = 0$, the constraint reduces to the trivial bound.

2.6 Soundness and Completeness of the ILP Model

Theorem 1. *Let $\mathcal{E}$ be an instance of the Evolomino puzzle on a board C with arrow set $\mathcal{A}$, and $S(\mathcal{E})$ be the linear system consisting of constraints (1)–(21), with all variables intended to be binary or integral. Then an assignment to the cell variables $\{x_i\}_{i\in C}$ corresponds to a valid solution of $\mathcal{E}$ if and only if there exists an integral assignment to the auxiliary variables that satisfies $S(\mathcal{E})$.*

Proof. ($\Rightarrow$) *Every valid Evolomino solution satisfies the ILP model.*

Given a valid Evolomino solution, set $x_i = 1$ on drawn squares and assign values to the auxiliary variables as follows. Index the blocks along each arrow and set the variables y_i^{ak}, b_{ak}, and N_{ak} accordingly. The constraints (1)–(11) follow directly from the definition of blocks and the rules of Evolomino: each square belongs to exactly one block (1); blocks are ordered along arrows (5)–(6); adjacent squares belong to the same block (7)–(8); and successive blocks differ in size by one (10)–(11).

For the flow variables, root a spanning tree of each connected polyomino block at its unique arrow-path cell s and direct all edges away from s. Assign to each directed edge (u,v) a flow equal to the number of cells in the subtree rooted at v. With this assignment, constraints (12)–(18) are satisfied by standard flow conservation.

For the translation variables, the evolutionary rule guarantees that there exists a unique two-dimensional shift (p^*, q^*) that maps block $k-1$ to a sub-shape of block k. By setting $t^{ak}_{p^*q^*} = 1$ and $t^{ak}_{pq} = 0$ for all $(p, q) \neq (p^*, q^*)$, we satisfy constraints (19)–(21).

(⇐) Every integer-feasible point of the model (1)–(21) *defines a valid solution of the Evolomino puzzle.*

Let (x, y, b, t, N, f, F) be an integer-feasible point. We verify Evolomino rules. Constraint (1) ensures that every square $x_i = 1$ belongs to exactly one block, and (4) guarantees that each block contains exactly one square on the pre-drawn arrow. The conditions $b_{a1} = b_{a2} = 1$ together with (5) ensure at least two blocks per arrow. Constraints (7)–(8) prevent orthogonally adjacent squares from belonging to different blocks, so blocks are geometrically separated.

Connectivity of each polyomino block follows from the flow model: any disconnected component would contain a cell unreachable from the unique source, which contradicts flow conservation (17)–(18), since (13)–(15) confine flow only to the cells of the block ($y^{ak}_i = 1$).

Finally, (19) selects exactly one translation (p^*, q^*), constraints (20)–(21) together enforce that every square i of block $k-1$ must reappear in block k at the translated position $\text{shift}(i, p^*, q^*)$, and (10)–(11) fix $N_{ak} = N_{a,k-1} + 1$; hence block k is exactly the translated copy of block $k-1$ plus one new square.

Thus, the integer-feasible point (x, y, b, t, N, f, F) induces a partition of the drawn squares into disjoint polyomino blocks $B_{ak} = \{i \in \mathcal{C} : y^{ak}_i = 1\}$, where each B_{ak} is connected, contains exactly one arrow cell, is obtained from $B_{a,k-1}$ by a single translation plus one additional square, and not touching other blocks orthogonally. Hence (x, y, b, t, N, f, F) defines a valid Evolomino solution. □

2.7 Model Size

We estimate how the number of variables and constraints in the ILP model (1)–(21) scales with the size of the Evolomino board, in order to determine the range of instances that can be solved in practice.

For a given rectangular Evolomino board, let $|\mathcal{C}|$ denote the number of cells on the board, and let K be the maximum possible number of polyomino blocks in a solution, where

$$K = \sum_{a \in \mathcal{A}} K_a.$$

It is easy to see that $K = O(|\mathcal{C}|)$ and, in fact, $K < |\mathcal{C}|$. In typical instances, the number of polyomino blocks in a solution is significantly smaller than the board size. However, it is technically possible to construct an Evolomino instance whose solution contains a number of blocks that grows linearly with the size of the board (for example, by densely placing short arrows of length 3, each supporting two blocks of sizes 1 and 2, respectively).

We do not attempt to compute the exact number of variables and constraints. Instead, we estimate their overall magnitude and identify which components contribute the most.

Variables. Among all variables, the dominant contribution comes from the block variables y_i^{ak}, the flow variables f_{ij}^{ak}, and the translation variables t_{pq}^{ak}. The size of each of these classes is $O(K \cdot |\mathcal{C}|)$. Consequently, the total number of variables in the model is also $O(K \cdot |\mathcal{C}|)$.

Constraints. Among constraints, the most significant contributions come from:

- the constraints preventing adjacent cells from belonging to different blocks (7)–(8), whose total count is $O(K^2 \cdot |\mathcal{C}|)$;
- the connectivity constraints for polyomino blocks enforced via flows (12)–(18), amounting to $O(K \cdot |\mathcal{C}|)$ (with the constant hidden in the big-O being substantially larger than for other constraints of the same asymptotic order);
- the correspondence constraints that maintain square-by-square consistency during block evolution along the arrow direction (21), whose total number is $O(K \cdot |\mathcal{C}|^2)$.

Thus, the overall number of constraints in the proposed ILP model for the Evolomino puzzle is $O(K \cdot |\mathcal{C}|^2)$, with the most expensive constraints being those that preserve the geometric shape during block evolution.

As an example, the ILP model for the 5×5 puzzle shown in Fig. 1 contains 509 variables and 1 709 constraints. In contrast, the model for the more complex 10×10 puzzle from Jonathan Chaffer's YouTube video [6] contains 6 334 variables and 55 057 constraints.

3 Puzzle Generation

To construct the dataset for the computational experiments, we developed an algorithm that generates Evolomino puzzles with a unique solution. Its pseudocode is presented in Algorithm 1.

The algorithm starts from an empty rectangular board and gradually fills it while attempting to reach a prescribed density (the total length of arrows and the total size of polyomino blocks). At each step, a new arrow is drawn using a biased random walk, where continuing straight is preferred over turning. The process stops when the arrow cannot grow further or when a probabilistic stopping criterion is triggered, implemented throughout the algorithm via repeated Bernoulli trials. Failed arrows are rolled back, and a new attempt is made.

For every admissible arrow, the block-placement procedure is executed. It first selects a random anchor position on the arrow for the initial block, prioritizing positions closer to the beginning of the arrow (weighted random sampling by Efraimidis and Spirakis [9]). From this anchor, a random polyomino block is grown. Subsequent blocks are placed afterward, with their number treated as a random variable, again drawn from a geometric distribution. For each candidate, the procedure enumerates all feasible evolution shapes, anchor positions on the arrow, and anchor-cell choices within the polyomino. Conflicts trigger a rollback and a new attempt. A successful sequence of blocks is committed to the board.

Algorithm 1. Evolomino puzzle generation

```
procedure GenerateEvolomino(m, n, P)
    Board ← empty m × n grid; Tries ← 0                          ▷ We start with an empty grid
    while Fill(Board) < P.Target and Tries < P.Tries do
        if TryAddArrow(Board, P) = fail then
            Backtrack(Board); Tries ← Tries + 1                  ▷ Undo the last arrow or reset
        end if
    end while
    return CarveToUnique(Board)                                  ▷ Final clue reduction and shading
end procedure
procedure TryAddArrow(Board, P)
    path ← [c ← rand cell]; d ← rand free dir
    repeat
        c ← c + d; path.push(c)
        d ← GetNextDir(c, d, Board, P)                           ▷ Biased random walk
    until IsStuck(c, Board) or Bernoulli(P.p_stop_arrow)
    if |path| < P.min_arrow then
        return fail
    end if
    res ← PlaceBlocks(Board, path, ∅, 0)
    if res ≠ ∅ then
        Commit(Board, res); return success                       ▷ Apply successful placement
    end if
    return fail
end procedure
procedure PlaceBlocks(Board, path, blocks, idx)
    if |blocks| ≥ P.min_blocks and Bernoulli(P.p_stop_blocks) then
        return blocks
    end if
    Cands ← GetValidIndices(path, idx, blocks)
    for each j ∈ WeightedShuffle(Cands) do                       ▷ Weighted random order
        Shapes ← GetEvolutions(blocks.last)
        for each S ∈ Shuffle(Shapes) do
            for each anchor ∈ Shuffle(S) do
                B ← Align(S, anchor, path[j])                    ▷ Align anchor to path[j]
                if CanPlace(B, Board) then
                    res ← PlaceBlocks(Board, path, blocks ∪ {B}, j)
                    if res ≠ ∅ then
                        return res
                    end if
                    Undo(B, Board)                               ▷ Backtracking: revert failed placement
                end if
            end for
        end for
    end for
    return ∅
end procedure
procedure CarveToUnique(Board)                                   ▷ Greedy clue removal and shading
    Sol ← ShadeAllFreeCells(Board)                               ▷ Fix a unique solution
    Cells ← Shuffle(AllCells)                                    ▷ Random order to avoid bias
    for all c ∈ Cells do
        Old ← Board[c]
        if Old ∈ {SHADED, SQUARE} then
            Board[c] ← EMPTY and UNSHADED
            if HasDifferentSolution(Board, Sol) then
                Board[c] ← Old                                   ▷ Revert if an alternative solution exists
            end if
        end if
    end for
    return Board                                                 ▷ Resulting puzzle with a unique solution
end procedure
```

In the final stage, after all arrows and blocks have been placed and the required density has been achieved, the algorithm shades all remaining empty cells, thereby fixing a unique solution. It then iterates through the cells in random order (classic Fisher–Yates shuffle [16]), greedily attempting to remove clue squares and shaded cells. If a removal does not introduce an alternative solution, as verified by calling the ILP solver, where the current solution x^* is excluded by the additional constraint

$$\sum_{i \in C} |x_i - x_i^*| \geq 1,$$

the change is accepted; otherwise, it is rolled back. The resulting puzzle contains a minimal set of clues that still uniquely determines the constructed solution.

4 Computational Experiments

Using the generator described in Algorithm 1, we constructed a custom benchmark dataset (available at [23]) that includes:

- 500 puzzles with unique solutions, ranging in size from 5×5 to 14×14, with 50 instances for each grid size;
- 200 puzzles with guaranteed, but not necessarily unique, solutions, ranging from 15×15 to 18×18, again evenly distributed across grid sizes.

For larger instances, we had to replace the CarveToUnique procedure, which guarantees solution uniqueness, with random clue removal and obstacle placement, since repeated calls to the ILP solver became computationally expensive.

The proposed ILP model was evaluated using the open-source software suite Google OR-Tools v9.15 [26]. All experiments were performed on a machine equipped with an Intel Core i7–13620H CPU and 16 GB of DDR5–5400 RAM, running Windows 11 Pro.

We conducted two series of experiments. In the first, using 50 puzzle instances of size 10×10, we compared four solvers available out-of-the-box in OR-Tools:

- CP-SAT: an award-winning constraint programming and SAT solver [25];
- SCIP: a high-performance academic mixed-integer programming solver [11];
- CBC: an open-source branch-and-cut solver from the COIN-OR initiative [19];
- BOP: a specialized SAT-based solver for pure Boolean problems [26].

This comparison was performed to identify the most efficient solver for our ILP model of the Evolomino puzzle.

The performance profile of this experiment is presented in Fig. 6, with each point representing two instances. CP-SAT emerges as the clear winner, being the only solver to successfully handle all 50 benchmark instances within the 180-second time limit. In comparison, SCIP solved 49 instances, while both BOP and CBC completed 48.

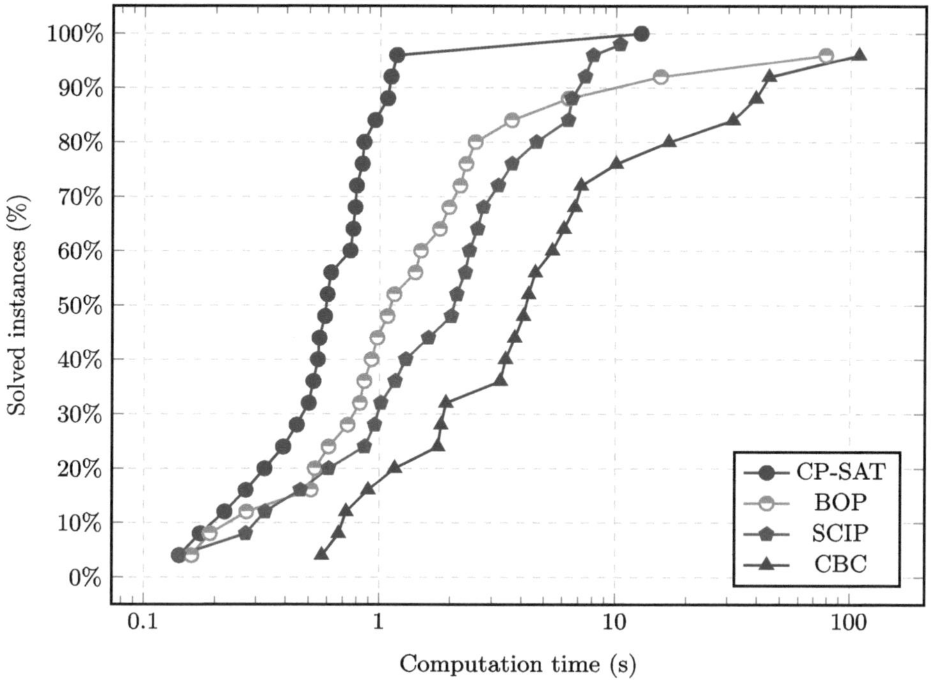

Fig. 6. Performance profile of the solvers on a 10 × 10 grid (180-s time limit).

Within a one-second cutoff, CP-SAT solved 43 out of 50 instances, BOP solved 22, SCIP 14, and CBC only 9. A notable outlier for CP-SAT is *sample39*, which required 12 s to solve. This instance proved particularly challenging: under the 3-minute limit, only CP-SAT and BOP managed to solve it at all.

In the second series of tests, the CP-SAT solver that demonstrated the best performance was evaluated on the entire dataset of 700 benchmarks (comprising 500 instances with unique solutions and 200 with guaranteed solutions). The results of these computational experiments are shown in Fig. 7 and in more detail in Table 1. The solid line indicates the median runtime, the shaded area represents the interquartile range (IQR), and individual points denote Tukey outliers ($Q_1 - 1.5\,\text{IQR}$ and $Q_3 + 1.5\,\text{IQR}$ [7]).

The plot demonstrates that solving time grows exponentially with the size of the puzzle grid and, consequently, with the number of linear constraints in the ILP model. For smaller grids up to 10 × 10, the median runtime remains below one second; for grids up to 13 × 13, it stays within five seconds. Starting from 15 × 15, the IQR expands and outliers become more frequent, indicating that the instances become less homogeneous and the solver's behavior less predictable. We stopped at puzzles of size 18 × 18, where the median runtime per instance (54 seconds) still remained under one minute.

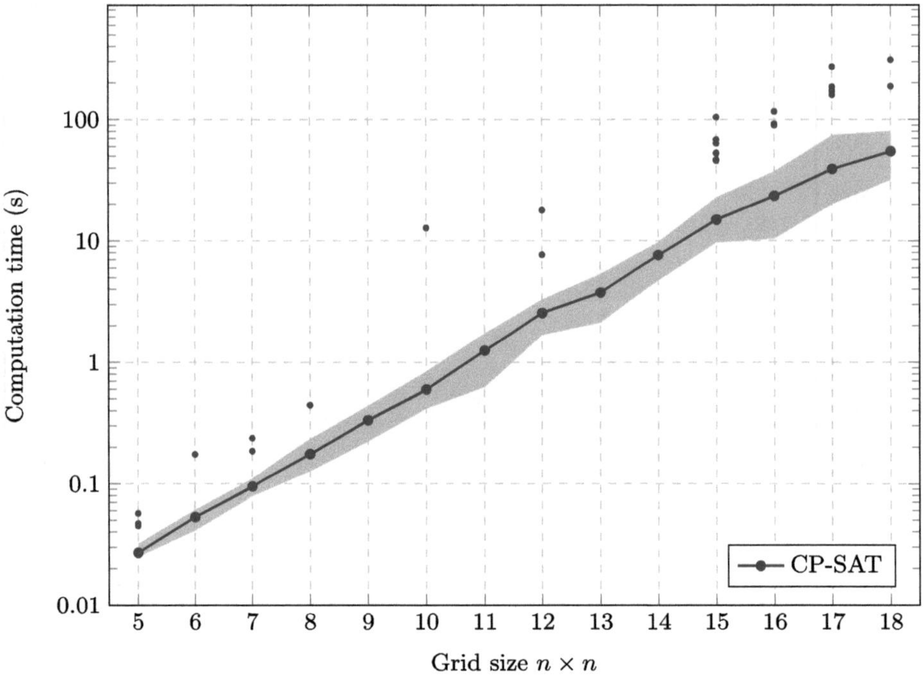

Fig. 7. CP-SAT performance: median (line), IQR (shaded), and outliers (dots).

Table 1. ILP model complexity and CP-SAT solver runtime on the benchmark dataset [23] (50 instances per grid size)

Grid size	Problem size (Med.)		Running time (s)		
	Variables	Constraints	$Q1$	Median	$Q3$
5×5	450	1 429	0.025	0.027	0.032
6×6	876	3 615	0.041	0.053	0.061
7×7	1 525	7 556	0.079	0.095	0.111
8×8	2 471	14 326	0.127	0.175	0.234
9×9	3 819	26 729	0.223	0.334	0.438
10×10	5 352	41 905	0.415	0.597	0.842
11×11	8 080	74 248	0.626	1.246	1.736
12×12	11 287	129 946	1.681	2.547	3.301
13×13	14 940	192 501	2.127	3.761	5.327
14×14	20 351	289 132	4.727	7.643	9.741
15×15	27 648	459 073	9.662	14.952	22.832
16×16	31 893	525 008	10.409	23.543	37.457
17×17	39 675	740 220	20.165	39.130	74.549
18×18	53 307	1 121 290	31.961	54.739	80.257

Given the complexity of the ILP model and the cubic growth $O(K \cdot |\mathcal{C}|^2)$ of the number of linear constraints relative to the grid size, these results are quite promising and highlight the efficiency of modern state-of-the-art solvers. In particular, for successfully solved 18×18 puzzles, the model contained approximately 50 000 variables and about one million linear constraints.

5 Conclusion

Evolomino is a relatively new pencil-and-paper logic puzzle that remains largely unexplored from a theoretical computer science perspective. To the best of our knowledge, the only existing result is a proof establishing that determining whether a given Evolomino instance has at least one valid solution is NP-complete and that counting the number of distinct solutions is **#**P-complete [22].

In this paper, we propose a computational approach to solving the puzzle within the transform-and-conquer paradigm by formulating the Evolomino rules as a set of constraints for an integer linear programming model. Furthermore, we introduce an algorithm for generating random Evolomino instances, where solution uniqueness is guaranteed via the proposed ILP framework.

Computational experiments conducted on a custom benchmark dataset demonstrate that the state-of-the-art CP-SAT solver from the OR-Tools library successfully solves puzzle instances of size up to 10×10 within one second, and up to 18×18 within one minute.

Future research could explore alternative approaches to solving the Evolomino puzzle, such as backtracking, constraint programming, SAT-based models [30], or a reduction to the exact cover problem and applying the DLX (Dancing Links) algorithm [17], which is particularly effective for placing polyominoes on a grid.

Acknowledgments. The research of Andrei V. Nikolaev was supported by Alfa-Bank JSC under the "Alfa-Future: Grants for Faculty" program. We thank the anonymous reviewers for their insightful and constructive comments, which helped improve the clarity and presentation of this paper.

Disclosure of Interests. The authors have no competing interests to declare that are relevant to the content of this article

References

1. Puzzle communication Nikoli **182**, (2023)
2. Bartlett, A., Chartier, T.P., Langville, A.N., Rankin, T.D.: An integer programming model for the Sudoku problem. J. Online Math. Appl. **8**(1), (2008)
3. Bitgen Sungur, B.: An integer programming formulation for the Futoshiki puzzle. Int. Rev. Econ. Manage. **10**(2), 38–49 (2022)
4. Bosch, R.A.: Painting by numbers. Optima **65**, 16–17 (2001)
5. Burkardt, J., Garvie, M.R.: An integer linear programming approach to solving the Eternity puzzle. Theoret. Comput. Sci. **975**, 114138 (2023). https://doi.org/10.1016/j.tcs.2023.114138

6. Chaffer, J.: Solving YOUR Evolmino puzzle! Pathologic 6-3 (2025). https://www.youtube.com/watch?v=yV7GDJo_8zo
7. Dekking, F., Kraaikamp, C., Lopuhaä, H., Meester, L.: A Modern Introduction to Probability and Statistics. Springer, London (2005). https://doi.org/10.1007/1-84628-168-7
8. Demaine, E.D., Okamoto, Y., Uehara, R., Uno, Y.: Computational complexity and an integer programming model of Shakashaka. IEICE Trans. Fund. Electron. Commun. Comput. Sci. **E97.A**(6), 1213–1219 (2014). https://doi.org/10.1587/transfun.E97.A.1213
9. Efraimidis, P.S., Spirakis, P.G.: Weighted random sampling with a reservoir. Inf. Process. Lett. **97**(5), 181–185 (2006). https://doi.org/10.1016/j.ipl.2005.11.003
10. Gavish, B., Graves, S.C.: The travelling salesman problem and related problems. Working Paper OR-078-78, Operations Research Center, Massachusetts Institute of Technology, Cambridge, MA (1978)
11. Hojny, C., et al.: The SCIP optimization suite 10.0. Technical report, Optimization Online (2025). https://optimization-online.org/2025/11/the-scip-optimization-suite-10-0/
12. Hurlimann, T.: Puzzles and Games, A Mathematical Modeling Approach. Lulu Press, Inc. (2025)
13. Karp, R.M.: Reducibility among combinatorial problems. In: Miller, R.E., Thatcher, J.W., Bohlinger, J.D. (eds.) Complexity of Computer Computations. The IBM Research Symposia Series. pp. 85–103. Springer US, Boston, MA (1972). https://doi.org/10.1007/978-1-4684-2001-2_9
14. Keçeci, B.: A mixed integer programming formulation for Smashed Sums puzzle: generating and solving problem instances. Entertain. Comput. **36**, 100386 (2021). https://doi.org/10.1016/j.entcom.2020.100386
15. Kino, F., Uno, Y.: An integer programming approach to solving Tantrix on fixed boards. Algorithms **5**(1), 158–175 (2012). https://doi.org/10.3390/a5010158
16. Knuth, D.E.: The Art of Computer Programming, Volume 2 (3rd ed.): Seminumerical Algorithms. Addison-Wesley Longman Publishing Co., Inc., USA (1997). https://doi.org/10.5555/270146
17. Knuth, D.E.: Dancing links (2000). https://arxiv.org/abs/cs/0011047
18. Koch, T., Berthold, T., Pedersen, J., Vanaret, C.: Progress in mathematical programming solvers from 2001 to 2020. EURO J. Comput. Optim. **10**, 100031 (2022). https://doi.org/10.1016/j.ejco.2022.100031
19. Lougee-Heimer, R.: The common optimization interface for operations research: Promoting open-source software in the operations research community. IBM J. Res. Dev. **47**(1), 57–66 (2003). https://doi.org/10.1147/rd.471.0057
20. Meuffels, W.J.M., den Hertog, D.: Puzzle—solving the battleship puzzle as an integer programming problem. INFORMS Trans. Educ. **10**(3), 156–162 (2010). https://doi.org/10.1287/ited.1100.0047
21. Mingote, L., Azevedo, F.: Colored nonograms: an integer linear programming approach. In: Lopes, L.S., Lau, N., Mariano, P., Rocha, L.M. (eds.) Progress in Artificial Intelligence, pp. 213–224. Springer, Heidelberg (2009). https://doi.org/10.1007/978-3-642-04686-5_18
22. Nikolaev, A.V.: Evolomino is NP-complete. Siberian Electron. Math. Rep. **22**(2), C54–C67 (2025). https://doi.org/10.33048/semi.2025.22.C05
23. Nikolaev, A.V., Myasnikov, Y.A.: Evolomino benchmark dataset. https://github.com/hsaruJ/evolomino-samples (2026)
24. Nikoli Co., Ltd.: Evolomino (2025). https://www.nikoli.co.jp/en/puzzles/evolomino/

25. Perron, L., Didier, F.: CP-SAT. https://developers.google.com/optimization/cp/cp_solver/
26. Perron, L., Furnon, V.: OR-Tools. https://developers.google.com/optimization/
27. Ravi, S., Knight, K.: Attacking letter substitution ciphers with integer programming. Cryptologia **33**(4), 321–334 (2009). https://doi.org/10.1080/01611190903030920
28. Simonis, H.: Kakuro as a constraint problem. In: Proceedings of the 7th International Workshop on Constraint Modelling and Reformulation (2008)
29. Sungur, B., Madenoğlu, F.S.: An integer programming formulation for generating and solving Survo puzzle. Erciyes Üniversitesi İktisadi ve İdari Bilimler Fakültesi Dergisi (71), 41–44 (2025). https://doi.org/10.18070/erciyesiibd.1666419
30. Weber, T.: A SAT-based Sudoku solver. In: Logic for Programming, Artificial Intelligence, and Reasoning (LPAR), pp. 11–15 (2005)

Approximation Algorithms for Minimizing Task Execution Cost in the Cloud

Yury Semenov[1(✉)] and Oleg Sukhoroslov[2]

[1] HSE University, Moscow, Russia
ydsemenov@hse.ru
[2] Institute for Information Transmission Problems of the Russian Academy of Sciences, Moscow, Russia
sukhoroslov@iitp.ru

Abstract. In this paper, we consider the problem of executing a set of computational tasks on virtual machines (VMs) leased from a cloud provider to minimize the total VM leasing cost. This problem is a case of the unrelated machines scheduling problem with the objective function that depends on the machine usage time. We exploit the relationship between the cloud scheduling problem and the uncapacitated facility location problem to devise algorithms with approximation guarantees and lower bounds on such. Our results include an $O(\log n)$ approximation algorithm, a corresponding $\Omega(\log n)$ lower bound, $2 - 1/m + \varepsilon$ approximation algorithms for important special cases when the number of machines is constant or the machines are uniformly related, and algorithms for obtaining tradeoffs between cost and makespan. We also present the results of computational experiments with those of our algorithms that are viable to be implemented in practice. The results show that the algorithm for the constant number of machines is within 5% of optimum in most cases.

Keywords: Unrelated Machines Scheduling · Cloud Scheduling · Approximation Algorithms · Facility Location Problem

1 Introduction

The problem of scheduling tasks in the cloud involves assigning tasks to virtual machines to minimize the total VM lease cost. The cost is paid for the total usage time of a VM, rounded up to the integer multiple of the billing interval. This may be seen as a case of the machine load balancing problem, that is, the problem of minimizing a function F of machine completion times. However, most of the existing results for the load balancing problem make some assumptions about F (e.g. F is convex [1], F is the l_p norm [4,5], F is the sum of task assignment costs and machine activation cost [14], F is a symmetric monotone norm [9]) that do not hold in our case. One exception we are aware of is [17] where a $(1 + o(1))\psi \ln n$ approximation is developed for non-decreasing, piecewise linear

A. Kazakov et al. (Eds.): MOTOR 2026, LNCS 16687, pp. 134–150, 2026.
https://doi.org/10.1007/978-3-032-31390-4_9

and left-continuous F with the condition that $F(2x) \leq \psi F(x)$. Our choice of F satisfies these properties with $\psi = 2$.

Our goal is to design algorithms with provable upper bounds on the approximation ratio and to prove lower bounds on the said ratio. So far, we are not aware of any results in this direction, except for the already mentioned $2(1+o(1)) \ln n$ approximation in [17] and the 6/5 lower bound in [21] for the case of two uniformly related machines. There are, however, results for some closely related problems. Makespan minimization on unrelated machines is impossible to approximate with ratio $3/2 - \varepsilon$, and there is a 2-approximate algorithm [16]. If the number of machines is constant, there is a PTAS [12], and in addition, it is possible to find ε-approximate tradeoffs between the makespan and the sum of individual task costs [2,13]. If the machines are identical, the problem of minimizing sum or maximum of convex functions of machine completion time admits a PTAS under some relatively mild conditions [1]. If the machines are unrelated, the problem of minimizing the l_p norm admits a 2-approximate algorithm for any p, and a PTAS if the number of machines is constant [5]. Minimizing an arbitrary symmetric monotone norm of the machine load admits a polynomial time randomized algorithm with $O(\log n)$ approximation factor [9]. For the problem of minimizing the sum of task assignment costs and machine activation costs it is possible to find a $(2, (1 + o(1)) \ln n)$-approximate makespan-cost tradeoff [14].

The rest of the paper is structured as follows. Section 2 formally defines the problem. Section 3 presents the approximation algorithms and lower bounds. Section 4 provides the experimental evaluation of the approximation algorithms. Section 5 concludes the work done.

2 Problem Formulation

A set of n computational tasks is to be executed on m virtual machines leased from a cloud provider. For each task, the scheduler chooses one of the m given machines to execute the task and determines the time to start executing the task on the chosen machine. Task i is non-preemptively executed on machine j for the known processing time p_{ij}, and no two tasks can simultaneously execute on the same machine. In general, the processing times are unrelated. However, we also consider a special case where p_{ij} is obtained by dividing task size by machine speed (uniformly related machines). As for the identical machines, the problem is trivially solvable in polynomial time, and therefore we do not consider such case.

If machine j is used for duration t_j, it incurs a monetary cost of $\lceil \frac{t_j}{I} \rceil \cdot f_j$ where f_j is the price of machine j, and I is the billing interval of the cloud provider. Machine usage time is computed as the time between the start of the first job and the end of the last job executed on that machine; machines that do not execute any jobs have usage time 0. The scheduling goal is to minimize $\sum_{j=1}^{m} \lceil \frac{t_j}{I} \rceil \cdot f_j$.

Throughout the rest of the paper, we refer to this problem as *Bag-of-Tasks Scheduling in the Cloud (BoTSC)*. The mathematical model of BoTSC is presented below. Variables x_{ij} denote the assignment of task i to machine j and variables l_j denote the number of used billing intervals on machine j.

$$
\begin{aligned}
\min & \sum_{j=1}^{m} f_j l_j \\
\text{s.t.} & \sum_{j=1}^{m} x_{ij} = 1 \quad \forall i \in \{1, \ldots, n\} \\
& \sum_{i=1}^{n} x_{ij} p_{ij} \leq I \cdot l_j \quad \forall j \in \{1, \ldots, m\} \\
& x_{ij} \in \{0, 1\} \quad \forall i \in \{1, \ldots, n\} \forall j \in \{1, \ldots, m\} \\
& l_i \in \mathbb{Z}_{\geq 0} \quad \forall j \in \{1, \ldots, m\}
\end{aligned}
\tag{1}
$$

3 Approximation Algorithms

In this section, we present approximation algorithms and inapproximability results for BoTSC. It goes without saying that all inapproximability results presented here are valid only if one assumes **P≠NP**.

3.1 Algorithms

To find approximate schedules, we will use the relation between BoTSC and the uncapacitated facility problem.

Uncapacitated Facility Location (FL) [1] [15]	
Input:	A set of n customers and m facilities. Customer i has the cost $c_{ij} \in \mathbb{R}_{\geq 0}$ of being assigned to facility j. Facility j has the cost $f_j \in \mathbb{R}_{>0}$ of being used for at least one customer.
Objective:	Find a subset S of used facilities and the assignment A of customers to facilities from S to minimize the total cost $\sum_{i=1}^{n} c_{iA(i)} + \sum_{j \in S} f_j$.

Lemma 1. *For an instance of BoTSC with n tasks and m machines, consider an instance of the uncapacitated facility location problem with n customers and m facilities where:*

1. *Task i corresponds to customer i, machine j corresponds to facility j.*

[1] The facility location problem is usually defined with non-negative costs as opposed to strictly positive facility costs in our definition. However, all results for the facility location problem used here remain valid if there are no facilities with zero cost.

2. *The cost of using facility j is f_j.*
3. *The cost of assigning customer i to facility j is $f_j \cdot p_{ij}/I$.*

For any assignment $A : \{1, \ldots, n\} \to \{1, \ldots, m\}$ the following holds:

$$cost_{sched}(A) \le cost_{FL}(A) \le 2cost_{sched}(A)$$

.

Proof. For assignment A and $j \in \{1, \ldots, m\}$, define A_j to be the set of all tasks (customers) assigned to machine (facility) j. Given assignment A of tasks to machines (or customers to facilities), the cost of the assignment equals $\sum_{j=1}^{m} \lceil \frac{\sum_{i \in A_j} p_{ij}}{I} \rceil \cdot f_j$ for scheduling, and $\sum_{j=1}^{m} (\sum_{i \in A_j} f_j \cdot p_{ij}/I + f_j \cdot [A_j \neq \varnothing]) = \sum_{j=1}^{m} (\frac{\sum_{i \in A_j} p_{ij}}{I} \cdot f_j + f_j \cdot [A_j \neq \varnothing])$ for facility location.

If $A_j = \varnothing$, $\lceil \frac{\sum_{i \in A_j} p_{ij}}{I} \rceil \cdot f_j = \frac{\sum_{i \in A_j} p_{ij}}{I} \cdot f_j + f_j \cdot [A_j \neq \varnothing] = 0$, otherwise $\lceil \frac{\sum_{i \in A_j} p_{ij}}{I} \rceil \cdot f_j \le \frac{\sum_{i \in A_j} p_{ij}}{I} \cdot f_j + f_j = \frac{\sum_{i \in A_j} p_{ij}}{I} \cdot f_j + f_j \cdot [A_j \neq \varnothing]$. Therefore, $cost_{sched}(A) \le cost_{FL}(A)$.

On the other hand, if $A_j \neq \varnothing$, we have $\lceil \frac{\sum_{i \in A_j} p_{ij}}{I} \rceil \cdot f_j \ge f_j$, therefore $\frac{\sum_{i \in A_j} p_{ij}}{I} \cdot f_j + f_j \cdot [A_j \neq \varnothing] = \frac{\sum_{i \in A_j} p_{ij}}{I} \cdot f_j + f_j \le 2\lceil \frac{\sum_{i \in A_j} p_{ij}}{I} \rceil \cdot f_j$, implying $cost_{FL}(A) \le 2cost_{sched}(A)$. □

The construction in Lemma 1 can be done in the opposite direction:

Lemma 2. *For an instance of the uncapacitated facility location problem with n customers and m facilities, consider an instance of BoTSC with n tasks and m machines where:*

1. *Task i corresponds to customer i, machine j corresponds to facility j.*
2. *The cost of using machine j is f_j.*
3. *The processing time of task i on machine j is c_{ij}/f_j.*
4. *The billing interval is 1.*

For any assignment $A : \{1, \ldots, n\} \to \{1, \ldots, m\}$ the following holds:

$$cost_{sched}(A) \le cost_{FL}(A) \le 2cost_{sched}(A)$$

.

Proof. The construction in Lemma 1 reverses the current construction, producing the same instance of the facility location problem. The result therefore follows from Lemma 1. □

The relation between BoTSC and the uncapacitated facility location problem described in the lemmas above allows us to relate the approximation algorithms for the two problems:

Lemma 3. *If there is an $r(n)$-approximate algorithm for the facility location problem, there is a $2r(n)$-approximate algorithm for BoTSC. If there is an $r(n)$-approximate algorithm for BoTSC, there is a $2r(n)$-approximate algorithm for the facility location problem.*

Proof. Whenever we do the reduction in Lemma 1 or the opposite reduction in Lemma 2, we always have $OPT_{sched} \leq OPT_{FL} \leq 2OPT_{sched}$. We have $OPT_{sched} \leq OPT_{FL}$ because the optimal assignment for the facility location corresponds to a feasible schedule A such that $cost_{sched}(A) \leq OPT_{FL}$, and $OPT_{sched} \leq cost_{sched}(A)$. Now consider the optimal schedule. It corresponds to some feasible facility assignment B such that $cost_{FL}(B) \leq 2OPT_{sched}$. Since we have $OPT_{FL} \leq cost_{FL}(B)$, we have $OPT_{FL} \leq 2OPT_{sched}$.

For an instance of BoTSC, apply the reduction in Lemma 1. If for some assignment A we have $cost_{FL}(A) \leq r(n)OPT_{FL}$, then we have $cost_{sched}(A) \leq cost_{FL}(A) \leq r(n)OPT_{FL} \leq 2r(n)OPT_{sched}$.

For an instance of the facility location problem, apply the reduction in Lemma 2. If for some assignment A we have $cost_{sched}(A) \leq r(n)OPT_{sched}$, then we have $cost_{FL}(A) \leq 2cost_{sched}(A) \leq 2r(n)OPT_{sched} \leq 2r(n)OPT_{FL}$. □

This lemma allows us to produce approximation algorithms for BoTSC from the approximation algorithms for the facility location problem.

Theorem 1. *For BoTSC, the following results hold:*

1. *There is an $O(\log n)$-approximate algorithm.*
2. *If the number of machines is a fixed constant, there is a 2-approximate algorithm.*
3. *If the machines are uniformly related, the algorithm that places all tasks on one machine and chooses the machine with the least total cost is 2-approximate.*

Proof. 1. There is an $O(\log n)$-approximate algorithm for the facility location problem [11], which by Lemma 3 gives an $O(\log n)$-approximate algorithm for BoTSC. The algorithm works by reformulating the facility location as the minimum weighted set cover problem as follows. The elements to be covered are the customers. The family of sets is the family of all subsets of customers, the cost of set S is $\min_{j=1}^{m}(f_j + \sum_{i \in S} c_{ij})$. It is known that a simple greedy algorithm which iteratively picks the set with the largest ratio of the number of newly covered elements to the cost achieves an $O(\log n)$ approximation ratio for the set cover problem. The only caveat is that the instance of the set cover problem has exponential size. However, it is possible to implement the greedy algorithm in such a way that it runs in $O(n^2m)$ time by noticing that the search can

be limited to the sets consisting of k customers with the smallest assignment cost for some facility.
This algorithm offers essentially the same $2(1+o(1))\ln n$ approximation ratio as the algorithm in [17], but with the additional advantage of being much simpler.
2. The uncapacitated facility location problem can be solved exactly if the number of facilities is a fixed constant. This is done by bruteforcing through each possible set S of active facilities and assigning customer i to $\arg\min_{j\in S} c_{ij}$. By Lemma 3, this gives a 2-approximate algorithm for the scheduling problem.
3. Consider what happens in the bruteforce above if the machines are uniformly related. For each set S of active facilities $\arg\min_{j\in S} c_{ij}$ always corresponds to the machine from S with the least price to speed ratio, regardless of the task. Therefore, the algorithm will place all tasks on one machine.

□

The 2-approximate algorithms can be improved to $2-1/m+\varepsilon$-approximate algorithms at the expense of the running time.

Theorem 2. *For any $\varepsilon > 0$ there are $2-1/m+\varepsilon$ approximate algorithms for BoTSC with a fixed constant number of machines and BoTSC with uniformly related machines. The algorithms have running time $O(2^m \cdot ((n^2\log n)/\varepsilon + n^2/\varepsilon^2))$ and $O(m^2 \cdot ((n^2\log n)/\varepsilon + n^2/\varepsilon^2))$ respectively.*

Proof. To improve the approximation ratio, we need to look at the 2-approximate algorithm for a fixed number of machines from a different angle. For a subset of machines M denote the optimal cost when using only machines from M as OPT^M. It's not hard to prove that when we assign task i to $\arg\min_{j\in M} f_j p_{ij}/I$, we actually solve the LP relaxation of the ILP BoTSC model 1 limited to machines from M. By summing up constraints $\sum_{i=1}^{n} x_{ij}p_{ij} \le I \cdot l_j$ with appropriate coefficients we get $\sum_{j\in M} f_j l_j \ge \frac{1}{I}\sum_{j\in M} f_j \sum_{i=1}^{n} x_{ij}p_{ij}$, and for any choice of x_{ij} it is always possible to choose l_j so that $\sum_{j\in M} f_j l_j = \frac{1}{I}\sum_{j\in M} f_j \sum_{i=1}^{n} x_{ij}p_{ij}$. Finally, $\frac{1}{I}\sum_{j\in M} f_j \sum_{i=1}^{n} x_{ij}p_{ij}$ attains minimum when i is assigned to $\arg\min_{j\in M} f_j p_{ij}/I$.

When we round variables l_j up, we obtain a feasible solution with cost at most $OPT^M_{LP} + \sum_{j\in M} f_j \le OPT^M + \sum_{j\in M} f_j$. Clearly, there is at least one set M' such that $OPT^{M'} = OPT$ and every machine from M' is used by at least one task. For such M' we have $\sum_{j\in M'} f_j \le OPT^{M'} = OPT$, and the cost is at most $2OPT$.

If $|M'| = 1$, then after rounding we found $OPT^{M'}$ and OPT exactly. Now suppose that $|M'| > 1$. Consider the most expensive machine j_{max} from M'.

We have $\sum_{j \in M' \setminus \{j_{max}\}} f_j \le \frac{|M'|-1}{|M'|} \sum_{j \in M'} f_j \le \frac{m-1}{m} \sum_{j \in M'} f_j \le \frac{m-1}{m} OPT$. If the optimal solution uses j_{max} for at least $\lceil 1/\varepsilon \rceil$ billing intervals, we have $\sum_{j \in M'} f_j = f_{j_{max}} + \sum_{j \in M' \setminus \{j_{max}\}} f_j \le \varepsilon OPT + \frac{m-1}{m} OPT$, and the approximation ratio of the rounded solution is $2 - 1/m + \varepsilon$.

Now consider the case when j_{max} is used for less than $\lceil 1/\varepsilon \rceil$ billing intervals. For $1 \le k < \lceil 1/\varepsilon \rceil$, define $OPT^{M',k}$ to be the optimal cost when only machines from M' are used and j_{max} is used for at most k billing intervals. For fixed k we build an instance of the 0–1 min-knapsack problem as follows. Task i corresponds to an item i with weight $w_i = p_{ij_{max}}$ and cost $c_i = \min_{j \in M' \setminus \{j_{max}\}} f_j p_{ij}/I$. The capacity of the knapsack is kI. The goal is to pack the items in the knapsack to minimize the total cost of the unpacked items. There is an FPTAS for the 0–1 min-knapsack problem that computes a $(1+\varepsilon)$-approximate solution in $O(n^2 \log n + n^2/\varepsilon)$ time and $O(n^2/\varepsilon)$ space [10]. For each k we run the FPTAS to determine which tasks go to machine j_{max} and place all other tasks on $M' \setminus \{j_{max}\}$ as before. For some k' we have $OPT^{M',k'} = OPT^{M'}$ and j_{max} is used for exactly k' intervals. For such k' the FPTAS-based schedule has cost at most $k' f_{j_{max}} + (1+\varepsilon)(OPT^{M',k'} - k' f_{j_{max}}) + \frac{m-1}{m} OPT^{M',k'} \le (2-1/m+\varepsilon) OPT^{M',k'} = (2 - 1/m + \varepsilon) OPT^{M'} = (2 - 1/m + \varepsilon) OPT$.

To sum up, the algorithm works as follows. It iterates over all non-empty sets of machines and for each set computes a schedule by placing task i on $\arg\min_{j \in M} f_j p_{ij}/I$. If the set has more than one machine, it computes $\lceil 1/\varepsilon \rceil$ additional schedules by running the min-knapsack FPTAS. The algorithm returns the cheapest schedule among the $O(2^m/\varepsilon)$ computed schedules. The running time is $O(2^m \cdot ((n^2 \log n)/\varepsilon + n^2/\varepsilon^2))$. If the machines are uniformly related, it is sufficient to check the sets with ≤ 2 machines, reducing the running time to $O(m^2 \cdot ((n^2 \log n)/\varepsilon + n^2/\varepsilon^2))$. □

Practitioners are often interested in schedules that achieve some tradeoff between the cost and the makespan. Similar ideas as above may be used to produce such tradeoffs.

Corollary 1. *Given makespan T, there is a polynomial-time algorithm that computes a schedule with makespan at most $2T$ and cost at most $2C_{opt}(T)(1 + o(1)) \ln n$, given that there exists a schedule with makespan at most T, and $C_{opt}(T)$ is the minimum cost of a schedule with makespan at most T.*

Proof. We use the same reduction to the facility location problem as before. Consider a hybrid of facility location and unrelated machine scheduling problems where in addition to assignment costs each customer has processing time p_{ij} on each of the facilities. This problem has been considered before under the name of machine activation with assignment cost. In [17], a polynomial-time algorithm is presented that finds for makespan T a schedule with makespan at most $2T$ and cost at most $C_{opt}(T)(1 + o(1)) \ln n$, given that there exists a schedule with makespan at most T. Modifying the reduction in Lemma 3 by adding processing times, we obtain a schedule with makespan at most $2T$ and cost at most $2C_{opt}(T)(1 + o(1)) \ln n$. □

It is possible to obtain a better tradeoff if the number of machines is constant.

Corollary 2. *Given makespan T, there is an $O(2^m n(n/\varepsilon)^m)$ algorithm that computes for any $\varepsilon > 0$ a schedule with makespan at most $(1 + \varepsilon)T$ and cost at most $2C_{opt}(T)$, given that there exists a schedule with makespan at most T, and $C_{opt}(T)$ is the minimum cost of a schedule with makespan at most T.*

Proof. We use the same reduction to the facility location problem as before. Bruteforce over all $2^m - 1$ non-empty sets of facilities and for each set S consider the problem of minimizing the total cost of assigning the customers to the facilities from S such that for each facility the sum of processing times of the customers is at most T. This problem has been addressed in [2] where an algorithm is proposed that finds an assignment with makespan at most $(1+\varepsilon)T$ and cost at most $C_{opt}(T)$ in time $O(n(n/\varepsilon)^m)$. We compute one such assignment for each set S of facilities, increase the cost by $\sum_{i \in S} f_i$, and find the minimum cost assignment among them. The corresponding schedule has makespan at most $(1 + \varepsilon)T$ and cost at most $2C_{opt}(T)$. □

3.2 Lower Bounds

The results from the previous section allow us to prove an $\Omega(\log n)$ lower bound in addition to the $O(\log n)$ upper bound.

Corollary 3. *Any polynomial-time approximation algorithm for BoTSC has $\Omega(\log n)$ approximation ratio.*

Proof. Follows from Lemma 3 and the fact that the facility location problem has the set cover problem as a special case where all assignment costs are either 0 or ∞. For the set cover, the $\Omega(\log n)$ lower bound is known [20]. □

Unlike the general case, it is impossible to use the relationship between FL and BoTSC to produce lower bounds in the case when the number of machines is fixed, because the corresponding facility location problem can be solved to optimality. Nevertheless, there are some non-trivial lower bounds, and no PTAS is possible. For the uniformly related machines, the bound of 6/5 is proved in [21]. For the unrelated machines, there is a simple 3/2 lower bound:

Proposition 1. *BoTSC on two unrelated machines is inapproximable with any approximation ratio less than 3/2.*

Proof. Consider an instance $a_1, \ldots, a_n$ of the partition problem with $\sum_{i=1}^{n} a_i = 2T$. This instance is transformed into an instance of BoTSC as follows. There are two machines, each has price 1. Each integer a_i is transformed into a task with duration a_i on both machines. In addition, there are two extra tasks. The first of them has duration 1 on the first machine and $3T + 3$ on the second machine, and the other is the opposite: 1 on the second machine and $3T + 3$ on the first machine. The billing interval equals $T + 1$.

If extra task 1 is placed on machine 2 or extra task 2 is placed on machine 1, the corresponding machine is used for at least 3 intervals and the cost is at least 3. Otherwise, each machine is used for at least 1 billing interval, so the cost is at least 2. If there is a partition of $a_1, \ldots, a_n$ into two parts with sum T, it's possible to execute each part on one of the machines so that each machine is used exactly for 1 interval, and the cost is 2. The opposite is also true: if the cost is 2, it means that each machine has the corresponding extra task and a subset of $a_1, \ldots, a_n$ with sum T.

Therefore, an algorithm with approximation ratio $\rho < 3/2$ can be used to distinguish between the feasible and infeasible instances of the **NP**-hard partition problem, which contradicts the **P**≠**NP** assumption. □

For two machines this bound is almost tight since Theorem 2 gives a $3/2+\varepsilon$-approximate algorithm.

3.3 A Remark on Dependency-Aware Scheduling

An important practical extension of task scheduling is dependency-aware scheduling, also known in the scheduling community as scheduling with communication delays. In this setting, the tasks form a DAG, and the presence of an edge (v, u, w) implies that task u must start after v finishes, and if the tasks are on different machines, at least w seconds must pass to transfer the data between the machines. Recently there has been a surge in the research on approximation algorithms for problems with communication delays (see e.g. [7,8,18,19]). However, the existing algorithms are developed for classic objective functions, such as makespan and weighted sum of completion times, and often make rather limiting assumptions, such as uniform communication delay. When it comes to optimizing machine usage cost, there is an additional difficulty: the strategy for sending data between the machines becomes important because a machine can't be switched off during the data transfers even if no more tasks are scheduled on that machine. In the existing research it is usually assumed that some simple strategy is used, e.g. the data is transferred immediately after the parent task is completed or the data is transferred only when the child task is about to start.

All lower bounds are naturally extensible to the case of dependencies, since a graph without edges is still a graph. The approximate algorithms, however, are mostly inapplicable in the case of data dependencies, with a notable exception of the uniformly related machines:

Observation 1 . *In the case of uniformly related machines the algorithm that places all tasks on one machine, and chooses the machine with the least total cost, is 2-approximate, even in presence of data dependencies between the tasks.*

This is due to the fact that taking any single-machine schedule for BoTSC and reordering the tasks in the topological order of the DAG yields a feasible schedule with the same cost for any DAG. The approximation ratio is still 2 because $cost_{sched} \leq 2OPT_{BoTSC} \leq 2OPT_{DAG}$ (the latter inequality is due to the fact that the optimal dependency-aware schedule remains feasible if we remove the dependencies).

This gap in our results reminds of a similar situation in makespan scheduling, where no nontrivial approximation algorithm is known in the case of unrelated machines and precedence constraints, except for some special cases, such as trees [3] and chains [22]. It has been conjectured recently [7] that a certain variant of the unrelated machines scheduling, called "unique machines precedence constraint scheduling", is **NP**-hard to approximate within a factor of n^{ε} for some $\varepsilon < 1$. This conjecture in fact implies a similar lower bound for our problem.

Proposition 2. *Consider the unique machines precedence constraint scheduling problem (UMPS) where a set of n jobs with precedence relations is to be non-preemptively scheduled on m machines to minimize the makespan, and job j can only be scheduled on machine $M(j)$. If BoTSC with precedence constraints can be approximated to the factor $r(n)$, the UMPS problem can be approximated to the factor $r(n) + \delta$ for any $\delta > 0$.*

Proof. To model the constraint forbidding certain tasks from executing on certain machines, one can set $p_{ij} = \infty$ for forbidden pairs (i, j). To convert makespan into cost, add two extra tasks s and t with processing time λ (will be defined later) as well as a new machine $m + 1$. Tasks s and t are only allowed on machine $m + 1$, and no other task is allowed on that machine. Add precedence relations (s, v) and (v, t) for every task v in the initial instance. The billing interval is λ. The cost of machine $m + 1$ is λ, and the cost of machines $1, \ldots, m$ is $\lambda^2/(m \cdot \max(1, U))$ where U is some upper bound on the optimal makespan (obtained e.g. by some heuristic).

Now consider a $r(n)$-approximate algorithm for cost minimization. Suppose it produced a schedule S. If we consider only the initial tasks, they are placed on the first m machines, and therefore the subset $\widehat{S}$ of S corresponding to those tasks is a feasible schedule for the UMPS. The makespan of this subset is not less than the cost of S, because the edges from s and to t ensure that the usage time of the last machine covers the makespan, and the cost of the last machine is at least its usage time. So, we have $makespan(\widehat{S}) \leq cost(S) \leq r(n)OPT_{cost}$.

It remains to prove that $r(n)OPT_{cost} \leq (r(n) + \delta)OPT_{UMPS}$. Consider the optimal schedule for the UMPS and extend it to a feasible schedule S' as follows. Shift all tasks by λ to account for dependency on s, start task s at time 0, and task t at time $OPT_{UMPS} + \lambda$. We have $OPT_{cost} \leq cost(S')$. By the construction of S', the cost of machine $m + 1$ is at most $OPT_{UMPS} + 3\lambda$. As for the first m machines, their total cost is at most $\lambda^2(OPT_{UMPS}/\lambda + 1)/\max(1, U) \leq \lambda + \lambda^2$. Therefore, $r(n)OPT_{cost} \leq r(n)cost(S') \leq r(n)OPT_{UMPS} + r(n)(4\lambda + \lambda^2)$. It remains to ensure that $r(n)(4\lambda + \lambda^2) \leq \delta OPT_{UMPS}$. This can be done by setting $\lambda = \min(1, \frac{\delta L}{5r(n)})$ where L is some lower bound on OPT_{UMPS} (obtained by e.g. solving the LP relaxation). □

The proof can be extended without much difficulty to the case of communication delays for all common data transfer strategies. We, however, decided to limit it to precedence constraints to omit ugly details.

4 Experimental Evaluation

The results proved in the previous section provide worst-case approximation guarantees. However, it's also desirable to estimate the approximation ratio encountered in practice, since it is often noticeably less than in the worst case. With this in mind, we have performed computational experiments with three of our algorithms: $O(\log n)$ general-case approximation, 2-approximation for the fixed number of machines, and single-machine 2-approximation for the uniformly related machines. The other algorithms are excluded because of their runtime and implementation complexity. The single-machine algorithm is also tested on the unrelated machines even though there are no worst-case guarantees in that case.

4.1 Input Data

Table 1. Virtual Machines

VM type	performance s_j	CPU cores C_j	price f_j
m5.large	3100	2	0.096
m5.xlarge	3100	4	0.192
m5.2xlarge	3100	8	0.384
m5.4xlarge	3100	16	0.768
c5n.large	3400	2	0.108
c5n.xlarge	3400	4	0.216
c5n.2xlarge	3400	8	0.432
c5n.4xlarge	3400	16	0.864

We test the algorithms on a set of 8 VMs with parameters taken from AWS EC2. The parameters are provided in Table 1. For VM j we have, in addition to price f_j, performance s_j and the number of CPU cores C_j. These parameters are used to estimate task execution time. The billing interval is 1 hour.

The tasks are generated randomly based on the model proposed in [6]. For each task we generate its runtime on the slowest machine (m5.large in our case) as a random number in $[0, T]$ where T is either 10 min, 1 hour, or 1 day. The runtime on any other machine T_j is generated based on the type of the task. There are 4 types of tasks:

1. **I/O bound.** These tasks have the same execution time regardless of the machine: $T_j = T_{slowest}$.
2. **CPU bound.** These tasks have execution time scaled proportionally to the CPU speed and the number of cores: $T_j = T_{slowest} \cdot \frac{C_{slowest} s_{slowest}}{C_j s_j}$.
3. **Memory bound.** For these tasks we set $T_j = T_{slowest} \cdot (\frac{C_{slowest} s_{slowest}}{C_j s_j})^2$.
4. **Mixed.** For these tasks T_j is randomly generated in $[0, T]$, regardless of $T_{slowest}$.

We consider datasets with $n \in \{100, 250, 500\}$ tasks, and for each n we consider three types of datasets:

1. **Balanced**. 10% I/O bound, 30% CPU bound, 30% memory bound, 30% mixed.
2. **Mixed**. All tasks are mixed.
3. **Uniform**. All tasks are CPU bound. This dataset type corresponds to the uniform machines case.

4.2 Results

The results for 10 min, 1 hour, and 1 day $T_{slowest}$ limit are presented in Tables 2, 3, and 4 respectively. For each dataset type and number of tasks we generated 100 instances, and the tables contain the minimum, mean, and maximum approximation ratio among these instances, rounded to 4 digits after the decimal point. The approximation ratio is computed w.r.t the solution obtained by solving ILP model 1 with a MILP solver.

On average, the empirical approximation ratio is much less than the theoretical bounds. The main trend here is that the ratio of the general and the fixed machines algorithms decreases as the instance size and the task duration grow; the same holds for the single-machine algorithm if one limits the analysis to the uniform data. The task duration effect is particularly noticeable. For the fixed machines and the single-machine algorithm it follows directly from the LP rounding viewpoint found in the proof of Theorem 2. Indeed, we have a set of machines M' such that $OPT^{M'} = OPT$, all VMs in M' are used, and the solution for this set has cost at most $OPT_{LP}^{M'} + \sum_{j \in M'} f_j \leq OPT + \sum_{j \in M'} f_j$. Approximation ratio 2 is obtained by noticing that each machine in M' is used for at least one interval, meaning that $\sum_{j \in M'} f_j \leq OPT^{M'} = OPT$. However, the longer are the tasks, the more billing intervals are used, so $\sum_{j \in M'} f_j$ decreases with respect to OPT. Preliminary experiments with 1-second billing interval also confirm this effect.

As for the disparity between the three types of datasets, it comes as no surprise that uniform datasets are the easiest with the mean ratio of all three algorithms being under 1.05 even for 100 tasks and $T_{slowest} \leq 10$ min. Balanced datasets also turned out to be quite easy: the fixed machines algorithm deviates from the optimum by $< 5\%$ on average (in fact, the ratio of this algorithm on balanced datasets is even less than on uniform). Finally, mixed datasets are the hardest. While the theoretical upper bound is not reached on mixed datasets, the average result is still noticeably worse than on the other two types.

Table 2. Empirical approximation ratio, $T_{slowest} \leq 10$ min

Dataset		general	fixed machines	single-machine
balanced 100 tasks	min	1.0503	1.0059	1.1667
	mean	1.1557	1.0343	1.2560
	max	1.2726	1.1009	1.4096
balanced 250 tasks	min	1.0350	1.0060	1.1713
	mean	1.0784	1.0203	1.2453
	max	1.1157	1.0391	1.3240
balanced 500 tasks	min	1.0299	1.0040	1.1774
	mean	1.0396	1.0120	1.2420
	max	1.0540	1.0215	1.3081
mixed 100 tasks	min	2.9535	1.0204	1.3061
	mean	5.5853	1.1715	1.5081
	max	6.6341	1.4146	1.7561
mixed 250 tasks	min	2.8241	1.0100	1.6162
	mean	3.0816	1.1403	1.7891
	max	3.5357	1.2717	2.0000
mixed 500 tasks	min	1.8586	1.0688	1.7486
	mean	1.9970	1.1541	1.8737
	max	2.1897	1.2368	1.9887
uniform 100 tasks	min	1.0000	1.0000	1.0000
	mean	1.0404	1.0387	1.0387
	max	1.1228	1.1077	1.1077
uniform 250 tasks	min	1.0000	1.0000	1.0000
	mean	1.0183	1.0183	1.0183
	max	1.0435	1.0435	1.0435
uniform 500 tasks	min	1.0000	1.0000	1.0000
	mean	1.0080	1.0080	1.0080
	max	1.0218	1.0218	1.0218

Table 3. Empirical approximation ratio, $T_{slowest} \leq 1$ hour

Dataset		general	fixed machines	single-machine
balanced 100 tasks	min	1.0008	1.0004	1.0251
	mean	1.0043	1.0017	1.0413
	max	1.0087	1.0036	1.0727
balanced 250 tasks	min	1.0010	1.0004	1.0281
	mean	1.0020	1.0009	1.0409
	max	1.0029	1.0017	1.0553
balanced 500 tasks	min	1.0005	1.0001	1.0304
	mean	1.0009	1.0005	1.0407
	max	1.0013	1.0009	1.0523
mixed 100 tasks	min	1.2239	1.0293	1.5484
	mean	1.7864	1.1158	1.8833
	max	2.0372	1.1895	2.3820
mixed 250 tasks	min	1.1718	1.0199	1.8462
	mean	1.2729	1.0832	2.0179
	max	1.3574	1.1371	2.2568
mixed 500 tasks	min	1.0610	1.0429	1.9125
	mean	1.1272	1.0787	2.0875
	max	1.1820	1.1219	2.2963
uniform 100 tasks	min	1.0000	1.0000	1.0000
	mean	1.0066	1.0066	1.0066
	max	1.0213	1.0213	1.0213
uniform 250 tasks	min	1.0000	1.0000	1.0000
	mean	1.0027	1.0027	1.0027
	max	1.0074	1.0074	1.0074
uniform 500 tasks	min	1.0000	1.0000	1.0000
	mean	1.0012	1.0012	1.0012
	max	1.0035	1.0035	1.0035

Table 4. Empirical approximation ratio, $T_{slowest} \leq 1$ day

Dataset		general	fixed machines	single-machine
balanced 100 tasks	min	1.0000	1.0000	1.0010
	mean	1.0000	1.0000	1.0017
	max	1.0000	1.0000	1.0030
balanced 250 tasks	min	1.0000	1.0000	1.0012
	mean	1.0000	1.0000	1.0017
	max	1.0000	1.0000	1.0023
balanced 500 tasks	min	1.0000	1.0000	1.0013
	mean	1.0000	1.0000	1.0017
	max	1.0000	1.0000	1.0022
mixed 100 tasks	min	1.0000	1.0000	1.6604
	mean	1.0113	1.0097	2.0965
	max	1.0307	1.0229	2.6255
mixed 250 tasks	min	1.0011	1.0011	1.9204
	mean	1.0065	1.0065	2.1266
	max	1.0145	1.0145	2.3853
mixed 500 tasks	min	1.0010	1.0010	1.9687
	mean	1.0039	1.0039	2.1416
	max	1.0078	1.0078	2.3856
uniform 100 tasks	min	1.0000	1.0000	1.0000
	mean	1.0003	1.0003	1.0003
	max	1.0008	1.0008	1.0008
uniform 250 tasks	min	1.0000	1.0000	1.0000
	mean	1.0001	1.0001	1.0001
	max	1.0003	1.0003	1.0003
uniform 500 tasks	min	1.0000	1.0000	1.0000
	mean	1.0000	1.0000	1.0000
	max	1.0001	1.0001	1.0001

5 Conclusion

In this work, we provide approximation algorithms for the bag-of-tasks scheduling problem in the cloud, its special cases, and the bi-objective problem of computing cost-makespan tradeoffs. We also identify the corresponding lower bounds, except for the bi-objective problem. Closing the gap between the algorithms and the lower bounds is a direction for future work.

We also perform an experimental evaluation of those algorithms whose runtime and implementation complexity allow for a practical use. The results show that in practice the approximation ratio of the proposed algorithms is much less than in theory, with the fixed machines algorithm being within 5% of optimum in most cases.

Acknowledgments. This work was supported by the Russian Science Foundation under grant No. 25-11-00349, https://rscf.ru/en/project/25-11-00349/.

Disclosure of Interests. The authors have no competing interests to declare that are relevant to the content of this article.

References

1. Alon, N., Azar, Y., Woeginger, G.J., Yadid, T.: Approximation schemes for scheduling on parallel machines. J. Sched. **1**(1), 55–66 (1998)
2. Angel, E., Bampis, E., Kononov, A.: A FPTAS for approximating the unrelated parallel machines scheduling problem with costs. In: auf der Heide, F.M. (ed.) Algorithms — ESA 2001, pp. 194–205. Springer, Heidelberg (2001)
3. Anil Kumar, V.S., Marathe, M.V., Parthasarathy, S., Srinivasan, A.: Scheduling on unrelated machines under tree-like precedence constraints. Algorithmica **55**(1), 205–226 (2009)
4. Awerbuch, B., Azar, Y., Grove, E., Kao, M.Y., Krishnan, P., Vitter, J.: Load balancing in the l_p norm. In: Proceedings of IEEE 36th Annual Foundations of Computer Science, pp. 383–391 (1995). https://doi.org/10.1109/SFCS.1995.492494
5. Azar, Y., Epstein, A.: Convex programming for scheduling unrelated parallel machines. In: Proceedings of the Thirty-Seventh Annual ACM Symposium on Theory of Computing, pp. 331–337. STOC '05, Association for Computing Machinery, New York (2005)
6. Convolbo, M.W., Chou, J.: Cost-aware dag scheduling algorithms for minimizing execution cost on cloud resources. J. Supercomput. **72**(3), 985–1012 (2016)
7. Davies, S., Kulkarni, J., Rothvoss, T., Sandeep, S., Tarnawski, J., Zhang, Y.: On the hardness of scheduling with non-uniform communication delays. In: Proceedings of the 2022 Annual ACM-SIAM Symposium on Discrete Algorithms (SODA), pp. 316–328 (2022)
8. Davies, S., Kulkarni, J., Rothvoss, T., Tarnawski, J., Zhang, Y.: Scheduling with communication delays via LP hierarchies and clustering II: Weighted completion times on related machines. In: Proceedings of the 2021 ACM-SIAM Symposium on Discrete Algorithms (SODA), pp. 2958–2977 (2021)
9. Deng, S., Li, J., Rabani, Y.: Generalized unrelated machine scheduling problem. In: Proceedings of the 2023 Annual ACM-SIAM Symposium on Discrete Algorithms (SODA), pp. 2898–2916 (2023)

10. Gens, G., Levner, E.: Complexity of approximation algorithms for combinatorial problems: a survey. SIGACT News **12**(3), 52–65 (1980)
11. Hochbaum, D.S.: Heuristics for the fixed cost median problem. Math. Program. **22**(1), 148–162 (1982)
12. Hochbaum, D.S., Shmoys, D.B.: Using dual approximation algorithms for scheduling problems theoretical and practical results. J. ACM **34**(1), 144–162 (1987)
13. Jansen, K., Porkolab, L.: Improved approximation schemes for scheduling unrelated parallel machines. In: Proceedings of the Thirty-First Annual ACM Symposium on Theory of Computing, pp. 408–417. STOC '99, Association for Computing Machinery, New York (1999)
14. Khuller, S., Li, J., Saha, B.: Energy efficient scheduling via partial shutdown. In: Proceedings of the 2010 Annual ACM-SIAM Symposium on Discrete Algorithms (SODA), pp. 1360–1372 (2010)
15. Korte, B., Vygen, J.: Combinatorial Optimization: Theory and Algorithms, 6th edn. Springer Publishing Company, Incorporated (2018)
16. Lenstra, J.K., Shmoys, D.B., Tardos, É.: Approximation algorithms for scheduling unrelated parallel machines. Math. Program. **46**(1), 259–271 (1990)
17. Li, J., Khuller, S.: Generalized machine activation problems. In: Proceedings of the Twenty-Second Annual ACM-SIAM Symposium on Discrete Algorithms, pp. 80–94. SODA '11. Society for Industrial and Applied Mathematics, USA (2011)
18. Liu, Q.C., Purohit, M., Svitkina, Z., Vee, E., Wang, J.R.: Scheduling with communication delay in near-linear time. In: Symposium on Theoretical Aspects of Computer Science (2021)
19. Maiti, B., Rajaraman, R., Stalfa, D., Svitkina, Z., Vijayaraghavan, A.: Scheduling precedence-constrained jobs on related machines with communication delay. In: 2020 IEEE 61st Annual Symposium on Foundations of Computer Science (FOCS), pp. 834–845 (2020)
20. Raz, R., Safra, S.: A sub-constant error-probability low-degree test, and a sub-constant error-probability PCP characterization of NP. In: Proceedings of the Twenty-Ninth Annual ACM Symposium on Theory of Computing, pp. 475–484. STOC '97. Association for Computing Machinery, New York (1997)
21. Semenov, Y., Sukhoroslov, O.: MIP models and complexity results for DAG scheduling in the cloud. In: Kochetov, Y., Khachay, M., Eremeev, A., Pardalos, P. (eds.) Mathematical Optimization Theory and Operations Research, pp. 317–331. Springer, Cham (2025)
22. Shmoys, D.B., Stein, C., Wein, J.: Improved approximation algorithms for shop scheduling problems. SIAM J. Comput. **23**(3), 617–632 (1994)

Stochastic Optimization

Stochastic Facility Location Problem with a Guaranteed Income-Based Objective Function

Sergey V. Ivanov(✉) and Valentina N. Akmaeva

Moscow Aviation Institute (National Research University), Volokolamskoe shosse, 4, Moscow 125993, Russia
sergeyivanov89@mail.ru

Abstract. The customer preference-based facility location problem is considered. It is assumed that a company produces products of several types. This problem can be considered as a bi-level problem, at the upper level of which the company selects from a finite set of facility locations, and at the lower level, customers select facilities to be served by based on their preference relations. The article uses an equivalent single-level problem in which customer preferences are set by a system of constraints. The concept of guaranteed income is introduced as the income received by the company if the income from the sale of each type of product is equal to the income from the sale of the least profitable type of product. The problem of minimizing the Value-at-Risk (VaR) of the guaranteed income taken with the opposite sign is formulated, defined as a guaranteed income that cannot be exceeded with a given probability. To solve the said problem, an algorithm is used based on the optimization of a set of realizations of random factors selected from a certain ball with an estimation of the resulting sets using the Monte Carlo method. A sample size sufficient to obtain a solution with a given accuracy is provided. The problem of minimizing the VaR of the company's total losses, which are the lower bound for the criterion function of the problem, is solved. It is shown that this problem reduces to a deterministic linear integer programming problem. The Conditional Value-at-Risk (CVaR) of the company's total losses is used as the upper bound for this criterion function. It is shown that the problem of finding the upper bound is reduced to a mixed integer linear programming problem. The results of numerical experiments are discussed.

Keywords: Facility location · Value-at-Risk · Conditional Value-at-Risk

1 Introduction

Facility location problems, which arise during company planning, involve selecting locations for establishing facilities to maximize income. The classic facility

A. Kazakov et al. (Eds.): MOTOR 2026, LNCS 16687, pp. 153–165, 2026.
https://doi.org/10.1007/978-3-032-31390-4_10

location problem is formulated in [1]. If the income received from customers is random, stochastic programming can be used to describe the mathematical model for facility location. One of the first works to take into account the stochastic nature of the facility location problem was [2]. Guaranteed income can be described using a quantile of the income level, which defines the income value guaranteed with a given probability, typically close to one. This approach to the facility location problem was proposed in [3]. Then it was subsequently generalized to competitive facility location problems [4,5], the deterministic formulations of which were previously studied in [6,7]. In [8] a competitive facility location problem with uncertain demand and two criterion functions is studied, one of which is guaranteed income and the other is the income quantile level. Guaranteed income in a two-stage facility location problem, in which a company can open additional locations based on the realization of random demand, was taken into account in [9]. A distinctive feature of [9] is the description of the methodology for selecting the confidence level in the quantile criterion. In [10], a two-stage competitive location problem is studied with an objective function in the form of company income, guaranteed with probability of one. In this article, the concept of guaranteed income with a given probability is extended to the case of multiple products. Guaranteed income is understood as the income received when the income from the sale of each product is equal to the income from the sale of the least profitable product. The problem of optimizing guaranteed income according to the quantile criterion (Value-at-Risk) is posed. To solve this problem, the algorithm proposed in the authors' work [11] is applied. The guaranteed income optimization problem is compared to the total income optimization problem. A lower bound on the guaranteed income (an upper bound on losses) is also constructed based on the CVaR criterion [12]. The CVaR, taken with the opposite sign, describes the average income in the unfavorable case, when income does not exceed the guaranteed value based on the probability criterion. The problem under consideration is posed within a facility location model, taking into account customers preferences. The resulting problem can be viewed as a bi-level problem, in which the first level determines which facilities to open, and at the second level, customers select the facilities they will serve based on their preferences. This article uses a single-level re-formulation of this problem, in which the optimality conditions for customers strategies are replaced by a system of constraints.

2 Mathematical Model and Problem Statement

Let I be a finite set of possible locations for the facilities, J is a finite set of customers, and K is a finite set of products.

It is assumed that the vector of costs for opening facilities is known and equal to $f = (f_i)$, $i \in I$.

We assume that the income D_{ij}^k, $j \in J$, received by facility $i \in I$ from customer $j \in J$ through the consumption of product $k \in K$, is a normal random variable with μ_{ij}^k and standard deviation σ_{ij}^k. The realizations of these random variables will be denoted by d_{ij}^k. The random three-dimensional matrix composed of the D_{ij}^k, will be denoted by D, and its realizations by d. In what follows, we will assume that $\sigma_{ij}^k = \sigma_j^k$, that is, the variance of income depends on the customer but is independent of the facility. This assumption states that all the facilities in the area pursue the compatible pricing policy. Thus, we do not take into account the case in which the facilities pricing policy differs so vastly that the variance does not depend on the facility. We will also assume that the D_{ij}^k values are independent. The proposed approach can be extended to the case of dependent random variables, but taking into account the dependencies will complicate its description.

Let a linear ordering $\succ_j$ be defined on a set I, describing the preferences of a customer $j \in J$. For this linear ordering, $i \succ_j k$ means that of two open facilities $i \in I$ and $k \in I$ consumer j will choose the facility i.

The optimization strategy is the vector $u = (u_i)$, $i \in I$, in which $u_i = 1$, if the facility $i \in I$ opens, and $u_i = 0$, if the facility i doesn't open. The variable v_{ij} indicates whether facility $i \in I$ is assigned to serve customer $j \in J$: $v_{ij} = 1$, if assigned, and $v_{ij} = 0$, if not assigned. We will use the notation $v = (v_{ij})$, $i \in I$, $j \in J$.

We will consider the income received by the company from selling the least profitable product:

$$\Psi(v, d) = \min_{k \in K} \sum_{i \in I} \sum_{j \in J} d_{ij}^k v_{ij}.$$

As a guaranteed income, we will consider the value $|K|\Psi(v, d)$, which is the lower bound on income, where $|K|$ is the number of elements in the set of goods K. Then the loss function can be introduced as

$$\Phi(u, v, d) = f^\top u - |K|\Psi(v, d) = \max_{k \in K} \left\{ f^\top u - |K| \sum_{i \in I} \sum_{j \in J} d_{ij}^k v_{ij} \right\}.$$

This function determines the company's losses when the income from sales of all product types is equal to the income from sales of the least profitable product type. Let's define the quantile function:

$$\Phi_\alpha(u, v) = \min\{\varphi \in \mathbb{R} \mid \mathbf{P}\{\Phi(u, v, D) \le \varphi\} \ge \alpha\}, \tag{1}$$

where $\alpha \in (0; 1)$ is a pre-selected probability level close to 1.

The article solves the problem of quantile optimization (also known as Value-at-Risk):

$$\Phi_\alpha(u, v) \to \min_{(u,v) \in W}, \tag{2}$$

where W is a set of pairs $(u, v) \in \{0; 1\}^m \times \{0; 1\}^{m \times n}$, for which the following restrictions are met:

$$\sum_{i \in I} v_{ij} = 1,\ j \in J; \tag{3}$$

$$u_i \geq v_{ij},\ i \in I,\ j \in J; \tag{4}$$

$$u_i + \sum_{l | i \succ_j l} v_{lj} \leq 1,\ i \in I,\ j \in J. \tag{5}$$

Constraint (3) requires that only one facility be assigned to serve the j customer. Constraint (4) allows only open facilities to be assigned to serve customers. Constraint (5) prohibits assigning facilities to serve the j customer that are less preferable to him than facility i, if it is open.

Remark 1. Let the utility of the i facility for the j customer be described by the value $a_{ij} \in \mathbb{R}$. We assume that $a_{ij} > a_{kj}$ if and only if $i \succ_j k$. Then v_{ij} satisfying constraint (5) are the solution to the problems

$$\sum_{i \in I} a_{ij} v_{ij} \to \max_{v_{1j}, \ldots v_{|I|j}} \tag{6}$$

subject to (3) and (4). Problems (6) are formulated for fixed $j \in J$ and $u_i \in \{0, 1\}$. These problems can be considered as lower-level problems. Therefore, constraints (3)–(5) essentially implement a bi-level decision-making structure, whereby the upper level determines which facilities to open, and at the lower level, customers select the most favorable facilities.

Guaranteed income optimization ensures not only maximum income from the sale of all products but also a uniform distribution of income across all products sold. Furthermore, the situation in which a particular product is unprofitable is excluded.

3 Algorithm for Solving the Problem

To solve the problem, we will use the algorithm proposed in [11] for the general stochastic linear programming problem. The base idea of the algorithm is sequential solution of parameterized convex (integer) problem for different values of this parameter. By using the solution of these problems, sets of random parameters values are constructed. Then, the measure of these sets is estimated by the Monte-Carlo method. If the measure is more or equal to α, then the solution to this problem provides an upper bound for the optimal objective value. A value of α is found using the dichotomy method. We will show that the application of this algorithm to the problem under consideration reduces to the sequential solution of mixed integer linear programming problems with the Monte Carlo method used to estimate the set measure. First, we represent the random variables D_{ij}^k in terms of independent standard normal random variables X_{ij}^k in the form

$$D_{ij}^k = \mu_{ij}^k + \sigma_{ij}^k X_{ij}^k.$$

Then the loss function can be written as

$$\Phi(u, v, d) = \tilde{\Phi}(u, v, x) = \max_{k \in K} \left\{ f^\top u - |K| \sum_{i \in I} \sum_{j \in J} (\mu_{ij}^k v_{ij} + \sigma_{ij}^k x_{ij}^k v_{ij}) \right\}.$$

Remark 2. If the random variables D_{ij}^k form a Gaussian random vector with dependent components, then they can be expressed in terms of independent standard normal random variables X_{ij}^k, but in this case the form of the function $\tilde{\Phi}(u, v, x)$ becomes somewhat more complex. In this case,

$$D_{ij}^k = \mu_{ij}^k + \sum_{k' \in K} \sum_{i' \in I} \sum_{j' \in J} b_{iji'j'}^{kk'} X_{i'j'}^{k'},$$

where coefficients $b_{iji'j'}^{kk'}$ are chosen with known methods in probability theory. However, in general, this form does not allow us to get constraint (9) given below.

As shown in [11], the upper and lower bounds for the optimal value of the criterial function $\Phi_\alpha(u)$ can be found by solving the problems

$$\psi(r) = \min_{(u,v) \in W} \max_{x \in B_r} \tilde{\Phi}(u, v, x) \tag{7}$$

to the number of cases D_{ij}^k with non-zero variance. It is easy to verify that the problem of finding the values $\psi(r)$ can be written in the form

$$\varphi \to \min_{\varphi \in \mathbb{R},\ (u,v) \in W}$$

under the constraints

$$f^\top u - |K| \sum_{i \in I} \sum_{j \in J} \mu_{ij}^k v_{ij} + |K| r \sqrt{\sum_{i \in I} \sum_{j \in J} (\sigma_{ij}^k)^2 v_{ij}} \le \varphi, \quad k \in K. \tag{8}$$

Constraint (8) concerns the general case when σ_{ij}^k depend on the facility i. These constraint is complex nonlinear. These constraints can be reduced to quadratic constraints by squaring the both parts. However, if we take into account the assumption $\sigma_{ij}^k = \sigma_j^k$, then by constraint (3)

$$\sum_{i \in I} \sum_{j \in J} (\sigma_{ij}^k)^2 v_{ij} = \sum_{j \in J} (\sigma_j^k)^2 \sum_{i \in I} v_{ij} = \sum_{j \in J} (\sigma_j^k)^2.$$

Therefore, nonlinear constraints (8) can be rewritten as linear constraints

$$f^\top u - |K| \sum_{i \in I} \sum_{j \in J} \mu_{ij}^k v_{ij} + |K| r \sqrt{\sum_{j \in J} (\sigma_j^k)^2} \le \varphi, \quad k \in K. \tag{9}$$

We denote the solution to problem (7) by $(u(r), v(r))$ and introduce the set

$$C_r = \left\{ x \in \mathbb{R}^{|I||J||K|} \mid \tilde{\Phi}(u(r), v(r), x) \le \psi(r) \right\}. \tag{10}$$

We denote the standard Gaussian measure of the set C_r by $h(r)$. To estimate this measure, we will use the Monte Carlo method. Therefore, we need to specify two parameters: the parameter $\varepsilon \in (0, 1-\alpha)$ of the accuracy of the measure calculation and the confidence probability $p \in [\alpha, 1)$. We also specify the parameter $\delta > 0$, describing the accuracy of the calculation of r. We denote by ρ_α the quantile of the level α of the standard normal distribution, and by $\chi^2_\alpha(m)$ and $F_{\chi^2(m)}$ the quantile of the level α and the chi-square distribution function with m degrees of freedom, respectively. We will use the notation $\lceil\cdot\rceil$ for rounding up to the nearest integer. As proved in [11], for a sample of size

$$N = \left\lceil \frac{\ln(1/(1-\sqrt[L]{p}))}{2\varepsilon^2} \right\rceil, \tag{11}$$

where

$$L = \left\lceil \log_2 \frac{|\bar{R}_\alpha - \rho_\alpha|}{\delta} \right\rceil,$$

a strategy is guaranteed to provide an upper bound for the criterion function of the problem with probability no less than p. This strategy is called a guaranteeing strategy. We present an adaptation of the algorithm proposed in [11], for finding a guaranteeing strategy for the problem under consideration.

Algorithm

1. Set
$$r_1 := \rho_\alpha,\ r_2 := \min\left\{\sqrt{\chi^2_\alpha(|I||J||K|)}, \rho_\beta\right\},$$
where
$$\beta = 1 - \frac{1-\alpha}{|K|}.$$
2. Find the lower bound for the solution $\psi(r_1)$ and the upper bound $\psi(r_2)$ of the optimal value of the criterial function, as well as the initial guaranteeing solution $u(r_2)$, by solving problem (7) for $r = r_1$ and $r = r_2$.
3. Repeat the following steps until $|r_1 - r_2| > \delta$:
   (a) Set
   $$r := \frac{r_1 + r_2}{2}.$$
   (b) Calculate $u(r)$ and $\psi(r)$ by solving the problem (7).
   (c) Simulate N independent realizations of the random vector X, where N is defined in (11).
   (d) Calculate
   $$\mu(r) \triangleq \mathbf{P}\{X \in B_r\} = F_{\chi^2(m)}(r^2).$$
   (e) Find $\hat{h}(r)$, the estimate of the measure of the set C_r, defined by formula (10):
   $$\hat{h}(r) = \mu(r) + \frac{s(r)}{N},$$
   where $s(r)$ is the number of sample elements in the set $C_r \setminus B_r$.
   (f) If $\hat{h}(r) \geq \alpha + \varepsilon$, then $r_2 := r$. Otherwise, $r_1 := r$.
4. Use $u(r_2)$ as the guaranteed solution.

The value of $\mu(r)$ can calculated by closed-form expression given in [15, p.210]. In [11], it was also proved that the lower bound for the optimal value of the criterial function is $\psi(\rho_\alpha)$.

4 Total Income Quantile Optimization Problem

In this section, we compare problem (2) solved in the article with the problem of minimizing the quantile of a company's total income. The total income received by a company can be written as

$$\Psi^g(v,d) = \sum_{k\in K}\sum_{i\in I}\sum_{j\in J} d_{ij}^k v_{ij}.$$

Then the loss function takes the form

$$\Phi^g(u,v,d) = f^\top u - \Psi(v,d).$$

We define the quantile functions:

$$\Phi_\alpha^g(u,v) = \min\{\varphi \in \mathbb{R} \mid \mathbf{P}\{\Phi^g(u,v,D) \leq \varphi\} \geq \alpha\},$$

where $\alpha \in (0;1)$ is a pre-selected probability level close to 1.

The problem of optimizing the total income quantile is

$$\Phi_\alpha^g(u,v) \to \min_{(u,v)\in W}$$

where W is the set of pairs $(u,v) \in \{0;1\}^m \times \{0;1\}^{m\times n}$, defined by constraints (3)–(5).

As above, we represent the random variables D_{ij}^k through independent standard normal random variables X_{ij}^k in the form

$$D_{ij}^k = \mu_{ij}^k + \sigma_{ij}^k X_{ij}^k.$$

Then the loss function can be written as

$$\Phi^g(u,v,d) = \tilde{\Phi}^g(u,v,x) = f^\top u - \sum_{k\in K}\sum_{i\in I}\sum_{j\in J}\left(\mu_{ij}^k v_{ij} + \sigma_{ij}^k x_{ij}^k v_{ij}\right).$$

It is known [15, Corollary 3.13], that for a linear loss function, the quantile function has the form

$$\Phi_\alpha^g(u,v) = \max_{v\in K_\alpha} \tilde{\Phi}^g(u,v,x),$$

where K_α is the kernel of the probability measure of level α, which for the standard normal distribution is a ball of radius ρ_α, where, as before, ρ_α is the quantile of level α of the standard normal distribution. Since the function $v \mapsto \Phi^g(u,v,x)$

is linear, its maximum on the ball, given the assumption $\sigma_{ij}^k = \sigma_j^k$ and constraint (3), can be found analytically:

$$\max_{v\in K_\alpha} \tilde{\Phi}^g(u,v,x) = f^\top u - \sum_{k\in K}\sum_{i\in I}\sum_{j\in J} \mu_{ij}^k v_{ij} + \rho_\alpha \sqrt{\sum_{k\in K}\sum_{j\in J}(\sigma_j^k)^2 \sum_{i\in I} v_{ij}} =$$
$$= f^\top u - \sum_{k\in K}\sum_{i\in I}\sum_{j\in J} \mu_{ij}^k v_{ij} + \rho_\alpha \sqrt{\sum_{k\in K}\sum_{j\in J}(\sigma_j^k)^2}.$$

Note that the last term obtained is independent of the optimization strategy.

Thus, the total revenue quantile optimization problem is reduced to an integer linear programming problem.

$$f^\top u - \sum_{k\in K}\sum_{i\in I}\sum_{j\in J} \mu_{ij}^k v_{ij} \to \min_{(u,v)\in W}. \tag{12}$$

If we denote its solution by (u^*, v^*) then the minimum value of the loss quantile will be

$$\Phi_\alpha^g(u^*,v^*) = f^\top u^* - \sum_{k\in K}\sum_{i\in I}\sum_{j\in J} \mu_{ij}^k v_{ij}^* + \rho_\alpha \sqrt{\sum_{k\in K}\sum_{j\in J}(\sigma_j^k)^2}.$$

Note that the solution to the optimization problem does not depend on the chosen α, and the choice of α only affects the value of the criterion function of the problem.

By the definition of the total loss function, the inequality

$$\Phi^g(u,v,d) \le \Phi(u,v,d)$$

always holds. This means that the total loss cannot be greater than the loss with guaranteed income. Therefore, the inequality

$$\Phi_\alpha^g(u,v) \le \Phi_\alpha(u,v),$$

which shows that the solution to the problem $\Phi_\alpha^g(u^*,v^*)$ provides a lower bound for the criterial function $(u,v) \mapsto \Phi_\alpha(u,v)$ in the quantile minimization problem (2). The difference between the total income and the guaranteed income can be considered a measure of the unevenness of the income distribution between the types of goods sold.

5 Constructing an Upper Bound

To construct an upper bound, we will use the CVaR (Conditional Value-at-Risk) of the loss function. It has several equivalent definitions, one of which is as follows [12]:

$$\phi_\alpha(u,v) = \min_{\varphi\in\mathbb{R}} \mathbf{E}\left[\varphi + \frac{1}{1-\alpha}\max\{0, \Phi(u,v,D) - \varphi\}\right], \tag{13}$$

where $\mathbf{E}$ is the expectation. Under certain regularity conditions, CVaR also has the meaning of average losses under the assumption that losses exceed VaR. This means that CVaR is of independent interest and not just serves as an upper bound for VaR. Moreover, the optimal value of φ in definition (13) coincides with the value of $\varphi_\alpha(u,v)$, if the minimum of φ in definition of VaR (1) is unique. Since the loss function in this work is linear and the random parameters are normally distributed, this condition will be satisfied.

It is known [12], that CVAR provides an upper bound for the quantile function:

$$\phi_\alpha(u,v) \geq \Phi_\alpha(u,v).$$

Therefore, to construct an upper bound for the quantile function, we will solve the problem

$$\phi_\alpha(u,v) \to \min_{(u,v)\in W}.$$

Using representation (13), this problem can be rewritten as

$$\mathbf{E}\left[\varphi + \frac{1}{1-\alpha}\max\{0, \Phi(u,v,D) - \varphi\}\right] \to \min_{\varphi\in\mathbb{R},\ (u,v)\in W}.$$

To solve this problem, we will use the method of sample approximations. Let $D_1, D_2, \ldots, D_N$ be a sample of random matrices D. We will replace the objective function in the resulting problem with its sample approximation, and obtain the problem

$$\varphi + \frac{1}{(1-\alpha)N}\sum_{\nu=1}^{N}\max\{0, \Phi(u,v,D_\nu) - \varphi\} \to \min_{\varphi\in\mathbb{R},\ (u,v)\in W}.$$

Let us fix the real samples $d_1, d_2, \ldots, d_N$. Taking into account the definition of the loss function, we obtain the problem

$$\varphi + \frac{1}{(1-\alpha)N}\sum_{\nu=1}^{N}\max_{k\in K}\left\{0, f^\top u - |K|\sum_{i\in I}\sum_{j\in J}(D_{ij}^k)_\nu v_{ij} - \varphi\right\} \to \min_{\varphi\in\mathbb{R},\ (u,v)\in W},$$

where $(d_{ij}^k)_\nu$ denotes the elements of the random matrix d_ν. The resulting problem, by introducing a vector of additional variables $z = (z_1, \ldots, z_N)$, is reduced to a mixed integer linear programming problem

$$\varphi + \frac{1}{(1-\alpha)N}\sum_{\nu=1}^{N} z_\nu \to \min_{\varphi\in\mathbb{R},\ (u,v)\in W, z\in\mathbb{R}^N} \tag{14}$$

under constraints

$$f^\top u - |K|\sum_{i\in I}\sum_{j\in J}(d_{ij}^k)_\nu v_{ij} - \varphi \leqslant z_\nu,$$

$$z_\nu \geqslant 0, \quad \nu \in \{1,2,\ldots,N\},\ k \in \{1,2,\ldots,K\}.$$

The convergence of the constructed approximations follows from the result proved in [14]. This paper demonstrates the almost-sure epiconvergence of expectation functions under uncomplicated assumptions, which are satisfied for the loss function under study. Epiconvergence implies the convergence of minima on a compact set, as well as the fact [13] that all partial limits of the sequence of solutions to the approximation problem are also optimal in the original problem. Although the variable φ is unbounded in the problem under consideration, adding the constraint $|\varphi| \leqslant C$ for some sufficiently large constant C does not change the solution to the problem but compactifies the set of admissible values of the variables.

6 Calculation Results

Example 1. There are 4 facilities, 3 customers, and 2 types of products. Let the vector of costs for opening the facilities have the form

$$f = (10, 20, 30, 40)^\top.$$

The goods are characterized by the quantities

$$(\mu^1_{ij}) = \begin{pmatrix} 40 & 50 & 70 \\ 30 & 40 & 50 \\ 30 & 40 & 60 \\ 40 & 60 & 70 \end{pmatrix}, \; (\mu^2_{ij}) = \begin{pmatrix} 40 & 50 & 30 \\ 30 & 80 & 50 \\ 70 & 40 & 60 \\ 30 & 40 & 70 \end{pmatrix},$$

$$\sigma^1_1 = 1, \sigma^1_2 = 2, \sigma^1_3 = 1, \sigma^2_1 = 4, \sigma^2_2 = 4, \sigma^2_3 = 1.$$

Customer preferences are defined by the relations

$$1 \succ_1 2 \succ_1 3 \succ_1 4, 4 \succ_2 3 \succ_2 2 \succ_2 1, 2 \succ_3 4 \succ_3 1 \succ_3 3.$$

The problem was solved for the $\alpha = 0{,}95$ level. The following algorithm parameters were specified:

$$\varepsilon = 10^{-3},\; \delta = 10^{-2},\; p = 0.99.$$

For these parameters values, the sample size $N = 3\,105\,297$. It took five iterations for the dichotomy method to solve the problems. One mixed integer programming problem (7) is solved per iteration. The solution time is 2121 s. The estimation of measure took the largest part (more than 99%) of the calculation time.

As a result of applying the algorithm, a guaranteed solution was found:

$$u = (0, 0, 1, 1).$$

In this case,

$$r_2 = 1.6645,\; \psi(r_2) = -241.8454,\; \rho_\alpha = 1.6449,\; \psi(\rho_\alpha) = -241.9419.$$

This means that the minimum value of the criterion function of the problem lies between -241.9419 and -241.8454. Considering that the costs of opening the facilities are 70, each facility is guaranteed to generate income approximately equal to $(241.9 + 70)/2 = 155.95$.

For comparison, we present the solution to the problem of minimizing the total income quantile:

$$u^* = (1, 0, 0, 1),$$

obtained as the solution to problem (12). The objective function quantile at the 0.95 level is -260.6953. The opening costs of these facilities are 50. Therefore, the total revenue of the two facilities is approximately $260.7 + 50 = 310.7$.

Thus, in both cases, two facilities should be opened, including a fourth one, and the next facility to be opened depends on the income being considered. In the case of guaranteed income, the optimal strategy is to additionally open a third facility, while in the case of shared income, the optimal strategy is to open the first facility. In the second case, the costs of opening the facilities were lower, and the income in both scenarios was approximately the same. However, in the case of guaranteed income, the income of one of the facilities may exceed the guaranteed income, which is equal to 155.95.

To find an upper bound for the criterion function, we also solved the CVaR loss minimization problem (14). This yielded the same strategy

$$u = (0, 0, 1, 1),$$

as when applying the developed algorithm. For the sample size of $N = 20000$, across ten different samples, the upper bound for the quantile varied from -240.0078 to -239.8119, and the value of the variable φ, which can be taken as the optimal value of the quantile function, varied from -242.0368 to -241.8757. It should be noted that the estimates of the objective functions using the sample approximation method are biased, so the true optimal value of the quantile may be higher than the obtained one.

Example 2. There are 10 facilities, 20 customers, and 2 types of products. Input values of parameters and customer preferences were generated randomly:

$$f_i \in [4, 13],\ \mu_{ij}^1 \in [25, 55],\ \mu_{ij}^2 \in [35, 65], \sigma_j^k \in [1, 5].$$

For this exmample, we reduced the accuracy parameters:

$$\varepsilon = 2 \cdot 10^{-3},\ \delta = 10^{-2},\ p = 0.98.$$

These parameters require the sample size $N = 689\,175$.

In this case,

$$r_2 = 1.6744,\ \psi(r_2) = -1\,692.7,\ \rho_\alpha = 1.6449,\ \psi(\rho_\alpha) = -1\,693.5.$$

This means that the optimal objective value of problem (2) belongs to the interval $(-1\,693.5, -1\,692.7)$. In the found solution, 3 facilities are opened.

It also took five iterations for the dichotomy method to solve the problems. The solution time is 3191 s. As in the previous example, the estimation of measure took the largest part (more than 99%) of the calculation time. This means that the solution time essentially depends on the sample size N defined by the accuracy parameters. By reducing the accuracy parameters, we can solve the problem in acceptable time, because solving problem (7) requires less than 1 s at each of 5 iterations.

Let us notice that the sample average approximation method for finding the quantile requires introducing a high number of additional binary variables. This leads us to a mixed integer problem of a very high dimension which requires a lot of machine memory. However, in the proposed algorithm, the dimension of integer programming problem (7) does not depend on the number of realizations. The estimation of measure does not require a lot of machine memory, but it can take a lot of calculation time as we can see in the examples considered. Moreover, for the sample size $N = 689\,175$ or $N = 3\,105\,297$, this is practically impossible to solve the sample approximation of the problem.

7 Conclusion

This article proposes a model for the location of facilities with a quantile criterion and guaranteed income. Guaranteed income is defined in terms of the quantile of the profit function obtained for the least profitable product. An algorithm is described that reduces the original problem to an integer linear programming problem with Monte Carlo estimation of the set measure. The guaranteed income problem is compared with a problem in which the quantile of the company's total income is optimized. An upper bound on the loss (lower bound on the guaranteed income) is also obtained in the form of a CVaR loss function. A sufficient sample size for constructing this estimate and studying its accuracy require further study. Further development of this work could also include the study of two-stage and bi-level competitive facility location problems with guaranteed income. It is of interest to study various criteria in these problems, such as the probability of achieving a given income (probability criteria) and the average income in the worst case (CVaR), as well as various risk measures.

Acknowledgments. The study was supported by the Russian Science Foundation (project No. 25-21-00487), https://rscf.ru/project/25-21-00487/.

Disclosure of Interests. The authors have no competing interests to declare that are relevant to the content of this article.

References

1. Beresnev, V.: Discrete Location Problems and Polynomials in Boolean Variables (in Russ). Publishing House of the Institute of Mathematics, Novosibirsk (2005)
2. Sheppard, E.S.: A conceptual framework for dynamic location-allocation analysis. Environ. Plan. A **6**, 547–564 (1974)

3. Chen, G., Daskin, M.S., Shen, Z.-J.M., Uryasev, S.: The α-reliable mean-excess regret model for stochastic facility location modeling. Nav. Res. Logist. **53**(7), 617–626 (2006)
4. Ivanov, S.V., Morozova, M.V.: Stochastic problem of competitive location of facilities with quantile criterion. Autom. Remote Control. **77**(3), 451–461 (2016)
5. Melnikov, A., Beresnev, V.: Upper bound for the competitive facility location problem with quantile criterion. In: Kochetov, Y., Khachay, M., Beresnev, V., Nurminski, E., Pardalos, P. (eds.) Discrete Optimization and Operations Research. DOOR 2016. LNCS, vol. 9869, pp. 373–387. Springer, Cham (2016). https://doi.org/10.1007/978-3-319-44914-2_30
6. Beresnev, V.: Upper bounds for goal functions of discrete competitive facility location problems. J. Appl. Ind. Math. **3**(4), 419–432 (2009)
7. Beresnev, V., Melnikov, A.A.: Approximation algorithms for the competitive facility location problem. J. Appl. Ind. Math. **5**(2), 180–190 (2011)
8. Beresnev, V., Melnikov, A.A.: ε-constraint method for bi-objective competitive facility location problem with uncertain demand scenario. EURO J. Comput. Optim. **8**, 33–59 (2020)
9. Ivanov, S.V., Akmaeva, V.N.: Two-stage stochastic facility location model with quantile criterion and choosing reliability level. Bull. South Ural State Univ. Ser. Math. Model. Program. Comput. Softw. (Bulletin SUSU MMCS) **14**(3), 5–17 (2021)
10. Beresnev, V., Melnikov, A.A.: Computing an upper bound in the two-stage bi-level competitive location model. J. Appl. Ind. Math. **16**(3), 377–386 (2022)
11. Ivanov, S.V., Kibzun, A.I., Akmaeva, V.N.: Parametric algorithm for finding a guaranteed solution to a quantile optimization problem. Autom. Remote Control. **84**(8), 947–957 (2023)
12. Rockafellar, R.T., Uryasev, S.: Conditional value-at-risk for general loss distributions. J. Bank. Financ. **26**, 1443–1471 (2002)
13. Rockafellar, R.T., Wets, R.J.-B.: Variational Analysis. Springer, Berlin (1998). https://doi.org/10.1007/978-3-642-02431-3_9
14. Artstein, Z., Wets, R.J.-B.: Consistency of minimizers and the SLLN for stochastic programs. J. Convex Anal. **2**, 1–17 (1995)
15. Kibzun, A.I., Kan, Y.S.: Stochastic programming problems with probabilistic criteria (in Russ). Fizmatlit, Moscow (2009)

Robust Batched Version of Thompson Sampling for the Multi-armed Bandit Problem

Alexander Kolnogorov(✉) and Konstantin Kunin

Yaroslav-the-Wise Novgorod State University, Veliky Novgorod, Russia
kolnogorov53@mail.ru

Abstract. The multi-armed bandit problem is considered in an application to batch processing of big data, if several alternative methods with different a priori unknown efficiencies are available for processing. During processing, it is necessary to determine a more efficient method and ensure its preferential use. Batch processing means that the methods are applied to the batches of processed data, and the cumulative results of processing the batch data are used for control, which have approximately normal distributions, i.e. a Gaussian multi-armed bandit is considered. Two equivalent robust algorithms, which are versions of the well-known Thompson Sampling algorithm, are proposed as a control strategy. For these algorithms, invariant descriptions are obtained on a control horizon equal to one that depend only on the batch processing parameters and ensure the order $N^{1/2}$ of maximum values of the regret, where N is the amount of processed data. Numerical experiments show the high efficiency of the proposed algorithms.

Keywords: Gaussian multi-armed bandit · Game against nature · Thompson Sampling · Robust Algorithms · Batch processing

1 Introduction

We consider the multi-armed bandit (MAB) problem [1–4], which is a classic example of solving the dilemma of "Information vs control" arising in conditions of a priori uncertainty. The name comes from a slot machine with multiple arms (hereinafter referred to as actions), the choice of each of which is accompanied by a random player's income. The distribution of one-step income depends only on the selected action, is fixed during the game but unknown to the player. The number of games against the MAB is called the control horizon and is known to the player. The player's goal is to maximize the mathematical expectation of his/her total income. This requires observing the statistics of the game to determine the most advantageous action and ensure its preferential use. The problem has numerous applications in behavior modeling, adaptive control in a random environment, medicine, Internet technologies and data processing [5–8].

A. Kazakov et al. (Eds.): MOTOR 2026, LNCS 16687, pp. 166–180, 2026.
https://doi.org/10.1007/978-3-032-31390-4_11

We consider the problem in an application to batch processing of big data using alternative methods with several a priori unknown efficiencies. It is assumed that this is a rather complex processing that can be performed with errors, for example, image classification. The probability of error varies for different methods, and different methods may be more effective for different types of data. In this case, the arms of the slot (actions) correspond to the processing methods, and the incomes correspond to the amounts of successfully processed data. Let us assume that data of the same type is offered for processing. In this case, all data is divided into batches of equal size, the same processing method is applied to all batch data, and the successfully processed batch data is summarized and used for control. If the size of the batches is large enough, then, by virtue of the central limit theorem, the incomes in them have approximately Gaussian distributions, therefore, the problem of a Gaussian MAB is considered.

Formally, a Gaussian MAB is a controlled random process ξ_{k+1}, $k = 0, 1, \ldots, K-1$, the values of which depend only on the currently selected actions $y_{k+1} = j$, ($j \in \{1, \ldots, J\}$, $J \geq 2$) and have a Gaussian probability density $f_{D_j}(x|m_j) = (2\pi D_j)^{-1/2} \exp\left(-(x - m_j)^2/2D_j)\right)$, if the action $y_{k+1} = j$ was selected. We assume the variances $D_1, \ldots, D_J$ to be a priori known. Thus, the Gaussian MAB is completely described by the parameter $\theta = (m_1, \ldots, m_J)$ which value is a priori unknown. The set of admissible parameter values is the following $\Theta = \{\theta : m_j = u + v_j;\ -\infty < u < \infty,\ |v_j| \leq C,\ j = 1, 2, \ldots, J\}$.

Control strategy σ at the point of time $(k + 1)$ determines the choice of action depending on the currently known history of the process. Instead of the entire history, we use sufficient statistics which represent the current cumulative incomes for the application of all actions. In this article, we impose additional restrictions on the class of strategies and consider the Thompson Sampling (TS) which was first proposed in [9]. Some new results are presented in [10–12].

Let us define a regret. If the parameter θ were known, it would always be necessary to apply an action corresponding to the largest of the one-step expected incomes $m_1, \ldots, m_J$, the mathematical expectation of total income in this case is $K \max(m_1, \ldots, m_J)$. In the case of using the strategy σ, the mathematical expectation of total income is less than the maximum one by the amount

$$L_K(\sigma, \theta) = K \max(m_1, \ldots, m_J) - \mathbf{E}_{\sigma,\theta}\left(\sum_{k=1}^{K} \xi_k\right), \tag{1.1}$$

which is called a regret and is caused by incomplete information. Here, $\mathbf{E}_{\sigma,\theta}$ denotes the mathematical expectation relative to the measure generated by the strategy σ and the parameter θ. By regret $L_K(\sigma, \theta)$ the minimax risk is defined as

$$R_K(\Theta)) = \inf_{\Sigma} \sup_{\Theta} L_K(\sigma, \theta). \tag{1.2}$$

If Σ is the most general class of strategies, then (1.2) provides the principle lower bound for the maximum regret values. A minimax approach to the MAB problem was proposed in [13]. In [14], an asymptotic minimax risk estimate of $R_K(\Theta)$ of the order of $K^{1/2}$ was obtained for a Bernoulli two-armed bandit. A number of algorithms that ensure the order of growth of minimax risk $K^{1/2}$ or close to it were proposed in [3,8,12,15]. An asymptotic minimax risk estimate for a Gaussian two-armed bandit was obtained in [16]:

$$\lim_{K\to\infty} (DK)^{-1/2} R_K(\Theta) = r. \tag{1.3}$$

Here $r \approx 0.637$, and $D = D_1 = D_2$ is the common variance value, which is assumed to be known a priori. It is worth noting that finding corresponding minimax strategy requires solving backwards the dynamic programming recursive Bellman equation or second order partial differential equation in the limiting case. Hence, there is a need for easy-to-implement and effective strategies.

Now let us explain why the variances can be often assumed to be a priori known. If the batch sizes are large, then unknown variances can be estimated with the required accuracy at the initial stage of control, when actions are applied in turn. Since the minimax risk changes little with a small change in variances, these estimates can then be used for control. Therefore, in the case of big data, the assumption of a priori known variances is reasonable, but it can be replaced by the requirement of possibility to estimate variances during the control process.

Note that a similar approach, consisting in estimating mathematical expectations with the required accuracy at the initial stage, is impossible. This is due to the fact that the maximum values of the regret are reached in the domain of close distributions satisfying the condition $|m_i - m_j| \le cK^{-1/2}$, $i, j = 1, \dots, J$. In the domain of close distributions, the error in determining the largest m_i, and, consequently, in determining the most advantageous action, is always at least some $q_e > 0$. Therefore, one-step regret is at least $L_1 = cK^{-1/2} q_e$, and regret over the entire control horizon is at least $K \times L_1$, i.e. it is of the order of $K^{1/2}$. Close distributions are discussed in more detail in [16].

On the other hand, if it is known that $|m_i - m_j| > \delta > 0$, $i \neq j$, then the asymptotic order of the regret is $\log(K)$. Strategies which ensure the order $\log(K)$ of the regret are presented, for example, in [10,11,17,18].

Batch processing was first proposed for the treatment of groups of patients with alternative drugs. Note that it is not possible to treat a large group of patients one at a time, as it takes a considerable amount of time to achieve treatment results. Therefore, it was suggested that different drugs should be given to relatively small test groups of patients first, and then the best one according to test results should be given to all the remaining ones. The results and a literature review can be found in [7]. In the application to batch data processing, this problem was addressed in [16,19].

The rest of the article is organized as follows. In Sect. 2, we describe Thompson Sampling algorithm. In Sects. 3 and 4 we present a robust and equivalent alternative robust version of the Thompson Sampling algorithm. Invariant descriptions of algorithms on a control horizon equal to one are also presented here. Invariant descriptions allow one to obtain upper bounds of minimax risk provided by TS-algorithm. Section 5 discusses the results of numerical experiments and presents the upper estimates of minimax risk obtained using Monte-

Carlo simulations. For example, if data is divided into 100 batches, then the normalized maximum regret, with respect to $(DK)^{1/2}$, in the domain of close distribution is approximately equal to 0.79. Although this is more than $r \approx 0.637$ in (1.3), one can see that TS-algorithm is quite efficient and much easier for implementation. Section 6 contains the conclusion.

2 Thompson Sampling

Thompson Sampling (TS) is a variant of the Bayesian control approach. Let us consider the case when only the mathematical expectations of one-step incomes are unknown. The Bayesian approach assumes the assignment of independent prior distribution densities $\lambda_{j,0}(m_j)$ of the components of the parameter θ at the start of the control. Here $m_j \in \Theta_j$, $j = 1, \dots, J$.

Let us assume that by the point of time k actions have been applied k_ℓ times $(k_1 + \dots, +k_J = k)$. Then independent random variables $\mu_{1,k}, \dots, \mu_{J,k}$ are generated, having distribution densities $\lambda_{1,k_1}(m_1), \dots, \lambda_{J,k_J}(m_J)$. To control at the next point of time $k+1$, an action is selected that corresponds to the maximum of $\mu_{1,k}, \dots, \mu_{J,k}$; if there are several such values (the probability of such an event is 0), then any of the corresponding actions can be selected.

Recalculation of densities $\lambda_{1,k_1}(m_1), \dots, \lambda_{J,k_J}(m_J)$ is implemented as follows. If at the point of time $k+1$ the action j was selected for control and income x_{k_j+1} was received in response, then the density $\lambda_{j,k_j}(m_j)$ is recalculated using the Bayes formula

$$\lambda_{j,k_j+1}(m_j) = \frac{f_{D_j}(x_{k_j+1}|m_j)\lambda_{j,k_j}(m_j)}{\int_{\Theta_j} f_{D_j}(x_{k_j+1}|m_j)\lambda_{j,k_j}(m_j)dm_j}, \tag{2.1}$$

and k_j increases by 1. Here $\lambda_{j,k_j+1}(m_j)$ is the posterior distribution density with respect to $\lambda_{j,k_j}(m_j)$. The remaining densities $\{\lambda_{i,k_i}(m_i)\}$ and the numbers of applications of actions $\{k_i\}$ do not change.

This procedure of selecting the actions and recalculating the distribution densities is cyclically repeated for $k = 0, 1, \dots, K-1$. If the prior densities $\lambda_{j,0}(m_j)$ are non-degenerate, then the sequences of posterior densities $\lambda_{j,k_j}(m_j)$ converge as $K \to \infty$ to degenerate densities concentrated at points m_j ($j = 1, \dots, J$) and, therefore, the algorithm ensures the preferential application of the best actions (corresponding to the maximum mathematical expectations of one-step incomes).

TS is most conveniently applied using conjugate distributions. A prior distribution $\lambda_{j,k_j}(m_j)$ is called conjugate to $f_{D_j}(x|m_j)$ if a posterior distribution $\lambda_{j,k_j+1}(m_j)$ belongs to the same type as a prior one but with different values of parameters. It is well known (see, for example, [20]) that the conjugate of a normal distribution is also a normal distribution.

Let us consider the TS-algorithm for a MAB with normally distributed incomes using conjugate distributions. At the beginning of the control, at $k_1 = \dots = k_J = 0$, the prior distributions are independent and are characterized by mathematical expectations and variances $(m^*_{j,0}, D^*_{j,0})$, $j = 1, \dots, J$. Further control is carried out according to the algorithm.

Thompson Sampling. Let us put $k = k_1 + k_2 + \dots + k_J$. In the loop, at the step number $k = 0, 1, 2, \dots, K-1$, we perform the following operations:

1. Sample from current distributions with parameters $(m^*_{1,k_1}, D^*_{1,k_1}), \dots, (m^*_{J,k_J}, D^*_{J,k_J})$ random variables $\mu_{1,k}, \dots, \mu_{J,k}$.
2. If $\mu_{j,k} = \max(\mu_{1,k}, \mu_{2,k}, \dots, \mu_{J,k})$, then increase k_j by 1, apply the action j, get a random income x_{j,k_j+1} and recalculate the parameters of the jth posterior distribution according to the formulas

$$\frac{1}{D^*_{j,k_j+1}} = \frac{1}{D^*_{j,k_j}} + \frac{1}{D_j}, \quad \frac{m^*_{j,k_j+1}}{D^*_{j,k_j+1}} = \frac{m^*_{j,k_j}}{D^*_{j,k_j}} + \frac{x_{j,k_j+1}}{D_j}, \tag{2.2}$$

which are equivalent to (2.1). The remaining $\{k_i\}$ and the parameters $\{(m^*_{i,k_i}, D^*_{i,k_i})\}$ of the posterior distributions do not change. If there are several maximum values $\{\mu_{j,k}\}$ (the probability of such an event is 0), then choose any of them. Increase k by 1.

Remark 1. It is easy to see that in the case under consideration, this algorithm is equivalent to the following. Let the action j be applied k_j times and the cumulative income $X_{j,k_j} = \sum_{i=1}^{k_j} x_{j,i}$ be received in response. Then the average income $\overline{X}_{j,k_j} = X_{j,k_j}/k_j$ has a normal distribution density with expectation m_j and variance D_j/k_j, and the parameters of the corresponding posterior density $\lambda_{j,k_j}(m_j)$ should be calculated using the formulas

$$\frac{1}{D^*_{j,k_j}} = \frac{1}{D^*_{0,j}} + \frac{1}{(D_j/k_j)}, \quad \frac{m^*_{j,k_j}}{D^*_{j,k_j}} = \frac{m^*_{0,j}}{D^*_{0,j}} + \frac{\overline{X}_{j,k_j}}{(D_j/k_j)}, \tag{2.3}$$

$j = 1, 2, \dots, J$, which corresponds to (2.2).

TS-algorithm was first proposed in [9] for Bernoulli two-armed bandit. It has recently been significantly improved in [10–12]. For a fixed parameter θ in [10, 11], it was found that the TS-algorithm provides a regret order of $\log(K)$ that cannot be improved [17]. In [10], the so-called dynamic bandits were considered, the distributions of one-step incomes of which can change during the control process, and it was shown that the TS-algorithm performs well in this case. In [12], various MAB were considered, for example, contextual MAB, linear models, adversarial MAB, and the regret in the worst case was estimated. The order of this estimate is close to $K^{1/2}$, like in the minimax setting

3 Robust Version of Thompson Sampling

Let us obtain a robust version of the algorithm. Consider a set of parameters of the form $\Theta = \{(m_1, m_2, \dots, m_J) : m_j = u + v_j;\ -\infty < u < \infty, |v_j| \le C,\ j = 1, 2, \dots, J\}$. We assume that the values of a controlled random process depend only on the currently selected actions and have normal distributions with parameters (m_j, D_j) if an action with the number j is selected for control

$(j = 1, 2, \dots, J)$. The variances of the distributions are assumed to be known and the mathematical expectations are unknown. In (2.3), let us set

$$\frac{1}{D^*_{j,0}} = 0, \quad \frac{m^*_{j,0}}{D^*_{j,0}} = 0, \quad j = 1, \dots, J. \tag{3.1}$$

This choice corresponds to $D^*_{j,0} = \infty$, $j = 1, \dots, J$ and characterizes the lack of information about the initial values of $m_1, \dots, m_J$. The algorithm has the form.

Robust Version of Thompson Sampling At the beginning of control, we apply all actions once in turn, receiving incomes $x_{1,1}, \dots, x_{J,1}$, forming cumulative incomes $X_{1,1} = x_{1,1}, \dots, X_{J,1} = x_{J,1}$ and set $k_1 = \dots = k_J = 1$.

Next, in the loop, at the step number $k = J, J+1, \dots, K-1$, we perform the following operations:

1. Given k_j, X_j, $j = 1, 2, \dots, J$, and taking into account (2.3), (3.1) we set

$$\frac{1}{D^*_{j,,k_j}} = \frac{k_j}{D_j}, \quad \frac{m^*_{j,k_j}}{D^*_{j,k_j}} = \frac{X_{j,k_j}}{D_j}, \quad j = 1, 2, \dots, J. \tag{3.2}$$

   Sample from the posterior distributions with parameters $(m^*_{1,k_1}, \kappa_1 D^*_{1,k_1}), \dots,$ $(m^*_{J,k_J}, \kappa_J D^*_{J,k_J})$, random variables $\mu_{1,k}, \dots, \mu_{J,k}$, respectively $(k = k_1 + \dots + k_J)$.
2. If $\mu_{j,k} = \max(\mu_{1,k}, \dots, \mu_{J,k})$, then we increase k_j by 1, apply the action j, get a random income x_{j,k_j+1} and recalculate the cumulative income $X_{j,k_j+1} = X_{j,k_j} + x_{j,k_j+1}$. The remaining $\{k_i\}$ and the parameters $\{(m^*_{i,k_i}, D^*_{i,k_i})\}$ of the posterior distributions do not change. If there are several such $\{\mu_{j,k}\}$ (the probability of such an event is 0), then we choose any of them. Increase k by 1.

Remark 2. In the standard TS-algorithm, $\kappa_1 = \dots = \kappa_J = 1$. However, using these coefficients as parameters, it is possible to reduce the maximum regret values.

Let us obtain an invariant description of the algorithm on a control horizon equal to one, depending only on the batch processing parameters. The mathematical expectation and variance of income received when processing a batch in the case of applying the jth action are represented as $m_j = u + \alpha_j (D/K)^{1/2}$, $D_j = \beta_j D$ and, hence, $v_j = \alpha_j (D/K)^{1/2}$. Suppose there is $N = MK$ of data for processing, divided into K batches of M units of data per batch. Then the income received for processing a unit of data when applying the jth action has normal distribution with mathematical expectation $\tilde{m}_j = m_j/M = u/M + \alpha_j (D/K)^{1/2}/M = u/M + \alpha_j (\tilde{D}/N)^{1/2}$ and the variance $\tilde{D}_j = \beta_j \tilde{D} = D_j/M$, $\tilde{D} = D/M$. In turn, $m_j = \tilde{m}_j M$, $D_j = \tilde{D}_j M$.

Let us denote $\varepsilon = 1/K$, $t_j = k_j/K$, $x_{j,i} = m_j + D_j^{1/2} \zeta_{j,i}$, where $\{\zeta_{j,i}\}$ are i.i.d. normally distributed random variables with a standard distribution (i.e.

with zero mathematical expectation and unit variance). Taking into account (3.2), we get

$$\frac{1}{D^*_{j,k_j}} = \frac{t_j K}{D_j}, \quad \frac{m^*_{j,k_j}}{D^*_{j,k_j}} = \frac{k_j u + \alpha_j t_j (DK)^{1/2} + D_j^{1/2} \sum_{i=1}^{k_j} \zeta_{j,i}}{D_j}$$
$$= \left(\frac{K}{D_j}\right) u t_j + \left(\frac{K}{D_j}\right)^{1/2} (\alpha_j \beta_j^{-1/2} t_j + t_j^{1/2} \eta_{j,k_j}),$$

where $\eta_{j,k_j} = k_j^{-1/2} \sum_{i=1}^{k_j} \zeta_{j,i}$. Therefore,

$$\begin{aligned} D^*_{j,k_j} &= (M/K)\tilde{D}\beta_j t_j^{-1}, \\ m^*_{j,k_j} &= u + (M/K)^{1/2}\tilde{D}^{1/2}\left(\alpha_j + \beta_j^{1/2} t_j^{-1/2} \eta_{j,k_j}\right), \end{aligned} \tag{3.3}$$

$j = 1, \ldots, J$. Denote by $P_{j,k} = \Pr(\mu_{j,k} = \max(\mu_{1,k}, \ldots, \mu_{1,J}))$ the probability of choosing the jth action on $(k+1)$th control step. Then for $\tilde{\mu}_{j,k} = (\mu_{j,k} - u)(K/M)^{1/2}$, obtained as a result of linear transformation of the initial random variables $\{\mu_{j,k}\}$, the equalities hold

$$\Pr(\tilde{\mu}_{j,k} = \max(\tilde{\mu}_{1,k}, \ldots, \tilde{\mu}_{1,J})) = P_{j,k}, \tag{3.4}$$

$j = 1, \ldots, J$; where $\{\tilde{\mu}_{j,k}\}$ have mathematical expectations and variances

$$\begin{aligned} \tilde{D}^*_{j,k_j} &= \kappa_j \tilde{D}\beta_j t_j^{-1}, \\ \tilde{m}^*_{j,k_j} &= \tilde{D}^{1/2}\left(\alpha_j + \beta_j^{1/2} t_j^{-1/2} \eta_{j,k_j}\right), \quad j = 1, \ldots, J, \end{aligned} \tag{3.5}$$

i.e. depend only on the characteristics of batch processing and do not depend on u.

Let us describe the regret. Let us assume for certainty that $m_1 = \max(m_1, \ldots, m_J)$. Then the regret is

$$L_N(\sigma, \theta) = (\tilde{D}N)^{1/2} l_K(\sigma, \theta). \tag{3.6}$$

where

$$l_K(\sigma, \theta) = \varepsilon \sum_{k=1}^{J-1} (\alpha_1 - \alpha_{k+1}) + \sum_{j=2}^{J} (\alpha_1 - \alpha_j) \left(\sum_{k=J}^{K-1} \varepsilon P_{j,k} \right). \tag{3.7}$$

Indeed, the regret consists of losses at the first J steps, when actions are applied in turn, and at the remaining horizon, $k = J, \ldots, K-1$, when Thompson Sampling is implemented:

$$L_N(\sigma, \theta) = \sum_{k=1}^{J-1} (m_1 - m_{k+1}) + \sum_{k=J}^{K-1} \sum_{j=2}^{J} (m_1 - m_j) P_{j,k}.$$

Assuming here $m_j = \alpha_j (D/K)^{1/2}$, changing the order of summation in the second term, and taking into account that $DK = \tilde{D}N$ we get (3.7). By virtue of (3.4), (3.5), the probabilities of $\{P_{j,k}\}$ depend only on the batch processing parameters. Let us present the obtained results as theorem.

Theorem 1. *The robust version of Thompson Sampling has an invariant description over a control horizon equal to one. In this case, the regret* (1.1) *is given by the expression* (3.6), *in which* $l_K(\sigma,\theta)$ *depends only on the batch processing parameters, does not depend on* u, *and is given by the expression* (3.7).

Remark 3. Control horizon one still assumes K rounds of the game against MAB with a one round duration of $1/K$. It allows to use the estimate (3.6) with the same constant $l_K(\sigma,\theta)$ for any sufficiently large amount of data N provided that the data is divided into K batches. Invariant descriptions with control horizon one can be obtained for various strategies, e.g., mirror descent algorithm [21] and UCB [22]. It is very likely to obtain such a description for ε-greedy strategy, as suggested by an anonymous reviewer.

4 Alternative Approach to a Robust TS-Algorithm

Let us describe another approach to implementing a robust batch version of Thompson Sampling, equivalent to the one described in Sect. 3. Denote $k'_j = k_j/D_j$, $j=1,\dots,J$. Let all actions be applied $k_1,\dots,k_J$ times, with $k_j \ge 1$, $j=1,\dots,J$. If average incomes $\overline{X}_{j,k_j} = X_{j,k_j}/k_j$, $j=1,\dots,J$, were obtained as a result, then their joint distribution density has the form:

$$\frac{1}{(2\pi)^{J/2}}\left(\prod_{j=1}^{J} k'_j\right)^{1/2} \exp\left(-\frac{1}{2}\sum_{j=1}^{J} k'_j(\overline{X}_j - m_j)^2\right). \tag{4.1}$$

In what follows, we use a formula that is straightforward to verify:

$$\sum_i k'_j(u-a_j)^2 = \left(\sum_j k'_j\right)\left(u - \frac{\sum_j a_j k'_j}{\sum_j k'_j}\right)^2 + \frac{\sum_j \sum_{i>j} k'_i k'_j (a_i - a_j)^2}{\sum_j k'_j}.$$

Let us put $m_i = u + v_i$, $v_{ij} = v_i - v_j$, $i,j = 1,\dots,J$, $j = 1,\dots,J$, $i \ne j$. Denote $k' = k'_1 + \dots + k'_J$ and $a_j = \overline{X}_j - v_j$. Taking into account the above formula, distribution (4.1) has the form

$$\begin{aligned}&\left(\frac{k'}{2\pi}\right)^{1/2} \exp\left(-\frac{k'}{2}\left(u - \frac{\sum_j(\overline{X}_j - v_j)k'_j}{k'}\right)^2\right)\\ &\times\left(\frac{\prod_j k'_j}{(2\pi)^{J-1}k'}\right)^{1/2} \exp\left(-\frac{\sum_j\sum_{i>j} k'_i k'_j(\overline{X}_i - \overline{X}_j - v_{ij})^2}{2k'}\right).\end{aligned} \tag{4.2}$$

According to Remark 1, the density (4.1) is the posterior joint distribution density of mathematical expectations $\{m_j\}$ in the robust version of Thompson Sampling. Then (4.2) describes the joint distribution density of $\{u, v_{ij}\}$, where, for example, $u, v_{12}, v_{23}, \dots, v_{J-1,J}$ can be selected as independent variables. In this case $v_{ij} = \sum_{l=i}^{j-1} v_{l,l+1}$ if $j > i+1$ and $v_{ij} = -v_{ji}$ if $i > j$. Note that only parameters $\{v_{ij} = m_i - m_j\}$ can be used to compare mathematical expectations $\{m_i\}$. Therefore, a marginal distribution of parameters $\{v_{ij}\}$ is used for Thompson Sampling

$$\left(\frac{\prod_j k'_j}{(2\pi)^{J-1}k'}\right)^{1/2} \exp\left(-\frac{\sum_j \sum_{i>j} k'_i k'_j (\overline{X}_i - \overline{X}_j - v_{ij})^2}{2k'}\right) \tag{4.3}$$

which can be obtained by integrating the density (4.2) over u.

Next, let us consider the algorithm in two special cases $J = 2$ and $J = 3$. For $J = 2$, the distribution (4.3) takes the form

$$\left(\frac{k'_1 k'_2}{2\pi(k'_1 + k'_2)}\right)^{1/2} \exp\left(-\frac{k'_1 k'_2 (\overline{X}_1 - \overline{X}_2 - v_{12})^2}{2(k'_1 + k'_2)}\right)$$

and characterizes the posterior distribution of the parameter $v_{12} = m_1 - m_2$. One can see that the mathematical expectation and variance of this parameter are

$$\mathbf{E}(v_{12}) = \overline{X}_1 - \overline{X}_2, \quad \mathbf{D}(v_{12}) = (k'_1 + k'_2)/(k'_1 k'_2). \tag{4.4}$$

In addition, let us put $k''_1 = k_1/(\kappa_1 D_1)$, $k''_2 = k_2/(\kappa_2 D_2)$ and obtain parameters of the normal distribution used for sampling

$$v^*_{12} = \overline{X}_1 - \overline{X}_2, \quad D^*_{12} = (k''_1 + k''_2)/(k''_1 k''_2), \tag{4.5}$$

which differ from the parameters in (4.4) by the factors of the variances D_1, D_2.

Alternative Version of Thompson Sampling for $J = 2$. At the beginning of control, apply both actions once in turn, receiving incomes $x_{1,1}, x_{2,1}$, forming cumulative incomes $X_{1,1} = x_{1,1}, X_{2,1} = x_{2,1}$ and set $k_1 = k_2 = 1$.

Next, in the loop, at the step number $k = 2, 3, \dots, K-1$, perform the following operations:

1. Given current values (k_1, X_1), (k_2, X_2), sample a random variable μ_{12} from the current posterior distribution with parameters v^*_{12}, D^*_{12} presented by formulas (4.5),
2. If $\mu_{12} > 0$, then increase k_1 by 1, apply action 1, get a random income x_{1,k_1+1} and recalculate the total income $X_{1,k_1+1} = X_{1,k_1} + x_{1,k_1+1}$.
   If $\mu_{12} < 0$, then increase k_2 by 1, apply action 2, get a random income x_{2,k_2+1} and recalculate the total income $X_{2,k_2+1} = X_{2,k_2} + x_{2,k_2+1}$.
   If $\mu_{12} = 0$ (the probability of such an event is 0), then choose the action j arbitrarily, apply the action j, get a random income x_{j,k_j+1} and recalculate the total income $X_{j,k_j+1} = X_{,k_j} + x_{j,k_j+1}$. Increase k by 1.

For $J = 3$, we put $y_1 = v_{12} - (\overline{X}_1 - \overline{X}_2)$, $y_2 = v_{23} - (\overline{X}_2 - \overline{X}_3)$. Then the exponent degree in (4.3) takes the form

$$-\frac{k'_1 k'_2 y_1^2 + k'_2 k'_3 y_2^2 + k'_3 k'_1 (y_1 + y_2)^2}{2k'}$$
$$= -\frac{(k'_1 k'_2 + k'_3 k'_1) y_1^2 + 2k'_3 k'_1 y_1 y_2 + (k'_2 k'_3 + k'_3 k'_1) y_2^2}{2k'}.$$

Hence, v_{12}, v_{23} have a joint two-dimensional normal distribution. Their mathematical expectations are

$$\mathbf{E}v_{12} = v_{12}^* = \overline{X}_1 - \overline{X}_2, \quad \mathbf{E}v_{23} = v_{23}^* = \overline{X}_2 - \overline{X}_3, \tag{4.6}$$

the variances $\mathbf{D}v_{12} = D_{12}$, $\mathbf{D}v_{23} = D_{23}$ and the correlation coefficient $\rho = \text{cov}(v_{12}, v_{23})(D_{12}D_{23})^{-1/2}$ can be determined as follows. Consider the covariance matrix of the distribution and its inverse:

$$\Sigma = \begin{pmatrix} D_{12} & (D_{12}D_{23})^{1/2}\rho \\ (D_{12}D_{23})^{1/2}\rho & D_{23} \end{pmatrix},$$

$$\Sigma^{-1} = \frac{1}{D_{12}D_{23}(1-\rho^2)} \begin{pmatrix} D_{23} & -(D_{12}D_{23})^{1/2}\rho \\ -(D_{12}D_{23})^{1/2}\rho & D_{12} \end{pmatrix}$$

The unknowns D_{12}, D_{23}, ρ can be found from the condition

$$\frac{(k_1'k_2' + k_3'k_1')y_1^2 + 2k_3'k_1'y_1y_2 + (k_2'k_3' + k_3'k_1')y_2^2}{k'} = (y_1\ y_2)\ \Sigma^{-1} \begin{pmatrix} y_1 \\ y_2 \end{pmatrix}.$$

Therefore, the equalities hold

$$\frac{1}{D_{12}(1-\rho^2)} = \frac{k_1'k_2' + k_3'k_1'}{k'}, \quad \frac{1}{D_{23}(1-\rho^2)} = \frac{k_2'k_3' + k_3'k_1'}{k'},$$
$$-\frac{\rho}{(D_{12}D_{23})^{1/2}(1-\rho^2)} = \frac{k_3'k_1'}{k'},$$

whence we obtain

$$\begin{gathered} D_{12} = \frac{k'}{(k_1'k_2' + k_3'k_1')(1-\rho^2)}, \quad D_{23} = \frac{k'}{(k_2'k_3' + k_3'k_1')(1-\rho^2)}, \\ \rho = -\frac{k_3'k_1'}{(k_1'k_2' + k_3'k_1')^{1/2}(k_2'k_3' + k_3'k_1')^{1/2}}. \end{gathered} \tag{4.7}$$

Next, for $k_1'' = k_1/(\kappa_1 D_1)$, $k_2'' = k_2/(\kappa_2 D_2)$ we put

$$\begin{gathered} D_{12}^* = \frac{k''}{(k_1''k_2'' + k_3''k_1'')(1-\rho^2)}, \quad D_{23}^* = \frac{k''}{(k_2''k_3'' + k_3''k_1'')(1-\rho^2)}, \\ \rho^* = -\frac{k_3''k_1''}{(k_1''k_2'' + k_3''k_1'')^{1/2}(k_2''k_3'' + k_3''k_1'')^{1/2}}. \end{gathered} \tag{4.8}$$

which differ from parameters in (4.7) by the factors of the variances D_1, D_2, D_3.

Alternative Version of Thompson Sampling $J = 3$. At the beginning of control, apply actions once in turn, receiving incomes $x_{1,1}, x_{2,1}, x_{3,1}$, forming cumulative incomes $X_{1,1} = x_{1,1}, X_{2,1} = x_{2,1}, X_{3,1} = x_{3.1}$ and set $k_1 = k_2 = k_3 = 1$.

Next, in the loop, at the step number $k = 3, 4, \ldots, K-1$, perform the following operations:

1. Given current values (k_1, X_1), (k_2, X_2), (k_3, X_3) sample random variables μ_{12}, μ_{23} from the current posterior two-dimensional normal distribution with mathematical expectations v_{12}^*, v_{23}^*, variances D_{12}^*, D_{23}^* and correlation coefficient ρ^* presented by formulas (4.6), (4.8). We assume $\mu_{13} = \mu_{12} + \mu_{23}$.
2. If $\mu_{12} > 0$, $\mu_{13} > 0$, then increase k_1 by 1, apply action 1, get a random income x_{1,k_1+1} and recalculate the cumulative income $X_{1,k_1+1} = X_{1,k_1} + x_{1,k_1+1}$. If $\mu_{12} < 0$, $\mu_{23} > 0$, then increase k_2 by 1, apply action 2, get a random income x_{2,k_2+1} and recalculate the cumulative income $X_{2,k_2+1} = X_{2,k_2} + x_{2,k_2+1}$. If $\mu_{13} < 0$, $\mu_{23} < 0$, then increase k_3 by 1, apply action 3, get a random income x_{3,k_3+1} and recalculate the cumulative income $X_{3,k_3+1} = X_{3,k_3} + x_{3,k_3+1}$. If some conditions are fulfilled in the form of equalities (the probability of such an event is 0), then choose the action j arbitrarily, apply the action j, get a random income x_{j,k_j+1} and recalculate the cumulative income $X_{j,k_j+1} = X_{,k_j} + x_{j,k_j+1}$. Increase k by 1.

Remark 4. To sample v_{12}, v_{23}, one can first sample i.i.d. standard Gaussian ξ, η. Then ξ and $\zeta = \rho^*\xi + (1-\rho^{*2})^{1/2}\eta$ are standard Gaussian with correlation coefficient ρ^*. Then

$$\mu_{12} = \overline{X}_1 - \overline{X}_2 + (D_{12}^*)^{1/2}\xi, \quad \mu_{23} = \overline{X}_2 - \overline{X}_3 + (D_{23}^*)^{1/2}\zeta.$$

Let us obtain an invariant description of the distribution density (4.3). Denote $\varepsilon = 1/K$, $t_j = k_j/K$, $t'_j = t_j/\beta_j$, $t' = t'_1 + \dots + t'_J$. Therefore, $t'_j = k'_j(D/K)$, $t' = k'(D/K)$. Next, $x_{j,i} = m_j + D_j^{1/2}\zeta_{j,i}$, where $m_j = u + \alpha_j(D/K)^{1/2}$, and $\{\zeta_{j,i}\}$ are i.i.d.r.v. with a standard normal distribution. Therefore,

$$\overline{X}_j = \frac{\sum_{i=1}^{k_j} D_j^{1/2}\zeta_{j,i}}{Kt_j} + \alpha_j(D/K)^{1/2} + u = \left(\frac{D}{K}\right)^{1/2} \chi_{l,k_j} + u, \tag{4.9}$$

where

$$\chi_{l,k_j} = \alpha_j + \beta_j^{1/2} t_j^{-1/2} \eta_{l,k_j}, \quad \eta_{l,k_j} = k_j^{-1/2} \sum_{i=1}^{k_j} \zeta_{j,i}. \tag{4.10}$$

Let us assume $v_{ij} = (D/K)^{1/2} w_{ij}$. Then, taking into account (4.9), (4.10) and the Jacobian of the transformation of variables $\partial(v_{12}, \dots, v_{J,J-1})/\partial(w_{12}, \dots, w_{J,J-1}) = (D/K)^{(J-1)/2}$ the distribution density (4.3) relative to the parameters $w_{12}, \dots, w_{J,J-1}$ takes the form

$$\left(\frac{\prod_j t'_j}{(2\pi)^{J-1} t'}\right)^{1/2} \times \exp\left(-\frac{\sum_i \sum_{j>i} t'_i t'_j \left(\chi_{i,k_i} - \chi_{j,k_j} - w_{ij}\right)^2}{2t'}\right) \tag{4.11}$$

Distribution (4.11) depends only on the batch processing parameters.

Remark 5. Since $\Pr(w_{ij} > 0) = \Pr(v_{ij} > 0)$ for all i, j, formula (4.11) gives another method to calculate probabilities $P_{j,k}$ in (3.4).

5 Numerical Experiments

The algorithm was explored using Monte-Carlo method; all estimates of the regret were averaged over 2×10^6 simulations. We considered the most important case of equal variances $D_j = D = 1$, $j = 1, \ldots, J$. For example, if incomes are Bernoullian, i.e., equal to 0 or 1, then in the domain of close distributions, where the regret reaches its maximum values, the variances are also close to each other and after normalization are close to 1. For the two-armed bandit, the results are shown in Fig. 1. Bold solid lines 1, 2, 3 and 4 represent the normalized regrets in the case of $K = 30, 50, 100$ and 150 batches, respectively. Thin dashed lines 1, 2, 3 and 4 correspond to normalized regrets for the same number of batches but without taking into account losses at the initial stage of control when actions are applied once in turn. We selected $m_1 = d(D/K)^{1/2}$, $m_2 = -m_1$, $\kappa = 1$. Here and after calculations were performed in the range of $0.1 < d < 15$ in increments of 0.1. The corresponding normalized regrets $(DK)^{-1/2} L_K(\sigma, \theta)$ are indicated by $l_K(d)$. For selected values of K, determined points $d_0 = \text{argmax}_d l_K(d)$, at which maximum normalized regrets $l_K(d)$ are reached, and maximum $l_K(d_0)$ are presented in the first multi-column of Table 1.

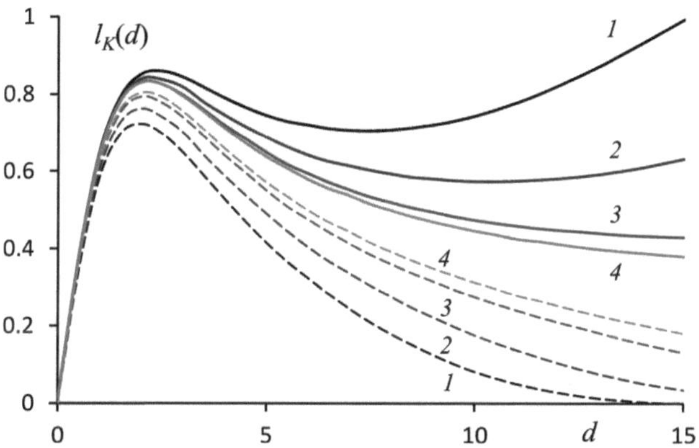

Fig. 1. TS-algorithm for two-armed bandit with $\kappa = 1$; $K = 30$, 50, 100, 150.

Figure Fig. 2 shows the same results for the two-armed bandit with optimization by κ, i.e., the maximum normalized regrets were minimized by κ. Note that the algorithm is not very sensitive to the choice of κ. The maximum regret icreases by no more than 1% if κ changes by 10%. The values of K, κ, d_0 and $l_K(d_0)$ are represented in the second multi-column of Table 1. One can see that maxima in the second multi-column of Table 1 are smaller than those in the first one.

Finally, the results of optimization by κ for various J are shown in Fig. 3 in the case of $K = 50$. Bold solid and thin dashed lines 1, 2, 3 and 4 represent the normalized regrets in the case of $J = 3, 4, 5$ and 6 with the latter representing

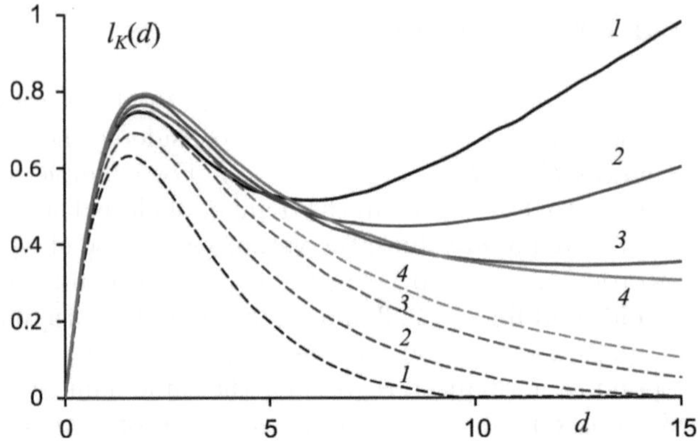

Fig. 2. TS-algorithm for three-armed bandit with $\kappa = 1$; $K = 30$, 50, 100, 150.

regrets without losses at the initial stage of control. We applied $m_1 = d(D/K)^{1/2}$, $m_j = -m_1/(J-1)$, $j = 2, \ldots, J$ (it was checked by simulations, that in the case of equal variances the maxima of regret are reached for the parameters with one larger value of m_j and other equal smaller ones). The values of J, κ, d_0 and $l_K(d_0)$ are represented in the third multi-column of Table 1.

Table 1. Parameters of TS-algorithms.

$J = 2$ $(\kappa = 1)$			$J = 2$				$K = 50$			
K	d_0	$l_K(d_0)$	K	κ	d_0	$l_K(d_0)$	J	κ	d_0	$l_K(d_0)$
30	2.3	0.86	30	0.2	2.0	0.74	3	0.20	2.9	1.26
50	2.2	0.84	50	0.4	2.0	0.76	4	0.15	3.8	1.64
100	2.2	0.83	100	0.5	2.0	0.79	5	0.10	4.4	1.97
150	2.2	0.83	150	0.6	2.0	0.79	6	0.08	4.9	2.27

We also note that all simulations in the case of $J = 2$ and $J = 3$ were performed using both methods and the same results were obtained.

Let us compare these results with previously obtained for batched versions of other algorithms. In [21] the batched version of the mirror descent algorithm (MDA) was investigated for the two-armed bandit. It turned out that by tuning the appropriate parameters the value of the maximum normalized regret $l_K(d)$ can be made approximately equal to 1.1 (for $d \approx 2.5$) and it changes little for $K = 50, 200, 1000$. In [22] the batched version of UCB strategy was explored. By tuning the parameters, for $J = 2, 3, 4, 6$ the maximum normalized regrets $l_K(d)$ were made approximately equal to 0.73, 1.22, 1.61, 2.23 (corresponding to our $d \approx 1.8$, 2.8, 3.6, 5) respectively if $K = 50$. One can see that the batched version of MDA is less efficient while the UCB batch strategy is slightly more efficient than the TS-algorithm.

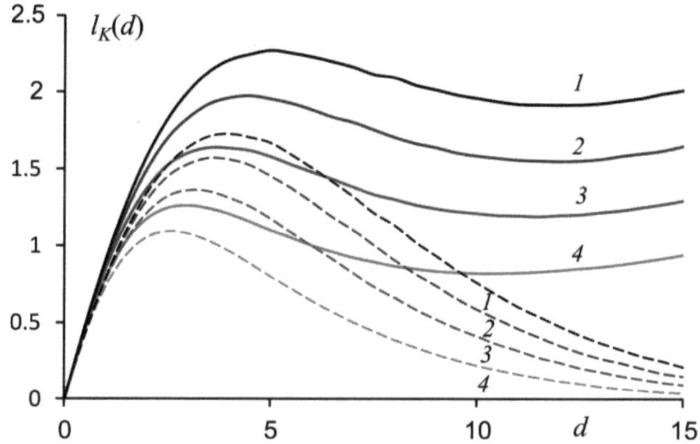

Fig. 3. TS-algorithm for two-armed bandit optimized by κ; $K = 30, 50, 100, 150$.

6 Conclusion

Two equivalent versions of robust TS-algorithms for the multi-armed bandit problem are proposed. The results show that TS-algorithm is quite effective and easy to implement. As the future research we indicate the analysis of processing incomes with different variances, the dynamically varied value of κ in order to minimize the maximum regrets and the usage of smaller sizes of batches at the beginning of the control in order to minimize the losses caused by initial stage.

Acknowledgments. The research was carried out during the implementation of the research "Mathematical modeling of natural processes" performed within the framework of the state assignment in the field of scientific activity.

Disclosure of Interests. The authors have no competing interests to declare that are relevant to the content of this article.

References

1. Berry, D.-A., Fristedt, B.: Bandit Problems: Sequential Allocation of Experiments. Chapman and Hall, London, New York (1985)
2. Presman, E.L., Sonin, I.-M.: Sequential Control with Incomplete Information. Academic Press, New York (1990)
3. Lattimore, T., Szepesvari, C.: Bandit Algorithms. Cambridge University Press, Cambridge (2020)
4. Slivkins, A.: Introduction to multi-armed bandits. arXiv:1904.07272v7 (2022). https://arxiv.org/pdf/1904.07272
5. Tsetlin, M.-L.: Automation Theory and Modeling of Biological Systems. Academic Press, New York (1973)
6. Sragovich, V.-G.: Mathematical Theory of Adaptive Control. Interdisciplinary Mathematical Sciences, vol. 4. World Scientific, New Jersey, London (2006)

7. Perchet, V., Rigollet, P., Chassang, S., Snowberg, E.: Batched bandit problems. Ann. Statist. **44**(2), 660–681 (2016)
8. Cesa-Bianchi, N., Lugosi, G.: Prediction, Learning, and Games. Cambridge University Press, Cambridge (2006)
9. Thompson, W.-R.: On the likelihood that one unknown probability exceeds another in view of the evidence of two samples. Biometrika **25**(3)–(4), 285–294 (1933)
10. Gupta, N., Granmo, O.-C., Agrawala, A.: Thompson Sampling for dynamic multi-armed bandits. In: Proceedings of the 2011 10th International Conference on Machine Learning and Applications, pp. 484–489 (2011). https://ieeexplore.ieee.org/abstract/document/6147024
11. Agrawal, S., Goyal, N.: Analysis of Thompson Sampling for the multi-armed bandit problem. In: Proceedings of the 25th Annual Conference on Learning Theory (COLT), vol. 23, pp. 39.1–39.26 (2012). https://proceedings.mlr.press/v23/agrawal12/agrawal12.pdf
12. Russo, D., Van Roy, B.: Learning to optimize via posterior sampling. Math. Oper. Res. **39**(4), 1221–1243 (2014)
13. Robbins, H.: Some aspects of the sequential design of experiments. Bull. AMS **58**(5), 527–535 (1952)
14. Vogel, W.: An asymptotic ninimax theorem for the two-armed bandit problem. Ann. Math. Stat. **31**, 444–451 (1960)
15. Juditsky, A., Nazin, A.-V., Tsybakov, A.-B., Vayatis, N.: Gap-free bounds for stochastic multi-armed bandit. In: Proceedings of the 17th World Congress IFAC, Seoul, Korea, 6–11 July 2008, pp. 11560–11563 (2008). https://www.sciencedirect.com/science/article/pii/S147466701640827X
16. Kolnogorov, A.-V.: On a limiting description of robust parallel control in a random environment. Autom. Remote. Control. **76**(7), 1229–1241 (2015)
17. Lai, T.-L., Robbins, H.: Asymptotically efficient adaptive allocation rules. Adv. Appl. Math. **6**, 4–22 (1985)
18. Auer, P., Cesa-Bianchi, N., Fischer, P.: Finite time analysis of multi-armed bandit problem. Mach. Learn. **27**(2)–(3), 235–256 (2002)
19. Kolnogorov, A.-V.: Determination of the minimax risk for the normal two-armed bandit. IFAC Proc. Vol.**43**(10), 231–236 (2010). https://www.sciencedirect.com/science/article/pii/S1474667015323703
20. DeGroot, M.-H.: Optimal Statistical Decisions. McGraw Hill Company, New York (1970)
21. Kolnogorov, A.-V., Nazin, A.-V., Shiyan, D.-N.: Two-armed bandit problem and batch version of the mirror descent algorithm. Autom. Remote. Control. **83**, 1288–1307 (2022)
22. Garbar, S.-V., Kolnogorov, A.-V.: Customization of J. Bather's UCB strategy for a Gaussian multiarmed bandit. Autom. Remote. Control. **83**, 1857–1873 (2022)

Primal–Dual LinUCB for Budget-Constrained Contextual Bandits with Delayed Feedback

Mariia Pavlova[1(✉)], Ekaterina Pavlova[2], and Vladimir Korkhov[3]

[1] Saint Petersburg State University, Saint Petersburg, Russia
st094853@student.spbu.ru
[2] Skolkovo Institute of Science and Technology, Moscow, Russia
ekaterina.pavlova@skoltech.ru
[3] Saint Petersburg State University, Saint Petersburg, Russia
v.korkhov@spbu.ru

Abstract. We study contextual bandits with a global budget constraint and delayed reward feedback. We consider a simple primal–dual variant of Disjoint LinUCB in which an adaptive dual variable acts as the resource price, while model updates follow a delayed-feedback heuristic. For reproducible evaluation, we build a scalable semi-synthetic benchmark on the Criteo Attribution dataset and use a stop-at-budget protocol that guarantees zero budget violations at execution time.

In the benchmark, rewards are generated from an arm-specific clipped-linear model fitted on a temporal training split. We compare the method against LinUCB and fixed cost-penalty heuristics of the form UCB $-\gamma c$. The experiments show that the fixed penalty is highly sensitive to both γ and the budget regime, whereas the adaptive primal–dual mechanism is more robust across tested budgets. These conclusions should be interpreted in the context of a delayed-feedback simulator with important modeling limitations.

Keywords: contextual bandits · delayed feedback · bandits with knapsacks · primal–dual methods · LinUCB

1 Introduction

Contextual multi-armed bandits and online stochastic optimization arise in sequential decision-making problems where actions must be chosen based on the observed context x_t while balancing exploration and exploitation. Typical applications include ad allocation, recommendation systems, online experimentation, and resource allocation, where each action yields a random reward and consumes a limited resource [9,22].

In practice, two factors make this problem substantially more challenging. First, actions incur costs, so the optimization objective must account for a

Code: https://github.com/Marypvlva/pd-linucb-delayed-budgets.

A. Kazakov et al. (Eds.): MOTOR 2026, LNCS 16687, pp. 181–195, 2026.
https://doi.org/10.1007/978-3-032-31390-4_12

resource constraint, as in Bandits with Knapsacks (BwK) and contextual BwK (CBwK) [2,3]. Second, rewards are often observed with delay (for example, conversions may occur hours or days after the action), leading to delayed-feedback bandit problems [10,24]. Together, these effects create a mismatch between immediately observed costs and delayed rewards, which may make standard methods such as LinUCB [13] unstable or require manual tuning of cost penalties.

We consider a sequential decision process over rounds $t = 1, \dots, T$. At round t, the learner observes a context $x_t \in \mathbb{R}^d$ and selects an action $a_t \in \{1, \dots, K\}$. This choice yields a reward $r_t \in [0,1]$ and incurs a cost $c(a_t) \geq 0$. The objective is to maximize the total reward subject to a total budget constraint

$$\sum_{t=1}^{\tau} c(a_t) \leq B,$$

where τ denotes the stopping time induced by the stop-at-budget protocol, that is, an action is not executed if it would cause the cumulative cost to exceed the budget. Reward feedback is delayed: the reward r_t becomes available to the algorithm only at time $t + D_t$, where $D_t \in \{0, 1, 2, \dots\}$.

Prior work studies these ingredients mostly separately: BwK and CBwK address resource constraints [2,3], delayed-bandit work studies feedback delays [10,24], and primal–dual methods provide adaptive resource pricing under constraints [14,17]. Our focus is the practical combination of these ingredients in a scalable delayed benchmark with strict budget feasibility.

In this work, we take Disjoint LinUCB as the base contextual method and augment it with an adaptive dual variable $\lambda_t \geq 0$ that penalizes costly actions online. Actions are selected using a Lagrangian score of the form $\mathrm{UCB}_t(x_t, a) - \lambda_t c(a)$, while λ_t is updated from the deviation between realized cost and a dynamic spending target based on the remaining budget and remaining horizon. Under delayed feedback, we use a practical design-now, reward-later update scheme: the design matrix is updated immediately after action selection, whereas the response vector is updated only when the delayed reward arrives. Budget feasibility is enforced by the common stop-at-budget protocol. Full details are given in Sect. 4.

Contributions.

- We study a contextual bandit setting with a global budget constraint, delayed reward feedback, and a stop-at-budget protocol that guarantees zero budget violations.
- We evaluate a simple primal–dual variant of Disjoint LinUCB with a dynamic spending target and delayed design-now, reward-later updates.
- We build a reproducible semi-synthetic benchmark on the Criteo Attribution dataset.
- We compare PD-LinUCB against LinUCB, cost-aware heuristics, and a context-free primal–dual baseline, including budget sweeps, delay ablations, and feedback-observation diagnostics.

2 Problem Statement

We consider a horizon $T \in \mathbb{N}$, a set of actions (arms) of size K, and a total budget $B > 0$. At each round $t = 1, \ldots, T$, the algorithm observes a context $x_t \in \mathbb{R}^d$ and selects an action $a_t \in \mathcal{A} = \{1, \ldots, K\}$. Selecting an action yields a stochastic reward $r_t \in [0,1]$ and incurs a cost $c(a_t) \geq 0$. We assume arm-dependent costs known at the time of selection. Reward r_t is observed with delay $D_t \in \{0, 1, 2, \ldots\}$ and can be used for learning only at time $t + D_t$.

Let $\text{spent}_t = \sum_{s=1}^{t} c(a_s)$ denote the cumulative cost (with $\text{spent}_0 = 0$). We use the stop-at-budget protocol, which prohibits executing an action if it would exceed the budget. At round t, action a_t is executed only if

$$\text{spent}_{t-1} + c(a_t) \leq B.$$

Let τ denote the number of actions executed before stopping. Then

$$\tau = \max\Big\{t : \text{spent}_t \leq B\Big\}.$$

The objective is to maximize the expected cumulative reward until stopping:

$$R(\pi) = \mathbb{E}\Big[\sum_{t=1}^{\tau(\pi)} r_t\Big], \qquad \tau(\pi) = \max\Big\{\tau : \sum_{t=1}^{\tau} c(a_t) \leq B\Big\}.$$

We use an arm-specific linear conditional mean model

$$\mu_a(x) = \mathbb{E}[r \mid x, a] = \text{clip}(\theta_a^\top x,\ 0,\ 1), \tag{1}$$

where $\text{clip}(u, 0, 1) = \min\{1, \max\{0, u\}\}$. Rewards are generated as

$$r_t \sim \text{Bernoulli}\big(\mu_{a_t}(x_t)\big). \tag{2}$$

The parameters θ_a are calibrated on the logged training split via ridge regression:

$$\widehat{\theta}_a = \arg\min_{\theta} \sum_{i:A_i=a} (R_i - \theta^\top X_i)^2 + \lambda \|\theta\|_2^2. \tag{3}$$

The main benchmark uses this clipped-linear simulator. In addition, we fit an optional arm-wise logistic simulator on the same training split and report held-out calibration diagnostics for it.

3 Related Work

Our setting combines three lines of work: bandits with knapsacks and contextual resource constraints, primal–dual methods for online constrained optimization, and bandits with delayed feedback.

BwK and CBwK formalize learning under limited resources and provide the main constrained-bandit framework [2,3,6,8]. More general resource formulations include concave rewards and convex knapsack constraints [1], while contextual extensions with multiple resources include packing and covering constraints [23,26].

Primal–dual methods treat the dual variable as the shadow price of a resource and update it online according to constraint violation [14,15,17–19]. Related contextual constrained optimization and reduction-based approaches include [4, 20]. Our method follows this perspective in a simple LinUCB-style setting.

Delayed feedback has been studied in both general online learning and bandits, including delayed conversions and delayed linear bandits [10,12,21,24,25]. Closest to our setting are recent works combining delayed feedback with budgeted decision-making [11,16]. We compare an adaptive primal–dual multiplier against a static cost penalty in a reproducible delayed benchmark.

4 Algorithms

All methods assume that the cost $c(a)$ is known at the time of action selection, while the reward r_t may be observed with a delay D_t.

Stop-at-Budget Protocol (Common to all Algorithms). Let $\text{spent}_{t-1} = \sum_{s=1}^{t-1} c(a_s)$ denote the cumulative cost before round t. At round t, the feasible set of actions is defined as

$$\mathcal{A}_t = \{a \in \mathcal{A} : \text{spent}_{t-1} + c(a) \leq B\}.$$

If $\mathcal{A}_t = \varnothing$, the process stops and $\tau = t - 1$. Otherwise, the algorithm selects an action $a_t \in \mathcal{A}_t$ according to its decision rule (maximizing its score or index over the feasible set), after which the cost is incurred. This protocol guarantees $\text{spent}_\tau \leq B$ and avoids the "early stopping" artifact that would arise if an infeasible action were selected while feasible alternatives still existed.

4.1 Disjoint LinUCB (Baseline)

Disjoint LinUCB maintains a separate linear ridge regression model for each action $a \in \{1, \ldots, K\}$ and selects actions using an upper confidence bound (UCB) criterion.

Parameter Estimation. For each action a, the algorithm maintains a design matrix and response vector:

$$A_{a,t} = \lambda I + \sum_{i \in \mathcal{P}_t : a_i = a} x_i x_i^\top, \qquad b_{a,t} = \sum_{i \in \mathcal{O}_t : a_i = a} x_i r_i,$$

where $\mathcal{P}_t$ is the set of rounds whose actions have already been committed by time t (the contexts x_i are known immediately), and $\mathcal{O}_t \subseteq \mathcal{P}_t$ is the subset of rounds for which the reward has already been observed by time t (accounting for delays).

The parameter estimate is

$$\widehat{\theta}_{a,t} = A_{a,t}^{-1} b_{a,t}.$$

This estimator is not the ordinary ridge solution on only the observed labeled samples unless $\mathcal{O}_t = \mathcal{P}_t$. In fact, it is equivalent to the minimizer of

$$\widehat{\theta}_{a,t} = \arg\min_{\theta} \left[\lambda \|\theta\|_2^2 + \sum_{i \in \mathcal{O}_t :\, a_i = a} (r_i - \theta^\top x_i)^2 + \sum_{i \in \mathcal{P}_t \setminus \mathcal{O}_t :\, a_i = a} (\theta^\top x_i)^2 \right].$$

Thus, pending unlabeled rounds act as temporary zero-imputed pseudo-observations. We use this form for implementation simplicity and immediate Sherman–Morrison updates.

UCB Index and Action Selection. Given context x_t, the algorithm computes

$$\mathrm{UCB}_t(x_t, a) = \widehat{\theta}_{a,t}^\top x_t + \alpha \sqrt{x_t^\top A_{a,t}^{-1} x_t}, \tag{4}$$

and selects

$$a_t = \arg\max_{a \in \mathcal{A}_t} \mathrm{UCB}_t(x_t, a).$$

Delayed Feedback: design-now, reward-later. Since contexts are available immediately, the design matrix is updated right after action selection:

$$A_{a_t, t+1} = A_{a_t, t} + x_t x_t^\top.$$

The response vector is updated only when the reward arrives:

$$b_{a_t, t+D_t+1} = b_{a_t, t+D_t} + x_t r_t,$$

with the case $D_t = 0$ included.

Thus, the uncertainty term in (4) is used heuristically rather than as a delay-correct confidence bound [10,12].

For computational efficiency, we maintain $A_{a,t}^{-1}$ explicitly and apply the Sherman–Morrison update:

$$A_{a_t,t+1}^{-1} = A_{a_t,t}^{-1} - \frac{A_{a_t,t}^{-1} x_t x_t^\top A_{a_t,t}^{-1}}{1 + x_t^\top A_{a_t,t}^{-1} x_t}, \qquad A_{a_t,t+1} = A_{a_t,t} + x_t x_t^\top. \tag{5}$$

4.2 Primal–Dual LinUCB (PD-LinUCB)

PD-LinUCB augments LinUCB with a nonnegative dual variable λ_t that penalizes costly actions online.

Action Selection Rule (Lagrangian Score). At round t, the action is selected according to the score

$$a_t = \arg\max_{a \in \mathcal{A}_t} \Big(\mathrm{UCB}_t(x_t, a) - \lambda_t \, c(a) \Big), \tag{6}$$

where UCB_t is computed as in (4), and $c(a)$ denotes the known cost of action a.

Stop-at-Budget (Feasible Set). At each round t, the feasible set of actions is

$$\mathcal{A}_t = \{a \in \mathcal{A} : \mathrm{spent}_{t-1} + c(a) \leq B\}.$$

If $\mathcal{A}_t = \varnothing$, the process stops.

Dual Update with a Dynamic Spending Target. We define a dynamic spending target based on the remaining budget and horizon:

$$b_t = \frac{B - \mathrm{spent}_{t-1}}{T - t + 1}. \tag{7}$$

The dual variable is then updated via a projected gradient step onto $\mathbb{R}_+$:

$$\lambda_{t+1} = \big[\lambda_t + \eta \, (c(a_t) - b_t)\big]_+. \tag{8}$$

If $c(a_t) > b_t$, the penalty increases; if $c(a_t) < b_t$, it decreases.

Interpretation in Simplified Regimes. In the no-delay case, PD-LinUCB can be viewed as a primal–dual method for cumulative reward maximization under a budget constraint. Its decision rule combines an optimistic reward estimate with a cost penalty, where the dual variable acts as an adaptive price of the remaining budget. If the budget is being spent too quickly, this price increases; if spending is below the target level, it decreases. Thus, the algorithm simultaneously favors high reward and controls the spending rate, reducing the risk of premature stopping.

Under bounded delays, the budget-control mechanism remains unchanged, because the cost is observed immediately and budget feasibility is checked at the moment of action selection. Delays mainly affect reward estimation, but not the enforcement of the budget constraint.

Delayed Feedback. PD-LinUCB uses the same design-now, reward-later update scheme as the LinUCB baseline: the design matrix is updated immediately after action selection, while the response vector is updated only when the reward becomes available. Thus, the dual update and stop-at-budget rule operate online from immediately observed costs, whereas reward learning proceeds from delayed labels.

4.3 Cost-Aware Baselines

To separate the effect of adaptive budget control from simple cost awareness, we compare PD-LinUCB with two heuristics based on the same UCB index (4).

CostNormUCB[ratio].

$$a_t = \arg\max_{a \in \mathcal{A}_t} \frac{\mathrm{UCB}_t(x_t, a)}{c(a) + \varepsilon},$$

where $\varepsilon > 0$ prevents division by zero.

CostNormUCB[sub] (fixed penalty).

$$a_t = \arg\max_{a \in \mathcal{A}_t} \left(\mathrm{UCB}_t(x_t, a) - \gamma\, c(a)\right),$$

where $\gamma \geq 0$ is a fixed multiplier (a static Lagrangian penalty) that must be tuned. In the experiments, we perform a sweep over γ and report how both total reward and budget utilization vary with this fixed penalty.

Both heuristics use the same design-now, reward-later delayed-update scheme for (A_a, b_a) and are evaluated under the same stop-at-budget protocol.

4.4 Context-Free Primal–Dual Baseline (CF-PD-BwK)

To isolate the contribution of contextual information from that of budget control, we include a baseline that ignores the context x_t and learns only the average reward of each action. Let $\widehat{\mu}_{a,t}$ denote the empirical mean reward of action a (based on observed rewards, accounting for delays), and let $n_{a,t}$ denote the number of observed rewards for action a by time t. The action is then selected using a UCB-style index with a Lagrangian penalty:

$$a_t = \arg\max_{a \in \mathcal{A}_t} \left(\widehat{\mu}_{a,t} + \alpha\sqrt{\tfrac{\log(t+1)}{\max(1, n_{a,t})}} - \lambda_t\, c(a)\right).$$

The multiplier λ_t is updated using the same dynamic rule (8) with the spending target b_t from (7). This baseline helps separate the effect of budget control from that of contextual adaptation.

5 Experiments

All methods are evaluated on identical context sequences and random seeds; results are reported as means with 95% confidence intervals.

5.1 Data and Feature Generation

We use the Criteo Attribution dataset [5] (n = 16,468,027 rows) as a source of realistic contexts, costs, and delays. To avoid leakage between simulator calibration and evaluation, we use a temporal 80/20 split: the semi-synthetic environment is calibrated on the earlier 80% of rows, while online evaluation uses contexts from the later 20%.

This split defines the environment and does not pre-train the evaluated policies. During evaluation, the policies learn online on held-out contexts. The fitted environment parameters are not assumed to be universal; they are treated as a fixed approximation of the reward-generation process under approximate stationarity. Policies may also be warm-started from a smaller sample and then updated online, while the simulator parameters remain fixed for fair comparison.

The context is a 65-dimensional vector consisting of 64 hashed binary features and one numerical feature

$$\log(1 + \max\{0, \texttt{time\_since\_last\_click}\}).$$

Actions are defined as hash buckets of campaign identifiers rather than original campaigns. We aggregate campaigns into $K = 50$ bandit arms via

$$a = h(\text{campaign}) \bmod K,$$

where $h(\cdot)$ is a fixed hash function, so multiple campaigns may map to the same action.

5.2 Online Environment and Reward Generation

The online environment is a simulator: at each round t, a held-out context x_t is sampled, the algorithm selects a_t, and the environment returns $(r_t, c(a_t), D_t)$.

The delay D_t is sampled from the empirical distribution of positive conversion delays (for $r = 1$). For $r = 0$, we use a censored delay

$$D_t = \lceil W/\Delta \rceil,$$

where W is the observation-window length and Δ is the discretization step.

The conversion probability is generated according to an arm-specific linear conditional mean model

$$\mu_a(x) = \text{clip}(\theta_a^\top x, 0, 1), \qquad r \sim \text{Bernoulli}(\mu_a(x))$$

where the parameters θ_a are calibrated using ridge regression on the training split for each arm (Sect. 2). This yields a controlled semi-synthetic environment with realistic context, cost, and delay distributions and reproducible rewards. We additionally fit an arm-wise logistic simulator on the same training split and evaluate its held-out calibration, but the main policy comparisons in this paper use the default clipped-linear simulator unless noted otherwise.

An important consequence of the default delayed setup is that when $T = 5000$ and *censor_steps* $= 5000$, censored zero labels do not arrive within the same run under the current scheduler. The delayed linear benchmark is therefore close to a positive-only censored-feedback regime: positive rewards may arrive before the horizon, while most negative labels remain pending.

Costs and Delays. We derive simulator costs and delays from the logged `cost`, `timestamp`, and `conversion_timestamp` fields. For each arm a, the online cost $c(a)$ is the average logged cost of events mapped to that arm, optionally normalized so that the mean arm cost is close to one. Delays are obtained by discretizing timestamp differences with step size Δ; converted events use the observed discretized delay, while non-converted events use observation-window censoring. In the online simulator, positive delays are sampled from the empirical pool of converted events and negative outcomes use the censored delay.

5.3 Observed Feedback Before Stopping

For interpreting the delayed benchmark, a more important diagnostic than held-out fit quality is how much feedback is actually observed before the run stops. Under the default delayed setup, positive rewards may arrive within the horizon, while censored zero labels typically do not, so the learner operates in a near positive-only censored-feedback regime.

Table 1 reports the fraction of executed actions whose labels are observed as positives, observed as zeros, or remain unobserved by stopping, under both delayed and no-delay regimes. In the delayed setting, only about 5% of executed labels are observed as positives before stopping and essentially none are observed as zeros; in the no-delay ablation, all labels are observed immediately.

Table 1. Observed feedback before stopping, split into delayed and no-delay regimes. Values are fractions of executed actions τ. In the delayed setup, almost no zero labels are observed within horizon, while in the no-delay ablation all labels are observed immediately.

Method	Delayed						No-delay ($D_t \equiv 0$)					
	Pos./τ	CI	Zero./τ	CI	Pending/τ	CI	Pos./τ	CI	Zero./τ	CI	Pending/τ	CI
LinUCB	0.042662	0.002147	0.000000	0.000000	0.957338	0.002147	0.045264	0.003498	0.954736	0.003498	0.000000	0.000000
PD-LinUCB	0.042743	0.002411	0.000000	0.000000	0.957257	0.002411	0.043981	0.002129	0.956019	0.002129	0.000000	0.000000
CostNormUCB[ratio]	0.043227	0.002478	0.000000	0.000000	0.956773	0.002478	0.044534	0.002012	0.955466	0.002012	0.000000	0.000000
CF-PD-BwK	0.038360	0.001894	0.000000	0.000000	0.961640	0.001894	0.041460	0.002352	0.958540	0.002352	0.000000	0.000000

5.4 Experiment Protocol and Metrics

All experiments use the stop-at-budget protocol: an action at round t is executed only if

$$\text{spent}_{t-1} + c(a_t) \leq B,$$

otherwise the process terminates. For a given horizon T, we parameterize the budget regime as

$$\rho = \frac{B}{T} \qquad \Longleftrightarrow \qquad B = \rho T,$$

where ρ represents the average available resource per round at the experiment scale. (Importantly, PD-LinUCB uses the dynamic spending target $b_t = \frac{B-\text{spent}_{t-1}}{T-t+1}$ rather than the fixed ratio B/T).

We report total reward $R = \sum_{t=1}^{\tau} r_t$, budget utilization spent$/B$, and stopping time τ.

In addition, the evaluation code records the fraction of executed actions whose labels are observed as positives, observed as zeros, or remain unobserved by stopping; Table 1 summarizes these diagnostics for the delayed and no-delay regimes. These diagnostics are useful for interpreting the effective delayed-feedback regime.

Confidence intervals are computed across seeds using the Student-t interval.

5.5 Comparison of Methods in a Fixed Budget Regime

Table 2 reports results for $T = 5000$ and $\rho = 0.7$ under the stop-at-budget protocol, averaged over $N = 10$ seeds with 95% confidence intervals.

Figure 1 shows the mean cumulative reward trajectories, while Fig. 2 shows the running-average cost per step in an example run, highlighting how each method tracks the target spending rate $\rho = B/T$.

Table 2. Results for $T = 5000$ and $\rho = 0.7$ (stop-at-budget). Mean values over $N = 10$ seeds with 95% confidence intervals.

Method	Reward	95% CI	Spent/B	95% CI	τ	95% CI
LinUCB	190.1	11.6	0.999947	0.000022	3641.2	6.7
PD-LinUCB	253.1	11.6	0.999815	0.000080	4999.7	0.3
CostNormUCB[ratio]	220.5	14.7	0.999943	0.000031	4262.0	11.6
CF-PD-BwK	219.1	8.7	0.999728	0.000331	4993.3	7.0

5.6 Gamma Sweep for CostNormUCB[sub]

For the UCB $- \gamma c$ heuristic, we perform a sweep over γ with fixed T and ρ. In the updated evaluation protocol, γ is selected by nested tuning: we tune on a separate split/seed set and report the selected $\gamma^\star$ only on held-out evaluation seeds. Concretely, in the budget-sweep experiment we choose

$$\gamma^\star \in \arg\max_{\gamma} \widehat{R}_{\text{tune}}(\gamma) \quad \text{subject to} \quad \widehat{\text{spent}/B}_{\text{tune}}(\gamma) \geq 0.99,$$

with a fallback to the largest-spend candidate if no γ satisfies the utilization threshold. This is why the visually best reward point in Fig. 3 need not coincide with the $\gamma^\star$ reported later in Table 4: tuning does not optimize reward alone. Since a fixed penalty may cause the method to underutilize the available budget, we report both total reward and budget utilization across the γ sweep.

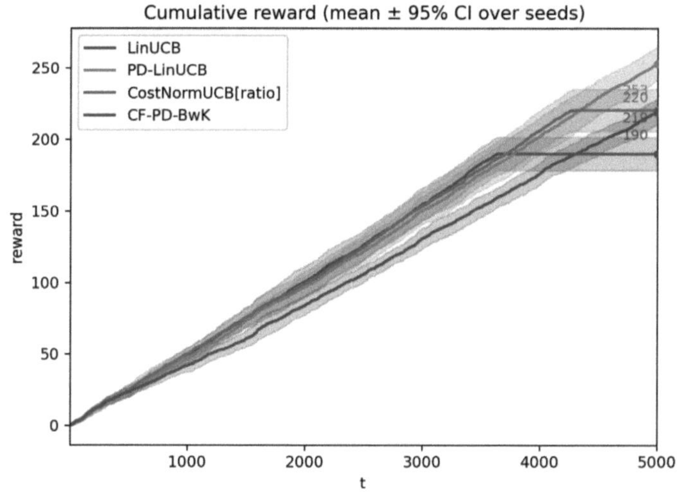

Fig. 1. Cumulative reward under the stop-at-budget protocol (mean ± 95% CI over seeds). Right-edge labels mark the final mean reward for each method.

Table 3. Delay ablation: empirical delays with censoring versus `no-delay`($D_t \equiv 0$)

Method	Regime	Reward	95% CI	Spent/B	95% CI	τ	95% CI
LinUCB	Delayed	163.6	9.0	0.999935	0.000031	3644.3	9.1
	No-delay	165.3	12.8	0.999971	0.000014	3652.0	3.8
PD-LinUCB	Delayed	220.3	11.3	0.999832	0.000094	4999.6	0.6
	No-delay	219.9	10.7	0.999793	0.000109	4999.9	0.2
CostNormUCB[ratio]	Delayed	191.1	11.9	0.999957	0.000023	4244.4	10.9
	No-delay	189.1	8.5	0.999940	0.000032	4246.2	8.9
CF-PD-BwK	Delayed	198.1	9.1	0.999513	0.000575	5000.0	0.0
	No-delay	207.3	11.8	0.999967	0.000024	5000.0	0.0

5.7 Delay Vs No-Delay

To assess the impact of delayed feedback, we compare two regimes: (i) empirical delays with observation-window censoring and (ii) the `no-delay` ablation ($D_t \equiv 0$).

We measure changes in total reward and reward variance, and also record the effect on stopping time τ and budget utilization spent/B. Because the delayed regime with $T = 5000$ and *censor_steps* $= 5000$ exposes little or no within-horizon negative feedback, this ablation should be interpreted as a change in information structure rather than as a small perturbation.

No-Delay Mode. The `no-delay` ablation is implemented by forcing all delays to zero: *censor_steps* $= 0$ and `delays_pos` $= \{0\}$, so the environment always returns $D_t \equiv 0$ for both $r = 1$ and $r = 0$.

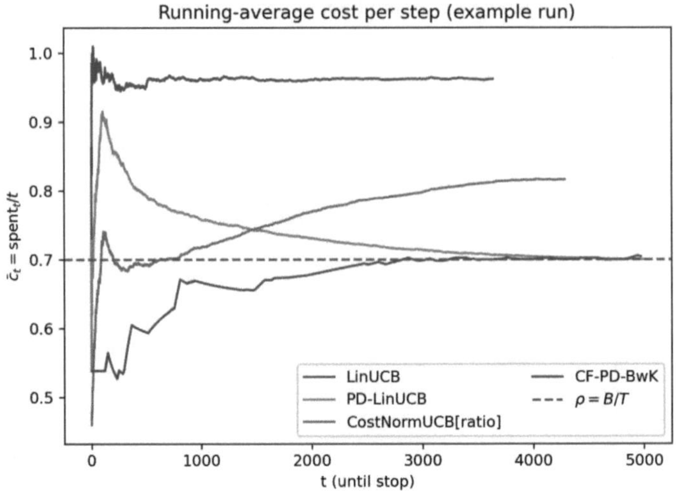

Fig. 2. Running-average cost per step in an illustrative run. The dashed line marks the target spending rate $\rho = B/T$; methods above it overspend early and stop sooner, while methods that adapt toward it sustain the horizon more evenly.

Table 4. Budget sweep over $\rho = B/T$: PD-LinUCB versus the tuned CostNormUCB[sub] baseline selected by nested tuning.

ρ	Method	Reward	95% CI	Spent/B	95% CI	τ	95% CI
0.40	PD-LinUCB	235.5	11.5	0.999896	0.000054	4001.2	1.8
0.40	CostNormUCB[sub] ($\gamma^\star = 10.0$)	222.7	11.2	0.999873	0.000055	4026.0	4.2
0.55	PD-LinUCB	273.0	18.3	0.999513	0.000223	5000.0	0.0
0.55	CostNormUCB[sub] ($\gamma^\star = 5.0$)	260.2	10.3	0.999906	0.000033	4823.7	8.7
0.70	PD-LinUCB	258.7	7.0	0.999818	0.000131	4999.9	0.2
0.70	CostNormUCB[sub] ($\gamma^\star = 2.0$)	237.7	13.5	0.999941	0.000029	4579.7	12.2
0.85	PD-LinUCB	253.9	9.4	0.999914	0.000049	4999.4	0.6
0.85	CostNormUCB[sub] ($\gamma^\star = 1.0$)	262.2	6.6	0.999973	0.000021	4966.0	11.4

Table 3 reports the quantitative comparison between the delayed and no-delay regimes.

5.8 Budget Sweep $\rho = B/T$

The key experiment sweeps the budget ratio ρ at fixed horizon T. We compare PD-LinUCB with the nested-tuned CostNormUCB[sub] baseline and show that the optimal fixed penalty $\gamma^\star$ depends strongly on ρ, whereas PD-LinUCB adapts automatically.

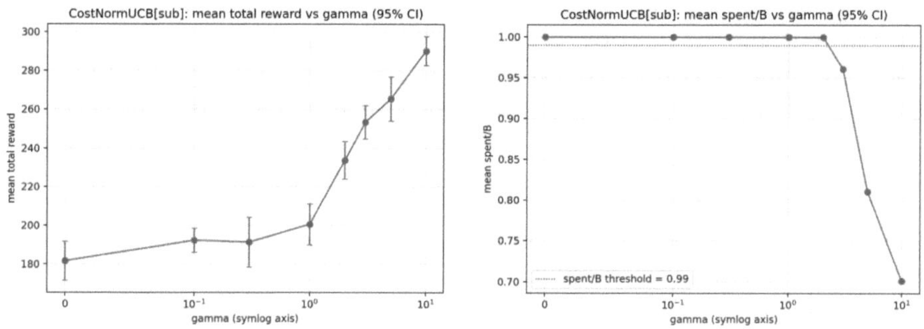

(a) Total reward (mean ± 95% CI).

(b) Budget utilization: spent/B (mean ± 95% CI).

Fig. 3. Sweep over the penalty γ for the UCB $- \gamma c$ baseline at $\rho = 0.7$ and $T = 5000$. Left: total reward. Right: budget utilization spent/B. The vertical marker shows the tuned $\gamma^\star$, and the horizontal line marks the utilization threshold 0.99.

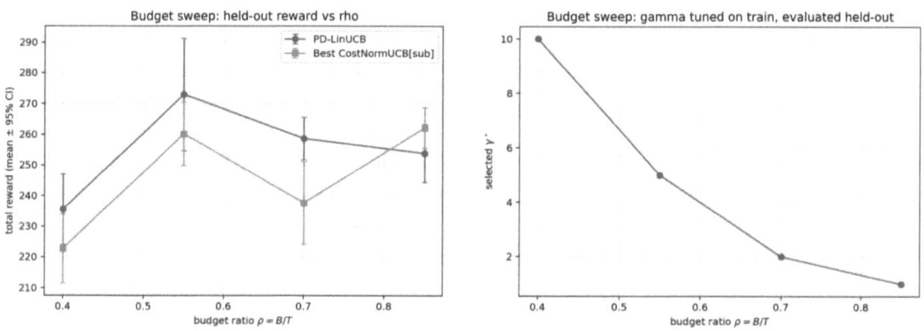

(a) Held-out reward versus ρ.

(b) Selected fixed penalty $\gamma^\star(\rho)$ from nested tuning.

Fig. 4. Budget sweep with nested tuning: the tuned fixed-penalty baseline remains regime-dependent, while PD-LinUCB does not require manual retuning across budget levels. The left panel shows held-out reward versus ρ, and the right panel shows the selected tuned penalty $\gamma^\star(\rho)$.

Table 4 reports the held-out numerical results across budget regimes, and Fig. 4 visualizes the corresponding reward trends together with the regime dependence of the tuned fixed penalty.

6 Conclusion

We studied contextual bandits with a global budget constraint and delayed rewards, where costs are observed immediately but rewards arrive later. To evaluate this setting, we implemented a primal–dual extension of Disjoint LinUCB and built a semi-synthetic benchmark on the Criteo Attribution dataset.

Across the experiments, PD-LinUCB was generally more reliable than standard LinUCB and fixed cost-penalty heuristics. Replacing a fixed penalty with an adaptive multiplier reduced sensitivity to the budget regime and the need for manual retuning, whereas the static UCB $-\gamma c$ baseline often underused the budget and depended strongly on γ. These findings are consistent with the primal–dual view of budgeted online learning: a fixed heuristic uses a static resource price, while PD-LinUCB updates this price online through the dual variable. Delays affect reward learning rather than feasibility, since costs are observed immediately and the stop-at-budget rule is enforced at execution time. We do not interpret the results as a delayed-regret guarantee, because the confidence term remains heuristic under delayed updates. The benchmark remains semi-synthetic: rewards are generated from fitted arm-specific models, campaigns are hashed into arms, and held-out contexts are sampled IID rather than replayed chronologically. Future work includes context-dependent costs, multiple-resource constraints, more expressive reward models such as GLM-UCB [7], and non-stationary delayed environments.

References

1. Agrawal, S., Devanur, N.R.: Bandits with concave rewards and convex knapsacks. In: Proceedings of the 15th ACM Conference on Economics and Computation (EC) (2014)
2. Agrawal, S., Devanur, N.R., Li, L.: An efficient algorithm for contextual bandits with knapsacks, and an extension to concave objectives. In: Proceedings of The 29th Conference on Learning Theory (COLT) (2016)
3. Badanidiyuru, A., Kleinberg, R., Slivkins, A.: Bandits with knapsacks. J. ACM (2018)
4. Bouvier, L., Prunet, T., Leclère, V., Parmentier, A.: Primal-dual algorithm for contextual stochastic combinatorial optimization. arXiv preprint arXiv:2505.04757 (2025)
5. Diemert, E., Meynet, J., Galland, P., Lefortier, D.: Attribution modeling increases efficiency of bidding in display advertising. In: Proceedings of the ADKDD and TargetAd Workshop at KDD (2017). https://dl.acm.org/doi/10.1145/3124749.3124752
6. Elumar, E.C., Tekin, C., Yagan, O.: Bandits with anytime knapsacks. arXiv preprint arXiv:2501.18560 (2025)
7. Filippi, S., Cappé, O., Garivier, A., Szepesvári, C.: Parametric bandits: the generalized linear case. In: Advances in Neural Information Processing Systems (NeurIPS) (2010). https://sites.ualberta.ca/~szepesva/papers/GenLinBandits-NeurIPS2010.pdf
8. Guo, H., Liu, X.: On stochastic contextual bandits with knapsacks in small budget regime. In: International Conference on Learning Representations (ICLR) (2025)
9. Harsha, P., et al.: Practical contextual bandits for large-scale structured discrete constrained optimization problems. In: Proceedings of the 31st ACM SIGKDD Conference on Knowledge Discovery and Data Mining (KDD), pp. 850–859 (2025)
10. van der Hoeven, D., Zierahn, L., Lancewicki, T., Rosenberg, A., Cesa-Bianchi, N.: A unified analysis of nonstochastic delayed feedback for combinatorial semi-bandits, linear bandits, and MDPs. In: Proceedings of the 36th Conference on Learning Theory (COLT) (2023). https://proceedings.mlr.press/v195/hoeven23a.html

11. Ji, S.: Inventory-constrained online learning for revenue management with delayed feedback. Transp. Res. Part E: Logist. Transp. Rev. **208**, 104649 (2026)
12. Joulani, P., Gyorgy, A., Szepesvári, C.: Online learning under delayed feedback. In: International Conference on Machine Learning, pp. 1453–1461. PMLR (2013)
13. Li, L., Chu, W., Langford, J., Schapire, R.E.: A contextual-bandit approach to personalized news article recommendation. In: Proceedings of the 19th International Conference on World Wide Web (WWW), pp. 661–670. ACM (2010). https://doi.org/10.1145/1772690.1772758
14. Li, X., Sun, C., Ye, Y.: The symmetry between arms and knapsacks: a primal-dual approach for bandits with knapsacks. In: Proceedings of the 38th International Conference on Machine Learning (ICML) (2021). https://proceedings.mlr.press/v139/li21s.html
15. Liu, H.Y., Xiao, X.T., Zhang, L.W.: Augmented Lagrangian methods for time-varying constrained online convex optimization. J. Oper. Res. Soc. China **13**(2), 364–392 (2025)
16. Liu, K., Maghsudi, S.: Budgeted recommendation with delayed feedback. arXiv preprint arXiv:2405.11417 (2024)
17. Mahdavi, M., Jin, R., Yang, T.: Trading regret for efficiency: online convex optimization with long term constraints. In: Advances in Neural Information Processing Systems (NeurIPS) (2012)
18. Miao, S., Wang, Y., Zhang, J.: A primal-dual approach toward resource-constrained revenue management with demand learning and large action space. Oper. Res. (2025)
19. Rivera Cardoso, A., Wang, H., Xu, H.: The online saddle point problem and online convex optimization with knapsacks. Math. Oper. Res. **50**(1), 1–39 (2025)
20. Sarkar, D., Sinha, A.: A simple reduction scheme for constrained contextual bandits with adversarial contexts via regression. arXiv preprint arXiv:2602.05019 (2026)
21. Schlisselberg, O., Lancewicki, T., Auer, P., Mansour, Y.: Improved best-of-both-worlds regret for bandits with delayed feedback. arXiv preprint arXiv:2505.24193 (2025)
22. Simchi-Levi, D., Xu, Y., Zhao, J.: Blind network revenue management and bandits with knapsacks under limited switches. Oper. Res. **73**(5), 2496–2514 (2025)
23. Slivkins, A., Zhou, H., Sankararaman, K., Foster, D.J.: Contextual bandits with packing and covering constraints: a modular Lagrangian approach via regression. J. Mach. Learn. Res. **25** (2024). https://www.jmlr.org/papers/volume25/24-1220/24-1220.pdf
24. Vernade, C., Cappé, O., Perchet, V.: Stochastic bandit models for delayed conversions. In: Proceedings of the 33rd Conference on Uncertainty in Artificial Intelligence (UAI) (2017). https://auai.org/uai2017/proceedings/papers/137.pdf
25. Vernade, C., Carpentier, A., Lattimore, T., Zappella, G., Ermis, B., Brückner, M.: Linear bandits with stochastic delayed feedback. In: Proceedings of the 37th International Conference on Machine Learning (ICML), pp. 9712–9721 (2020). https://proceedings.mlr.press/v119/vernade20a.html
26. Zhao, Y., Yang, L.: Constrained contextual bandit algorithm for limited-budget recommendation system. Eng. Appl. Artif. Intell. **128**, 107558 (2024)

Scheduling

En-Route Processing in the Two-Machine Routing Open Shop: Complexity and Approximation

Ilya Chernykh[1(✉)] and Alexandra Shigina[2]

[1] Sobolev Institute of Mathematics, Koptyug ave. 4, 630090 Novosibirsk, Russia
idchern@gmail.com

[2] Novosibirsk State University, Pirogova st. 2, 630090 Novosibirsk, Russia
a.shigina@g.nsu.ru

Abstract. We introduce and study en-route processing in the two-machine routing open shop, a generalization of the classical routing open shop problem in which jobs may be located on the edges of the transportation network, rather than exclusively at its nodes. A machine visiting such a job may process it while traversing the edge on which it resides, entering from either endpoint and emerging at the other upon completion; the edge remains available for transit by other machines throughout. This model is further motivated by a recent instance reduction technique for the two-machine routing open shop with the preservation of the standard lower bound, whose reduced form naturally gives rise to edge-located jobs with en-route processing. The objective is to minimize the makespan. The problem is NP-hard in general, as it simultaneously generalizes the two-machine routing open shop and the Rural Postman Problem. We identify a polynomially solvable special case arising from natural restrictions on the network topology. We establish NP-hardness for a minimal structurally non-trivial configuration. For the general two-machine problem we present a 2-approximation algorithm.

Keywords: Routing open shop · En-route processing · Approximation algorithm · Optima localization

1 Introduction

The open shop scheduling problem, introduced in [11], is concerned with scheduling of a set of n jobs on a set of m machines when each job must be processed by every machine exactly once, but the order of such operations is unrestricted, provided that two operations of the same job/machine are not processed simultaneously. The processing times of all operations are known in advance. The objective is typically to minimize the makespan—the completion time of the last

The research was supported by Russian Science Foundation grant N 22-71-10015-П, https://rscf.ru/en/project/22-71-10015-II/.

A. Kazakov et al. (Eds.): MOTOR 2026, LNCS 16687, pp. 199–213, 2026.
https://doi.org/10.1007/978-3-032-31390-4_13

operation, denoted as $C_{\max}$. According to the classical three-field notation for scheduling problems [13], open shop with m machines is denoted as $Om||C_{\max}$. The $O2||C_{\max}$ problem is known to be polynomially solvable, while $O3||C_{\max}$ being NP-hard [11]. It is still an open question, whether $Om||C_{\max}$ is strongly NP-hard for some constant $m \geqslant 3$. When m is a part of input, the problem is denoted as $O||C_{\max}$ and is strongly NP-hard, and cannot be approximated with worst-case performance ratio less than 5/4, unless $P = NP$ [18].

This classical problem assumes that machines can switch between jobs instantaneously. However, in real-world applications significant delay between consecutive operations of a machine could be unavoidable due to technological restrictions, switching tools, cleaning workspace etc. There is a number of ways one can introduce such delays in a scheduling model (see *e.g.* [3,15,17]). In another approach jobs are distributed across nodes of a *transportation network*, which has to be traversed by a fleet of mobile machines, travel times represent the delays mentioned above. The machines start from the common predefined *depot* and have to return there after processing all operations. The goal is to minimize the makespan $R_{\max}$, *i.e.* the completion time of the last activity of the machine, which is either the processing of some operation or traveling to the depot. This routing open shop problem was introduced in [1,2], where its NP-hardness was established even for two machines on simplest network topology with two nodes and first approximation algorithms were developed. Subsequent work has refined these approximation ratios [6,12,19], explored structural properties of optimal schedules, and identified polynomially solvable special cases [4,5,7]. We denote m-machine open shop routing problem as $ROm||R_{\max}$, or $ROm|G = X|R_{\max}$ in case we want to specify the structure of the transportation network G.

A further generalization idea comes from a natural question: what if jobs can also reside on the edges of the network? Such a generalization is motivated by two distinct directions. Firstly, in broader literature on arc routing, including the Chinese Postman Problem [9] and Rural Postman Problem [14], demands or tasks are assumed to be located on edges rather than on nodes. In those problems, a vehicle must traverse certain edges to service them, but no scheduling structure (such as machine assignment or operation ordering) is imposed. Secondly, and more directly relevant to the routing open shop, recent work [5] establishes an instance reduction technique that transforms any two-machine routing open shop instance into a canonical form with at most six jobs. While this reduced form has a simple transportation network structure (see Fig. 1), it introduces a new feature absent from the original routing open shop problem: edge-located jobs. This provides strong motivation for studying the two-machine routing open shop with en-route processing as an independent model, discussed in this paper.

In this model, jobs may be located either at nodes or on edges of the transportation network. An edge contains at most one job, while nodes can contain a number of jobs. In order to process an edge-located job, a machine traverses the edge, entering from either endpoint and emerging at the opposite endpoint upon completion. The time required to traverse an edge while processing a job located on it is the sum of the edge's travel time and the job's processing time for that

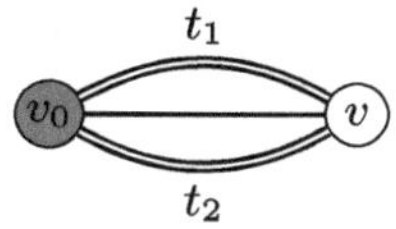

Fig. 1. Transportation network with two edge-located jobs.

machine. Crucially, the edge remains available for traversal by the other machine throughout, but no simultaneous processing of the job is allowed. The standard two-machine routing open shop can be seen as a special case where all jobs are node-located and no en-route processing occurs. We denote the m-machine routing open shop with en-route processing as $\overline{\overline{R}}Om||R_{\max}$ (or $\overline{\overline{R}}Om|G = X|R_{\max}$ to specify the structure of the transportation network G), where $\overline{\overline{R}}$ represents routing with en-route processing on tunnels.

This model arises from applications in geophysical surveying with incompatible sensor systems. Two mobile units equipped with different sensor types (e.g., electromagnetic and seismic) must collect data from a network of locations, but cannot operate simultaneously at the same site due to physical interference between their measurement modalities. Point measurements at fixed locations correspond to node jobs, while linear features such as pipelines or geological transects requiring continuous data collection correspond to tunnel jobs processed en-route. Both vehicles operate from a central depot (base camp), and the objective is to minimize the total survey time while ensuring each location receives measurements from both sensor types.

Our main contributions are as follows:

1. **Polynomially solvable case.** We identify a special case (star structure with central depot) for which an optimal schedule achieving makespan $R^*_{\max} = \overline{\overline{R}}$ can be constructed in $O(n)$ time.
2. **Minimal NP-hard case.** We establish NP-hardness for the case of two edge-located jobs on a parallel edges, demonstrating a tight boundary between tractable and intractable instances.
3. **2-approximation algorithm.** We present a 2-approximation algorithm for the general two-machine routing open shop with en-route processing.

The remainder of the paper is organized as follows. Section 2 provides formal definitions and notation. Section 3 presents a polynomially solvable special case and establishes NP-hardness for a minimal hard case. Section 4 presents the 2-approximation for the general two-machine problem. The conclusion follows in Sect. 5.

2 Preliminaries

2.1 Problem Formulation

We now give a formal description of the two-machine routing open shop with en-route processing.

Two machines M_1 and M_2 are to process jobs from $\mathcal{J} = \{J_1, \ldots, J_n\}$. Each machine M_i must perform an operation O_{ji} on each job J_j. The processing times p_{ji} are given for each operation O_{ji}. The operations can be processed in any partial order, provided that operations of the same job, or two operations of the same machine, are not processed simultaneously. To be precise, the intersection of the processing time intervals must not contain any interior point of either interval.

The transportation network is represented by a connected graph $G = (V, E)$ with a predefined depot $v_0 \in V$. Each edge e has a weight $w(e)$ which represents the travel time over e. Machines are mobile, and are allowed to traverse the transportation network without any restriction: nodes can be visited multiple times, edges can be used for simultaneous traveling of both machines in any direction. We assume shortest paths in G are precomputed; the travel time between $v, u \in V$ via a shortest path is denoted $\mathbf{dist}(v, u)$.

Jobs from $\mathcal{J}$ are located either at nodes or on edges of G. Nodes can contain multiple of jobs; the set of jobs located at node $v \in V$ is denoted as $\mathcal{J}(v)$. A node v is *active* if it contains at least one job ($\mathcal{J}(v) \neq \emptyset$) or is the depot ($v = v_0$). The set of active nodes is denoted as $V_A \subseteq V$.

Each edge contains at most one job. Edges containing jobs are called *tunnels* and form the subset $\mathcal{T} \subseteq E$. Tunnel $t \in \mathcal{T}$ has a *near endpoint* $v^-(t)$, which is closest to the depot, and a *far endpoint* $v^+(t)$ (with ties broken arbitrary). For any $t \in \mathcal{T}$, let $j(t)$ denote the index of the job located on t, so the job is $J_{j(t)}$. An example transportation network is shown at Fig. 2.

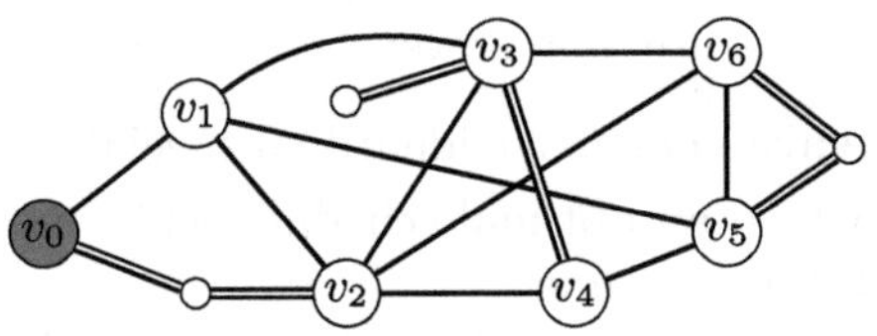

Fig. 2. An example of a transportation network G. Named nodes are active, tunnels are represented as double lines.

Both machines are initially located at the depot v_0 and can traverse the network G with unit speed. An operation on a node-located job can be performed only when the machine is positioned at the same node. Operations on edge-located jobs are performed while the machine traverses the edge. The machine may enter from either endpoint of the edge and then it exits at the other endpoint upon completing the operation. The time required to complete such an operation is the sum of the processing time p_{ji} and the travel time $w(t)$ of edge t. Note that en-route processing by one machine does not prevent the other machine from traversing the same edge.

At any time, each machine is in one of four states: (i) traversing an edge without processing, (ii) processing an operation at a node, (iii) processing an operation while traversing an edge (en-route), or (iv) idle.

A schedule $\sigma = (S, \Theta)$ consists of:

- starting times $S = \{s_{ji}(\sigma) \mid i \in \{1,2\},\ j \in \{1,\dots,n\}\}$, and
- entry endpoints $\Theta = \{\theta_{j(t),i}(\sigma) \in \{v^-(t), v^+(t)\} \mid i \in \{1,2\},\ t \in \mathcal{T}\}$ for en-route operations.

For a job $J_j = J_{j(t)}$ on tunnel t and machine M_i, the *exit endpoint* $\theta'_{j(t),i}(\sigma)$ is chossen to satisfy

$$\{\theta_{j(t),i}(\sigma), \theta'_{j(t),i}(\sigma)\} = \{v^-(t), v^+(t)\}.$$

For consistency we define entry and exit endpoints for operations of node-located jobs: for each $J_j \in \mathcal{J}(v)$ and each M_i

$$\theta_{ji}(\sigma) = \theta'_{ji}(\sigma) = v.$$

The completion time of operation O_{ji} in schedule σ is defined as

$$c_{ji}(\sigma) = \begin{cases} s_{ji}(\sigma) + p_{ji}, & \text{if } J_j \in \mathcal{J}(v) \text{ for some } v \in V, \\ s_{ji}(\sigma) + p_{ji} + w(t), & \text{if } J_j = J_{j(t)} \text{ for some } t \in \mathcal{T}. \end{cases}$$

When the schedule is clear from context, we omit the argument σ and write c_{ji}, s_{ji}, etc.

A schedule $\sigma = (S, \Theta)$ is *feasible* if the following conditions hold:

1. *Dependent operations are not processed simultaneously.* The processing time intervals $I_1 = [s_{ji}, c_{ji}]$ and $I_2 = [s_{j'i'}, c_{j'i'}]$ of any two different operations O_{ji} and $O_{j'i'}$ with $i = i'$ or $j = j'$ must satisfy:

$$I_1 \cap \mathrm{int}(I_2) = I_2 \cap \mathrm{int}(I_1) = \emptyset,$$

   where $\mathrm{int}(I)$ is the interior of interval I.
2. *Initial positioning.* For each machine M_i and each operation O_{ji}:

$$s_{ji} \geq \mathbf{dist}(v_0, \theta_{ji}).$$

3. *Travel time between consecutive operations.* For each machine M_i and any two operations O_{ji} and $O_{j'i}$ such that no operation $O_{j''i}$ satisfies $c_{ji} < s_{j''i} < s_{j'i}$:

$$s_{j'i} \geq c_{ji} + \mathbf{dist}(\theta'_{ji}, \theta_{j'i}).$$

After completing all the operations, both machines must return to the depot. The completion time R_i of machine M_i is the maximum of its last operation completion time C_i and its arrival time back to the depot. The objective is to construct a feasible schedule minimizing the makespan $R_{\max} = \max\{R_1, R_2\}$. We use $R_{\max}$ rather than the standard $C_{\max}$ for the makespan to emphasize that the objective is the maximum completion time of the machines (including return to depot), not the maximum completion time of individual operations.

2.2 Notation and Definitions

Throughout the paper, we use the following notation.

For machine M_i, let

$$\ell_i = \sum_{j=1}^{n} p_{ji}$$

denote the total processing time (*machine load*), and for job J_j, let

$$d_j = p_{j1} + p_{j2}$$

denote the total processing time across both machines (*job duration*).

Let $\ell_{\max} = \max\limits_{i=1,2} \ell_i$ be the maximum machine load and $d_{\max}(v) = \max\limits_{j \in \mathcal{J}} d_j$ be the maximum job duration at node v.

For active node $v \in V_A$, let

$$\Delta(v) = \sum_{j \in \mathcal{J}(v)} d_j$$

denote the *load* of node v. *Total load* of the instance is

$$\Delta = \sum_{j \in \mathcal{J}} d_j = \sum_{i=1,2} \ell_i.$$

The routing component of our problem is closely related to the *Rural Postman Problem* (RPP) [14], defined as follows: given a graph $G = (V, E)$ with edge weights, and a subset of *required edges* $E_R \subseteq E$, find a minimum-length closed walk that traverses every edge in E_R at least once.

In our setting, the underlying RPP instance, on top of required edges $\mathcal{T}$ (the tunnels), also has *required nodes* V_A (active nodes), which must be visited by each machine at least once. Note that we can reduce this problem to the standard RPP by adding zero-weight required loops to each active node. We denote the optimal solution value of this RPP instance by T^*_{RPP}.

Proposition 1. *For any instance of* $\overline{\overline{R}}O2||R_{\max}$*, the optimal makespan satisfies*

$$R^*_{\max} \geq \overline{\overline{R}} = \max\Big\{ \ell_{\max} + T^*_{RPP}, \max_{v \in V_A} \left(d_{\max}(v) + 2\mathbf{dist}(v_0, v) \right), \max_{t \in \mathcal{T}} \left(d_{j(t)} + 2w(t) + 2\mathbf{dist}(v_0, v^-(t)) \right) \Big\}. \quad (1)$$

Proof. The first component $\ell_{\max} + T^*_{\text{RPP}}$ is a lower bound on the optimal makespan because each machine must visit every tunnel and every active node (to process its operations on all jobs). Thus, each machine must complete a routing tour of length at least T^*_{RPP}. Additionally, one machine must process at least $\ell_{\max}$ time worth of operations. Therefore, the machine with maximum load requires time at least $\ell_{\max} + T^*_{\text{RPP}}$.

To justify the second component $\max_{v \in V_A}(d_{\max}(v) + 2\mathbf{dist}(v_0, v))$, consider a job $J_j \in \mathcal{J}(v)$ with maximum duration $d_j = d_{\max}(v)$. A machine must first travel from v_0 to v (time $\mathbf{dist}(v_0, v)$), then both operations on J_j must be completed (total time d_j), and finally at least one machine must return to v_0 (time $\mathbf{dist}(v_0, v)$). Thus $R^*_{\max} \geq d_j + 2\mathbf{dist}(v_0, v)$.

For the third component, consider tunnel $t \in \mathcal{T}$ with job $J_{j(t)}$. A machine must travel from v_0 to one of the tunnel's endpoints (needs at least $\mathbf{dist}(v_0, v^-(t))$), traverse the tunnel while processing (time $p_{j(t),i} + w(t)$ for machine M_i), and the other machine M_{3-i} must also process $J_{j(t)}$ (time $p_{j(t),3-i} + w(t)$) and eventually return to v_0 (at least $\mathbf{dist}(v_0, v^-(t))$ time). These sums to $d_{j(t)} + 2w(t) + 2\mathbf{dist}(v_0, v^-(t))$, making it another lower bound. □

Note that $\Delta = \ell_1 + \ell_2 \leqslant 2\ell_{\max}$, therefore

$$\Delta \leqslant 2\left(\overline{\overline{R}} - T^*_{RPP}\right). \tag{2}$$

3 Complexity

3.1 Polynomial Case: Star with Central Depot

We identify a restricted class of instances that can be solved optimally in polynomial time. The restriction relies on a structure of the transportation network, defined as follows.

Definition 1. *A transportation network G has* tunnel star structure *if G is a star with center v_0 (the depot), all edges are tunnels ($\mathcal{T} = E$), and the depot is the only active node ($V_A = \{v_0\}$). This structure is denoted as St'.*

Theorem 1. *The problem $\overline{\overline{R}}O2|G = St'|R_{\max}$ admits an $O(n)$-time algorithm that computes an optimal schedule with makespan $R^*_{\max} = \overline{\overline{R}}$.*

Proof. Consider an instance of $\overline{\overline{R}}O2|G = St'|R_{\max}$. Perform the following **Tunnel reduction**:

For each tunnel $t \in \mathcal{T}$ satisfying

$$d_{j(t)} + 4w(t) \leq \overline{\overline{R}} \tag{3}$$

perform:

- Remove tunnel t from G and job $J_{j(t)}$ from $\mathcal{J}$.
- Add a new job J^t at depot v_0 with processing times

$$p^t_i = p_{j(t),i} + 2w(t) \quad \text{for } i \in \{1, 2\}.$$

Note that this transformation does not increase the standard lower bound $\overline{\overline{R}}$: the duration of the new job J^t contributes is exactly $d_{j(t)} + 4w(t)$, which by (3) is at most $\overline{\overline{R}}$ itself.

Moreover, this transformation is reversible: any feasible schedule for the modified instance can be converted into a feasible schedule for the original instance with the same makespan. Specifically, the processing of an operation of the transformed job J^t can be viewed as the respective machine performing the en-route operation on $J_{j(t)}$ (time $p_{j(t),i} + w(t)$) and returning to the depot (time $w(t)$).

Case 1: All Tunnels are Transformed

If condition (3) holds for every tunnel $t \in \mathcal{T}$, then after the transformation we obtain a classical two-machine open shop instance $O2||C_{\max}$ with all jobs located at the depot. For this problem, the lower bound is $\max\{\ell_{\max}, d_{\max}\}$, and the Gonzalez-Sahni algorithm [11] computes an optimal schedule achieving this bound in $O(n)$ time.

Since the tunnel transformation can be performed in $O(n)$ time, the total running time for Case 1 is $O(n)$. By the reversibility of the transformation, this yields a schedule for the original instance with makespan $R_{\max} = \max\{\ell_{\max}, d_{\max}\} = \overline{\overline{R}}$, which is therefore optimal.

Case 2: At Least One Tunnel Remains

Let's prove that in this case the transformed instance has exactly one remaining tunnel. Suppose otherwise, and there are tunnels t_1 and t_2 for which condition (3) does not hold, i.e., $d_{j(t_1)} + 4w(t_1) > \overline{\overline{R}}$ and $d_{j(t_2)} + 4w(t_2) > \overline{\overline{R}}$. Due to the star structure,

$$T^*_{RPP} = 2\sum_{t\in\mathcal{T}} w(t). \tag{4}$$

Then adding these inequalities and (4), we have

$$d_{j(t_1)} + d_{j(t_2)} > 2\overline{\overline{R}} - 4w(t_1) - 4w(t_2) \geqslant 2\overline{\overline{R}} - 2T^*_{RPP},$$

which contradicts (2).

We now construct an optimal schedule for the transformed instance with the single remaining tunnel t. Denote

$$A = \ell_1 - p_{j(t),1}, \quad B = \ell_2 - p_{j(t),2}.$$

Machine M_1 processes operation $O_{j(t),1}$ en-route through tunnel t (time $p_{j(t),1} + w(t)$), returns to the depot ($w(t)$), and at time $\max\{B, p_{j(t),1} + 2w(t)\}$ continues processing the remaining jobs at the depot in arbitrary order, taking A time units.

Meanwhile, machine M_2 processes the jobs at the depot in arbitrary order (taking time B), traverses through the tunnel t ($w(t)$), and at time $\max\{B + w(t), p_{j(t),1} + w(t)\}$ performs en-route processing of its operation on job $J_{j(t)}$ (time $p_{j(t),2} + w(t)$).

The schedule is feasible, because neither processing of jobs at the depot nor the en-route processing of job $J_{j(t)}$ overlap by construction.

Machine M_2 finishes at time $R_2 = \max\{B + w(t), p_{j(t),1} + w(t)\} + p_{j(t),2} + w(t) = \max\{\ell_2 + 2w(t), d_{j(t)} + 2w(t)\} \leqslant \overline{\overline{R}}$.

Machine M_1 finishes at time $R_1 = \max\{B, p_{j(t),1} + 2w(t)\} + A = \max\{\ell_1 + 2w(t), A + B\}$. Let's prove that $A + B \leqslant \overline{\overline{R}}$.

Using (2), (4), and the fact, that (3) does not hold for tunnel t:

$$A + B = \Delta - d_{j(t)} < 2\overline{\overline{R}} - 2T^*_{RPP} - (\overline{\overline{R}} - 4w(t)) = \overline{\overline{R}}.$$

The makespan of the schedule built is $\overline{\overline{R}}$ and it is constructed explicitly in $O(n)$ time.

By the reversibility of the tunnel transformation, this yields an optimal schedule for the original instance with makespan $R^*_{\max} = \overline{\overline{R}}$. The schedule can be constructed in $O(n)$ time. □

3.2 NP-Hardness: Two Parallel Tunnels

We now establish NP-hardness for a minimal configuration.

Consider the following structure (see Fig. 3):

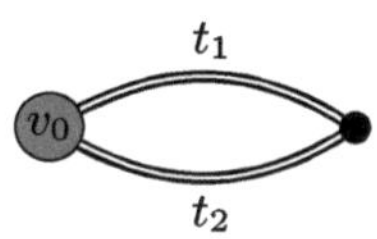

Fig. 3. The structure PT of the transportation network.

Definition 2. *A transportation network G has* parallel tunnel structure, *if it consists of two nodes v_0 (depot) and v, connected by two tunnels $t_1, t_2 \in \mathcal{T}$, with only v_0 being an active node. This structure is denoted as PT.*

Theorem 2. *The problem $\overline{\overline{R}}O2|G = PT|R_{\max}$ is NP-hard.*

Proof. The proof is by reduction from the well-known Partition problem.

Given an instance of Partition with positive integers $a_1, \ldots, a_N$ satisfying $\sum_{i=1}^{N} a_i = 2K$, we construct an instance of $\overline{\overline{R}}O2|G = PT|R_{\max}$ as follows:

- There are N jobs $J_1, \ldots, J_N$ at the depot v_0 with processing times $p_{j1} = 0$ and $p_{j2} = a_j$, $j = 1, \ldots, N$.
- The tunnels t_1 and t_2 have jobs $J_{j(t_1)} = J_{N+1}$ and $J_{j(t_2)} = J_{N+2}$ with identical processing times $p_{N+1,1} = p_{N+2,1} = 2K$ and $p_{N+1,2} = p_{N+2,2} = K$.
- Travel times over t_1 and t_2 are equal: $w(t_1) = w(t_2) = K$.

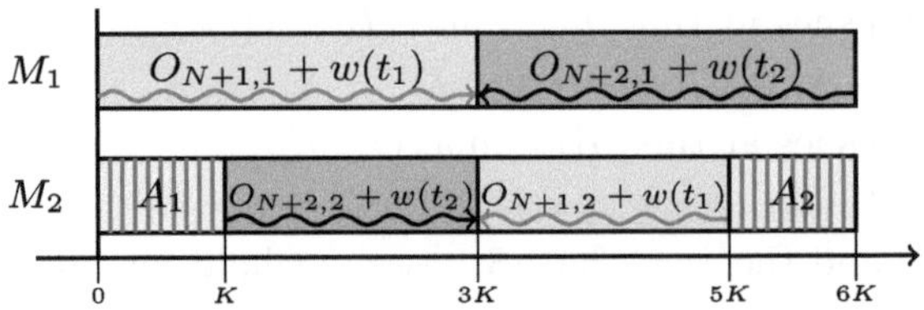

Fig. 4. The sketch of feasible schedule of makespan $6k$ in case the Partition exists. Wavy arrows indicate en-route processing, direction correspond to the starting endpoint.

We claim that the Partition instance has a solution if and only if there exists a feasible schedule for the constructed instance with makespan at most $6K$.

($\Rightarrow$): Suppose the Partition instance has a solution: $A_1 \cap A_2 = \emptyset$, $A_1 \cup A_2 = \{1, \ldots, N\}$ and $\sum_{i \in A_1} a_i = \sum_{i \in A_2} a_i = K$.

Then we construct a feasible schedule as follows (see the sketch of Gantt diagram at Fig. 4):

- Machine M_1 starts with en-route processing of job J_{N+1} from the v_0 (finishes at time $3K$) and immediately after that processes job J_{N+2} en-route from the far endpoint.
- Machine M_2 starts processing of jobs with indices from A_1 in arbitrary order without idle time, then en-route processes J_{N+2}, immediately after that processes job J_{N+1} en-route from the far endpoint and continues at the depot processing jobs with indices from A_2.

The makespan is exactly $6K$. The feasibility of this schedule follows from the construction: depot jobs are processed sequentially by M_2 without overlap, and the tunnel jobs are processed in opposite order by the two machines, ensuring their processing intervals are disjoint.

($\Rightarrow$): Suppose there exists a feasible schedule with makespan at most $6K$. Note that for the constructed instance, $\ell_1 = \ell_2 = 4K$ and $T^*_{\mathrm{RPP}} = 2K$, giving a lower bound of $\ell_{\max} + T^*_{RPP} = 6K$. Therefore, to achieve makespan exactly $6K$, each machine must process jobs without any idle time and must not traverse any tunnel without performing en-route processing. Without loss of generality we assume that M_1 processes J_{N+2} after J_{N+1} (otherwise we reverse the schedule). Then the schedule for M_1 coincides with the one from Fig 4. For M_2, feasibility dictates that the en-route processing of J_{N+2} must be contained in the interval $[0, 3K]$ (to avoid overlap with M_1's processing of J_{N+2} during $[3K, 6K]$), and the en-route processing of J_{N+1} must be contained in $[3K, 6K]$ (to avoid overlap with M_1's processing of J_{N+1} during $[0, 3K]$). Moreover, since M_2 cannot be idle or traverse edges without en-route processing, it must start en-route processing of J_{N+1} immediately after completing its operation on J_{N+2}, i.e., exactly at time $3K$ (consistent with Fig. 4). This leaves intervals $[0, K]$ and $[5K, 6K]$ for processing operations of jobs $J_1, \ldots, J_N$ by machine M_2. Since the machine cannot be idle, these intervals are tightly filled with processing operations, and the corresponding sets A_1 and A_2 of job indices satisfy $\sum_{j\in A_1} a_j = K$ and $\sum_{j\in A_2} a_j = K$, yielding a solution to the Partition instance.

□

4 A 2-Approximation Algorithm

We present a 2-approximation algorithm for the general $\overline{\overline{R}}O2||R_{\max}$ problem, adapting the framework of Averbakh et al. [2] for the classical routing open shop. Their algorithm constructs an approximate TSP tour using the Christofides-Serdyukov algorithm [8,16], which is used by both machines to traverse in opposite directions. We replace the TSP subroutine with Frederickson's 3/2-approximation for the Rural Postman Problem [10] to handle both required (active) nodes and required edges (tunnels). The algorithmic structure otherwise remains unchanged.

Algorithm A.

1. Construct the job sequence π and entry endpoints Θ:
   (a) Construct an instance of the Rural Postman Problem by adding a required loop of weight 0 to each active node.
   (b) Run Frederickson's algorithm on this instance, resulting in a tour R.
   (c) Enumerate jobs following tour R starting from v_0: when R visits node v, list all jobs at v in arbitrary order; when R first traverses tunnel t, record job $J_{j(t)}$ with entry endpoint defined by the direction of traversal in R.
2. Construct a schedule σ as follows:
   - Machine M_1 processes jobs in the order given by π, using entry endpoints from Θ.
   - Machine M_2 processes jobs in the reverse order π^{-1}, using reversed entry endpoints Θ^{-1} (i.e., if $\theta_{ji} = v^-(t)$ in Θ, then M_2 uses $v^+(t)$, and vice versa).
3. If the schedule σ is feasible, return σ and terminate.
4. Otherwise, due to the opposite processing orders, there exists exactly one job J_c whose operations overlap in σ. We call J_c the *conflict job*. Resolve the conflict by constructing two candidate schedules:
   (a) In schedule σ_1: Machine M_1 waits at the entry point of its operation on J_c until M_2 completes O_{c2}, then processes O_{c1} and continues with the remaining jobs.
   (b) In schedule σ_2: Machine M_2 waits at the entry point of its operation on J_c until M_1 completes O_{c1}, then processes O_{c2} and continues with the remaining jobs.
5. Return the schedule with smaller makespan: $\sigma_A = \arg\min\{R_{\max}(\sigma_1), R_{\max}(\sigma_2)\}$.

Note that the running time is dominated by Frederickson's algorithm for the Rural Postman Problem, which runs in $O(|E|^3)$ time [10]. The remaining steps (job enumeration and schedule construction) require $O(n)$ time.

Lemma 1. *Let σ_A be the schedule returned by Algorithm A. Then*

$$R_{\max}(\sigma_A) \leqslant \frac{3}{2}T^*_{RPP} + \ell_{\max} + \begin{cases} 0, & \text{if there is no conflict job,} \\ \frac{d_c}{2}, & \text{if } J_c \text{ is node-located,} \\ \frac{d_c}{2} + w(t), & \text{if } c = j(t),\ t \in \mathcal{T}. \end{cases} \quad (5)$$

Proof. **Case 1.** There is no conflict in schedule σ.

Then there are no idle intervals on both machines and $R_{\max}(\sigma_A) = \ell_{\max} + w(R) \leqslant \ell_{\max} + \frac{3}{2}T^*_{RPP}$.

Case 2. There is a conflict job $J_c \in \mathcal{J}(v)$ for some v in schedule σ. Without loss of generality assume that M_1 arrives at its respective entry endpoint of job J_c first, and M_2 is late by x time units (Fig. 5).

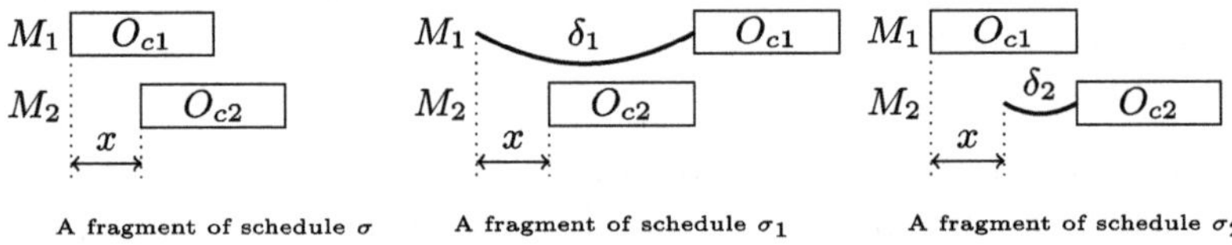

Fig. 5. Avoiding conflict at node v in Algorithm A. δ_1 and δ_2 denote respective idle times.

Then

$$R_{\max}(\sigma_1) \leqslant \ell_1 + \frac{3}{2}T^*_{RPP} + \delta_1 = \ell_1 + \frac{3}{2}T^*_{RPP} + p_{c2} + x,$$

$$R_{\max}(\sigma_2) \leqslant \ell_2 + \frac{3}{2}T^*_{RPP} + \delta_2 = \ell_2 + \frac{3}{2}T^*_{RPP} + p_{c1} - x,$$

$$\min\{R_{\max}(\sigma_1), R_{\max}(\sigma_2)\} \leqslant \frac{1}{2}(R_{\max}(\sigma_1) + R_{\max}(\sigma_2)) \leqslant \ell_{\max} + \frac{3}{2}T^*_{RPP} + \frac{d_c}{2}.$$

Case 3. There is a conflict job J_c, $c = j(t)$ for some tunnel t in schedule σ. Without loss of generality assume that M_1 arrives at its respective entry endpoint of job J_c first, and M_2 is late by x time units (Fig. 6).

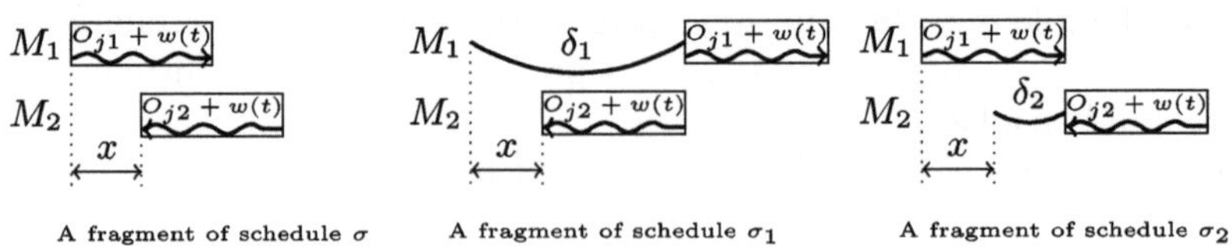

Fig. 6. Avoiding conflict at an edge-located job.

Then

$$R_{\max}(\sigma_1) \leqslant \ell_1 + \frac{3}{2}T^*_{RPP} + \delta_1 = \ell_1 + \frac{3}{2}T^*_{RPP} + p_{c2} + w(t) + x,$$

$$R_{\max}(\sigma_2) \leqslant \ell_2 + \frac{3}{2}T^*_{RPP} + \delta_2 = \ell_2 + \frac{3}{2}T^*_{RPP} + p_{c1} + w(t) - x,$$

$$\min\{R_{\max}(\sigma_1), R_{\max}(\sigma_2)\} \leqslant \ell_{\max} + \frac{3}{2}T^*_{RPP} + \frac{d_c}{2} + w(t).$$

This completes the proof of the Lemma. □

Theorem 3. *For any instance of $\overline{\overline{R}}O2||R_{\max}$, Algorithm A produces a feasible schedule with makespan $R_{\max}(\sigma_A) \leq 2\overline{\overline{R}}$ in polynomial time. This yields a 2-approximation since $\overline{\overline{R}} \leq R^*_{\max}$.*

Proof. From (5) and (1) we have

$$R_{\max}(\sigma_A) \leqslant \frac{3}{2}T^*_{RPP} + \ell_{\max} + \frac{1}{2}\overline{\overline{R}} \leqslant \frac{3}{2}(T^*_{RPP} + \ell_{\max}) + \frac{1}{2}\overline{\overline{R}} \leqslant 2\overline{\overline{R}}.$$

This concludes the proof of the Theorem. □

5 Conclusion

We introduced and studied the two-machine routing open shop with en-route processing. In this new model, jobs can be allocated not only on nodes, but on edgers of transportation network as well. Hence the considered problem is a generalization of a classical routing open shop problem where jobs are associated with exclusively with nodes.

Our contribution includes

1. A polynomially solvable special case with $G = St'$ (Theorem 1). It is proved that in the case when transportation network is a star with center depot, all edges of which are tunnels and all node-associated jobs are located at the depot, an optimal schedule with makespan $\overline{\overline{R}}$ can be constructed in linear time.
2. Minimal NP-hard special case with $G = PT$ (Theorem 2). It is shown be reduction from Partition that the special case with two parallel tunnels (Fig. 3) is NP-hard.
3. 2-approximation algorithm for the general case (Theorem 3). For two-machine routing open shop problem with en-route processing and arbitrary transportation network a feasible schedule with makespan at most $2\overline{\overline{R}}$ can be constructed in polynomial time.

The following directions can be suggested for future research:

1. Investigate a strong NP-hardness for the special case $G = PT$ (Subsect. 3.2).
2. Approximation algorithm for $\overline{\overline{R}}O2||R_{\max}$ with better approximation ratio.
3. Approximation algorithms for $\overline{\overline{R}}O3||R_{\max}$ and $\overline{\overline{R}}Om||R_{\max}$ problems.

References

1. Averbakh, I., Berman, O., Chernykh, I.: A 6/5-approximation algorithm for the two-machine routing open-shop problem on a two-node network. Eur. J. Oper. Res. **166**(1), 3–24 (2005). https://doi.org/10.1016/j.ejor.2003.06.050
2. Averbakh, I., Berman, O., Chernykh, I.: The routing open-shop problem on a network: complexity and approximation. Eur. J. Oper. Res. **173**(2), 531–539 (2006). https://doi.org/10.1016/j.ejor.2005.01.034
3. Brucker, P., Knust, S., Cheng, T.C.E., Shakhlevich, N.V.: Complexity results for flow-shop and open-shop scheduling problems with transportation delays. Ann. Oper. Res. **129**, 81–106 (2004). https://doi.org/10.1023/B:ANOR.0000030683.64615.c8
4. Chernykh, I.D., Lgotina, E.V.: Two-machine routing open shop on a tree: instance reduction and efficiently solvable subclass. Optim. Methods Softw. **36**(4), 821–841 (2020)
5. Chernykh, I.: Two-machine routing open shop: how long is the optimal makespan? In: Pardalos, P., Khachay, M., Kazakov, A. (eds.) MOTOR 2021. LNCS, vol. 12755, pp. 253–266. Springer, Cham (2021). https://doi.org/10.1007/978-3-030-77876-7_17
6. Chernykh, I., Kononov, A., Sevastyanov, S.: Efficient approximation algorithms for the routing open shop problem. Comput. Oper. Res. **40**(3), 841–847 (2013). https://doi.org/10.1016/j.cor.2012.01.006
7. Chernykh, I., Pyatkin, A.: Irreducible bin packing and normality in routing open shop. Ann. Math. Artif. Intell. **89**(8-9), 899–918 (2021). https://doi.org/10.1007/s10472-021-09759-x
8. Christofides, N.: Worst-case analysis of a new heuristic for the travelling salesman problem. Report 388, Graduate School of Industrial Administration, Carnegie-Mellon University, Pittsburg, PA (1976)
9. Edmonds, J., Johnson, E.L.: Matching, Euler tours and the Chinese postman. Math. Program. **5**(1), 88–124 (1973). https://doi.org/10.1007/BF01580113
10. Frederickson, G.N.: Approximation algorithms for some postman problems. J. ACM **26**(3), 538–554 (1979). https://doi.org/10.1145/322139.322150
11. Gonzalez, T.F., Sahni, S.: Open shop scheduling to minimize finish time. J. ACM **23**(4), 665–679 (1976). https://doi.org/10.1145/321978.321985
12. Kononov, A.: $o(\log m)$-approximation for the routing open shop problem. RAIRO - Oper. Res. **49**(2), 383–391 (2015). https://doi.org/10.1051/ro/2014051
13. Lawler, E.L., Lenstra, J.K., Rinnooy Kan, A.H.G., Shmoys, D.B.: Chapter 9. Sequencing and scheduling: algorithms and complexity. In: Logistics of Production and Inventory, Handbooks in Operations Research and Management Science, vol. 4, pp. 445–522. Elsevier (1993). https://doi.org/10.1016/S0927-0507(05)80189-6
14. Orloff, C.S.: A fundamental problem in vehicle routing. Networks **4**(1), 35–64 (1974). https://doi.org/10.1002/net.3230040105
15. Rebaine, D., Strusevich, V.A.: Two-machine open shop scheduling with special transportation times. J. Oper. Res. Soc. **50**, 756–764 (1999). https://doi.org/10.1057/palgrave.jors.2600769
16. Serdyukov, A.: On some extremal routes in graphs (in Russian). Upravlyaemye Sistemy **17**, 76–79 (1978)
17. Strusevich, V.A.: A heuristic for the two-machine open-shop scheduling problem with transportation times. Discret. Appl. Math. **93**(2–3), 287–304 (1999). https://doi.org/10.1016/S0166-218X(99)00115-8

18. Williamson, D.P., et al.: Short shop schedules. Oper. Res. **45**(2), 288–294 (1997). https://doi.org/10.1287/opre.45.2.288
19. Yu, V.F., Lin, S.W., Chou, S.Y.: Routing open shop and flow shop scheduling problems. Eur. J. Oper. Res. **213**(1), 261–269 (2011). https://doi.org/10.1016/j.ejor.2011.03.006

Scheduling Under in-Tree Precedence Constraints to Minimise Maximum Lateness

Julia Memar(✉)

University of Technology Sydney, Sydney, Australia
julia.memar@uts.edu.au

Abstract. This paper presents a worst-case performance guarantee for a modification of the Brucker-Garey-Johnson algorithm applied to a problem on parallel machines, where a set of jobs is subject to in-tree precedence constraints and jobs have arbitrary processing times. The objective is to minimise maximum lateness. We show that this performance guarantee is tight for any maximum processing time of all jobs, that is assumed to be positive integer.

Keywords: Scheduling · Approximation algorithms · Arbitrary Processing times · Maximum lateness

1 Introduction

Scheduling problem of minimising maximum lateness on parallel machines - and its particular case - minimising maximum completion time or makespan - are among the most studied problems in scheduling theory [3,15]. Despite decades of research, these scheduling problems continue to attract the attention of the academic community [11,13,14,17]. In fact, the complexity of makespan problem on identical parallel machines when the number of machines is fixed and greater than two is one of the only two problems that are still open [1] among the open problems mentioned in the classical work on the complexity of scheduling problems [8]. The new result presented in this paper is derived for the problem that has been in a focus of researchers' scrutiny since 1960's.

The problem discussed in this paper can be stated as follows. A set of jobs N is to be processed on m identical parallel machines. Each machine can process any job, and only one job at a time. Once a job's processing has commenced, the job must be completed, no pre-emtions are allowed. Each job $i \in N$ is assigned a processing time p_i, and a due date d_i - the desirable time the job must be completed by. The processing times and due dates are assumed to be integer. The jobs are subject to precedence constraints in the form of in-tree, i.e. a job cannot have more than one successor. Let $C_i(\sigma)$ be the completion time of a job i in schedule σ. The objective is to minimise maximum lateness

A. Kazakov et al. (Eds.): MOTOR 2026, LNCS 16687, pp. 214–230, 2026.
https://doi.org/10.1007/978-3-032-31390-4_14

$L_{max}(\sigma) = \max_{i \in N} C_i(\sigma) - d_i$. In a three-field notation [3,15] the problem can be stated as

$$P|intree|L_{max} \tag{1}$$

The problem above is NP-hard in a strong sense, as even its particular cases $P2|chains|L_{max}$ and $P||C_{max}$ are NP-hard in a strong sense due to [6,7], correspondingly. The motivation for this paper is the observation that the precedence constraints in the form of in-tree or chains could make the problems easier - there are sub-cases of the problem (1) that can be solved in a polynomial time, for example, $P|intree, p_i = p|L_{max}$ [4] and $P|chains, p_i = 1|L_{max}$ [2,5]. In this paper, we investigate whether the worst-case performance guarantee for the classical approximation algorithm - the Brucker-Garey-Johnston algorithm for the general problem $P|prec|L_{max}$ [14] can be improved, when the algorithm is applied to the problem (1).

The problem considered (1) has been a topic of research since the 1960s. The Hu's longest path algorithm [10], which solves $P2|intree, p_i = 1|C_{max}$, was applied to the problem with a set of jobs with arbitrary processing times $P|intree|C_{max}$ in [12] . It has been shown that the algorithm is "almost optimal" and the worst-case bound cannot be improved:

$$C_{max}(s^{Hpmtn}) \leq C_{max}(s^*) \leq C_{max}(s^{LP}) \leq C_{max}(s^{Hpmtn}) + p_{max} - \frac{p_{max}}{m}, \tag{2}$$

where s^{Hpmtn} - is the schedule obtained by Hu's longest path algorithm with preemptions allowed after each unit of time, s^* is an optimal schedule without preemptions and s^{LP} is the schedule obtained by the longest path (LP) algorithm without pre-emptions; p_{max} signifies the maximum processing time.

The Brucker-Garey-Johnson algorithm [4] (hereafter referred to as BGJ algorithm) solves $P|intree, p_i = p|L_{max}$. The BGJ algorithm is a generalisation of Hu's algorithm to the problem with the objective of maximum lateness. In [16] the following tight upper bound for the worst-case performance is derived when the BGJ algorithm is applied to the problem with general precedence constraints $P|prec, p_i = 1|L_{max}$:

$$L_{max}(s^{BGJ}) \leq \left(2 - \frac{1}{m}\right) L_{max}(s^*) + \left(1 - \frac{1}{m}\right) \max_{1 \leq j \leq n} d_j - \left(\frac{m-1}{m}\right), \tag{3}$$

where s^{BGJ} is the schedule constructed by the BGJ algorithm, and s^* is an optimal schedule.

The performance of the BGJ algorithm for $P|prec|L_{max}$ was investigated in [14]. The tight performance guarantee was obtained for cases when the maximum processing time does not exceed the number of machines:

$$L_{max}(s^{BGJ}) \leq \left(2 - \frac{1}{m}\right) L_{max}(\sigma^*) + \left(1 - \frac{1}{m}\right) d + \frac{p_{max}}{m} - p_{min}, \tag{4}$$

where s^{BGJ} is the schedule constructed by the BGJ algorithm, s^* is an optimal schedule, d is the maximum due date, p_{max} and p_{min} are the maximum and minimum processing times. It was demostrated that (4) does not hold if the maximum processing time is strictly greater than the number of machines.

In this paper, we bridge the gap in the literature and derive the worst-case performance guarantee for the BGJ algorithm applied to the problem (1):

$$L_{max}(\sigma) \leq L_{max}(\sigma^*) + p_{max} - \frac{p_{min}}{m}, \quad (5)$$

thus improving (4) for the cases when the precedence constraints are in the form of in-tree. We show that (5) holds for any maximum processing time, that is assumed to be positive integer, and that the bound is tight - we construct instances for which (5) becomes an equality - and hence the bound cannot be improved.

In what follows, Sect. 2 summarises the framework for analysis of a schedule, Section 3 provides insights into the structure of the schedule constructed by the BGJ algorithm. In Sect. 4 tight worst-case performance guarantee (5) is derived, and we show that all results from Sects. 3 and 4 also hold for the particular case of problem (1) - $P|chains|L_{max}$ - when the precedence constraints are in the form of chains. The paper concludes with a summary of results.

2 Framework for Analysis of a Schedule

To obtain the results presented in this paper, we will use the framework for analysis of a schedule developed in [14]. For the convenience of the reader, the summary is provided below.

When determining performance bounds for approximation algorithms, it is preferable to replace the maximum lateness problem with an equivalent problem with a "delivery times criterion" [9]. Hence such a criterion $G_{max}(\sigma)$ is introduced as follows. For job $i \in N$ denote by $K(i)$ the set of all successors of i: $K(i) = \{j : i \rightarrow j\}$ and by $Q(i)$ the set of all predecessors of i: $Q(i) = \{j : j \rightarrow i\}$. Let d_{max} be the maximum due date among all jobs. For each job $i \in N$, an auxiliary priority μ_i is computed: $\mu_i = d_{max} - d_i, if K(i) = \emptyset$, otherwise

$$\mu_i = \max\left\{d_{max} - d_i, \max_{j \in K(i)} p_j + \mu_j\right\} \quad (6)$$

A priority β_i is calculated as $\beta_i = p_i + \mu_i$. A schedule σ is then constructed by a list algorithm in non-increasing order of the βs. Algorithms such as Hu's longest path algorithm [10,12] or the BGJ algorithm [4,16], were scheduling jobs with a longer chain of successors earlier in a schedule to shorten the length of the schedule. As these algorithms considered problems where all jobs have the same processing times, the processing time of a job did not affect its priority. However, if jobs' processing times are arbitrary, a job's processing time affects the longest path that includes the job and its successors, and, as proposed in [14], is included in the calculation of βs. The difference between the schedules constructed in non-increasing order of μs and βs - denoted as σ^μ and σ^β, correspondingly, is demonstrated in Fig. 1.

Denote by $S_i(\sigma)$ and by $C_i(\sigma)$ the starting and the completion times of job $i \in N$ in σ, correspondingly, and let $G_{max}(\sigma) = \max\limits_{i \in N}(S_i(\sigma) + \beta_i)$. It has been shown in [14], that for any schedule σ,

$$G_{max}(\sigma) = L_{max}(\sigma) + d_{max}. \tag{7}$$

The following notation and definitions are used to analyse a schedule's structure:

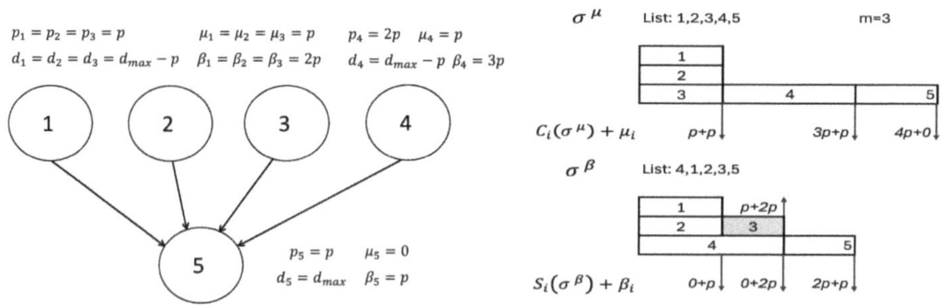

Fig. 1. σ^μ vs. σ^β.

- a time-slot t is the interval $[t-1, t]$; a job $i \in N$ is *processed* in time-slot t if $S_i(\sigma) < t \leq C_i(\sigma)$;
- let g be the job with the earliest starting time among all jobs j such that $S_j(\sigma) + \beta_j = G_{max}(\sigma)$; then a job $i \in N$ is a *complete* job, if $\beta_i \geq \beta_g$, otherwise i is *incomplete*;
- a time-slot $t \leq S_g(\sigma)$ is *complete*, if $|\{j : S_j(\sigma) = t and \beta_j \geq \beta_g\}| = m$; otherwise t is *incomplete*;
- an incomplete time-slot t is ***Type I***, if at least one of the conditions is satisfied:
  a. in t, there is at least one idle machine;
  b. there exists an incomplete task j such that $S(\sigma) = t - 1$;
  c. all tasks j, processed in t, have the same starting times $S_j(\sigma)$.
- an incomplete t, which is not Type I, is a **Type II** time-slot. The definition implies that in any Type II time-slot all machines are busy.

The following results from [14] are used to derive the worst-case performance guarantee provided in this paper:

i. For any Type II time-slot t, there exists a Type I time-slot t' such that $t' < t$; any τ such that $t' < \tau \leq t$ is Type II time-slot; the Type I slot t', is the *supporting* time-slot for the Type II time-slot t.
ii. Any incomplete task, processed in t, is also processed in slot t'.

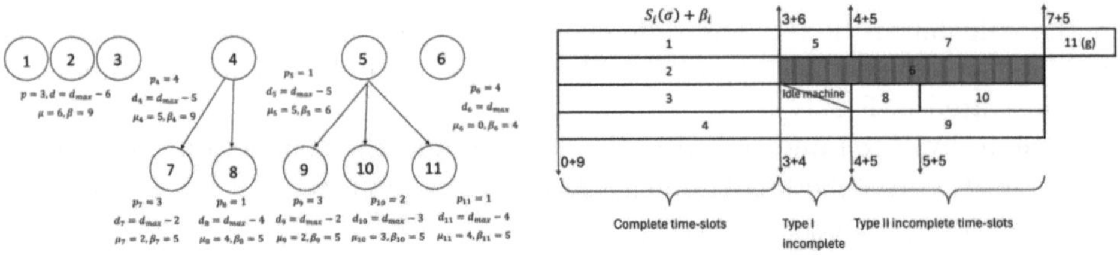

Fig. 2. Example instance.

iii. For any Type I time-slot t and any complete task j with $S_j(\sigma) \geq t$, there is a complete task i processed in t and $i \rightarrow j$.
iv. At least one complete task is processed in each t: $1 \leq t \leq S_g(\sigma)$.

To illustrate these concepts, let us consider the instance and the schedule σ^{BGJ} constructed by the BGJ algorithm presented in Fig. 2. The value of $G_{max}(\sigma^{BGJ})$ is achieved on job 11 with $\beta_{11} = 5$. Hence complete jobs are $\{1, 2, 3, 4, 5, 7, 8, 9, 10\}$, and the job 6 with $\beta_6 = 4$ is an incomplete job. Further, in σ^{BGJ}, time-slots $1-3$ are complete, time-slot 4 - is Type I incomplete time-slot, and the time-slots $5-7$ are Type II incomplete time-slots.

3 Structure of a Schedule

Lemma 1. *In a schedule σ constructed by the BGJ algorithm, all time-slots $t \leq S_g(\sigma)$ are complete.*

Proof. Assume that there is at least one incomplete time-slot $t \leq S_g(\sigma)$. Let τ be the largest Type I incomplete time-slot among all Type I incomplete time-slots $t : t \leq S_g(\sigma)$. If such τ does not exist, then According to [i.] there are also no incomplete Type II time-slots $t \leq S_g(\sigma)$, hence the statement of the lemma holds. Assume that such τ exists. According to [iii.], there is a predecessor $h \rightarrow g$ that is processed in τ, hence $\tau < S_g(\sigma)$, as otherwise h should be selected as the job with the earliest starting time among all jobs j such that $S_j(\sigma)+\beta_j = G_{max}(\sigma)$. Denote by N_c^t and N_{ic}^t the number of complete and incomplete jobs, respectively, that are processed in a time-slot $t \leq S_g(\sigma)$. According to [iv.] $N_c^t \geq 1$ for any $t \leq S_g(\sigma)$. Further, for any complete job j with $S_j \geq \tau$, by virtue of [iii.], there is $q \rightarrow j$, processed in τ. As the precedence constraints are in the form of an in-tree, $|K(i)| \leq 1$ for any i processed in τ. Hence for any $\tau < t \leq S_g(\sigma)$

$$m > N_c^\tau \geq N_c^t, \tag{8}$$

which implies that every $\tau < t \leq S_g(\sigma)$ is an incomplete time-slot. As τ is the largest Type I time-slot, every time-slot $\tau < t \leq S_g(\sigma)$ is Type II incomplete time-slot, and τ is the supporting time-slot for this group of incomplete Type II time-slots.

Now assume that there is an idle machine in the time-slot τ:

$$N_c^\tau + N_{ic}^\tau < m. \tag{9}$$

At the same time, by virtue of [ii.], any incomplete task, processed in incomplete Type II slot $t : \tau + 1 \leq t \leq S_g(\sigma)$, is also processed in τ. Thus for any such t

$$N_{ic}^t \leq N_{ic}^\tau. \tag{10}$$

By definition, there are no idle machines in Type II incomplete time-slots, and taking into account (8)–(10), we have that for any $t : \tau + 1 \leq t \leq S_g(\sigma)$

$$m = N_{ic}^t + N_c^t \leq N_{ic}^\tau + N_c^\tau < m,$$

which contradicts the assumption that there is an idle machine in τ. Hence there are no idle machines on the interval $[\tau - 1, S_g(\sigma)]$.

By the definition of the Type II incomplete time-slot, no incomplete job can start in the interval $[\tau, S_g(\sigma)]$. Assume that there are incomplete jobs j processed in τ with $C_j(\sigma) < S_g(\sigma)$, and let h be the job with the smallest completion time $C_h(\sigma) = t_h$ among all such jobs. Then

$$N_{ic}^t = N_{ic}^\tau forall \tau < t \leq t_h and N_{ic}^{t_h+1} < N_{ic}^\tau. \tag{11}$$

It has been established above that $m = N_{ic}^t + N_c^t$ for any time-slot $\tau \leq t \leq S_g(\sigma)$, then (11) implies that $N_{ic}^{t_h+1} > N_{ic}^\tau$, which contradicts (10). Thus, for all incomplete jobs j that are processed in τ, $C_j(\sigma) \geq S_g(\sigma)$.

Next we will show that for any complete job q processed in the time-slot τ, $C_q(\sigma) \geq S_g(\sigma)$. Assume that there is at least one complete job q processed in the time-slot τ, such that $C_q(\sigma) < S_g(\sigma)$. Denote by

$$t_1 = \min_{i:S_i(\sigma)<\tau\leq C_i(\sigma) \text{ and } \beta_i\geq\beta_g} C_i(\sigma).$$

Let h be a complete job processed in τ with the completion time $C_h(\sigma) = t_1$. As only complete jobs can start on the interval $[\tau, S_g(\sigma)]$, there exists a complete job $j : S_j(\sigma) = t_1$. Then by virtue of [iii.], $h \to j$, as for all other complete jobs i processed in τ, $C_i(\sigma) > t_1$. If there are $l \leq N_c^\tau$ complete jobs h_i, $i = 1..l$ processed in τ with the completion time $C_{h_i}(\sigma) = t_1$, then there are l complete jobs j_i with $S_{j_i}(\sigma) = t_1$, and $h_i \to j_i$. Iteratively, we will construct N_c^τ chains by determining

$$t_k = \min_{i:C_i(\sigma)>t_{k-1} \text{ and } \beta_i\geq\beta_g} C_i(\sigma), \quad k = 1..v.$$

Let a be the complete job with the largest completion time $C_a(\sigma) = \tilde{t} < S_g(\sigma)$ among complete jobs j with $C_j(\sigma) < S_g(\sigma)$. Denote by A the set of all complete jobs i: $C_i(\sigma) > \tilde{t}$. Then for $i \in A$, $C_i(\sigma) \geq S_g(\sigma)$. Due to selection of g for any $h \to g$, $C_h(\sigma) < S_g(\sigma)$. At the same time any $i \in A$ is a part of one of the N_c^τ chains constructed above. Thus g does not have a predecessor in the chains initiated by the jobs processed in τ, which contradicts property [vi.]. The initial assumption that there are incomplete time-slots $t \leq S_g(\sigma)$ does not hold, hence all time-slots $t \leq S_g(\sigma)$ are complete. The lemma is demonstrated by the example in Fig. 1, where in schedule σ^β job g is job 3, and all time-slots before $S_3(\sigma^\beta)$ are complete. □

4 Worst-Case Performance Guarantee

The following theorem derives the upper bound for the worst-case performance of the *BGJ* algorithm for problem (1). Let σ be the schedule constructed by the *BGJ* algorithm, and σ^* be an optimal schedule. Denote by p_{max} and p_{min} the maximum and minimum processing times, respectively, for the set of jobs N. In what follows, it is assumed that $p_{max} > p_{min}$, as $P|intree, p_i = p|L_{max}$ is solvable in polynomial time [4].

Theorem 1. *The upper bound for the worst-case performance of the BGJ algorithm applied to $P|intree|L_{max}$ scheduling problem is*

$$L_{max}(\sigma) \leq L_{max}(\sigma^*) + p_{max} - \frac{p_{min}}{m}, \tag{12}$$

and the bound is tight.

Proof. Denote by U the set of all jobs that started before $S_g(\sigma)$: $U = \{i : S_i(\sigma) < S_g(\sigma)\}$. According to Lemma 1, any $i \in U$ is a complete job. Let $a \in U \cup g$ be the job with the largest completion time among all jobs in $U \cup g$ in σ^*. Observe that a is a complete job and $\beta_a \geq \beta_g$, hence $p_a + \mu_a \geq p_g + \mu_g$. The following lower bound for $G_{max}(\sigma^*)$ is obtained:

$$\begin{aligned} G_{max}(\sigma^*) &\geq S_a(\sigma^*) + \beta_a = C_a(\sigma^*) + \mu_a \\ &\geq \left\lceil \frac{\Sigma_{i \in U \cup g} p_i}{m} \right\rceil + \mu_a \geq S_g(\sigma) + \frac{p_g}{m} + p_g + \mu_g - p_a \end{aligned} \tag{13}$$

According to (13) and taking into account that $G_{max}(\sigma) = S_g(\sigma) + \beta_g = S_g(\sigma) + p_g + \mu_g$, we have

$$\begin{aligned} & G_{max}(\sigma) - G_{max}(\sigma^*) \leq \\ & S_g(\sigma) + p_g + \mu_g - \left(S_g(\sigma) + \frac{p_g}{m} + p_g + \mu_g - p_a\right) = p_a - \frac{p_g}{m} \\ & \leq p_{max} - \frac{p_{min}}{m}. \end{aligned} \tag{14}$$

Finally, substituting (7) into (14), we show that (12) holds:

$$\begin{aligned} & L_{max}(\sigma) - L_{max}(\sigma^*) = L_{max}(\sigma) + d_{max} - (L_{max}(\sigma^*) + d_{max}) = \\ & = G_{max}(\sigma) - G_{max}(\sigma^*) \leq p_{max} - \frac{p_{min}}{m}. \end{aligned}$$

□

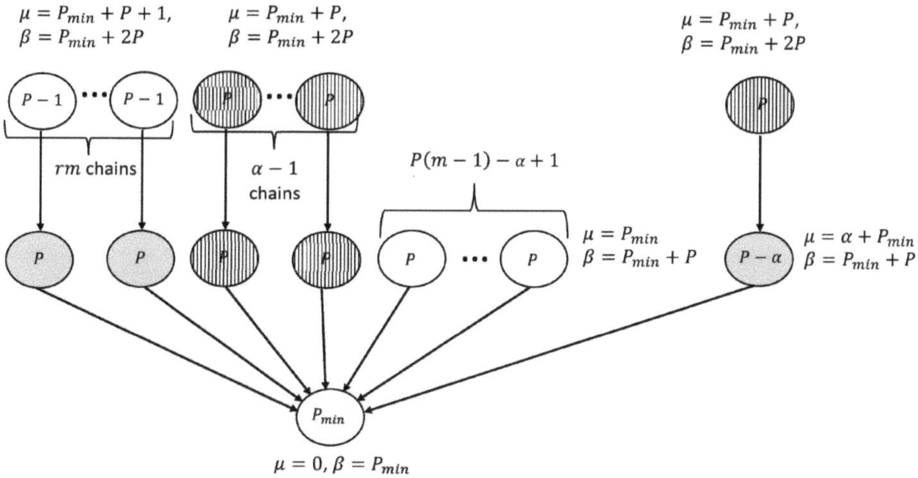

Fig. 3. Precedence graph, $p_{max} = rm + \alpha$, $r \geq 1$, $1 \leq \alpha < m$.

4.1 Tightness of the Worst-Case Performance Guarantee

In this section we will first demonstrate the tightness of (12) for any integer value of $p_{max} > 0$. Observe that as $p_{max} > p_{min}$, $p_{max} \geq 2$.

Let us assume that $p_{max} = rm + \alpha$, $r \geq 1$, $1 \leq \alpha < m$ and consider the instance presented in Fig. 3. The precedence graph is an in-tree, where $P = p_{max}$ and signifies a job with p_{max} processing time. There are rm chains, each consisting of two jobs: a job with $p_{max} - 1$ processing time (referred to as $P - 1$ job) that precedes a job with processing time P (referred to as P^s job). There are $\alpha - 1$ two-job chains, each comprised of a job P (referred to as P^α job) preceding a job P (referred to as $P^{\alpha s}$ job). Another two-job chain consists of a job P (referred to as P^β job) preceding a job with processing time $P - \alpha$ (referred to as $P - \alpha$ job). In addition, there are other $P(m-1) - \alpha + 1$ of P jobs (referred to as P^f jobs). Jobs P^s, $P^{\alpha s}$, P^f and the job $P - \alpha$ precede a root job with processing time p_{min}. Due dates d_i are set as follows:

- for jobs $P - 1$: $d_i = P - 1$;
- for jobs P^α, P^β: $d_i = P$;
- for jobs P^s, $P^{\alpha s}$, P^f: $d_i = 2P$;
- for job $P - \alpha$: $d_i = 2P - \alpha$.
- the root job has due date $p_{min} + 2P$

Priorities μ and β are calculated according to (6) and can be viewed in Fig. 3.

Figure 4 presents the schedule σ^{BGJ} constructed by the BGJ algorithm and an optimal schedule σ^*. In σ^{BGJ}, $P - 1$ jobs scheduled in the first r columns, followed by a column with all P^α jobs and job P^β and $m - \alpha$ P^f jobs. All remaining jobs, except for the root job, have equal priorities $\beta_i = p_{min} + P$. The

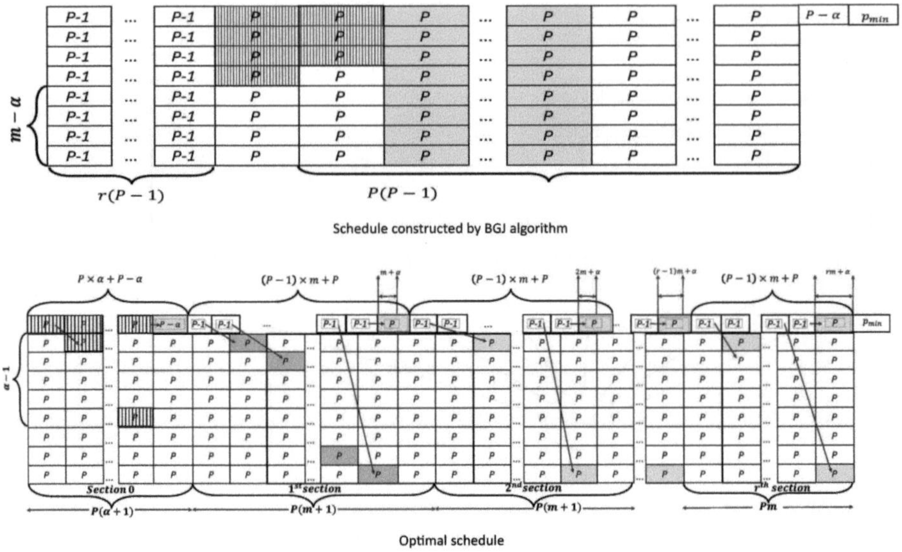

Fig. 4. BGJ and optimal schedules, $p_{max} = rm + \alpha$, $r \geq 1$, $1 \leq \alpha < m$

total number of the remaining jobs is

$$\underbrace{P(m-1) - \alpha + 1 - (m-\alpha)}_{P^f \text{ jobs}} + \underbrace{\alpha - 1}_{P^{\alpha s} \text{ jobs}} + \underbrace{rm}_{P^s \text{ jobs}} + \underbrace{1}_{P-\alpha \text{ job}} =$$

$$Pm - (rm + \alpha) - m + \alpha + rm + 1 = (P-1)m + 1,$$

hence these jobs occupy $P(P-1) + P - \alpha$ time-slots. Then

$$G_{max}(\sigma^{BGJ}) = \underbrace{r(P-1) + P + P^2 - \alpha}_{C_{P-\alpha}(\sigma^{BGJ})} + \underbrace{\alpha + p_{min}}_{\mu \text{ of job } P-\alpha}, \tag{15}$$

and the value is achieved on job the $P - \alpha$.

The optimal schedule σ^* consists of $r+1$ sections. In section 0, P^α and P^β jobs are scheduled on the 1st machine. A $P^{\alpha s}$ job from chain $i = 1..\alpha - 1$ is scheduled on the machine $i+1$ in the time interval $[iP, (i+1)]$. The $P - \alpha$ job is placed on the 1st machine in the interval $[\alpha P, (\alpha+1)P + P - \alpha]$. There are $2(m-1) + (\alpha-1)(m-2)$ P^f jobs that occupy the remaining time-slots on machines $2 - m$ within the interval $[0, P(\alpha+1)]$. Table 1 provides details of jobs' allocation within the section 0.

Sections $1 - r$ of σ^* have the same structure. Each section has m two-job chains of $P-1$ job preceding P^s job; there are $2(m-1)+(m-1)(m-2) = m(m-1)$ P^f jobs in each section $1-(r-1)$, and $(m-1)^2$ P^f jobs in section r. Tables 2 and 3 provide details of jobs' starting and completion times, and the machines they are assigned to within these sections. In these tables $\Delta = \alpha P + P - \alpha$.

In section $1 \leq j \leq r$, all $P-1$ jobs are assigned to the first machine, and P^s job from chain $i = 1..m-1$ is scheduled to machine $i+1$. P^s job from m^{th} chain is

Table 1. Section 0 of optimal schedule

Section 0	Job P		Job P^s		
Chain N	Starting time	Completion time	Starting time	Completion time	Machine
1	0	P	P	$2P$	2
2	P	$2P$	$2P$	$3P$	3
i	$(i-1)P$	iP	iP	$(i+1)P$	$i+1$
$\alpha-1$	$(\alpha-2)P$	$(\alpha-1)P$	$(\alpha-1)P$	$\alpha\, P$	α
α	$(\alpha-1)P$	$\alpha\, P$	$\alpha\, P$	$\alpha\, P+P-\alpha$	1
Jobs P^f are in the interval:		$[0,(\alpha+1)P]$		on machines:	$2-m$

placed on the 1st machine. P^f jobs occupy the remaining time-slots on machines $2-m$ within the interval $[(\alpha+1)P+(j-1)(m+1)P,(\alpha+1)P+j(m+1)P]$. Precedence constraints are satisfied for each chain $i=1...m-1$ of section $j=1...r$:

$$\underbrace{(P\alpha+1)P+(j-1)(m+1)P+iP}_{\text{starting time of } P^s}-\underbrace{(\Delta+(j-1)(mP-m+P)+i(P-1))}_{\text{completion time of } P-1}=$$
$$P\alpha+P-(P\alpha+P)+\alpha+i(P-P+1)+(j-1)(mP+P-mP+m-P)=$$
$$\alpha+i+(j-1)m>0.$$

Observe, that the difference between the completion time of P^s job of chain $m-1$ and the starting time of P^s job of chain m in section $j=1..r$ is $jm+\alpha$, hence in the last section r the jobs start at the same time. On machines $2-m$, each section $j=1...r-1$ occupies $P(m+1)$ time-slots, and the section r occupies Pm time-slots; while on the 1st machine each section $j=1...r$ occupies $(P-1)m+P$ time-slots. Thus,

$$G_{max}(\sigma^*)=P(\alpha+1)+(r-1)P(m+1)+Pm+p_{min}=$$
$$P\alpha+P+Prm+rP-Pm-P+Pm+p_{min}=P^2+rP+p_{min}. \quad (16)$$

The value is achieved in section r on the last P^f jobs, P^s job of chain $m-1$ and both jobs of chain m, and the root job. The total number of P^f jobs in σ^*:

$$\underbrace{2(m-1)+(\alpha-1)(m-2)}_{\text{section 0}}+\underbrace{m(m-1)(r-1)}_{\text{in sections } 1-(r-1)}+\underbrace{(m-1)^2}_{\text{section } r}=$$
$$2(m-1)+\alpha(m-1)-\alpha-(m-2)+rm(m-1)-m(m-1)+$$
$$m(m-1)-(m-1)=(m-1)(\alpha+rm)-\alpha+1=P(m-1)-\alpha+1,$$

the same number as the shown in Fig. 3. Finally, taking into account (15)–(16), we have:

$$G_{max}(\sigma^{BGJ})-G_{max}(\sigma^*)$$
$$=r(P-1)+P+P^2+p_{min}-(P^2+rP+p_{min})=P-r \quad (17)$$

Table 2. Sections $1 \leq j <$ r of the optimal schedule

Sections $1-(r-1)$	Job $P-1$		Job P^s		
Chain N	**Starting time**	**Completion time**	**Starting time**	**Completion time**	**Machine**
1	$\Delta+(j-1)(mP-m+P)$	$\Delta+(j-1)(mP-m+P)+P-1$	$(\alpha+1)P+(j-1)(m+1)P+P$	$(\alpha+1)P+(j-1)(m+1)P+2P$	2
i	$\Delta+(j-1)(mP-m+P)+(i-1)(P-1)$	$\Delta+(j-1)(mP-m+P)+i(P-1)$	$(\alpha+1)P+(j-1)(m+1)P+iP$	$(\alpha+1)P+(j-1)(m+1)P+(i+1)P$	$i+1$
$m-1$	$\Delta+(j-1)(mP-m+P)+(m-2)(P-1)$	$\Delta+(j-1)(mP-m+P)+(m-1)(P-1)$	$(\alpha+1)P+(j-1)(m+1)P+(m-1)P$	$(\alpha+1)P+(j-1)(m+1)P+mP$	m
m	$\Delta+(j-1)(mP-m+P)+(m-1)(P-1)$	$\Delta+(j-1)(mP-m+P)+m(P-1)$	$\Delta+(j-1)(mP-m+P)+m(P-1)$	$\Delta+j(mP-m+P)$	1
Jobs P^j are in the interval:		$[(\alpha+1)P+(j-1)(m+1)P, (\alpha+1)P+j(m+1)P]$		**on machines:**	$2-m$

Table 3. Section r of the optimal schedule

Section r	Job $P-1$		Job P^s		
Chain N	**Starting time**	**Completion time**	**Starting time**	**Completion time**	**Machine**
1	$\Delta+(r-1)(mP-m+P)$	$\Delta+(r-1)(mP-m+P)+P-1$	$(\alpha+1)P+(r-1)(m+1)P+P$	$(\alpha+1)P+(r-1)(m+1)P+2P$	2
i	$\Delta+(r-1)(mP-m+P)+(i-1)(P-1)$	$\Delta+(r-1)(mP-m+P)+i(P-1)$	$(\alpha+1)P+(r-1)(m+1)P+iP$	$(\alpha+1)P+(j-1)(m+1)P+(r+1)P$	$i+1$
$m-1$	$\Delta+(r-1)(mP-m+P)+(m-2)(P-1)$	$\Delta+(r-1)(mP-m+P)+(m-1)(P-1)$	$(\alpha+1)P+(r-1)(m+1)P+(m-1)P$	$(\alpha+1)P+(r-1)(m+1)P+mP$	m
m	$\Delta+(r-1)(mP-m+P)+(m-1)(P-1)$	$\Delta+(r-1)(mP-m+P)+m(P-1)$	$\Delta+(r-1)(mP-m+P)+m(P-1)$	$\Delta+r(mP-m+P)$	1
Jobs P^f are in the interval:		$[(\alpha+1)P+(r-1)(m+1)P, (\alpha+1)P+(r-1)(m+1)P+mP]$		**on machines:**	$2-m$

Recall, that $P = p_{max} = rm + \alpha$, $r \geq 1$, $\alpha \geq 1$. By virtue of (17) and letting $p_{min} = p_{max} - \alpha = rm$ we show that (14), and hence (12) are tight:

$$G_{max}(\sigma^{BGJ}) - G_{max}(\sigma^*) = P - r = p_{max} - \frac{p_{min}}{m}. \tag{18}$$

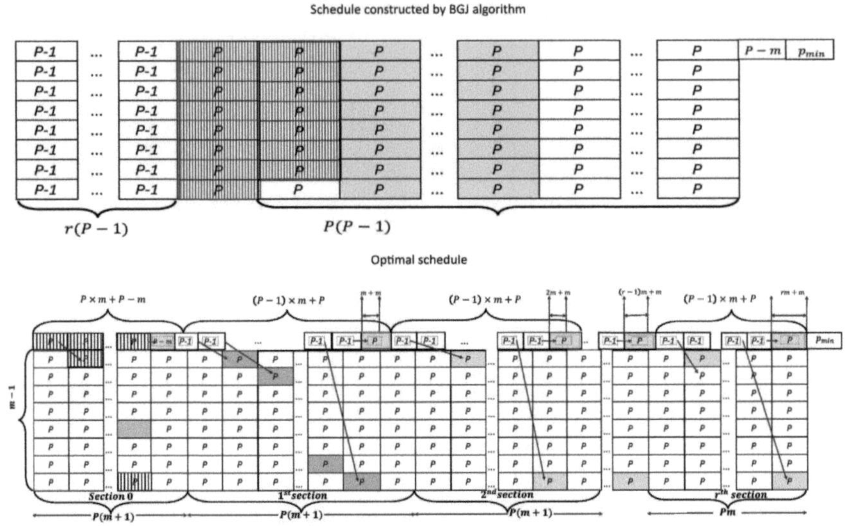

Fig. 5. Schedules, $p_{max} = km$, $k \geq 2$.

The tightness of the bound (12) for the case when $P = p_{max} = km$, $k \geq 2$, can be shown if we consider the instance in Fig. 3, and assume that $r = k - 1$, and $\alpha = m$. The schedules σ^{BGJ} and σ^* are presented in Fig. 5, and have the same structure and the values of $G_{max}(\sigma^{BGJ})$ and $G_{max}(\sigma^*)$ as in the previous case. Thus (12) also holds.

Finally, assume that $p_{max} \leq m$. Then (13) can be tightened:

$$G_{max}(\sigma^*) \geq S_a(\sigma^*) + \beta_a = C_a(\sigma^*) + \mu_a \geq S_g(\sigma) + 1 + p_g + \mu_g - p_a \tag{19}$$

Hence, if $p_{max} \leq m$,

$$G_{max}(\sigma) - G_{max}(\sigma^*) \leq p_{max} - 1, \text{ hence } L_{max}(\sigma) - L_{max}(\sigma^*) \leq p_{max} - 1 \tag{20}$$

Consider the instance presented in Fig. 6. The precedence graph is an in-tree, where P represents jobs with p_{max} processing time. There are P two-job chains, each consisting of $P-1$ job that precedes a P^s job. All P^s jobs precede the root job with the processing time p_{min} and the due date $2P + p_{min} - 1$. There are $P\,(m-2)+1$ of P^f jobs, each preceding the root job. Each P^f and P^s job has a due date $2P - 1$, and job $P - 1$ has a due date of $P - 1$. Priorities μ and β are calculated according to (6) and can be viewed in Fig. 6.

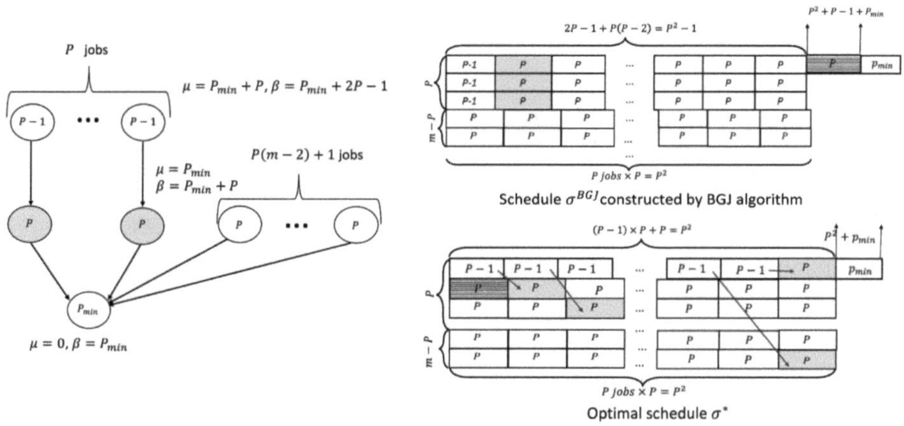

Fig. 6. Precedence graph and schedules, $p_{max} \leq m$.

The schedule σ^{BGJ} constructed by BGJ algorithm and an optimal schedule σ^* for this instance are depicted in Fig. 6. In σ^{BGJ}, all $P-1$ jobs are scheduled on the first P machines followed by the successors P^s on the same P machines. The remaining $m-P$ machines are occupied by $(m-P)P$ P^f jobs. Other $P\,(P-2)+1$ P^f jobs are scheduled on the first P machines after the P^s jobs. The root job is scheduled after all jobs. Observe that the total number of P^f jobs is

$$(m-P)P+(P-2)P+1 = mP-2P+1 = P(m-2)+1,$$

the same number as depicted on the precedence graph in Fig. 6. The maximum completion time of P^f jobs on the last $m-P$ machines is P^2. The maximum completion time of P^f on the first P machines is P^2-1+P. Hence the value of $G_{max}(\sigma^{BGJ})$ is achieved on the last P^f job and on the root job:

$$G_{max}(\sigma^{BGJ}) = \underbrace{P-1+P}_{P\,\text{chains}} + \underbrace{P\,(P-2)+P}_{P^f\,\text{jobs}} + \underbrace{p_{min}}_{\mu} = P^2+P-1+p_{min} \quad (21)$$

In σ^* all $P-1$ jobs are scheduled one by one on the first machine. On the next $P-1$ machines, the jobs with processing time P are scheduled as follows. A job P^s from the ith chain, $i=1..P-1$, is scheduled on a machine $i+1$, in the time interval $[iP,(i+1)P]$, and the P^f jobs are scheduled in all other time slots within the interval $[0,P^2]$. Observe that a $P-1$ job in the ith chain, $i=1..P-1$, is scheduled in the interval $[(i-1)(P-1), i(P-1)]$, thus the precedence constraints in the chains are satisfied. For the last chain P, the $P-1$ job is scheduled in the interval $[(P-1)(P-1), P(P-1)]$, and its successor P^s is scheduled in the interval $[P(P-1), P^2]$. In the equation below, we refer to the time-interval $[(i-1)P, iP]$, $i=1..P$ as to i^{th} *interval*. The remaining P^f jobs are scheduled on the last $m-P$ machines and occupy the time-interval $[0,P^2]$, as the total number of P^f jobs on the $m-P$ machines is

$$\underbrace{P\,(m-2)+1}_{\text{total number of } P^f\text{jobs}} - \left(\underbrace{P-1}_{\text{in } 1^{st}\text{Pinterval}} + \underbrace{(P-2)}_{\text{in}2^{nd}-P^{th}\text{intervals}} \times \underbrace{(P-1)}_{\text{N of intervals}} \right)$$
$$= P \times m - P^2,$$

The root job is scheduled after all jobs. Hence, $G_{max}(\sigma^*) = P^2 + p_{min}$, thus

$$G_{max}(\sigma^{BGJ}) - G_{max}(\sigma^*) = P^2 + P - 1 + p_{min} - (P^2 + p_{min}) = P - 1, \quad (22)$$

which confirms that the bound (20) is tight. Observe that p_{min} can be any integer from the interval $[1, p_{max} - 1]$.

4.2 $P|chains|L_{max}$

In what follows, we will restrict the precedence graph further from in-tree to chains and apply the BGJ algorithm to this particular case of (1):

$$P|chains|L_{max} \quad (23)$$

The problem (23) remains NP-hard in the strong sense due to [6]. Observe that Lemma 1 holds if the precedence constraints are in the form of chains. Indeed, in Lemma 1 the inequality (8) is derived using the property of in-tree - that $K(i) \leq 1$ for any $i \in N$. The precedence constraints in the form of chains have the same property. Thus Lemma 1 and, therefore, Theorem 1 hold for (23). The instances in Fig. 3 and 6 could be easily modified to show that (12) is tight for problem (23) - the modified instances are depicted in Fig. 7. The schedules constructed by BGJ algorithm and optimal schedules for these instances will be the same as shown in Fig. 4 - 6, but without the root job p_{min}. Thus, when $p_{max} > m$,

$$G_{max}(\sigma^{BGJ}) = \underbrace{r(P-1) + P + P^2 - \alpha +}_{C_{P-\alpha}(\sigma^{BGJ})} \underbrace{\alpha}_{\mu \text{of job } P-\alpha} \quad \text{and } G_{max}(\sigma^*) = P^2 + rP$$
$$G_{max}(\sigma^{BGJ}) - G_{max}(\sigma^*) = r(P-1) + P + P^2 - (r(P-1) + P + P^2) = P - r,$$

hence (12) is tight. When $p_{max} \leq m$,

$$G_{max}(\sigma^{BGJ}) = P^2 + P - 1 \text{ and } G_{max}(\sigma^*) = P^2$$
$$G_{max}(\sigma^{BGJ}) - G_{max}(\sigma^*) = P^2 + P - 1 - P^2 = P - 1,$$

and (12) is tight for this case as well.

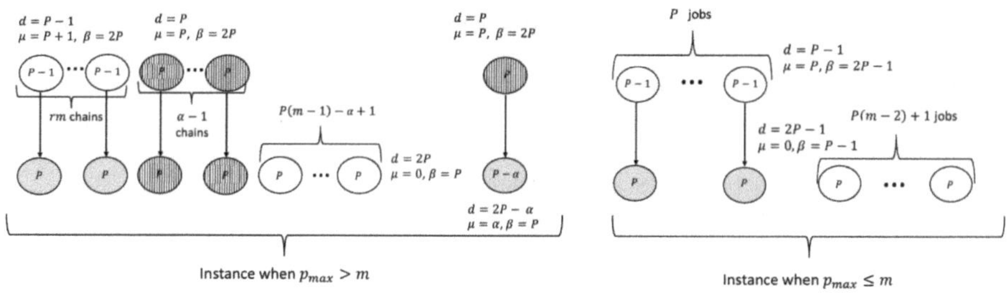

Fig. 7. Precedence graphs in a form of chains

5 Conclusion

This paper presented a worst-case performance guarantee for a modification of BGJ algorithm applied to the classical scheduling problem $P|intree|L_{max}$ and its particular case $P|chains|L_{max}$. This guarantee is tight for any positive integer value of maximum processing time, thus bridging the gap in the literature.

References

1. Agnetis, A., Billaut, J.C., Pinedo, M., Shabtay, D.: Fifty years of research in scheduling-theory and applications. Eur. J. Oper. Res. (2025)
2. Baptiste, P., Timkovsky, V.G.: Shortest path to nonpreemptive schedules of unit-time jobs on two identical parallel machines with minimum total completion time. Math. Methods Oper. Res. **60**(1), 145–153 (2004)
3. Brucker, P.: Scheduling Algorithms, 5th edn. (2006)
4. Brucker, P., Garey, M.R., Johnson, D.S.: Scheduling equal-length tasks under tree-like precedence constraints to minimize maximum lateness. Math. Oper. Res. **2**(3), 275–284 (1977)
5. Dror, M., Kubiak, W., Dall'Olmo, P.: "Strong"-"weak" chain constrained scheduling. Ricerca Oper. (1997/83) (1997)
6. Du, J., Leung, J.Y., Young, G.H.: Scheduling chain-structured tasks to minimize makespan and mean flow time. Inf. Comput. **92**(2), 219–236 (1991)
7. Garey, M.R., Johnson, D.S.: "Strong"NP-completeness results: motivation, examples, and implications. J. ACM (JACM) **25**(3), 499–508 (1978)
8. Garey, M.R., Johnson, D.S.: Computers and Intractability, vol. 174. Freeman, San Francisco (1979)
9. Hall, L.A., Shmoys, D.B.: Approximation schemes for constrained scheduling problems. In: 30th Annual Symposium on Foundations of Computer Science, pp. 134–139. IEEE Computer Society (1989)
10. Hu, T.C.: Parallel sequencing and assembly line problems. Oper. Res. **9**(6), 841–848 (1961)
11. Jiang, X., Lee, K., Pinedo, M.L.: Approximation algorithms for bicriteria scheduling problems on identical parallel machines for makespan and total completion time. Eur. J. Oper. Res. **305**(2), 594–607 (2023)
12. Kaufman, M.T.: An almost-optimal algorithm for the assembly line scheduling problem. IEEE Trans. Comput. **100**(11), 1169–1174 (1974)

13. Lee, M., Lee, K., Pinedo, M.: Tight approximation bounds for the LPT rule applied to identical parallel machines with small jobs. J. Sched. **25**(6), 721–740 (2022)
14. Memar, J., Zinder, Y., Kononov, A.V.: Worst-case analysis of a modification of the Brucker-Garey-Johnson algorithm. In: Eremeev, A., Khachay, M., Kochetov, Y., Pardalos, P. (eds.) OPTA 2018. CCIS, vol. 871, pp. 78–92. Springer, Cham (2018). https://doi.org/10.1007/978-3-319-93800-4_7
15. Pinedo, M.L.: Scheduling: Theory, Algorithms, and Systems. Springer (2016)
16. Singh, G., Zinder, Y.: Worst-case performance of two critical path type algorithms. Asia-Pac. J. Oper. Res. **17**(1), 101 (2000)
17. Wang, S., Cui, W.: Approximation algorithms for the min-max regret identical parallel machine scheduling problem with outsourcing and uncertain processing time. Int. J. Prod. Res. **59**(15), 4579–4592 (2021)

Heuristic Algorithms for Scheduling Multi-mode Jobs with Resource Constraints

Yulia Zakharova(✉) and Maria Sakhno

Sobolev Institute of Mathematics, Omsk Department, Omsk, Russia
yzakharova@ofim.oscsbras.ru

Abstract. We consider a problem of scheduling multi-mode jobs on processor cores, where job durations depend on data bus utilization, while memory capacity and the number of available processor cores are treated as additional resource constraints. Each job can be executed in several modes corresponding to different degrees of parallelism, affecting its processing time and resource consumption. The objective is to minimize the makespan. We investigate the computational complexity of the problem and propose greedy heuristics, several variants of the (1+1) evolutionary strategy, and an adaptive genetic algorithm. The algorithms are experimentally evaluated on synthetic and well-known PSPLIB benchmark instances, and the results are statistically validated.

Keywords: Schedule · Metaheuristic · Computational complexity · Experiment

1 Introduction

Modern scheduling problems that arise in computer systems are characterized by the use of renewable and non-renewable resources and complex structural constraints. The intensity of resource consumption can affect the duration of operations or jobs. Cache memory, data bus, and cores are typical renewable resources in computer systems. Although energy and finance are important non-renewable resources.

In literature, there exist a number of approaches to planning job assignments on processor cores considering variable job execution durations. Typically, such problems are solved using fast heuristics that operate in an online mode, i.e., jobs arrive sequentially and only a limited number of jobs are considered at each moment in time. The job scheduling heuristics proposed in [18,27] and several other articles are based on the principle that jobs should be placed on processor cores in a complementary manner, so that jobs with the most diverse resource usage requirements are executed simultaneously (in [18,27] such resources imply the usage of data bus bandwidth and cache usage at different levels).

In scheduling theory, similar problem formulations can be found in the area of scheduling with controllable processing times. Models and scheduling methods

A. Kazakov et al. (Eds.): MOTOR 2026, LNCS 16687, pp. 231–246, 2026.
https://doi.org/10.1007/978-3-032-31390-4_15

for the case where preemptions are allowed are considered in [21]. Since jobs slow each other down primarily due to competition for renewable resources (data bus bandwidth, cache), we refer to [2,15] for reviews of problems with renewable resources, where resource allocation can change over time. The case of discrete resources is considered in [2], while continuous resources are considered in [15]. The formulation from [15] assumes that the amount of resources allocated to each job is limited, but continuous and determined by the scheduler at each moment in time.

In [7], it is supposed that the execution speed of jobs is completely determined by the set of jobs co-scheduled on other cores (or machines in traditional scheduling terminology). A greedy algorithm is proposed, where job slowdown is calculated based on information about their data bus usage. A configuration mixed integer linear model is also proposed and tested on small-size instances (up to 10 modules) in [7].

The scheduling problem with co-run degradation is considered in papers [3,8, 14,22,23,25]. In this paper every subset of cores resides on a chip, sharing a cache (a job is assigned on core). Co-run degradation factor (equal to the increase in application execution time when jobs within it share cache, compared to running jobs in isolation) was calculated using machine learning techniques [25] or offline profiling runs and predictive models [3,8,22]. In [14] jobs are represented as vertices connected by edges, and edge weights are given by the sum of mutual degradation factors when two jobs co-run. Then for dual-core processors, under some simplifying assumptions, the optimal schedule can be found by solving a minimum-weight perfect matching problem. It is shown that the problem is NP-hard in the general case, and several approximate heuristic algorithms are proposed. Although the methodology from [14] and the corresponding algorithms would be too expensive for online use, they are acceptable for assessing the quality of other approaches.

The authors of [23,24] propose a new fairness-aware co-scheduling algorithm based on a non-cooperative game to reduce second-level cache misses. A certain amount of cache on a processor is shared among its cores, and a job's execution speed on a processor depends on which jobs are placed on the other cores of that processor. The number of jobs equals the number of cores of all processors and all jobs start at the same time. In [23] it is proven that this problem is NP-hard in the general formulation, but for a special case (dual-core processor without job migration between cores) the problem is solvable in $O(n^{2.5} \log\ n)$. Furthermore, a number of algorithms are presented in [23] for the general case, including a greedy algorithm, for finding optimal or approximate solutions.

In production scheduling applications, problem formulations with variable processing times are also important. For example, [17] considers a coking production scheduling problem, where jobs affect the execution time of other jobs due to an increase in the production unit's temperature. The authors of [17] build an integer programming model to minimize the makespan, propose several heuristics including a genetic algorithm, and compare their performance.

The project scheduling problem with renewable resources (which do not influence on durations of jobs) is actively investigated from the theoretical and experimental point of view (see, e.g. [10–12,20]). PSPLIB library (https://www.om-db.wi.tum.de/psplib/) contains instances of this problem. The computational complexity of the problem of assignment resources to cores with makespan criterion is investigated in [1]. In this paper assignment of jobs to processors and their order are fixed, but identical processors share a common renewable resource.

We consider a problem of scheduling multi-mode jobs with durations dependent on renewable resource consumption (processor cores, data bus and RAM). Our main result is greedy methods and evolutionary algorithms with problem specific encoding-decoding scheme and operators allowing to take into account renewable resources and technological constraints. Approximation guarantees are proved for the greedy algorithms proposed for the case of scheduling independent single-mode jobs. The scheduling problem with data bus constraint and without multi-mode options has been also considered in the previous research [7,26], where list-type and evolutionary algorithms were proposed.

2 Problem Statement and Computational Complexity

We consider the multi-machine scheduling problem with resource constraints. A set of jobs $J = \{1, ..., n\}$ must be scheduled on a set of parallel machines (processors, processor cores) $I = \{1, ..., m\}$ with RAM equal to V. Partial order between jobs is given by a directed graph $G = (J, A)$, where arc $(j, j') \in A$ means that job j' can be scheduled after completion of job j.

Each job can be processed in several modes. Each mode $i \in I_j$ of job $j \in J$ is characterized by $M_j = (V_j, size_{ji}, b_{ji}, p_{ji})$, where
I_j is the set of indexes of modes for job j;
$V_j \in Z^+$ is the required volume of RAM for job j;
$size_{ji} \in Z^+$ is the required number of machines for job j in mode i;
$b_{ji} \in Z^+$ is the required percentage of data bus under ideal conditions for job j in mode i;
$p_{ji} \in Z^+$ is the duration of job j in mode i under ideal conditions.

A data bus is a part of the system bus that is used to transfer data between computer components. The data bus capacity is a renewable resource. The parallel execution of jobs and their joint consumption of the data bus influences the speed and duration of the jobs. In practice, the distribution of data bus among the threads is very hardware-specific and depends on many factors (see e.g. [19]). As a simple approximation, we assume that the data bus bandwidth allocation to jobs is determined by the following rule. Let b_{ji} denote the percent of data bus consumed by job $j \in J$ in some mode $i \in I_j$. Initially we assign to each job $b'_{ji} := \frac{100\%}{m} size_{ji}$ of the data bus. If a job j has $b_{ji} < b'_{ji}$ then the remaining resource $(b'_{ji} - b_{ji})$ is divided between jobs $j' \in J_s$, for which $b_{j'i'} > b'_{j'i'}$. The processing time of job $j' \in J_s$ is equal to $\frac{b'_{j'i'}}{b_{j'i'}} p_{j'i'}$.

The goal is to schedule the jobs on the processor cores with the minimum makespan (i.e. the time of completion of the last job) taking into account bounds on resources, machines and precedence constraints for jobs. Preemptions of jobs are disallowed.

Using the reduction from the well known Partition problem [9] we can prove NP-hardness of the scheduling problem with renewable resources and makespan criterion in the following cases: (1) two processor cores, jobs of arbitrary length, no resource; (2) parallel jobs of the same duration with preemption and migration; (3) jobs of the same duration, RAM as the common resource. In the case of the partial order between jobs, the problem is strongly NP-hard even for the two processor cores [5]. The lower bound on the makespan can be calculated as follows

$$LB = \max\left\{P^{\max}, \frac{\sum_{j\in J} ps_j^{\min}}{m}, \frac{\sum_{j\in J} pv_j^{\min}}{V}, \frac{\sum_{j\in J} pb_j^{\min}}{100}\right\}, \tag{1}$$

where $ps_j^{\min} = \min_{i\in I_j} p_{ji} size_{ji}$, $pv_j^{\min} = \min_{i\in I_j} p_{ji} V_j$, $pb_j^{\min} = \min_{i\in I_j} p_{ji} b_{ji}$, $j \in J$, value $P^{\max}$ is the length of the critical path in the partial order graph with durations of jobs minimal over modes.

Now we investigate greedy algorithms for the single-mode jobs. We will denote by $size_j$ (p_j, b_j) the number of the required machines (duration, percentage of data bus) of job $j \in J$ in the case of single-mode statement.

Theorem 1. *For the problem of scheduling independent single-mode jobs, a 6-approximate solution can be constructed in polynomial time.*

Proof. We will consider schedules, for which the data bus requirement does not exceed 100% at each time moment (i.e. at each time moment t we have a subset of jobs J_t with $\sum_{j\in J_t} b_j \le 100\%$). We construct a schedule according to the rule: as soon as a job completes its execution, select the next job on the list to schedule from the set of jobs such that their RAM, data bus and processor requirements do not exceed the currently available (if this set is non-empty). Let $L(\sigma)$ denote the length (makespan) of schedule σ. Jobs can be preliminarily ordered in the list by decreasing (increasing) durations, sizes, RAM, data bus consumption and so on. We show that the schedule σ is a 6-approximate solution.

Suppose that at every moment in time, the schedule σ uses at least $\frac{m+1}{2}$ machines (the total length of such intervals is denoted by L_m) or at least $\frac{V}{2}$ RAM (the total length of such intervals is denoted by L_V) or at least $\frac{100\%}{2}$ data bus utilization (the total length of such intervals is denoted by L_b). We state that

$$L_m \frac{m}{2} \le \sum_{j\in J} p_j size_j \le LB\cdot m,\ L_m \le 2\cdot LB.$$

Similarly $L_V \le 2\cdot LB$ for RAM and $L_b \le 2\cdot LB$ for data bus. As a result, we obtain $L(\sigma) = L_m + L_V + L_b \le 6LB$.

Otherwise, we identify the last interval in σ of duration Δ, where $m' < \frac{m+1}{2}$ machines and $V' < \frac{V}{2}$ RAM and $b' < 50\%$ data bus are used. By the construction of σ, there exists a job j that runs throughout this interval with the specified property: at least m' machines and V' RAM, and b' percent of data bus are utilized. Note that the rest of the intervals in σ have the property that at least $(m - m' + 1)$ machines (the total duration of such intervals is denoted by L'_m) or $(V - V' + 1)$ RAM (the total duration of such intervals is denoted by L'_V) or $(100 - b' + 1)$ percentage of data bus (the total duration of such intervals is denoted by L'_b) are used in them. So, then the total workloads are

$$m' \cdot \Delta + (m - m' + 1) \cdot L'_m \leq LB \cdot m,$$

$$V' \cdot \Delta + (V - V' + 1) \cdot L'_V \leq LB \cdot V,$$

$$b' \cdot \Delta + (100 - b' + 1) \cdot L'_b \leq LB \cdot 100.$$

Using property $\Delta \leq p_j \leq LB$ we estimate

$$L'_m + \Delta \leq \frac{mLB + \Delta(m - 2m' + 1)}{(m - m' + 1)} \leq \frac{LB(2m - 2m' + 1)}{(m - m' + 1)} \leq 2LB,$$

and $L'_V + \Delta \leq 2 \cdot LB,\ L'_b + \Delta \leq 2 \cdot LB$.

And similarly to the previous case, we obtain $L(\sigma) = L'_m + L'_V + L'_b + \Delta \leq 6LB$. □

Theorem 2. *For the problem of scheduling single-mode jobs constrained by an arbitrary partial order, a* $\left(\frac{6-3q}{1-q}\right)$*-approximate solution can be constructed in polynomial time in the case, when* $size_j \leq qm$, $V_j \leq qV$, $b_j \leq 100q$ *for* $j \in J$, *and* $q \in (0, 1)$.

Proof. We construct a feasible schedule σ by taking into account the precedence constraints according to the rule: as soon as some job is completed, select the next job on the list to schedule from the set of jobs such that their RAM, data bus and processor requirements do not exceed currently available and all predecessors of the job are completed.

Let $t_1 < t_2 < \cdots < t_\mu < t_{\mu+1}$ be the points in time corresponding to the job completion times in schedule σ, sorted in ascending order, $\mu < n$. Define the subintervals $I_{i-1} = (t_{i-1}, t_i]$, $i = 1, \ldots, \mu + 1$, where $t_0 = 0$ is the start time of the first job, and $t_{\mu+1} = L_\sigma$ is the completion time of the last job in schedule σ.

We divide subintervals into two disjoin sets S_1 and S_2 by the following rule. We start from the end of the schedule and find the job j that completes at time L_σ. All subintervals where j executes are called covered by job j and putted in set S_1. Next, we check why job j did not start earlier (in the subinterval I' preceding its start subinterval) and, depending on the reason, we put the subinterval in S_1 or S_2. If job j does not start at I' because its immediate predecessor j' has not completed, we put I' and all subintervals covered by j' in set S_1. Then we check why job j' did not start earlier and so on. If job j does not start at I' because there are not enough resources available for its execution, we put I' in set S_2. Then we go to the subinterval immediately preceding I' and check why job j does not start at that subinterval and so on. We also determine the chain of preceding jobs that define the subintervals from set S_1 (the number of jobs in the chain is denoted by k).

The total length of subintervals from set S_1 is denoted by L_1, and the total length of subintervals from set S_2 is denoted by L_2. Then the makespan is $L_\sigma = L_1 + L_2$.

From the definition of the lower bound LB we state $LB \geq \sum_{l=1}^{k} p_{j_l} = L_1$.

In a subinterval from S_2 we have the property: the number of busy machines is at least $(1-q)m+1$ or the busy value of RAM is at least $(1-q)V+1$ or the utilized percentage of data bus is at least $(1-q)100+1$. We divide L_2 into three parts L_{2m}, L_{2V} and L_{2b} according to the lower bounds on the utilized resources. Also, at least one unit of each resource is busy during each subinterval from S_1. Consequently, the total workloads are

$$L_1 + ((1-q)m+1)\, L_{2\,m} \leq \sum_{j\in\mathcal{J}} p_j size_j \leq LB,$$

$$L_1 + ((1-q)V+1)\, L_{2V} \leq \sum_{j\in\mathcal{J}} p_j V_j \leq LB,$$

$$L_1 + ((1-q)100+1)\, L_{2b} \leq \sum_{j\in\mathcal{J}} p_j b_j \leq LB.$$

As a result, we obtain the following estimate for the ratio of parts of the length of the schedule to the lower bound LB

$$\frac{L_1 + L_{2\,m}}{LB} \leq \frac{L_1 + L_2}{\max\{L_1;\ L_1/m + (1-q+1/m)L_2\}} \leq \frac{2-q}{1-q},$$

$$\frac{L_1 + L_{2V}}{LB} \leq \frac{2-q}{1-q}, \frac{L_1 + L_{2b}}{LB} \leq \frac{2-q}{1-q},$$

and in total we have

$$L_1 + L_{2\,m} + L_{2V} + L_{2b} \leq \frac{6-3q}{1-q}.$$

□

Remark that excluding one of the resources (machines, RAM or data bus) improves the approximation bound by 2 for the empty partial order and by $\frac{2-q}{1-q}$ for the non-empty partial order.

3 Heuristics

We propose a greedy algorithm and evolutionary algorithms (EAs) for the considered problem. The EA simulates the evolution of a population comprising individuals that encode potential solutions to the given problem [13]. Solutions are encoded by permutations of jobs.

3.1 Greedy Algorithm

The proposed greedy method (GR) generalizes algorithms presented in the previous section to the data bus influencing and multi-mode options. The approximation guarantee of this method is an open question for further research.

Rule of the greedy algorithm is the following. Let some subset of jobs (configuration C_{cur}) is executed and the job $j_{\min}$ of the minimal duration is completed. Then we consider jobs, that requires RAM and processors not greater that the system has at this time and can be placed in accordance with the partial order (all predecessors have been completed). We select a job $j_{\max}$ with the maximum duration in the mode with the maximum number of processors. Then we test all modes of job $j_{\max}$ in configuration $C_{cur} \setminus \{j_{\min}\} \cup j_{\max}$ and select the best one in terms of duration of job j. And so on.

3.2 Schedule for Permutation

Under the given permutation of jobs, the optimal dividing them between the processors is NP-hard problem even in the case of identical data bus requirements for jobs and two machines (the proof is based on the reduction of the Ordered Partition Problem [9]). So, we propose a heuristic algorithm to construct a schedule for a permutation.

We use the following rule to construct a feasible schedule and compute objective for the given permutation of jobs:
We take the maximum subset J_0 of the first jobs from the permutation taking into account RAM and machine constraints and compute data bus consumption for them. Then calculate durations of jobs and identify the shortest job j_s. After the completion of this job we go to the next job the permutation and try to put it. If RAM constraint or machine constraint is not satisfied, then we are waiting for the completion time of the next job in the current schedule. A mode is selected as in the greedy algorithm. And so on.

Moreover, the permutation is preliminary corrected in accordance with the partial order. We take at each step the next job in the permutation, for which all predecessors are already placed and RAM and processor constraints are satisfied.

3.3 (1+1)-ES

We use the basic scheme of $(1+1)$-ES, where a new random solution is constructed from the current solution and the best one of these two solutions is selected for the next iteration. The scheme is presented in Algorithm 1.

We apply the classic mutation operators such as swap and shift [13], and such variations as k-swap and k-shift, where the operator is used k times, $k = 2,\ 4$. We also integrate scramble mutation and heavy-tailed scramble mutation [4].

Scramble mutation operator works as follows. The operator samples a number n_p according to a Poisson distribution with mean λ_p, selects a random set of n_p elements from $[1 \ldots n]$, and applies a random permutation to this set.

Algorithm 1 The (1 + 1) Evolutionary Strategy (ES).

1: Construct an initial permutation π.
2: While the stopping criterion is not satisfied do

  2.1: Select a mutation operator *mut*.
  2.2: Construct $\pi' = mut(\pi)$.
  2.3: Update π to the best permutation (according to the objective function) between π and π'.

We also apply a heavy-tailed scramble mutation operator. We say that an integer random variable X follows a power-law distribution with parameters β and u if

$$Pr[X = i] = \begin{cases} C_{\beta,u} i^{-\beta}, & \text{if } i \in [1 \dots u], \\ 0 & \text{otherwise.} \end{cases}$$

where $C_{\beta,u} = (\sum_{i=1}^{u} i^{-\beta})^{-1}$ denotes the normalization coefficient. We write $X \sim pow(\beta, u)$ and call u the range of X and β the power-law exponent.

The heavy-tailed scramble mutation (with power-law exponent β) is the mutation operator that first samples a number $n_p \sim pow(\beta, n)$, then selects a random subset of n_p elements from $[1 \dots n]$, and finally applies a random permutation on this set.

We also propose (1+1)-ES with several operators in the adaptive (ESA) and sequential (ESVO) variants.

In the adaptive scheme we have a set of mutation operators, and one of them is selected at each iteration using an (ε, δ)-learning scheme. A random operator is selected with probability ε, and the following rule is used with probability $(1 - \varepsilon)$. We select an operator that has the best weight with probability δ, and choose a random operator with probabilities proportional to weights otherwise (with probability $(1 - \delta)$).

The reward of operator *oper* is calculated as

$$r_{oper} = \begin{cases} 1, & \text{if the obtained solution becomes a new record,} \\ 0.3 & \text{otherwise.} \end{cases}$$

The weight of operator *oper* is updated as $R_{oper} = \lambda r_{oper} + (1 - \lambda) R_{oper}$. Before the main learning schema we introduce training period, where operators are selected with equals probabilities, but weights are updated

The operators are applied sequentially within a given number of iterations in the sequential variant of (1+1)-ES. The scheme is presented in Algorithm 2.

3.4 Genetic Algorithm

We propose an adaptive genetic algorithm (GA) outlined in Scheme 3. The individuals in the initial population are either generated randomly with probability p_{init} or constructed using the "arbitrary insertion" approach [6] with probability

Algorithm 2 (1+1)-ES with Variable Operators (ESVO).

1: Construct an initial permutation π.
2: While the global stopping criterion is not satisfied do
3: Select the next mutation operator *mut*.
4: While the local stopping criterion is not satisfied do

  4.1: Construct $\pi' = mut(\pi)$.
  4.2: Update π to the best permutation (according to the objective function) between π and π'.

$(1 - p_{\text{init}})$, similar to a technique used in the Traveling Salesman Problem. This latter method is employed with a predefined probability and operates as follows: r_{job} percent of jobs are arranged in a random order, and each subsequent job is selected randomly and inserted into the most favorable position among those already assigned, minimizing the objective function of the incomplete solution.

Algorithm 3 Elitist Genetic Algorithm (*GA*)

1: Create an initial population of k individuals. Preserve the top n_{elites} individuals from this initial set as elites, based on their objective function values.
2: Repeat the following steps until the termination condition is satisfied:

  2.1 Choose individuals from the current population to serve as parents for reproduction.
  2.2 Perform mutation and recombination on the selected individuals.
  2.3 Generate offspring individuals.
  2.4 Update the main population and the elite group.

3: Output the best individual discovered during the run.

At the reproduction stage, we select parent individuals using either rank-based or s-tournament selection. We implement a classic restart strategy: the algorithm restarts once the number of objective function evaluations since the best solution was found exceeds a predefined threshold n_{FC}. The algorithm follows an elitist strategy, and the number of elite individuals n_{elites} is determined during preprocessing. After applying reproduction operators, the resulting offspring replaces the least fit individual in the current population.

Crossover is applied with probability P_c. We employ such randomized operators as Order crossover (OX), One-point crossover (1PX), Cycle crossover (CX), and Partially mapped crossover (PMX) [16].

A variety of crossover operators is available, with one chosen per iteration using the same (ε, δ)-learning mechanism as in the (1+1)-ES. The reward for a given operator *oper* is computed as follows: The reward of operator *oper* is calculated as

$$r_{oper} = \begin{cases} w_1, & \text{if the obtained solution becomes a new record,} \\ w_2, & \text{if the obtained solution is a new record in the current restart,} \\ w_3, & \text{if the obtained solution is better than one of the parents,} \\ 0 & \text{otherwise.} \end{cases}$$

Mutation is applied with probability P_m. We use standard mutation operators, including swap, shift [13], scramble mutation, and heavy-tailed scramble mutation [4].

4 Computational Experiments

The computational experiment consists of three parts. Firstly, we compare the algorithms on single-mode test instances, where each job has only one execution mode. Secondly, we evaluate the performance of algorithms on multi-mode test series. Finally, we present the results obtained on instances from PSPLIB library.

Testing on Single-Mode Series. The test instances are generated similarly to those described in [7,26], where the job parameters were based on actual computational procedures from the oneMKL library. Each job j is characterized by volume (V_j) and one mode with the number of required CPU cores ($size_j$), duration (p_j), and databus bandwidth (b_j). These parameters are sampled from three specific intervals (or discrete sets) I_i^x, where $i \in \{1,2,3\}$ and $x \in \{V, SIZE, P, B\}$:

- Volume: $I_1^V = [0, 10), I_2^V = [10, 20), I_3^V = [20, 35)$.
- Cores: $I_1^{SIZE} = \{1\}, I_2^{SIZE} = \{2\}, I_3^{SIZE} = \{4\}$.
- Duration: $I_1^P = [2, 10), I_2^P = [10, 25), I_3^P = [25, 50)$.
- Databus: $I_1^B = [0, 30), I_2^B = [30, 80), I_3^B = [80, 100)$.

The generation process of the *standard* (STD) configurations consists of two steps. At the first step, each job is assigned to one of three classes $\mathcal{K} \in \{\text{Small}, \text{Medium}, \text{Big}\}$ according to a series-specific distribution $\mathbf{P}_{series} = (p_S, p_M, p_B)$, where $p_{\mathcal{K}}$ is the probability of selecting class $\mathcal{K}$. At the second step, once a class $\mathcal{K}$ is determined for a job, each parameter $x \in \{V, SIZE, P, B\}$ is drawn uniformly from an interval (or discrete set) I_i^x. The index i is selected with probability α_i specified by the class weight vector $\mathbf{W}_{\mathcal{K}} = (\alpha_1, \alpha_2, \alpha_3)$. Specifically, we use values: $\mathbf{W}_{\text{Small}} = (0.7, 0.2, 0.1)$, $\mathbf{W}_{\text{Medium}} = (0.15, 0.7, 0.15)$, and $\mathbf{W}_{\text{Big}} = (0.1, 0.2, 0.7)$.

Beyond the *standard* configurations, we define *random* (RND) series, where class and parameter selection follow a uniform distribution. We also evaluate *inverse* (INV) series, where the duration intervals are reversed: $I_1^{P'} = [25, 50)$, $I_2^{P'} = [10, 25)$, and $I_3^{P'} = [2, 10)$. Each series consists of 30 instances.

We compare solutions of the greedy heuristic (GR), the variants of $(1+1)$ evolutionary strategy (ES, ESA, ESVO) and the genetic algorithm (GA) found over the given CPU budget. Parameter values for the algorithms were presented in Tables 1-2 (IRACE package was used for tuning of parameters). The stopping criterion of the algorithms is the number of the objective function calculations equal to 50000.

Table 1. Configuration parameters for the ES, ESA, and ESVO algorithms.

Parameter	ES	ESA	ESVO
Total objective calculations	50,000	50,000	50,000
Mutation Types	shift	k-shift, k-swap $k \in \{1, 2, 4\}$	
Mutation Schemas	classic	classic	classic, scramble ($\lambda_p = 3$), heavy tailed ($\beta = 1.5$)
$\lambda, \varepsilon, \delta$	—	0.75, 0.0, 0.0	—
Training period	—	2500 iterations	—
Local stopping criterion	—	—	150 iterations

Table 2. Configuration parameters of GA.

Parameter	Value
Total objective calculations	50000
n_{FC}	10000
k	232
n_{elites}	8
p_{init}	0.76
r_{job}, %	19
Selection	6-tournament
P_c	0.83
Mutation Type	shift
Mutation Schema	classic
P_m	0.26
Training period	3000 objective calculations
$\lambda, \varepsilon, \delta$	0.75, 0.84, 0.58
w_1, w_2, w_3	1.0, 0.8, 0.49

We use the relative deviation from the best known record to evaluate the performance of the algorithms. For each test instance, this deviation is calculated as $\frac{f_a - f_{best}}{f_{best}} \times 100\%$, where f_a is the objective value obtained by algorithm a, and f_{best} is the best objective value achieved among all compared algorithms for that instance. Table 3 presents these deviations averaged over all instances within each series. We use the record as a baseline because relative deviations from the theoretical lower bound are large and hide the performance differences between algorithms. The column Δ_{LB} represents the average relative deviation of the records from the lower bound. The data is presented for $n = 100$, as results for $n = 50,\ 200$ jobs are similar. The best results are shown in bold, and asterisks (*) indicate statistically significant superiority (Wilcoxon test, p-value $= 0.05$).

Table 3. Average relative deviation from the best known record for $n = 100$ in single-mode series, %.

Series				Algorithms					
Type	p_S	p_M	p_B	**GR**	**ES**	**ESA**	**ESVO**	**GA**	Δ_{LB}
STD	0.7	0.2	0.1	6.25	1.25	1.14	0.64	**0.16***	53.43
STD	0.1	0.2	0.7	3.1	0.61	0.29	0.22	**0.11**	74.68
STD	0.33	0.33	0.34	4.15	0.51	0.28	0.22	**0.16**	67.65
STD	0.15	0.7	0.15	3.86	0.51	0.21	0.23	**0.13**	57.1
STD	0.5	0.0	0.5	4.55	1.04	0.45	0.53	**0.21**	65.58
STD	0.3	0.0	0.7	3.6	0.49	0.26	0.32	**0.08***	72.65
RND	–	–	–	4.9	0.59	0.34	0.31	**0.1***	86.46
INV	0.7	0.2	0.1	8.71	2.12	1.58	1.62	**0.06***	33.04
INV	0.1	0.2	0.7	4.17	0.69	0.73	0.85	**0.19**	70.88
INV	0.3	0.0	0.7	6.81	3.03	1.2	0.91	**0.22***	52.35
INV	0.7	0.0	0.3	10.73	1.73	1.27	1.3	**0.02***	36.44

The GA demonstrates statistically significantly better results over all other algorithms in several series, specifically those with a high proportion of small jobs. The ESA and ESVO algorithms demonstrate statistically significant improvements over the basic ES across all test series. It is worth noting that while the greedy algorithm runs faster than the metaheuristics, it still provides acceptable results. Its average relative deviations across all tested series with empty partial order do not exceed 11%. It demonstrates good results on series with a low proportion of small jobs and on the random series. The results also show that the gap between the found solutions and the lower bound remains significant. It is important to improve both upper and lower bounds in the future.

We introduced precedence constraints between jobs to further evaluate the algorithms under more complex conditions. Two types of partial orders were implemented: a random acyclic graph (rand) and a binary tree structure (bitree), arising in practice. For each order type, three test series were generated. Table 4 presents the average relative deviations (in percent) from the best known record for these series.

The GA remains the most effective approach, maintaining statistically significantly better results in all test series. Similarly, the ESA and ESVO variants show statistically significantly improvements over the baseline ES.

Table 4. Average relative deviation from the best known record for $n = 100$ in single-mode series with partial order, %.

Series					Algorithms					
Type	Order	p_S	p_M	p_B	**GR**	**ES**	**ESA**	**ESVO**	**GA**	Δ_{LB}
STD	rand	0.33	0.33	0.34	16.77	2.33	0.58	0.47	**0.11***	190.11
RND	rand	–	–	–	13.73	1.98	0.45	0.67	**0.08***	352.3
INV	rand	0.7	0.0	0.3	8.08	0.7	0.23	0.09	**0.04***	179.15
STD	bitree	0.33	0.33	0.34	18.72	2.26	1.01	0.73	**0.12***	92.53
RND	bitree	–	–	–	19.41	2.5	0.71	0.94	**0.12***	140.22
INV	bitree	0.7	0.0	0.3	9.68	1.19	0.46	0.38	**0.06***	97.03

Testing on Multi-mode Series. We now evaluate the algorithms on the same test instances but with the added ability to execute jobs in multiple modes. This means we can assign a smaller number of cores to a job j than its maximum requirement $size_j$. In this case, the job duration increases and the databus bandwidth decreases proportionally, while the memory volume remains unchanged. The results for this scenario are presented in Table 5.

The results show that while the GA often finds the best average results, it is not the absolute leader in all series. In several test series, other algorithms achieve better average values, though these differences are not statistically significant. Notably, no significant statistical difference is observed on STD and RND series. However, the GA frequently demonstrates statistically significantly better results on INV series and series with partial order. It is important to highlight that the crossover operator plays a key role in the GA. Without this operator, the algorithm performs worse than other evolutionary strategies.

Note that the multi-mode approach demonstrates statistically significantly better results than the results of the single-mode one in terms of records (the best results) for series with partial order. This confirms that the algorithms can successfully use the multiple modes to find better schedules.

Testing on PSPLIB Series. We also tested the algorithms using instances from the well-known PSPLIB library with n = 60, 90, and 120 jobs. In the original PSPLIB instances, four resources are given. In our experiment, we treat the first resource as a databus. This means that jobs can use less of this resource than originally requested, but they will run proportionally slower. The number of CPU cores for each instance is set equal to the width of its partial order. Table 6 presents the average relative deviations from the length of the critical path for these series.

The GA again shows statistically significantly better results (p-value = 0.05). There is no statistically significantly difference between results of ES-A and ES-VO, but both demonstrate statistically significantly better results than the basic ES. It is important to note that the average relative deviations from the length of the critical path for the GA were 15.65%, 13.94%, and 38.78% for n = 60,

Table 5. Average relative deviation from the best known record for $n = 100$ in multi-mode series, %.

Series					Algorithms					
Type	Order	p_S	p_M	p_B	**GR**	**ES**	**ESA**	**ESVO**	**GA**	Δ_{LB}
STD	no	0.7	0.2	0.1	6.76	1.63	0.62	0.5	**0.47**	53.95
STD	no	0.1	0.2	0.7	3.06	0.5	0.24	0.21	**0.2**	74.89
STD	no	0.33	0.33	0.34	4.38	0.61	**0.19**	0.29	0.32	67.81
STD	no	0.15	0.7	0.15	4.52	0.65	0.29	**0.17**	0.2	57.25
STD	no	0.5	0.0	0.5	5.83	0.95	0.39	0.44	**0.21**	65.94
STD	no	0.3	0.0	0.7	3.68	0.62	0.28	0.22	**0.21**	72.81
STD	no	0.7	0.0	0.3	8.01	1.06	0.68	0.52	**0.28**	61.14
RND	no	–	–	–	5.57	0.52	0.29	**0.22**	0.23	86.58
INV	no	0.7	0.2	0.1	10.32	1.63	1.07	1.15	**0.13***	34.3
INV	no	0.1	0.2	0.7	4.97	1.23	0.78	0.35	**0.32**	71.09
INV	no	0.3	0.0	0.7	9.78	3.3	0.73	1.17	**0.61***	53.08
INV	no	0.7	0.0	0.3	12.91	1.64	1.06	0.94	**0.08***	36.99
STD	rand	0.33	0.33	0.34	16.27	2.18	0.47	0.44	**0.16**	189.37
RND	rand	–	–	–	14.35	2.27	0.52	0.38	**0.14***	352.04
INV	rand	0.7	0.0	0.3	8.49	1.15	0.39	0.28	**0.04***	178.13
STD	bitree	0.33	0.33	0.34	20.01	2.49	0.65	0.86	**0.07***	92.46
RND	bitree	–	–	–	20.72	2.92	1.19	0.98	**0.17***	141.05
INV	bitree	0.7	0.0	0.3	12.77	0.96	0.43	0.42	**0.07***	95.88

Table 6. Average relative deviation from the length of the critical path for PSPLIB series, %.

	GR	ES	ESA	ESVO	GA
$n = 60$	206.62	16.17	14.35	14.4	**13.5***
$n = 90$	230.42	15.01	13.81	13.83	**12.67***
$n = 120$	557.84	42.96	40.59	40.54	**37.71***

90, and 120 respectively for the original PSPLIB instances, and our results are comparable to ones presented in [10–12]. The obtained results show that treating a resource as a databus (influencing on durations of jobs) allows the algorithms to find schedules of better quality.

5 Conclusion

We investigated a problem of scheduling multi-mode jobs with resource dependent durations and constraints, arising in planning of procedures (tasks) in computer systems. The computational complexity, approximation guarantees and lower bounds on the makespan were analyzed for particular cases. Greedy heuristics, several variants of the (1+1) evolutionary strategy, and an adaptive genetic algorithm were proposed and experimentally tested on series of instances of different structures. Further results can be undertaken to improve lower bounds on the makespan in the case of different types of resources.

Acknowledgements. The research was supported by Russian Science Foundation grant N 22-71-10015-П, https://rscf.ru/en/project/22-71-10015-П/.

References

1. Althaus, E., et al.: Scheduling shared continuous resources on many-cores. J. Sched. **21**(1), 77–92 (2018)
2. Blazewicz, J., Brauner, N., Finke, G.: Scheduling with discrete resource constraints. In: Leung, J.Y.-T. (ed.) Handbook of Scheduling: Algorithms, Models, and Performance Analysis, pp. 23-1-23-18. CRC Press, Boca Raton (2004)
3. Chandra, D., Guo, F., Kim, S., Solihin, Y.: Predicting inter-thread cache contention on a chip multi-processor architecture. In: 11th International Symposium on High-Performance Computer Architecture (HPCA), pp. 340–351. IEEE Computer Society, Washington (2005)
4. Doerr, B., Ghannane, Y., Ibn Brahim, M.: Runtime analysis for permutation-based evolutionary algorithms. Algorithmica **86**, 90–129 (2024)
5. Du, J., Leung, J.Y.-T., Young, G.H.: Scheduling chain-structured tasks to minimize makespan and mean flow time. Inf. Comput. **92**(2), 219–236 (1991)
6. Eremeev, A.V., Kovalenko, Y.V.: A memetic algorithm with optimal recombination for the asymmetric travelling salesman problem. Memet. Comput. **12**, 23–36 (2020)
7. Eremeev, A.V., Malakhov, A.A., Sakhno, M.A., Sosnovskaya, M.Y.: Multi-core processor scheduling with respect to data bus bandwidth. In: Olenev, N., Evtushenko, Y., Khachay, M., Malkova, V. (eds.) OPTIMA 2020. CCIS, vol. 1340, pp. 55–69. Springer, Cham (2020). https://doi.org/10.1007/978-3-030-65739-0_5
8. Fedorova, A., Seltzer, M., Smith, M.D.: Improving performance isolation on chip multiprocessors via an operating system scheduler. In: 16th International Conference on Parallel Architecture and Compilation Techniques (PACT 2007), pp. 25–38. IEEE Computer Society, Los Alamitos (2007)
9. Garey, M.R., Johnson, D.S.: Computers and Intractability: A Guide to the Theory of NP-Completeness. Freeman, San Francisco (1979)
10. Goncharov, E.N., Leonov, V.V.: Genetic algorithm for the resource-constrained project scheduling problem. Autom. Remote. Control. **78**(6), 1101–1114 (2017). https://doi.org/10.1134/S0005117917060108
11. Goncharov, E.N.: A local search algorithm for the resource-constrained project scheduling problem. Discrete Anal. Oper. Res. **29**(4), 15–37 (2022)
12. Goncharov, E.N.: A greedy algorithm for the resource-constrained project scheduling problem. J. Appl. Ind. Math. **18**(4), 679–685 (2024)
13. Holland, J.H.: Adaptation in Natural and Artificial Systems: An Introductory Analysis with Applications to Biology, Control, and Artificial Intelligence. The MIT Press, Cambridge (1992)

14. Jiang, Y., Shen, X., Chen, J., Tripathi, R.: Analysis and approximation of optimal co-scheduling on chip multiprocessors. In: Moshovos, A. (ed.) 17th International Conference on Parallel Architectures and Compilation Techniques (PACT 2008), pp. 220–229. ACM, New York (2008)
15. Jozefowska, J., Weglarz, J.: Scheduling with resource constraints - continuous resources. In: Leung, J.Y.-T. (ed.) Handbook of Scheduling: Algorithms, Models, and Performance Analysis, pp. 24-1–24-15. CRC Press, Boca Raton (2004)
16. Kellegoz, T., Toklu, B., Wilson, J.: Comparing efficiencies of genetic crossover operators for one machine total weighted tardiness problem. Appl. Math. Comput. **199**(2), 590–598 (2008)
17. Liu, M., Chu, F., He, J., Yang, D., Chu, C.: Coke production scheduling problem: a parallel machine scheduling with batch preprocessings and location-dependent processing times. Comput. Oper. Res. **104**, 37–48 (2019)
18. Merkel, A., Stoess, J., Bellosa, F.: Resource-conscious scheduling for energy efficiency on multicore processors. In: Morin, C. (ed.) 5th European Conference on Computer Systems (EuroSys 2010), pp. 153–166. ACM, New York (2010)
19. Nesbit, K.J., Aggarwal, N., Laudon, J., Smith, J.E.: Fair queuing memory systems. In: Conte, T., Zhou, H. (eds.) 39th Annual IEEE/ACM International Symposium on Microarchitecture (MICRO 2006), pp. 208–222. IEEE Computer Society, Los Alamitos (2006)
20. Servakh, V.V.: An efficiently solvable case of the renewable resource-constrained project scheduling problem. Discrete Anal. Oper. Res. **7**(1), 75–82 (2000)
21. Shioura, A., Shakhlevich, N.V., Strusevich, V.A.: Preemptive models of scheduling with controllable processing times and of scheduling with imprecise computation: a review of solution approaches. Eur. J. Oper. Res. **266**, 795–818 (2018)
22. Snavely, A., Tullsen, D.M.: Symbiotic job scheduling for a simultaneous multithreaded processor. In: 9th International Conference on Architectural Support for Programming Languages and Operating Systems (ASPLOS 2000), pp. 234–244. ACM, New York (2000)
23. Tian, K., Jiang, Y., Shen, X., Mao, W.: Optimal co-scheduling to minimize makespan on chip multiprocessors. In: Cirne, W., Desai, N., Frachtenberg, E., Schwiegelshohn, U. (eds.) JSSPP 2012. LNCS, vol. 7698, pp. 114–133. Springer, Heidelberg (2013). https://doi.org/10.1007/978-3-642-35867-8_7
24. Xiao, Z., Chen, L., Wang, B., Du, J., Li, K.: Novel fairness-aware co-scheduling for shared cache contention game on chip multiprocessors. Inf. Sci. **256**, 68–85 (2020)
25. Zacarias, F.V., Petrucci, V., Nishtala, R., Carpenter, P., Mossé, D.: Intelligent colocation of workloads for enhanced server efficiency. In: 31st International Symposium on Computer Architecture and High Performance Computing (SBAC-PAD), pp. 120–127. IEEE Press, New York (2019)
26. Zakharova, Y.V., Sakhno, M.Y.: Evolutionary algorithm with optimized operators for multi-processor scheduling with resources. SEMR **22**(2), 85–103 (2025)
27. Zhuravlev, S., Blagodurov, S., Fedorova, A.: Addressing shared resource contention in multicore processors via scheduling. In: 15th International Conference on Architectural Support for Programming Languages and Operating Systems (ASPLOS'10), pp. 129–142. ACM, New York (2010)

Applications in Machine Learning and Mathematical Economics KL-means C

KL-Means Clustering via Nonconvex Optimization

Tatiana V. Gruzdeva(✉), Maria S. Okhotina, and Anton V. Ushakov

Matrosov Institute for System Dynamics and Control Theory of SB RAS, 134 Lermontov Street, 664033 Irkutsk, Russia
{gruzdeva,mokhotina,aushakov}@icc.ru

Abstract. Cluster analysis is one of the central and most actively developed research directions in unsupervised machine learning. The Euclidean minimum-sum-of-squares clustering problem, referred to as the k-means problem, remains the most widely studied clustering model. A more general Bregman divergence-based clustering approach extends this classical model by replacing the squared Euclidean distance with a Bregman divergence for measuring the dissimilarity between data points and their nearest cluster centers. In the present work, we study this generalized clustering setting from the point of view of mathematical programming. In particular, we consider a practically important case of Bregman divergences, Kullback-Leibler divergence. We focus on the corresponding clustering problem as a nonconvex optimization problem and develop a local search method based on linearization of a nonconvex component in its DC representation. The developed approach is tested in a series of computational experiments and compared with conventional implementations of k-means and KL-means.

Keywords: Clustering · Bregman divergence · Kullback-Leibler divergence · KL-means · Nonconvex optimization · DC programming · Local search

1 Introduction

Cluster analysis constitutes one of the central and most actively developed research directions in unsupervised machine learning, combinatorial optimization, and multivariate statistics. The fundamental objective of clustering is to divide a given set of observations or data points into meaningful groups, referred to as clusters, in such a manner that intra-cluster similarity is maximized while inter-cluster similarity remains low. Despite its seemingly straightforward formulation, clustering represents a difficult computational and mathematical problem being of critically important in practice. Indeed, clustering routinely appears as a critical preprocessing or exploratory step in diverse application domains ranging

---

The research was funded by the Russian Science Foundation, project No 25–21-00304.

A. Kazakov et al. (Eds.): MOTOR 2026, LNCS 16687, pp. 249–261, 2026.
https://doi.org/10.1007/978-3-032-31390-4_16

from bioinformatics and medical image analysis to natural language processing, computer vision, and recommender systems.

Among the broad landscape of clustering methodologies, center-based approaches hold a particularly prominent position, both from theoretical and applied perspectives. The unifying principle of these methods is to represent each cluster by a single representative point, commonly called a cluster center or prototype, and to define cluster membership by proximity to these representatives with respect to some chosen dissimilarity measure. The resulting optimization problem—finding places for centers that minimize the total dissimilarity between data points and their assigned representatives—is inherently nonconvex and, in general, NP-hard [11]. Nevertheless, efficient algorithms have been developed that perform remarkably well in practice, making center-based methods dominant in large-scale clustering applications. Note that center-based clustering models are closely related to facility location problems [26].

The most extensively studied instance of center-based clustering is undoubtedly the minimum sum-of-squares clustering problem. Its objective is to find k cluster centers so as the total sum of squared Euclidean distances from data items to the closest centers is minimized. The corresponding optimization problem of finding such kind of clustering is NP-hard [1,18].

The k-means algorithm, introduced by Lloyd [13,17], remains to this day the most widely deployed algorithm for this problem, owing to its simplicity, low per-iteration computational cost, and surprisingly competitive empirical performance despite lacking global optimality guarantees. The algorithm alternates between two steps: assigning each data point to its nearest center, and updating each center as the arithmetic mean of its assigned points. A critical property that makes this procedure computationally attractive is the existence of a closed-form solution for the center-update step, which reduces to computing simple arithmetic means and requires only linear time in the number of data points. This property is a direct consequence of the first-order optimality conditions associated with squared Euclidean distance minimization.

A natural well-motivated generalization of the minimum sum-of-squares model arises upon replacing squared Euclidean distance with a Bregman divergence as the underlying dissimilarity measure. Bregman divergences constitute a broad family of dissimilarity measures, defined as the gap between a strictly convex differentiable function and its first-order Taylor approximation. This family encompasses numerous practically important measures, including the squared Euclidean distance, the Kullback-Leibler divergence, the Itakura-Saito distance, and the logistic loss, among others. The theoretical significance of Bregman divergences in the context of clustering was established in the seminal work [3], where it was proved that the class of Bregman divergences is precisely the class for which the arithmetic mean remains the optimal cluster representative. Thus, the closed-form center-update step is preserved, which remains k-means computationally efficient. It was also shown that such algorithms converge to a locally optimal solution in a finite number of steps.

Despite this theoretical foundation, the development of practically efficient clustering algorithms for general Bregman divergences remains an active and challenging research area. Existing algorithms either inherit the local convergence limitations of standard k-means, or sacrifice computational efficiency in pursuit of improved solution quality. In particular, the initialization sensitivity of k-means-type algorithms (their tendency to converge to poor local optima depending on the choice of initial centers) is well-known and constitutes a significant practical challenge.

Despite a huge number of publications devoted to conventional k-means, there are only quite limited amount of publications on clustering algorithms with Bregman divergences, especially considering the clustering problem as a mathematical optimization problem. For example, the Bergman divergences were applied as distance measures in a multi-task clustering [28], co-clustering [2,20], and agglomerative clustering [24] settings. In [15] the authors studied the conditions under which the k-means algorithm with any Bregman divergence converges to a locally optimal solution and propose simple modifications to the k-means algorithm which ensure local optimality. The total Bregman soft clustering algorithm was introduced in [16]. It is build upon the family of total Bregman divergences being an efficient and robust measure to quantify the dissimilarity between shapes. In [27] the authors introduced a data-driven approach based on data preprocessing to find initial centroids and improve computational speed of Bregman k-means algorithms.

In [9], the clustering problem with Bregman divergences was formulated as a nonconvex optimization problem. The DC approach for the simplest Bregman divergence induced by the function $F(x) = \|x\|^2$ resulting in squared Euclidean distance $D_F(x, y) = \|x - y\|^2$ is studied and successfully tested on clustering problems [6,7]. Another example of Bregman divergence that is useful in many applications is the squared Mahalanobis distance, $D_F(x, y) = \frac{1}{2}\langle (x-y), Q(x-y)\rangle$ induced by the convex function $F(x) = \frac{1}{2}\langle x, Qx\rangle$ with positive defined matrix Q of appropriate dimension was studied by the view-point of DC minimization approach as well [8].

In this paper we consider another practically important particular case of Bregman divergences, Kullback-Leibler divergence. We formulate the corresponding clustering problem (KL-means) as a nonconvex optimization problem. Next, we show its reduction to a DC programming problem with a nonconvex objective minimized over the canonical simplex. We then developed a local search method based on linearization of a nonconvex component in the DC representation. The developed approach is tested in a series of computational experiments on widely used test data instances and compared with conventional implementations of k-means.

2 Problem Statement

A Bregman divergence or Bregman distance [5] is a measure of difference between two points, defined in terms of a strictly convex function. Let $F(\cdot) : \Omega \to \mathbb{R}$ be a continuously differentiable, strictly convex function defined on a closed convex set Ω. The Bregman distance associated with function $F(\cdot)$ for points $p, q \in \Omega$ is the difference between the value of function $F(\cdot)$ at point p and the value of the first-order Taylor expansion of $F(\cdot)$ around point q evaluated at point p [5]:

$$D_F(p,q) = F(p) - F(q) - \langle \nabla F(q), p - q \rangle. \tag{1}$$

Bregman divergence has several well-known and useful properties [3]. In the paper, we use its feature of non-negativity: $D_F(p,q) \geq 0 \;\; \forall p, q$; as well as the convexity of $D_F(p,q)$ with respect to p (but not necessarily to the q [4]).

In [9], we formulated the Bregman clustering problem as a non-convex optimization problem. Given a finite set $J = \{1, \ldots, m\}$ of data items, where each item $j \in J$ is represented by a feature vector $a^j \in \mathbb{R}^n$. The objective is to find k cluster centers that minimize the total sum of dissimilarities (distances) between each data item and its nearest cluster's center. Obviously, one supposes that the number of data items m is greater than k, otherwise the problem is trivially solved.

Naturally, all items assigned to the same center constitute a cluster. As the measure of dissimilarity between a data item and its closest cluster center, we employ a Bregman divergence.

Let us introduce the following binary variables

$$x_{ij} = \begin{cases} 1, \text{ if data item } j \text{ is assigned to cluster } i, \\ 0, \text{ otherwise}, \end{cases} \quad i = 1, \ldots, k, \; j = 1, \ldots, m.$$

which are often referred to as assignment variables, and continuous variables $y^i \in I\!R^n$, $i = 1, \ldots, k$, corresponding to the unknown cluster centers' locations.

With these notations, we can formulate the following mixed integer programming (MIP) problem:

$$\sum_{i=1}^{k} \sum_{j=1}^{m} x_{ij} D_F(y^i, a^j) \downarrow \min_{(x,y)}, \tag{2}$$

$$\sum_{i=1}^{k} x_{ij} = 1 \qquad \forall\, j = 1, \ldots, m; \tag{3}$$

$$x_{ij} \in \{0, 1\} \qquad \forall i = 1, \ldots, k; \; \forall j = 1, \ldots, m. \tag{4}$$

The goal function (2) minimizes the sum of dissimilarities between data items and cluster centers, whereas constraints (3) guarantee that each data item is assigned to exactly one cluster.

For fixed centers y^i, thanks to data items are always assigned to the closest cluster centers, the assignment variables x_{ij} take binary values in the corresponding optimal solution. Consequently, in our approach we consider a natural relaxation of (2)–(4) where the binary constraints $x_{ij} \in \{0, 1\}$ are replaced with

$x_{ij} \in [0,1]$, $i = 1, \ldots, k$; $j = 1, \ldots, m$. The resultant problem is to minimize a nonconvex function over a convex feasible set [9]:

$$f(x, y) = \sum_{i=1}^{k} \sum_{j=1}^{m} x_{ij} D_F(y^i, a^j) \downarrow \min_{(x,y)}, \quad x \in S, \quad y \in I\!R^{k \times n}, \tag{5}$$

where $S = \{x_{ij} \in [0,1] : \sum_{i=1}^{k} x_{ij} = 1, \; j = 1, \ldots, m\}$, $S \subset I\!R^{k \times m}$.

If we intend to solve the nonconvex problem (5) by applying the global search theory [21,23], we need an explicit DC representation of the nonconvex objective function.

3 DC Representation and Local Search

It is well-known that DC representation is not unique, and different DC decompositions generate various auxiliary convex problems. Here we use a DC representation of the objective function in the problem (5) constructed by the following way [9].

Let us fix i and j, denote $x := x_{ij} \in I\!R$, $y := y^i \in I\!R^n$, $a := a^j \in I\!R^n$, and consider the following representation of one nonconvex term of the sum in the objective function $f(\cdot)$ of the problem (5):

$$x \, D_F(y, a) = \frac{1}{2}(x + D_F(y, a))^2 - \frac{1}{2}\left(x^2 + D_F^2(y, a)\right). \tag{6}$$

Since $D_F(y, a)$ is convex with respect to its first argument y, $D_F(y, a) \geq 0$ for all y, a, and $x + D_F(y, a) \geq 0$ for all $x \in [0,1]$ then

$$\begin{aligned} g(x, y) &= \frac{1}{2} \sum_{i=1}^{k} \sum_{j=1}^{m} (x_{ij} + D_F(y^i, a^j))^2, \\ h(x, y) &= \frac{1}{2} \sum_{i=1}^{k} \sum_{j=1}^{m} \left(x_{ij}^2 + D_F^2(y^i, a^j)\right) \end{aligned} \tag{7}$$

are convex functions.

The explicit view of the functions (7) allows us to consider the following DC minimization problem

$$f(x, y) = g(x, y) - h(x, y) \downarrow \min_{(x,y)}, \quad x \in S, \quad y \in I\!R^{k \times n}. \tag{$\mathcal{P}$}$$

In order to find a local solution of Problem ($\mathcal{P}$), we apply the well-known DC Algorithm [12,21,22]. It consists of linearizing, at a current point, the function $h(\cdot)$ which defines the basic nonconvexity of Problem ($\mathcal{P}$). The resultant convex approximation of the objective function $f(\cdot)$ obtained by replacing the nonconvex part with its linearization is then minimized. Obviously, such an approach allows finding local solutions by employing conventional convex optimization techniques [19].

The scheme of the DC Algorithm for Problem ($\mathcal{P}$) is the following. We start with an initial point $(x^0, y^0) : y^0 \in I\!R^{k\times n},\ x^0 \in S$. Suppose there is point $(x^s, y^s),\ x^s \in S$, already found. Then, we look for (x^{s+1}, y^{s+1}) as an approximate solution to the linearized problem

$$\Phi_s(x,y) = g(x,y) - \langle \nabla h(x^s,y^s), (x,y)\rangle \downarrow \min_{(x,y)},\ x \in S,\ y \in I\!R^{k\times n}. \qquad (\mathcal{PL}_s)$$

It means that the next iteration (x^{s+1}, y^{s+1}) satisfies the following inequality:

$$\begin{aligned} &g(x^{s+1},y^{s+1}) - \langle \nabla h(x^s,y^s), (x^{s+1},y^{s+1})\rangle \le \\ &\le \inf_{\substack{x\in S\\ y\in I\!R^{k\times n}}} \{g(x,y) - \langle \nabla h(x^s,y^s),(x,y)\rangle\} + \delta_s, \end{aligned} \qquad (8)$$

where $\delta_s \ge 0,\ s = 0,1,2,\ldots;\ \sum_{s=0}^{\infty} \delta_s < \infty$.

As it was proven in [22], the clustering solution $(x^*, y^*),\ x^* \in S,\ y^* \in I\!R^{k\times n}$, of the sequence $\{(x^s, y^s)\}$ generated by the method, is a solution to Linearized Problem ($\mathcal{PL}_*$), and stationary (critical) point of Problem ($\mathcal{P}$).

Note that Linearized Problem ($\mathcal{PL}_s$) is convex, whereas Problem ($\mathcal{P}$) is nonconvex.

As it was suggested in [21,22], one of the following inequalities can be employed as a stopping criterion:

$$\begin{aligned} f(x^s,y^s) - f(x^{s+1},y^{s+1}) &\le \frac{\tau}{2}, \\ \Phi_s(x^s,y^s) - \Phi_s(x^{s+1},y^{s+1}) &\triangleq g(x^s,y^s) - g(x^{s+1},y^{s+1}) + \\ &+ \langle \nabla h(x^s,y^s), (x^{s+1},y^{s+1}) - (x^s,y^s)\rangle \le \frac{\tau}{2}, \end{aligned} \qquad (9)$$

where τ is a given accuracy. Therefore, if $\delta_s \le \frac{\tau}{2}$, the point (x^s, y^s) is a τ-solution to Problem ($\mathcal{PL}_s$).

Given the DC representation (7). To find local solutions, we have to solve a series of the following linearized problems:

$$\left.\begin{aligned} \Phi_s(x,y) = \frac{1}{2}\sum_{i=1}^{k}\sum_{j=1}^{m}(x_{ij} + D_F(y^i,a^j))^2 - \langle \nabla h(x^s,y^s),(x,y)\rangle \downarrow \min_{(x,y)}, \\ x \in S,\ y \in I\!R^{k\times n}, \end{aligned}\right\} \qquad (10)$$

where $\nabla h(x^s,y^s) = (\nabla_x h(x^s,y^s), \nabla_y h(x^s,y^s)) \in I\!R^{k\times m + k\times n}$ with the following components

$$\begin{aligned} &[\nabla_x h(x^s,y^s)]_{ij} = x^s_{ij},\quad i = 1,\ldots,k,\quad j = 1,\ldots,m; \\ &[\nabla_y h(x^s,y^s)]_{il} = \sum_{j=1}^{m} D_F(y^{si},a^j)\left[\nabla_y D_F(y^{si},a^j)\right]_l, \\ &\qquad i = 1,\ldots,k,\quad l = 1,\ldots,n. \end{aligned} \qquad (11)$$

Next, we will show how the local search method (LSM) for DC minimization problems may be applied for a specific special case of Bregman divergence.

4 Case Study: Generalized I-Divergence

One of the widely used Bregman divergence is the so-called generalized I-divergence (the information or Kullback-Leibler divergence [14]) which is induced by the negative entropy function.

If $p \in I\!R^n_+$, the negative entropy $F(p) = \sum_{l=1}^{n} p_l \ln p_l = \langle p, \ln p \rangle$ is a convex function, and $\nabla F(p) = \ln p + 1\!\!1$, where $1\!\!1 = (1, \ldots, 1) \in I\!R^n$. The corresponding Bregman divergence is

$$\begin{aligned} D_F(p, q) &= \langle p, \ln p \rangle - \langle q, \ln q \rangle - \langle \nabla F(q), p - q \rangle = \\ &= (\langle p, \ln p \rangle - \langle p, \ln q \rangle) - \langle 1\!\!1, p - q \rangle = \left\langle p, \ln \frac{p}{q} \right\rangle - \langle 1\!\!1, p - q \rangle. \end{aligned} \tag{12}$$

Therefore, in the problem (5),

$$D_F(y^i, a^j) = \left\langle y^i, \ln \frac{y^i}{a^j} \right\rangle - \langle 1\!\!1, y^i - a^j \rangle, \quad i = 1, \ldots, k, \quad j = 1, \ldots, m,$$

$$\left[\nabla_y D_F(y^i, a^j)\right]_l = \ln \frac{y^i_l}{a^j_l}, \quad i = 1, \ldots, k, \quad l = 1, \ldots, n,$$

and, using (7), we get the following functions $g(\cdot)$ and $h(\cdot)$ in DC representation of the objective function $f(\cdot)$:

$$\begin{aligned} g(x, y) &= \frac{1}{2} \sum_{i=1}^{k} \sum_{j=1}^{m} \left(x_{ij} + \left\langle y^i, \ln \frac{y^i}{a^j} \right\rangle - \langle 1\!\!1, y^i - a^j \rangle \right)^2, \\ h(x, y) &= \frac{1}{2} \sum_{i=1}^{k} \sum_{j=1}^{m} \left(x_{ij}^2 + \left(\left\langle y^i, \ln \frac{y^i}{a^j} \right\rangle - \langle 1\!\!1, y^i - a^j \rangle \right)^2 \right). \end{aligned} \tag{13}$$

To construct the linearized problem (10) in this case, the components of the gradient $\nabla h(x^s, y^s) = (\nabla_x h(x^s, y^s), \nabla_y h(x^s, y^s))$ can be calculated as follows:

$$\begin{aligned} &[\nabla_x h(x^s, y^s)]_{ij} = x^s_{ij}, \quad i = 1, \ldots, k, \quad j = 1, \ldots, m; \\ &[\nabla_y h(x^s, y^s)]_{il} = \sum_{j=1}^{m} \left(\left\langle y^{si}, \ln \frac{y^{si}}{a^j} \right\rangle - \langle 1\!\!1, y^{si} - a^j \rangle \right) \ln \frac{y^{si}_l}{a^j_l}, \\ &i = 1, \ldots, k, \quad l = 1, \ldots, n. \end{aligned} \tag{14}$$

To solve the linearized problems (10), we apply the simplest projection gradient algorithm with the following iteration:

$$(x, y)^{r+1} = P_S\left[(x, y)^r - \alpha_r \nabla \Phi_s\left((x, y)^r\right)\right], \tag{15}$$

where α_r is a step size, P_S is a projection onto the canonical simplex S.

On the algorithm's iterations, we used the natural idea for holding the constraints (3)–(4): attaching data elements to the new nearest cluster center y^i. The step size α at r-iteration was found using the golden-section search on the interval $[0, 1]$. The descent direction $\nabla \Phi_s$ has the following components

$$
\begin{aligned}
&[\nabla_x \Phi_s(x,y)]_{ij} = x_{ij} + D_F(y^i,a^j) - x^s_{ij}, \;\; i=1,\ldots,k, \;\; j=1,\ldots,m;\\
&[\nabla_y \Phi_s(x,y)]_{il} = \sum_{j=1}^{m} \left(x_{ij} + D_F(y^i,a^j)\right) \left[\nabla_y D_F(y^i,a^j)\right]_l - \\
&\qquad - \sum_{j=1}^{m} D_F(y^{si},a^j) \left[\nabla_y D_F(y^{si},a^j)\right]_l, \;\; i=1,\ldots,k, \;\; l=1,\ldots,n.
\end{aligned} \tag{16}
$$

Next section describes the details of the implementation and the results of the computational experiments.

5 Computational Experiments

Here we describe computational experiments to test the proposed clustering algorithm, LSM. We compare our approach to the conventional KL-means clustering algorithm (k-means with Kullback-Leibler divergence as dissimilarity measure).

Our first test bed consists of several standard classification test instances from the KEEL dataset repository[1], which are widely used to test clustering algorithms. The second set of test problems is synthetically generated two-dimensional Gaussian mixture data instances of up to 1000 data items that widely used in testing solution algorithms for clustering and facility location problems [7,10,25]. For our experiments, we generated instances of Types I and III of different complexity, following the methodology from [29]. The problems of Type I are considered to be easier to solve and those of Type III are harder. The numbers of clusters vary from 4 to 49 and depend on the problem size.

Beside the value of the objective function, we also use the following external measures to assess accuracy of a clustering (pairwise precision Pr, recall R, and F-measure):

$$
Pr = \frac{TP}{TP+FP}, \quad R = \frac{TP}{TP+FN}, \quad F\text{-measure} = 2 \cdot \frac{Pr \cdot R}{Pr + R}.
$$

where TP are true positive pairs of data items (correctly clustered), FP — false positive pairs, and FN — false negative pairs. The pairwise F-measure is defined as harmonic mean of pairwise precision and recall.

Observe that we compare our algorithm with KL-means with respect to solution quality only. Our approach is not competitive against the algorithm with respect to run time. The initial solutions were taken at random from the data items and was the same for KL-means and LSM.

For tests from the KEEL dataset, we report the computational results in Table 1, where the following denotations are used:

- m is the number of data items;
- n is the number of features;

[1] https://sci2s.ugr.es/keel/datasets.php.

Table 1. Clustering comparison on KEEL-dataset

Dataset	m	n	k	F-measure		Objective value	
				LSM	KLM	LSM	KLM
zoo	101	16	7	**0.8373**	0.8125	**2258.43**	2382.08
iris	150	4	3	0.6612	0.6612	27.96	27.96
hayesroth	160	4	3	0.4276	0.4276	1313.41	1313.41
wine	178	13	3	0.8901	0.8901	138.68	138.68
glass	214	9	6	0.3952	**0.4024**	**222.42**	229.14
ecoli	336	7	8	0.4091	0.4091	234.15	234.15
bupa	345	6	2	0.5967	**0.6208**	**116.27**	122.80
ionosphere	351	33	2	0.6024	0.6024	8228.59	**8212.21**
monk-2	432	6	2	**0.6326**	0.6012	9906.64	**9647.80**
wbdc	569	30	2	**0.8654**	0.8587	**845.23**	859.15
balance	625	4	3	0.4093	**0.4170**	**3890.51**	3961.65
vehicle	846	18	4	0.3008	**0.3056**	**1081.19**	1095.20
yeast	1484	8	10	**0.3681**	0.3663	350.93	**135.07**
winequality-red	1599	6	3	0.2911	**0.3433**	**482.12**	616.15
titanic	2201	3	2	**0.7199**	0.6764	14877.50	**6995.24**
winequality-white	4898	7	3	0.2606	**0.2640**	**779.99**	812.80
spambase	4597	57	2	**0.6862**	0.6608	52448.00	**47142.30**
banana	5300	2	2	0.5090	**0.5175**	**227.52**	232.51
phoneme	5404	5	2	0.5879	**0.5966**	**679.11**	682.83
page-blocks	5472	10	5	**0.8953**	0.5901	2078.81	**741.49**
thyroid	7200	21	3	**0.9249**	0.8420	95764.60	**82296.70**
twonorm	7400	20	2	**0.9388**	0.8963	**2800.49**	2818.22
coil2000	9822	85	2	**0.9406**	0.9404	521464.00	**521316.00**
magic	19020	10	2	**0.7249**	0.5483	7561.05	**5758.80**

- k is the number of clusters;
- *Objective value* is the best value of the objective function (2) provided by the LSM and KL-means (KLM) algorithms, respectively;
- F-measure is the value of pairwise F-measure.

We report the computational results on the synthetically generated instances of Types I and III in Table 2 with the same notations.

The results on real-world datasets from the KEEL repository demonstrate an advantage of developed LSM over the standard KL-means algorithm. LSM achieves higher F-measure on 13 out of 23 datasets and ties on 6, meaning that KL-means fails to outperform LSM in clustering accuracy on the overwhelming majority of the test instances. The advantage of LSM is particularly compelling

Table 2. Clustering comparison on synthetically generated datasets

m	n	k	F-measure		Objective value	
			LSM	KLM	LSM	KLM
Type I						
50	2	4	**0.8801**	0.8430	22.59	**22.21**
200	2	16	0.6913	**0.7036**	110.24	**109.88**
300	2	16	**0.7505**	0.7086	**132.17**	141.04
400	2	16	**0.8175**	0.8116	189.01	**188.00**
1000	2	16	**0.6418**	0.6274	**514.51**	562.89
500	2	25	**0.8973**	0.8928	137.66	**136.96**
600	2	25	0.7635	**0.8195**	246.85	**209.65**
800	2	36	0.8452	0.8452	185.46	185.46
1000	2	49	0.7635	**0.7677**	**302.90**	303.65
Type III						
50	2	4	0.9295	0.9295	6.60	6.60
200	2	16	0.5600	0.5600	8.28	8.28
300	2	16	**0.7002**	0.6846	4.78	**4.53**
400	2	16	0.6499	0.6499	12.05	12.05
500	2	25	**0.6353**	0.6248	6.39	**5.97**
600	2	25	**0.6949**	0.6939	**3.23**	6.40
800	2	36	0.7054	0.7054	4.09	4.09
1000	2	49	**0.7114**	0.7105	4.26	**3.96**

on relatively large-scale datasets such as *magic*, *titanic*, and *thyroid*, where the improvement in F-measure is substantial. This confirms that LSM is more robust to premature convergence as dataset size and dimensionality increase. Although on the instances where KL-means outperforms LSM, the difference between computed metrics is relatively low, indicating almost similar performance for both approaches.

The experimental results on synthetically generated datasets of Type I, LSM demonstrates a better F-measure value in 5 out of 9 instances. On the datasets of Type III, LSM matches or outperforms KL-means in all eight problem instances, confirming its consistently superior cluster quality on this dataset type. Overall, across both Type I and Type III, LSM produces better F-measure results in 8 out of 17 instances and ties in 7, while KL-means wins in only 3 cases. All the latter instances are confined to Type I. These results indicate that LSM more reliably discovers cluster structures even in cases where KL-means achieves a marginally lower objective value. The trade-off between objective value and clustering accuracy further highlights that minimizing the KL-means criterion does not necessarily lead to better partitions, and that LSM's search strategy is better suited to finding solutions of higher quality.

It is interesting to note that solution quality according to objective value does not always correlate to that according to the F-measure.

6 Conclusion

In this work, we developed a KL clustering algorithm based on reformulating the objective function in the corresponding clustering model in DC form. We developed a local search algorithm which is based on solving a series of linearized convex problems with gradient descent procedures. Experimental evaluation on synthetic and real-world benchmarks confirms that the proposed approach outperforms conventional KL-means and k-means in terms of both objective value and external clustering measures across diverse datasets. The advantage is especially compelling on large and complex instances, where KL-means reliably falls into poor local minima. Crucially, cases where KL-means achieves a better objective value do not correspond to better clustering, which demonstrates that the KL-means criterion is not always a good measure of solution quality.

References

1. Aloise, D., Deshpande, A., Hansen, P., Popat, P.: NP-hardness of Euclidean sum-of-squares clustering. Mach. Learn. **75**, 245–248 (2009). https://doi.org/10.1007/s10994-009-5103-0
2. Banerjee, A., Dhillon, I., Ghosh, J., Merugu, S., Modha, D.S.: A generalized maximum entropy approach to bregman co-clustering and matrix approximation. J. Mach. Learn. Res. **8**(67), 1919–1986 (2007). http://jmlr.org/papers/v8/banerjee07a.html
3. Banerjee, A., Merugu, S., Dhillon, I.S., Ghosh, J.: Clustering with bregman divergences. J. Mach. Learn. Res. **6**, 1705–1749 (2005)
4. Bauschke, H., Borwein, J.: Joint and separate convexity of the bregman distance. Stud. Comput. Math. **8**, 23–36 (2001). https://doi.org/10.1016/S1570-579X(01)80004-5
5. Bregman, L.M.: The relaxation method of finding the common points of convex sets and its application to the solution of problems in convex programming. Comput. Math. Math. Phys. **7**, 200–217 (1967)
6. Gruzdeva, T.V., Ushakov, A.V.: A computational study of the dc minimization global optimality conditions applied to k-means clustering. In: Olenev, N., et al. (eds.) Optimization and Applications, Lecture Notes in Computer Science, vol. 13078, pp. 79–93. Springer, Cham (2021). https://doi.org/10.1007/978-3-030-91059-4_6
7. Gruzdeva, T.V., Ushakov, A.V.: K-means clustering via a nonconvex optimization approach. In: Pardalos, P., Khachay, M., Kazakov, A. (eds.) MOTOR 2021. LNCS, vol. 12755, pp. 462–476. Springer, Cham (2021). https://doi.org/10.1007/978-3-030-77876-7_31
8. Gruzdeva, T.V., Ushakov, A.V.: Mahalanobis-distance-based clustering via dc optimization approach. In: Olenev, N., et al. (eds.) Optimization and Applications. Lecture Notes in Computer Science, vol. 16426. Springer, Cham (2026)

9. Gruzdeva, T., Ushakov, A.: On a nonconvex distance-based clustering problem. In: Pardalos, P., Khachay, M., Mazalov, V. (eds.) Mathematical Optimization Theory and Operations Research, vol. 13367, pp. 139–152. Springer, Cham (2022). https://doi.org/10.1007/978-3-031-09607-5_10
10. Hansen, P., Brimberg, J., Urosević, D., Mladenović, N.: Solving large p-median clustering problems by primal-dual variable neighborhood search. Data Min. Knowl. Discov. **19**(3), 351–375 (2009). https://doi.org/10.1007/s10618-009-0135-4
11. Hansen, P., Jaumard, B.: Cluster analysis and mathematical programming. Math. Program. **79**(1–3), 191–215 (1997)
12. Hoai An, L.T., Tao, P.D.: The DC (difference of convex functions) programming and DCA revisited with dc models of real world nonconvex optimization problems. Ann. Oper. Res. **133**, 23–46 (2005)
13. Jain, A.K.: Data clustering: 50 years beyond k-means. Patt. Recogn. Lett. **31**(8), 651–666 (2010)
14. Kullback, S.: Information Theory and Statistics. A Wiley Publication in Mathematical Statistics, Dover Publications (1997)
15. Li, M., Metel, M.R., Takeda, A.: Modified k-means algorithm with local optimality guarantees (2025). https://arxiv.org/abs/2506.06990
16. Liu, M., Vemuri, B.C., Amari, S.I., Nielsen, F.: Shape retrieval using hierarchical total bregman soft clustering. IEEE Trans. Patt. Anal. Mach. Intell. **34**(12), 2407–2419 (2012). https://doi.org/10.1109/TPAMI.2012.44
17. Lloyd, S.: Least squares quantization in PCM. IEEE Trans. Inf. Theory **28**(2), 129–137 (1982). https://doi.org/10.1109/TIT.1982.1056489
18. Mahajan, M., Nimbhorkar, P., Varadarajan, K.: The planar k-means problem is np-hard. Theor. Comput. Sci. **442**, 13–21 (2012). https://doi.org/10.1016/j.tcs.2010.05.034. special Issue on the Workshop on Algorithms and Computation (WALCOM 2009)
19. Nocedal, J., Wright, S.J.: Numerical Optimization, 2nd edn. Operations Research and Financial Engineering, Springer, New York (2006)
20. Shen, G., Yang, W., Wang, W., Yu, M., Dong, G.: Agglomerative hierarchical co-clustering based on bregman divergence. In: Herawan, T., Ghazali, R., Deris, M.M. (eds.) Recent Advances on Soft Computing and Data Mining, pp. 389–398. Springer, Cham (2014)
21. Strekalovsky, A.S.: On solving optimization problems with hidden nonconvex structures. In: Rassias, T.M., Floudas, C.A., Butenko, S. (eds.) Optimization in Science and Engineering, pp. 465–502. Springer, New York (2014)
22. Strekalovsky, A.S.: On local search in d.c. optimization problems. Appl. Math. Comput. **255**, 73–83 (2015)
23. Strekalovsky, A.: On the minimization of the difference of convex functions on a feasible set. Comput. Math. Math. Phys. **43**, 380–390 (2003)
24. Telgarsky, M., Dasgupta, S.: Agglomerative bregman clustering. In: Proceedings of 29th International Conference on Machine Learning, pp. 1011–1018. Omnipress, Madison (2012)
25. Ushakov, A.V., Vasilyev, I.L., Gruzdeva, T.V.: A computational comparison of the p-median clustering and k-means. Int. J. Artif. Intell. **13**(1), 229–242 (2015)
26. Vasilyev, I., Ushakov, A.V.: Discrete facility location in machine learning. Diskretn. Anal. Issled. Oper **28**(4), 5–60 (2021). https://doi.org/10.33048/daio.2021.28.714
27. Xiao, Y., Li, H., Zhang, Y.: DBGSA: a novel data adaptive Bregman clustering algorithm. Eng. Appl. Artif. Intell. **131**, 107846 (2024). https://doi.org/10.1016/j.engappai.2024.107846

28. Zhang, J., Zhang, C.: Multitask Bregman clustering. Neurocomputing **74**(10), 1720–1734 (2011). https://doi.org/10.1016/j.neucom.2011.02.004
29. Zhang, T., Ramakrishnan, R., Livny, M.: Birch: a new data clustering algorithm and its applications. Data Min. Knowl. Discov. **1**(2), 141–182 (1997). https://doi.org/10.1023/A:1009783824328

A Hierarchical Data Envelopment Analysis Approach for Multicriteria Decision Making Problems with Multi-level Criteria

Chiang Kao[1,2](✉)

[1] Chung Kung Research Center, College of Management, National Cheng Kung University, Tainan, Taiwan
ckao@mail.ncku.edu.tw

[2] Department of Industrial and Information Management, National Cheng Kung University, Tainan, Taiwan

Abstract. The conventional multicriteria decision making (MCDM) problem concerns assessing alternatives based on a set of criteria. When the number of criteria is large, the criteria are usually grouped into categories at different levels to make the problem manageable. The problem is solved level by level independently. In this case, the relationship between the criteria at different categories is ignored. This paper proposes a common-weight hierarchical data envelopment analysis (DEA) model without inputs to solve this multi-level type of MCDM problem. The model treats each category as a subunit of the hierarchical system to properly take the interactions between the criteria at different levels into consideration. The model uses the average values, rather than the maximum values, to calculate efficiency, which avoids producing unreliable results caused by outliers. The case of evaluating the junior colleges of technology (JCTs) in Taiwan discussed in the literature is revisited. The idea of using average values as the basis obtains efficiencies with an average of one, which helps identify the weaknesses of the JCTs. Making improvements on the weaknesses raises the performance index in future evaluations.

Keywords: Data Envelopment Analysis · Multicriteria Decision Making · Hierarchical System

1 Introduction

Multicriteria decision making (MCDM) concerns with evaluating the performance of alternatives based on a set of criteria. In general, the importance, usually expressed in weights, of each criterion is firstly determined. The performances of an alternative on the criteria multiplied by the corresponding weights is then the aggregate performance of this alternative. Based on the aggregate performance, subsequent analyses are conducted and decisions are made. Numerous solution methods for the MCDM problem have been proposed and discussed in the literature [1–3].

A. Kazakov et al. (Eds.): MOTOR 2026, LNCS 16687, pp. 262–277, 2026.
https://doi.org/10.1007/978-3-032-31390-4_17

The conventional MCDM problem has only one level of criteria. The performance of an alternative is evaluated based on all criteria directly. In practice, there are cases where the performance of the alternative is evaluated from several aspects, and each aspect is expressed by a number of criteria. In other words, the criteria are grouped into different categories. This is especially common when the number of criteria is large. For example, in assessing the national competitiveness of Southeast Asian countries, Kao et al. [4] grouped sixteen criteria into four categories of economy, technology, human resources, and management, and calculated the national competitiveness index of a country indirectly via the four categories. In studying the collaboration partner selection for city logistics planning, Awasthi et al. [5] grouped twenty-one criteria into four categories of benefits, costs, opportunities, and risks. Theoretically, the aggregate performance of an alternative can be calculated from all criteria directly. However, due to the large number of criteria, it is difficult to determine the importance of a criterion in relative to other criteria. For problems with a substantial number of criteria, the criteria may even need to be grouped into categories of several levels such that each group of criteria is of a manageable size. The conventional way for dealing with multi-level criteria is to treat each group of criteria as an independent MCDM problem and aggregate the results of the categories level by level to obtain the final performance of the alternative. The analytical hierarchy process (AHP) technique [6] that formulates the problem in a hierarchical structure is a typical example. It solves the problem category by category and level by level independently. Solving each category of criteria independently loses the effect of the interactions between the criteria of different categories. A solution for this type of MCDM problem is the hierarchical data envelopment analysis.

Data envelopment analysis (DEA) is a technique proposed by Charnes et al. [7] for measuring the relative efficiency of a set of decision-making units (DMUs) that apply multiple inputs to produce multiple outputs. When there is no input, this technique can be applied to assessing alternatives by treating the criteria as outputs. The conventional DEA treats the production system as a whole, without considering the operations of the internal components. In reality, a production system is composed of several subunits operating interdependently, where the outputs of a subunit are the inputs of another subunit, with the relationship expressed in a network structure. Ignoring the operations of the subunits inside the system may produce misleading results [8]. The DEA for dealing with network production systems is named network DEA [9], and the hierarchical DEA deals with a specific network of hierarchical structure [10]. Chen et al. [11] applied a hierarchical DEA technique to evaluate the global food security indexes for 113 countries, where 9 sub-indicators and 19 indicators were grouped into three categories of affordability, availability, and quality and safety. Guo et al. [12] computed the digital economy development indexes for 30 provinces in China, where 32 sub-indicators and 10 indicators were grouped into four categories of data resourcing, digital industrialization, industrial digitalization, and digital economy development environment. The weights used by different countries and provinces to calculate the indexes were not the same, which made the results not comparable. More seriously, the variable substitution technique and the golden section search used in these two studies to calculate the indexes were not correct.

Aiming at the shortcoming of incompatible weights, Babaee et al. [13] proposed a common-weight hierarchical DEA model to calculate road user behavior indexes for 13 European countries, where the indicators were grouped into categories at three levels. Shahari et al. [14] measured the performance of the district health centers in Malaysia with the indicators grouped into categories at two levels. Yu and Rakshit [15] calculated a logistic performance index for 100 countries; the indicators were also grouped into categories at two levels. A common weakness of the methods used in these studies is that the original data for different indicators must be standardized to make the measures in different units and scales comparable and that bounds on the associated weights can be imposed. However, different methods of standardization yield different and probably misleading results. This will be explained in the next section. One of the merits of the DEA technique is that data standardization is not needed because the difference in units of the indicators can be absorbed by the associated multipliers, similar to the case in regression analysis. If bounds for the multipliers are necessary, they can be handled via assurance regions [16] or other similar ideas.

The conventional DEA approach uses the maximum point to construct frontiers as the basis for measuring efficiency. The frontiers are inclined to be affected by outliers. In this study we use the average point to construct hyperplanes as the basis for calculating efficiency. A common-weight hierarchical DEA model is proposed to solve the MCDM problem with multi-level indicators. The model has three characteristics. First, it is a common-weight approach where all DMUs apply the same set of weights for aggregation such that the results are comparable. Second, the original data is used for calculation, without standardization, to avoid producing misleading results. Third, the average-based hyperplane explains the trend of the DMUs better and obtains more representative results.

In the following sections, a common-weight average-based DEA model is firstly developed to evaluate a group of DMUs on the same basis. The model is then extended to the MCDM problem with multiple levels of criteria. Based on this model, the case of evaluating the junior colleges of technology (JCTs) in Taiwan studied in Kao [17] is revisited for illustrative purposes. Finally, some conclusions are drawn based on the discussions in the preceding sections.

2 Efficiency Measurement Based on Average Values

The idea of DEA is to use the maximum value as the basis for measuring efficiency. The position of a DMU in relative to the maximum point is the efficiency of this DMU. A DMU that is farther away from the maximum value is less efficient.

Let Y_{ij} denote the value of the ith output, $i = 1,\ldots, k$ of the jth DMU, $j = 1,\ldots, n$. Kao [17] formulated the following CCR model [7] without inputs to measure the performance of the JCTs in Taiwan:

$$\begin{aligned} E_0 = \ & \max \ \sum_{i=1}^{k} u_i Y_{i0} \\ & \text{s.t.} \ \sum_{i=1}^{k} u_i Y_{ij} \le 1, \quad j=1,\ldots, n \\ & \qquad u_i \ge \varepsilon, \qquad\qquad i=1,\ldots, k, \end{aligned} \tag{1}$$

where E_0 is the efficiency of the JCT, or DMU, being evaluated, with one as the highest attainable value, and ε is a small non-Archimedean number [18, 19] imposed upon the multipliers u_i to avoid the corresponding indicator from being ignored. When there is no further information for determining the values for ε, Färe and Hunsaker [20] suggested using one-half of the smallest u_i appeared in calculating the efficiency of all DMUs for ε.

Model (1) allows every DMU to select the most favorable multipliers u_i to calculate efficiency. The multipliers selected by every DMU may not be the same. The resulted efficiencies, in the sense of MCDM, are not comparable. The efficiency simply shows how far a DMU is from their perceived production frontier. To make the results comparable, all DMUs must use the same set of weights. Roll et al. [21] proposed an idea of finding the weights that produce the maximum efficiency for all DMUs in an aggregate sense. The model is:

$$\begin{aligned} \max \quad & \sum_{j=1}^{n}\sum_{i=1}^{k} u_i Y_{ij} \\ \text{s.t.} \quad & \sum_{i=1}^{k} u_i Y_{ij} \le 1, \quad j=1,\ldots,n \\ & u_i \ge \varepsilon, \qquad\qquad i=1,\ldots,k. \end{aligned} \tag{2}$$

This model is similar to Model (1), except that Model (1) is solved once for every DMU while this model is solved once for all DMUs. After the optimal values u_i^* are solved, the hyperplane $\sum_{i=1}^{k} u_i^* Y_i = 1$ is the production frontier for all DMUs to calculate efficiency, where Y_i is a variable denoting the value of the ith output. The efficiency of the jth DMU is $E_j = \sum_{i=1}^{k} u_i^* Y_{ij}, j = 1,\ldots, n$.

Note that the multiplier u_i is not the weight, i.e. the importance, of the ith output. It can be very large or very small, depending on the scale of the associated output Y_i. Larger u_i not necessarily implies that output Y_i is more important. The sum of all u_i is not required to equal one. To let u_i represent the importance of the ith output and the efficiency be a weighted average of the k outputs, the existing studies on the hierarchical DEA standardize the original data Y_{ij} to $X_{ij} \in [0, 1]$ to construct aggregate performance indexes to assess the DMUs. The sum of all u_i is required to equal one and bounds on u_i are added. Since X_{ij} is less than or equal to one, $\sum_{i=1}^{k} u_i X_{ij}$ will not exceed one, and the constraints $\sum_{i=1}^{k} u_i Y_{ij} \le 1$ in Model (2) are no longer needed. The model becomes:

$$\begin{aligned} \max \quad & \sum_{j=1}^{n}\sum_{i=1}^{k} u_i X_{ij} \\ \text{s.t.} \quad & \sum_{i=1}^{k} u_i = 1, \quad j = 1,\ldots,n \\ & \alpha_i \le u_i \ge \beta_i, \quad i = 1,\ldots,k. \end{aligned} \tag{3}$$

Notably, different data standardization methods produce different standardized data which leads to different or even misleading results. Consider two popular standardization methods of $X_j = (Y_j - Y_{min})/(Y_{max} - Y_{min})$ and $X_j = Y_j / Y_{max}$, where Y_j is the observation of the jth output, $Y_{min} = \min\{Y_j, j = 1,\ldots, n\}$, and $Y_{max} = \max\{Y_j, j = 1,\ldots, n\}$. Both methods produce a standardized value in the range of 0 and 1. Given the data of $\{95, 97, 98, 100\}$ for four DMUs. The standardized value obtained from these two methods are $\{0, 0.4, 0.6, 1\}$ and $\{0.95, 0.97, 0.98, 1\}$, respectively. Although the orders of the values standardized from these two methods are the same, the respective means and variances are different. They will very probably produce different rankings for the four DMUs after the subsequent efficiency measurement. In fact, the DEA technique allows the outputs measured in different units and scales to be aggregated directly, without the standardization process. This merit should be utilized. If restrictions on the multipliers u_i are required, the methods of absolute ranges [21], assurance regions [16], polyhedral cone ratios [22], etc. can be applied.

The common-weight model (2) seeks a hyperplane superimposed upon all DMUs that produces the maximum efficiency for all DMUs. The hyperplane is inclined to be dragged toward the points that are farther away from the hyperplane, regarded as outliers. The hyperplane that passes through the average of all DMUs, on the other hand, reflects the trend of the DMUs better. The idea of this paper is to construct a hyperplane that passes through the average of all DMUs as the basis for calculating efficiency. Let $\overline{\boldsymbol{Y}} = (\overline{Y}_1, \ldots, \overline{Y}_k)$ be the average DMU, where $\overline{Y}_i = \sum_{j=1}^{n} Y_{ij}/n$. This hyperplane is expressed as $\sum_{i=1}^{k} u_i \overline{Y}_i = 1$. The relative distance between DMU j and the hyperplane is s_j, where $s_j = 1 - \sum_{i=1}^{k} u_i Y_{ij}$. The multiplier u_i is calculated via the conventional least-squares technique formulated as the following quadratic program:

$$
\begin{array}{lll}
\min & \sum_{j=1}^{n} s_j^2 & \\
\text{s.t.} & \sum_{i=1}^{k} u_i Y_{ij} + s_j = 1, & j = 1,\ldots, n \\
& \sum_{i=1}^{k} u_i \overline{Y}_i = 1 & \\
& u_i \geq \varepsilon, & i = 1,\ldots, k \\
& s_j \text{ unrestricted in sign}, & j = 1,\ldots, n.
\end{array}
\tag{4}
$$

At optimality, the hyperplane obtained is $\sum_{i=1}^{k} u_i^* Y_i = 1$ and the efficiency of DMU j is calculated as $E_j = \sum_{i=1}^{k} u_i^* Y_{ij}$. Note that the efficiency in this case can exceed one, which is super efficiency in DEA terminology [23]. The efficiency calculated from the hyperplane passing through the average point will be more representative than that calculated from the hyperplane passing through the maximum point. Kao [24] proposed a model that used Y_i^{max} to determine the hyperplane, where $Y_i^{max} = \max\{Y_{ij}, j = 1,\ldots, n\}$. The hyperplane obtained from this model is similar to that obtained from the maximum-efficiency model (2), which is prone to be affected by outliers.

Similar to the case of regression analysis where the sum of the residuals is equal to zero, the sum of the slack variables s_j in Model (4) is equal to zero. This property is proven in the following proposition.

Proposition 1. The sum of the slack variables s_j in the average-based common-weight model (4) is zero.

Proof. The slack variable s_j is defined in the first constraint of Model (4). The sum of s_j in this set of constraint is $\sum_{j=1}^{n}\sum_{i=1}^{k} u_i Y_{ij} + \sum_{j=1}^{n} s_j = n$. Dividing both sides of this constraint by n obtains $\sum_{i=1}^{k} u_i \overline{Y}_i + \sum_{j=1}^{n} s_j/n = 1$. Since $\sum_{i=1}^{k} u_i \overline{Y}_i = 1$, as required by the second constraint of Model (4), we have $\sum_{j=1}^{n} s_j/n = 0$, which implies $\sum_{j=1}^{n} s_j = 0$. ◊

In the next section, the average-based model is extended from one level to multiple levels to handle the MCDM problem with the criteria grouped into categories of multiple levels.

3 Common-Weight Multi-level Hierarchical DEA Model

We start the discussion with two-level extensions. The problem of measuring the efficiency of the production units discussed in the preceding section is switched to evaluating the performance of alternatives. In this case Y_{ij} becomes the value of the ith indicator of the jth alternative. Figure 1 shows the structure of the two-level hierarchical system, where k indicators at the lower level are grouped into q categories at the upper level.

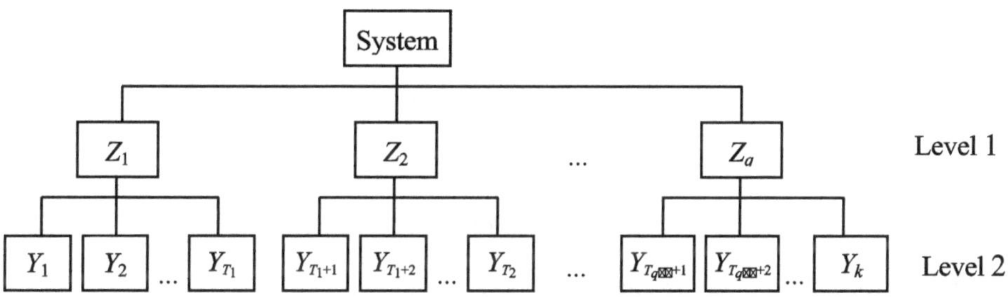

Fig. 1. The two-level hierarchical structure.

The pth category at the upper level comprises S_p indicators at the lower level. The notation of $T_p = S_1 + \ldots + S_p$ is introduced to indicate the cumulative number of indicators for ease of expression. Three specific cases are $T_0 = 0$, $T_1 = S_1$, and $T_q = k$. At level 2, the weight u_i is applied to the ith indicator to obtain $Z_{pj} = \sum_{i=T_{p-1}+1}^{T_p} u_i Y_{ij}$ as the value of the pth category of the jth alternative. At Level 1, the weight w_p is applied to the pth category to obtain $E_j = \sum_{p=1}^{q} w_p Z_{pj}$ as the performance index of the jth alternative. An alternative is seeking the most favorable weights u_i and w_p to calculate the performance index. Again, since the most favorable weights chosen by every DMU to calculate the index are different, a common-weight solution is desired, and the average-based common-weight model developed in the preceding section is employed.

3.1 Independent Two-Level Model

Conventionally, the value of every category $Z_{pj} = \sum_{i=T_{p-1}+1}^{T_p} u_i Y_{ij}$ is calculated independently via Model (4). Let $\overline{Z}_p = \sum_{j=1}^{n} Z_{pj}/n$ denote the average value of all alternatives of the pth category of indicators. The hyperplane for every alternative to calculate the performance index is $\sum_{p=1}^{q} w_p \overline{Z}_p = 1$. The distance between the jth alternative and this hyperplane is $s_j = 1 - \sum_{p=1}^{q} w_p Z_{pj}$. The average score $\overline{Z}_p$ has a property of equal to one as proven in the following proposition.

Proposition 2 The average value of all alternatives of every category $\overline{Z}_p$, $p = 1,\ldots, q$ is equal to one.

Proof. By definition, $\overline{Z}_p = \sum_{j=1}^{n} Z_{pj}/n = \sum_{j=1}^{n} \sum_{i=T_{p-1}+1}^{T_p} u_i Y_{ij}/n = \sum_{i=T_{p-1}+1}^{T_p} u_i \overline{Y}_i$. The second constraint of Model (4) requires $\sum_{i=T_{p-1}+1}^{T_p} u_i \overline{Y}_i = 1$ when the category is considered as a system. We thus have $\overline{Z}_p = \sum_{i=T_{p-1}+1}^{T_p} u_i \overline{Y}_i = 1$. ◊

Following Model (4), the model for calculating the performance index of the jth alternative at Level 1 independently from Level 2 is formulated as:

$$
\begin{aligned}
\min\quad & \sum_{j=1}^{n} s_j^2 \\
\text{s.t.}\quad & \sum_{p=1}^{q} w_p Z_{pj} + s_j = 1, && j=1,\ldots,n \\
& \sum_{p=1}^{q} w_p = 1 \\
& w_p \geq \varepsilon, && p=1,\ldots,q \\
& s_j \text{ unrestricted in sign}, && j=1,\ldots,n,
\end{aligned}
\tag{5}
$$

The performance index E_j has a property similar to Z_{pj} that the average of all alternatives is equal to one:

Proposition 3 The average value of the performance index of all alternatives calculated from Model (5), $\overline{E}$, is equal to one.

Proof. By definition, $\overline{E} = \sum_{j=1}^{n} E_j/n = \sum_{j=1}^{n} \sum_{p=1}^{q} w_p Z_{pj}/n = \sum_{p=1}^{q} w_p \overline{Z}_p$. Since $\overline{Z}_p = 1$, as proven in Proposition 2, we have $\overline{E} = \sum_{p=1}^{q} w_p = 1$, where the last equality is due to the second constraint of Model (5). ◊

Based on Proposition 2, a category with the score less than one indicates that this category performs below average. Similarly, according to Proposition 3, an alternative with the performance index less than one indicates that this alternative performs below the average of all alternatives. The categories with values less than one are the ones that cause the unsatisfactory performance of the alternative. These two properties help identify the weaknesses of an alternative. Making improvements on the weak points obtains more competitive alternatives.

3.2 Related Two-Level Model

Calculating the weights u_i and w_p independently ignores the relationship between the indicators at different levels. The u_i and w_p weights obtained are local optimal solutions. To obtain the global optimal solution, all weights should be calculated at the same time in one model. This is done by combining models (4) and (5) together.

The combination of Model (4) for the indicators at Level 2 and Model (5) for the categories at Level 1 is:

$$\begin{aligned}
\min \quad & \sum_{j=1}^{n} s_j^2 \\
\text{s.t.} \quad & \sum_{p=1}^{q} w_p Z_{pj} + s_j = 1, && j=1,\ldots,n \\
& \sum_{p=1}^{q} w_p = 1 \\
& Z_{pj} = \sum_{i=T_{p-1}+1}^{T_p} u_i Y_{ij}, && p=1,\ldots,q, \quad j=1,\ldots,n \\
& \sum_{i=T_{p-1}+1}^{T_p} u_i \overline{Y}_i = 1, && p=1,\ldots,q \\
& w_p, u_i \geq \varepsilon, && p=1,\ldots,q, \quad i=1,\ldots,k. \\
& s_j \text{ unrestricted in sign}, && j=1,\ldots,n.
\end{aligned} \tag{6}$$

This model is referred to as the related two-level hierarchical DEA model. The total weight assigned to the q categories is $\sum_{p=1}^{q} w_p = 1$. In other words, the efficiency is a weighted average of the q category values. By substituting s_j in the objective function by $s_j = 1 - \sum_{p=1}^{q} w_p Z_{pj}$ defined in the first constraint, and Z_{pj} by $Z_{pj} = \sum_{i=T_{p-1}+1}^{T_p} u_i Y_{ij}$ defined in the third constraint, Model (6) becomes the following quadruple program:

$$\begin{aligned}
\min \quad & \sum_{j=1}^{n} \Big(1 - \sum_{p=1}^{q} \sum_{i=T_{p-1}+1}^{T_p} w_p u_i Y_{ij}\Big)^2 \\
\text{s.t.} \quad & \sum_{p=1}^{q} w_p = 1 \\
& \sum_{i=T_{p-1}+1}^{T_p} u_i \overline{Y}_i = 1, \quad p=1,\ldots,q \\
& w_p, u_i \geq \varepsilon, \quad p=1,\ldots,q, \; i=1,\ldots,k.
\end{aligned} \tag{7}$$

Multiplying both sides of the second constraint and the $u_i \geq \varepsilon$ constraints by w_p and substituting $w_p u_i$ by v_{pi}, the quadruple program becomes the following quadratic program, which is easier to solve:

$$\begin{aligned} \min \quad & \sum_{j=1}^{n} \Big(1-\sum_{p=1}^{q} \sum_{i=T_{p-1}+1}^{T_p} v_{pi} Y_{ij}\Big)^2 \\ \text{s.t.} \quad & \sum_{p=1}^{q} w_p = 1 \\ & \sum_{i=T_{p-1}+1}^{T_p} u_i \overline{Y}_i = w_p, \quad p=1,\ldots,q \\ & v_{pi} \geq \varepsilon w_p, \quad p=1,\ldots,q,\ i=1,\ldots,k \\ & w_p \geq \varepsilon, \quad p=1,\ldots,q. \end{aligned} \tag{8}$$

At optimality, the performance index of the jth alternative is calculated as $E_j = \sum_{p=1}^{q} \sum_{i=T_{p-1}+1}^{T_p} v_{pi} Y_{ij}$. The optimal weight associated with the ith indicator is calculated as $u_i^* = v_{pi}^*/w_p^*$ and the optimal value of the pth category is calculated as $Z_{pj} = \sum_{i=T_{p-1}+1}^{T_p} u_i^* Y_{ij}$. A proof similar to Proposition 3 shows that the average of the performance indexes of all alternatives calculated from Model (8) is equal to one.

Comparing the related model (8) with the independent model (5), it is noted that they both produce performance indexes of average one. The difference is that the former produces smaller total squared distance $\sum_{j=1}^{n} s_j^2$. This is because the Z_{pj} in Model (5) are fixed values while that in Model (8) are variables, whose best values are to be determined and used. Smaller total squared distance implies that the corresponding hyperplane fits the observations better. The related model is thus preferred.

3.3 Three-Level Hierarchical DEA Model

The two-level model can be extended to three levels. The idea is straightforward, the notation, however, is annoying. Continuing the discussion in Sect. 3.2, suppose the q Z-type categories of indicators are grouped further into h categories of R-type categories at one level higher. Category R_f at the upper level comprises A_f Z-type categories of indicators at the lower level. The cumulative number of f categories of the Z-type category is $B_f = A_1 + A_2 + \ldots + A_f$. The R_f value for the jth alternative is $R_{fj} = \sum_{p=B_{f-1}+1}^{B_f} w_p Z_{pj}$. Let t_f denote the weight associated with R_f. The final performance index of the jth alternative is $E_j = \sum_{f=1}^{h} t_f R_{fj}$.

Following the idea of the average-based common-weight efficiency measurement expressed in Model (4), the model for calculating the performance index of the jth alternative for the three-level MCDM problem is formulated as:

$$
\begin{aligned}
\min \quad & \sum_{j=1}^{n} s_j^2 \\
\text{s.t.} \quad & \sum_{f=1}^{h} t_f R_{fj} + s_j = 1, && j=1,\ldots,n \\
& \sum_{f=1}^{h} t_f \overline{R}_f = 1 \\
& t_f \geq \varepsilon, && f=1,\ldots,h \\
& s_j \text{ unrestricted in sign}, && j=1,\ldots,n,
\end{aligned}
\tag{9}
$$

where $\overline{R}_f = \sum_{j=1}^{n} R_{fj}/n$. A proof similar to that in Proposition 2 obtains the following proposition:

Proposition 4 The average value of all alternatives of every category $\overline{R}_f, f = 1,\ldots, h$ is equal to one.

Based on this proposition, the constraint $\sum_{f=1}^{h} t_f \overline{R}_f = 1$ in Model (9) becomes $\sum_{f=1}^{h} t_f = 1$. Combining the models at the levels corresponding to Y_i, Z_p, and R_f together, one obtains the following three-level hierarchical DEA model:

$$
\begin{aligned}
\min \quad & \sum_{j=1}^{n} s_j^2 \\
\text{s.t.} \quad & \sum_{f=1}^{h} t_f R_{fj} + s_j = 1, \quad j=1,\ldots,n \\
& \sum_{f=1}^{h} t_f = 1 \\
& R_{fj} = \sum_{p=B_{f-1}+1}^{B_f} w_p Z_{pj}, \quad f=1,\ldots,h, \quad j=1,\ldots,n \\
& \sum_{p=B_{f-1}+1}^{B_f} w_p = 1, \quad f=1,\ldots,h \\
& Z_{pj} = \sum_{i=T_{p-1}+1}^{T_p} u_i Y_{ij}, \quad p=1,\ldots,q, \quad j=1,\ldots,n \\
& \sum_{i=T_{p-1}+1}^{T_p} u_i \overline{Y}_i = 1, \quad p=1,\ldots,q \\
& t_f, w_p, u_i \geq \varepsilon, \quad \forall f, p, i. \\
& s_j \text{ unrestricted in sign}, \quad j=1,\ldots,n,
\end{aligned}
\tag{10}
$$

Replacing s_j with $s_j = 1 - \sum_{f=1}^{h} t_f R_{fj}$, R_{fj} with $R_{fj} = \sum_{p=B_{f-1}+1}^{B_f} w_p Z_{pj}$, and Z_{pj} with $Z_{pj} = \sum_{i=T_{p-1}+1}^{T_p} u_i Y_{ij}$, Model (10) becomes:

$$\begin{aligned}
\min \quad & \sum_{j=1}^{n} \left(1 - \sum_{f=1}^{h} \sum_{p=B_{f-1}+1}^{B_f} \sum_{i=T_{p-1}+1}^{T_p} t_f w_p u_i Y_{ij}\right)^2 \\
\text{s.t.} \quad & \sum_{f=1}^{h} t_f = 1 \\
& \sum_{p=B_{f-1}+1}^{B_f} w_p = 1, f=1,\ldots,h \\
& \sum_{i=T_{p-1}+1}^{T_p} u_i \overline{Y}_i = 1, \quad p=1,\ldots,q \\
& t_f, w_p, u_i \geq \varepsilon, \qquad \forall f, p, i.
\end{aligned} \tag{11}$$

This model is a linearly constrained nonlinear program due to the cubic term $t_f w_p u_i$. Guo et al. [12] proposed a golden section method to solve a similar problem. They treated all $t_f, f = 1,\ldots, h$ variables as one variable to solve. The results are incorrect.

This idea can be extended to construct models for handling the problem with four or more levels in a similar manner. In the next section we use a case to explain how the idea of the multi-level hierarchical DEA model is applied to solve the MCDM problem with two-level criteria.

4 A Case

Kao [17] evaluated the industrial management department of eleven JCTs in Taiwan from the aspects of *instructors*, *curriculum*, *equipment*, and *administration*. The indicators for assessing *instructors* are the average degree of the instructors, average position of the instructors, and average number of refereed papers published by the instructors. For *curriculum*, they are the number of instructors per 100-student and the number of cooperation companies per 100-student. For *equipment*, they are the average expenditures per year per 100-student in 1,000 US dollars and the score of this aspect in the past evaluation. And for *administration*, they are the average numbers of financially supported workshops per instructor in the last three years for durations longer than one month and shorter than one month. Kao [17] had detailed descriptions of these indicators.

This problem is a typical two-level MCDM problem. In the evaluation, Kao [17] used the conventional CCR model (1) to aggregate the values of the indicators subordinated to each aspect to form the aspect scores. The four aspect scores were aggregated according to the pre-determined weights of 0.3, 0.2, 0.25, and 0.2 to form the performance index of the JCT. Note that the total weight of the four aspects was 0.95, instead of 1. The shortfall of 0.05 was assigned to the aspect of *objective* of the JCT. This aspect is difficult to find appropriate indicators to represent. Since the weight of 0.05 is small, this aspect was ignored in the study of Kao [17]. Table 1 shows the data for the eleven JCTs. The scores of the four aspects and the final performance indexes are shown in Table 2. The numbers in parentheses in the last column are the ranks of the corresponding efficiencies.

Following the suggestion of Färe and Hunsaker [20], 0.00005 = 0.0001/2 was assigned as the value for the non-Archimedean number ε since the smallest multiplier appeared in the calculation was 0.0001. As discussed in the preceding sections, the aspect scores of the eleven JCTs are not comparable because different weights were applied by each JCT.

Table 1. Data for the eleven JCTs in Taiwan.

JCT	Instructors			Curriculum		Equipment		Administration	
	Degree	Publi-cation	Posi-tion	Ins.-stu. ratio	Compa-nies	Expendi-tures	Past score	Workshops >1 mo.	<1 mo.
Taipei	2.00	1.47	2.60	2.76	1.84	25.24	16.5	0.67	0.60
Nan-Tai	2.00	0.67	2.33	2.22	3.96	8.96	16.5	0.38	1.29
Oriental	1.88	0.25	2.00	2.28	0.74	23.80	13.0	0.62	0.88
Chin-Yi	2.00	0.57	2.00	2.02	0.46	11.50	18.0	0.14	0.86
Ming-Chi	2.00	2.29	2.43	3.12	11.62	13.08	16.0	0.28	0.00
Van-Nung	1.90	0.33	2.19	2.26	1.50	11.62	15.0	0.10	1.03
Lien-Ho	1.83	1.00	2.08	2.14	0.96	8.00	14.0	0.17	1.00
St. M & J	1.94	0.06	2.00	1.88	1.42	11.28	16.0	0.39	0.06
Lung-Hwa	1.92	0.42	2.08	2.28	1.16	10.78	14.0	0.33	0.58
Shu-Teh	1.84	0.26	2.21	1.58	0.92	8.26	11.5	0.58	0.21
Tung-Fang	1.64	0.00	2.08	2.48	0.86	16.32	13.0	0.36	0.00

By applying the related two-level model (8) the global optimal results are obtained. Table 3 shows the results. When applying this model, the constraint $\sum_{p=1}^{q} w_p = 1$ is changed to $\sum_{p=1}^{q} w_p = 0.95$ to suit the situation of this example. The optimal weights obtained are $w_1 = 0.94306$, $w_2 = 0.00023$, $w_3 = 0.00541$, and $w_4 = 0.00130$, as shown in the last row of Table 3 where the instructors aspect dominates the other three aspects. Compared to the results of Kao [17] shown in Table 2, it is noted that the three undistinguishable efficient JCTs, Taipei, Nan-Tai, and Oriental, in the administration aspect have been differentiated. More importantly, none of the eight inefficient JCTs have been correctly ranked by Kao [17]. The rankings of the JCTs based on the performance index for these two models are quite different, with an average absolute difference of 1.8181 ranks. The largest difference occurs at Oriental, whose rank is dropped from third to eighth.

If the aspect weights in Model (8) are fixed at 0.3, 0.2, 0.25, and 0.2 to conform to that used in Kao [17], then the aspect scores and the performance indexes are calculated, as shown in Table 4. Comparing the results of this model to that of Kao [17], it is noted that the rankings are rather different. The average absolute difference is 0.9091 ranks, with the largest difference of three ranks occurring at Lung-Hwa. Since Kao [17] allowed every JCT to use different weights u_i to calculate the aspect score, the results are actually

Table 2. Results of the CCR model used in Kao [17].

JCT	Instructors	Curriculum	Equipment	Administration	Total (Rank)
Taipei	1.0000	0.8842	1.0000	1.0000 (1)	0.9268 (1)
Nan-Tai	0.9998	0.7113	0.9166	1.0000 (1)	0.8714 (2)
Oriental	0.9398	0.7304	0.9427	1.0000 (1)	0.8637 (3)
Chin-Yi	0.9998	0.6471	1.0000	0.6667 (7)	0.8127 (4)
Ming-Chi	1.0000	1.0000	0.9051	0.4179 (11)	0.8099 (5)
Van-Nung	0.9498	0.7240	0.8449	0.7984 (5)	0.8006 (6)
Lien-Ho	0.9149	0.6855	0.7778	0.7752 (6)	0.7610 (7)
St. M & J	0.9697	0.6023	0.8949	0.5821 (9)	0.7515 (8)
Lung-Hwa	0.9598	0.7304	0.7882	0.5898 (8)	0.7490 (9)
Shu-Teh	0.9198	0.5062	0.6441	0.8656 (4)	0.7113 (10)
Tung-Fang	0.8199	0.7945	0.7677	0.5373 (10)	0.7043 (11)
Weight	0.30	0.20	0.25	0.20 0.95	

Table 3. Results of the related two-level hierarchical DEA model with flexible weights.

JCT	Instructors	Curriculum	Equipment	Administration	Total (Rank)
Taipei	1.0705	1.0028	1.7616	1.8147 (1)	1.0217 (1)
Nan-Tai	1.0527	1.3472	0.7237	1.0658 (4)	0.9983 (3)
Oriental	0.9770	0.6583	1.6374	1.6917 (2)	0.9326 (8)
Chin-Yi	1.0309	0.5406	0.8995	0.4075 (10)	0.9778 (4)
Ming-Chi	1.0593	3.2132	0.9818	0.7487 (8)	1.0060 (2)
Van-Nung	0.9985	0.8197	0.8794	0.3071 (11)	0.9470 (7)
Lien-Ho	0.9598	0.6758	0.6394	0.4931 (9)	0.9094 (10)
St. M & J	1.0039	0.7194	0.8670	1.0452 (5)	0.9530 (5)
Lung-Hwa	1.0002	0.7499	0.8166	0.9047 (7)	0.9490 (6)
Shu-Teh	0.9728	0.5450	0.6329	1.5590 (3)	0.9230 (9)
Tung-Fang	0.8743	0.7281	1.1606	0.9626 (6)	0.8322 (11)
Weight	0.94306	0.00023	0.00541	0.00130	0.95

not comparable among all JCTs and the results of the common-weight model (8) are more reliable.

The results of the related two-level hierarchical DEA model also have managerial implications. The average performance index of all JCTs is 0.95. Thus, if a JCT whose

Table 4. Results of the related two-level hierarchical DEA model with fixed weights.

JCT	Instructors	Curriculum	Equipment	Administration	Total (Rank)
Taipei	1.0503	1.2132	1.1121	1.5930	1.1544 (1)
Nan-Tai	1.0501	0.9764	1.1089	1.3741	1.0624 (2)
Oriental	0.9870	1.0020	0.8770	1.6351	1.0428 (3)
Chin-Yi	1.0500	0.8877	1.2100	0.6968	0.9344 (5)
Ming-Chi	1.0504	1.3738	1.0762	0.5415	0.9672 (4)
Van-Nung	0.9976	0.9934	1.0088	0.7037	0.8909 (8)
Lien-Ho	0.9609	0.9405	0.9409	0.8242	0.8765 (9)
St. M & J	1.0185	0.8264	1.0758	0.7840	0.8966 (7)
Lung-Hwa	1.0081	1.0021	0.9415	0.9256	0.9233 (6)
Shu-Teh	0.9661	0.6945	0.7733	1.2258	0.8672 (10)
Tung-Fang	0.8610	1.0899	0.8755	0.6962	0.8344 (11)
Weight	0.30	0.20	0.25	0.20	0.95

performance index is less than 0.95, then its performance is below average. This JCT should find its weaknesses to make improvements in the next evaluation. As which aspects are the weaknesses of a JCT, one simply examines the aspects whose scores are less than one. For example, Chin-Yi has a performance index of 0.9344, which is less than 0.95. Its weaknesses are curriculum and administration because their scores of 0.8877 and 0.6968 are less than one. These two aspects are the weaknesses of this JCT. If the scores of these two aspects are increased, then the performance index of Chin-Yi will be increased, and its rank be raised.

5 Conclusion

For MCDM problems with a large number of indicators where the conventional solution methods are to group the indicators into categories of indicators at different levels and treat each category of indicators as an independent MCDM problem, this paper proposes a hierarchical DEA model without inputs to solve the problem by treating each category of indicators as a subunit of the hierarchical system. The proposed method has several characteristics.

First, the weights associated with the indicators and categories are objectively reflected from the data, instead of being determined by the experts. The doubts regarding subjectivity are removed. Second, different from the conventional DEA models that allow every DMU to use their favorable weights to calculate efficiency, this model requires all DMUs to use the same set of weights to calculate efficiency. The results are thus comparable. Third, the model uses the average values, rather than the maximum values, as the basis to calculate efficiency, which avoids the well-known shortcoming of the DEA of producing misleading results caused by outliers. The average-based method is similar to the regression analysis which reflects the trend of the observations better. Fourth, the average-based method produces an average of one for the category scores of all DMUs and the performance indexes of all DMUs. This property helps examine whether a DMU performs below average and identify the categories that cause the unsatisfactory performance of the DMU. Making improvements on these categories of indicators raises the performance index of the DMU. Fifth, the model is a network DEA model that takes the interactions between the indicators of different categories into consideration. The results are more reliable. Sixth, while the existing hierarchical DEA models require the data to be standardized, the proposed model does not need a data standardization process, the possibility of obtaining misleading results is avoided. These characteristics are also the merits of the proposed model.

The model developed in this paper is quadratic for two-level problems, which is easy to solve. For problems of three levels or higher, the model is nonlinear, which makes a solution difficult to obtain. How to transform the nonlinear model into a quadratic model or to devise easier solution methods is a direction for future study.

Acknowledgments. This research is partially supported by the National Science and Technology Council of the Republic of China (Taiwan), under grant No. NSTC 114-2410-H-006-088-MY3.

Disclosure of Interests. The author has no competing interests to declare that are relevant to the content of this article.

References

1. Hwang, C.L., Yoon, K.: Multiple Attribute Decision Making – Methods and Applications. A State-of-the-Art Survey. Springer-Verlag, New York (1981)
2. Triantaphyllou, E.: Multi-criteria Decision Making Methods: A Comparative Study. Springer, Boston (2000)
3. Zeleny, M.: Multiple Criteria Decision Making. McGraw-Hill, New York (1982)
4. Kao, C., Wu, W.Y., Hsieh, W.J., Wang, T.Y., Lin, C., Chen, L.H.: Measuring the national competitiveness of Southeast Asian countries. Eur. J. Oper. Res. **187**, 613–628 (2008)
5. Awasthi, A., Adetiloye, T., Crainic, T.C.: Collaboration partner selection for city logistics planning under municipal freight regulations. Appl. Math. Model. **40**, 510–525 (2016)
6. Saaty, T.L.: The Analytic Hierarchy Process. McGraw-Hill, New York (1980)
7. Charnes, A., Cooper, W.W., Rhodes, E.: Measuring the efficiency of decision making units. Eur. J. Oper. Res. **2**, 429–444 (1978)
8. Kao, C., Hwang, S.N.: Efficiency measurement for network systems: IT impact on firm performance. Decis. Support Syst. **48**, 437–446 (2010)
9. Kao, C.: Network Data Envelopment Analysis: Foundations and Extensions, 2nd edn. Springer Nature, Cham, Switzerland (2023)
10. Kao, C.: Efficiency measurement for hierarchical network systems. Omega **51**, 121–127 (2015)
11. Chen, P.C., Yu, M.M., Shih, J.C., Chang, C.C., Hsu, S.H.: A reassessment of the global food security index by using a hierarchical data envelopment analysis approach. Eur. J. Oper. Res. **272**, 687–698 (2019)
12. Guo, C., Song, Q., Yu, M.M., Zhang, J.: A digital economy development index based on an improved hierarchical data envelopment analysis approach. Eur. J. Oper. Res. **316**, 1146–1157 (2024)

13. Babaee, S., Toloo, M., Hermans, E., Shen, Y.: A new approach for index construction: the case of the road user behavior index. Comput. Indust. Eng. **152**, Article No. 106993 (2021). https://doi.org/10.1016/j.cie.2020.106993
14. Shahari, M.R., See, K.F., Mohammed, N.S., Yu, M.M.: Constructing the performance index of Malaysia's district health centers using effectiveness-based hierarchical data envelopment analysis. Socio-Econ. Plan. Sci. **89**, Article No. 101662 (2023). https://doi.org/10.1016/j.seps.2023.101662
15. Yu, M.M., Rakshit, I.: An alternative assessment approach to global logistics performance evaluation: COMMON weight H-DEA approach. Int. Trans. Oper. Res. **32**, 839–862 (2025)
16. Thompson, R.G., Singleton Jr., F.D., Thrall, R.M., Smith, B.A.: Comparative site evaluations for locating a high-energy physics lab. In Texas. Interfaces **16**(6), 35–49 (1986)
17. Kao, C.: Evaluation of junior colleges of technology: the Taiwan case. Eur. J. Oper. Res. **72**, 43–51 (1994)
18. Charnes, A., Cooper, W.W., Rhodes, E.: Short communication: Measuring the efficiency of decision making units. Eur. J. Oper. Res. **3**, 339 (1979)
19. Charnes, A., Cooper, W.W.: The non-Archimedean CCR ratio for efficiency analysis: a rejoinder to Boyd and Färe. Eur. J. Oper. Res. **15**, 333–334 (1984)
20. Färe, R., Hunsaker, W.: Notions of efficiency and their reference sets. Manage. Sci. **32**, 237–243 (1986)
21. Roll, Y., Cook, W.D., Golany, B.: Controlling factor weights in data envelopment analysis. IIE Trans. **23**, 2–9 (1991)
22. Charnes, A., Cooper, W.W., Wei, Q.L., Huang, Z.M.: Cone ratio data envelopment analysis and multi-objective programming. Int. J. Syst. Sci. **20**, 1099–1118 (1989)
23. Andersen, P., Petersen, N.C.: A procedure for ranking efficient units in data envelopment analysis. Manage. Sci. **39**, 1261–1264 (1993)
24. Kao, C.: Weight determination for consistently ranking alternatives in multiple criteria decision analysis. Appl. Math. Model. **34**, 1779–1787 (2010)

Importance Compression: A New Sparsification Mechanism for Distributed Learning

Fanis Khafizov[2,3](✉), Michael Ermoshin[2], Vadim Kasiuk[2], Daniil Medyakov[1,2,3], Dmitry Bylinkin[1,2,3], and Aleksandr Beznosikov[1,2,3,4]

[1] Federated Learning Problems Laboratory, Moscow, Russia
[2] Moscow Independent Research Institute of Artificial Intelligence, Moscow, Russia
khafizov.f@miriai.org
[3] Basic Research of Artificial Intelligence Laboratory, Shanghai, China
[4] Innopolis University, Innopolis, Russia

Abstract. Communication efficiency is critical for scalable distributed learning, but biased methods such as TopK prioritize gradient magnitude rather than their actual contribution to objective decrease. We introduce ImpK, a family of importance-based compressors that select coordinates by solving a local optimization subproblem on a shared trial function $\hat{f}$ built from a small proxy dataset. We prove that ImpK achieves linear convergence to a vicinity of the solution without error compensation, despite the operator being biased. We also propose `SCAM` (Sparsification with Compensation and Aggregation Method), which uses error-augmented signals for coordinate selection while transmitting clean reweighted gradients. On *ResNet-18* (*CIFAR-10*) and *GPT-2* (*WikiText2*) at a 1% compression ratio, ImpK with `SCAM` consistently outperforms TopK baselines.

Keywords: Distributed optimization · Gradient compression · Sparsification · Federated learning

1 Introduction

We consider the general distributed optimization problem

$$\min_{x\in\mathbb{R}^d}\left\{f(x) := \frac{1}{n}\sum_{i=1}^{n} f_i(x)\right\},$$

where $x \in \mathbb{R}^d$ represents the parameters of a statistical model to be optimized, n denotes the number of workers, $f_i(x)$ is the local loss function for worker i [14]. The local loss $f_i : \mathbb{R}^d \to \mathbb{R}$ is typically expressed as $f_i(x) := \mathrm{E}_{\xi\sim\mathcal{D}_i}[f_\xi(x)]$, where $\mathcal{D}_i$ is the data distribution associated with worker i.

While gradient computation on individual devices is relatively efficient, the exchange of information between the server and workers constitutes the principal bottleneck in distributed learning [2]. One powerful method to address

A. Kazakov et al. (Eds.): MOTOR 2026, LNCS 16687, pp. 278–293, 2026.
https://doi.org/10.1007/978-3-032-31390-4_18

this constraint is communication compression techniques that reduce the size of transmitted messages [2,4,7,8,13,15]. In practice, the most widely used class of compressors is biased [5], e.g., those whose expectation differs from their input. However, they require specialized optimization frameworks [8,12]. Traditional techniques such as error feedback introduce notable drawbacks: the descent direction incorporates compression errors from previous iterations, which can become problematic over long horizons as these errors accumulate. This naturally raises a fundamental question: *Is there a robust compression operator that guarantees convergence without optimization complications?*

1.1 Related Works

The challenge of communication bottleneck in distributed learning has been addressed through two main approaches in the literature. The first direction is the development of compression operators. Compression operators generally fall into two categories: sparsification and quantization techniques. *Sparsification methods* transmit only a subset of gradient coordinates. Examples include random sparsification (RandK) [13], TopK sparsification [1], and adaptive random sparsification [1]. *Quantization methods* reduce precision by mapping gradient values to a small discrete set. Examples include natural compression [7], sign compression [4], and power compression [15]. The second direction is the development of optimization methods. This line of work develops optimization methods that provably converge under biased compression. `Error Feedback` (`EF`) represents a significant improvement in this direction [5,8], enabling the use of biased compressors in distributed settings without compromising convergence guarantees. The core idea is simple but powerful: preserve the compression error in memory and add it to the gradient in the next iteration before compression. This mechanism effectively compensates for the bias introduced by compression, ensuring that no gradient information is permanently lost. Building on this foundation, `Error Feedback 21` (`EF21`) [12] introduced significant advances to the original `EF` framework. `EF21` maintains a slightly different error memory structure and updates it using a modified scheme that provides stronger theoretical guarantees for non-convex optimization problems [12].

Despite their theoretical appeal, approaches relying on error feedback and variance reduction mechanisms have notable practical limitations. Variance-reduced optimization methods, for instance, have been shown to perform poorly when applied to deep neural networks [6]. Similarly, in error feedback frameworks like `EF` and `EF21`, the compensation error accumulated throughout the training process can introduce deleterious staleness into the optimization trajectory, which negatively impacts convergence and generalization.

1.2 Notation and Definitions

We denote the standard inner product in $\mathbb{R}^d$ as $\langle x, y\rangle := \sum_{i=1}^d x_i y_i$, where x_i and y_i are the i-th components of vectors $x, y \in \mathbb{R}^d$, respectively. The induced ℓ_2 norm is written as $\|x\|_2 := \sqrt{\langle x, x\rangle}$, the ℓ_1 norm is defined as $\|x\|_1 = \sum_{i=1}^d |x_i|$. The number of non-zero coordinates of a vector x is denoted by $\|x\|_0 := \sum_{i=1}^{d} \mathbb{I}\{x_i \neq 0\}$. The support of a vector x is denoted by $\mathrm{supp}(x) := \{i : x_i \neq 0\}$. The all-ones vector is denoted by $\mathbf{1} = (1, \dots, 1)^\top \in \mathbb{R}^d$. We denote mathematical expectation by $\mathbb{E}[\cdot]$. The Hadamard product (element-wise product) of vectors $x, y \in \mathbb{R}^d$ is denoted by $x \odot y$, where $(x \odot y)_i = x_i y_i$.

2 Importance Compression

In this section, we introduce importance-based compression operators that prioritize coordinates according to their effect on a shared trial objective.

2.1 Trial Loss Function

The objective of this work is to construct a sparsification operator that selects coordinates of the local gradient based on their relative importance to the training process. To implement this idea, we assume that each device maintains a small subset of samples drawn from the server-side distribution $\mathcal{D}_1$ that approximates the local distributions in terms of their modes. The empirical loss function defined on a device using this subset is referred to as the trial function. This concept is well established in the literature. For instance, [10] employs the trial function to filter out clients that adversely affect the training process. Inspired by this, we adapt the trial function concept to estimate coordinate importance without accessing the full dataset. Formally, the trial function $\hat{f}$ is defined as:

$$\hat{f}(x) = \frac{1}{N} \sum_{i=1}^{N} f_1(x, \xi_i),$$

where N denotes the size of the trial dataset. This function is available on every device and can be used to guide the compression process locally. Although $\hat{f}$ may deviate from the local loss functions f_i depending on the sample size N from $\mathcal{D}$, it provides a sufficient approximation of the optimization landscape to determine coordinate importance. In our mechanism, the trial function is used to assess the importance of individual gradient coordinates. Informally, a client applies an update under a given sparsification pattern and evaluates the resulting change in the trial function. If the value increases, this indicates that an important coordinate has been discarded. Minimizing the trial function over different sparsification patterns constitutes the conceptual foundation of the proposed compression operator. For the details, see Sect. 2.2.

Assumption 1. *Each function f_i and the trial function $\hat{f}$ are L-smooth, i.e., $\|\nabla g(x) - \nabla g(y)\|_2 \leq L\|x - y\|_2$ for any $x, y \in \mathbb{R}^d$, where $g \in \{f_1, \dots, f_n, \hat{f}\}$.*

While Assumption 1 is generally unrealistic for neural networks, it is common for the optimization literature and necessary to construct any theory for deep learning [11].

Assumption 2. *Each function f_i and the trial function $\hat{f}$ are μ-strongly convex, i.e., for all $x, y \in \mathbb{R}^d$:*

$$g(y) \geq g(x) + \langle \nabla g(x), y - x \rangle + \frac{\mu}{2}\|y - x\|_2^2,$$

where $g \in \{f_1, \ldots, f_n, \hat{f}\}$.

Assumption 2 does not hold for deep learning models either. Nevertheless, it provides a coarse approximation of the behavior observed in neural networks and serves as a reasonable starting point for the development of methods that perform well in practice [5].

Assumption 3. *For all $x \in \mathbb{R}^d$, the following bound holds for the local gradients relative to the trial function:*

$$\frac{1}{n}\sum_{i=1}^{n} \|\nabla f_i(x)\|_2^2 \leq \|\nabla f(x)\|_2^2 + \sigma^2,$$

where $\sigma^2 \geq 0$ is a constant.

This assumption is not too restrictive, since we allow $\sigma > 0$, i.e., we do not require uniform boundedness of local gradients.

Our theory also relies on the following lemma.

Lemma 1 (Lemma C.1 in Appendix[10]). *For all $x \in \mathbb{R}^d$, the gradient of the approximation $\hat{f}$ satisfies:*

$$\|\nabla f(x) - \nabla \hat{f}(x)\|_2^2 \leq \zeta,$$

where $\zeta \geq 0$ is a constant representing the approximation error bound.

Theoretical results regarding this bound are provided by [10, Lemma C.1]. Specifically, as the dataset size N increases, the approximation error ζ behaves as $\tilde{\mathcal{O}}\left({}^1/_N\right)$. For our framework, this implies the existence of a public distribution $\mathcal{D}$ on the server. By drawing a sample of size N from it, the resulting trial function inherently exhibits similarity properties with the overall global distribution.

2.2 Importance Vector and Compressor Definition

To reduce per-round communication from $\mathcal{O}(d)$ to $\mathcal{O}(k)$, a sparsifier should transmit only $k \ll d$ coordinates. In our work, we select coordinates via a one-step objective-aware problem on $\hat{f}$ that maximizes the predicted decrease after a masked gradient step. Following intuition from the beginning of Sect. 2.1, we design a trial-based weighting of coordinates.

Definition 1. *Using the trial function $\hat{f}$, step size $\gamma > 0$ and a set $\Omega \subseteq \mathbb{R}^d$, we define the importance vector w_i^t at the t-th iteration on the i-th device as:*

$$w_i^t = \arg\min_{w \in \Omega} \hat{f}(x^t - \gamma w \odot \nabla f_i(x^t)).$$

The choice of Ω controls how coordinates are selected. To enforce selection of exactly k coordinates, we consider the ℓ_0-sparsified feasible set

$$\Omega_k := \Omega \cap \{w \in \mathbb{R}^d : \|w\|_0 = k\}.$$

In this case, w is k-sparse and $w \odot v$ keeps exactly k coordinates of v (possibly with reweighting).

Since optimization problems involving the ℓ_0 norm are NP-hard, we use continuous relaxations to solve (1) in practice. The following relaxations are considered for Ω:

1. Scaled simplex (corresponding importance compressor is denoted as ImpK_s):

$$\Delta_{d-1} = \left\{ w : \sum_{i=1}^{d} w_i = d, w_i \geq 0 \, \forall i = \overline{1,d} \right\}.$$

2. Cube (corresponding importance compressor is denoted as ImpK_c):

$$[0,2]^d = \left\{ w : 0 \leq w_i \leq 2 \, \forall i = \overline{1,d} \right\}.$$

We can apply optimization methods that operate on such sets. For instance, for Δ_{d-1}, we can use the `Mirror Descent` method [3] with KL-divergence as a regularizer, and for $[0,2]^d$, `Gradient Descent` with projection onto the cube. Thus, the proposed trial-based compression operator is as follows.

Definition 2. *The (ideal) importance compression operator* ImpK_i^t *on the i-th device at iteration t, based on the importance vector w_i^t (see Definition 1) and the trial function $\hat{f}$, acts on a vector $v \in \mathbb{R}^d$ as*

$$\mathrm{ImpK}_i^t(v) := w_i^t \odot v.$$

In particular, when Ω is a box constraint (e.g., $[0,2]^d$) and we use Ω_k, the update keeps exactly k coordinates.

3 Gradient Descent with Importance Compression

Instead of `EF`, we use `DCGD` (Distributed Compressed Gradient Descent) [5] to handle the proposed importance compressor:

$$x^{t+1} = x^t - \gamma \frac{1}{n} \sum_{i=1}^{n} \mathrm{ImpK}_i^t(\nabla f_i(x^t)),$$

where ImpK_i^t is the importance compressor on the i-th device at the t-th iteration, as defined in Definition 2.

Theorem 1. *Let Assumptions 1–3 hold. Consider solving (1) using* `DCGD` *with the importance compressor from Definition 2 instantiated with* $\Omega = [0, 2]^d$ *and* $\Omega_k = \Omega \cap \{w : \|w\|_0 = k\}$. *By setting the constant step size tuned to the horizon* T*:*

$$\gamma = \min\left\{\frac{1}{4L}, \frac{2d}{k\mu T}\log\left(\max\left\{1, \frac{k\mu^2 T(\hat{f}(x^0) - \hat{f}^*)}{2Ld\sigma^2}\right\}\right)\right\},$$

the iterates satisfy:

$$\hat{f}(x^T) - \hat{f}^* \leq \tilde{\mathcal{O}}\left(\exp\left(-\frac{k\mu}{8Ld}T\right)(\hat{f}(x^0) - \hat{f}^*) + \frac{Ld\sigma^2}{k\mu^2 T}\right) + \frac{3\zeta}{2\mu},$$

where $\tilde{\mathcal{O}}$ *hides logarithmic factors. If* $\zeta = 0$ *(ideal trial function), the method converges to the exact optimum of the trial function* $\hat{f}^*$.

This is an interesting result, as the ImpK operator is biased, and there are known examples where biased compressors diverge [5]. Here, however, convergence is obtained without error compensation.

The bound also clarifies the roles of σ^2 and ζ. The variance term σ^2 is suppressed by the horizon-tuned choice of γ, yielding the decaying contribution $\tilde{\mathcal{O}}(1/T)$. In contrast, in the one-step recursion the approximation term enters as a term linear in γ, hence after unrolling it produces a non-vanishing floor of order ζ/μ rather than a term that can be eliminated by step-size decay alone. Nevertheless, this floor is controllable in practice because $\zeta = \tilde{\mathcal{O}}(1/N)$ for a trial set of size N: increasing N improves $\hat{f}$ and reduces the residual term.

4 `SCAM`: Error-guided Coordinate Selection

Traditional biased compressors like TopK may diverge without error compensation [5]. Standard `Error Feedback` addresses this by adding accumulated errors to the gradient before compression. While this tracking mechanism drastically improves both theoretical convergence and practical scaling over uncompensated methods, it transmits a mixture of the fresh current gradient and past temporal errors. This can introduce deleterious staleness into the optimization trajectory, which can dampen final generalization.

Our ImpK operators converge without error compensation (Theorem 1), which opens a new possibility: we can use historical error information solely for *coordinate selection* while transmitting the *clean* reweighted gradient. This separation effectively combines the coverage benefits of error feedback with the freshness of uncompensated gradients. This separation forms the key insight behind `SCAM` (Sparsification with Compensation and Aggregation Method).

Algorithm 1 is the practical version of this idea and directly addresses the `EF` issues highlighted in the Introduction: accumulated errors are used only for support selection and no longer affect the update direction.

Algorithm 1 SCAM

1: **Input:** starting point x^0, learning rate γ, number of iterations T, number of workers n, initial errors $\varepsilon_i^0 = 0$ for $i = 1, \ldots, n$.
2: **for** $t = 0, 1, \ldots, T-1$ **do**
3: Master broadcasts x^t to all nodes
4: **for** each node $i = 1, \ldots, n$ **in parallel do**
5: $g_i^t = \nabla f_i(x^t)$
6: $S_i^t = \mathrm{supp}(\mathcal{C}(\varepsilon_i^t + g_i^t))$
7: $\tilde{g}_i^t = \mathcal{C}_{S_i^t}(w_i^t \odot g_i^t)$
8: $\varepsilon_i^{t+1} = \varepsilon_i^t + g_i^t - \tilde{g}_i^t$
9: **end for**
10: $x^{t+1} = x^t - \gamma \frac{1}{n} \sum\limits_{i=1}^{n} \tilde{g}_i^t$
11: **end for**

Here $\mathcal{C}_S(x) = \sum_{j \in S} x_j e_j$ denotes the projection onto coordinates in S, and w_i^t is the importance vector. The compressor $\mathcal{C}$ can be any sparsifying operator; when combined with ImpK, it solves the importance subproblem at each iteration. In Line 6, we select coordinates using the error-augmented gradient $\varepsilon_i^t + g_i^t$ (see Definition 2). In Line 7, we transmit the clean reweighted gradient restricted to these coordinates. In Line 8, we update the error buffer to account for the discarded coordinates.

The key distinction from `Error Feedback` is the following. In `EF`, the transmitted message is $\mathcal{C}(\varepsilon + g)$ — a mixture of current gradient and accumulated errors. In `SCAM`, we transmit $\mathcal{C}_S(w \odot g)$: the current gradient reweighted by importance, restricted to coordinates suggested by error-augmented selection. This ensures that coordinates neglected for many iterations get prioritized via ε, the transmitted gradient remains fresh without error contamination, and importance weighting w is preserved in the update direction.

Note that `SCAM` can be applied to any biased compressor $\mathcal{C}$, not only ImpK. When $w = \mathbf{1}$ and $\mathcal{C} =$ TopK, `SCAM` provides error-guided coordinate selection with clean gradient transmission for the standard TopK operator.

5 Experiments

5.1 Implementation Details

In our experiments, we fix the compression ratio $\alpha = k/d$ (set to 1%), where k is the number of selected coordinates and d is the total number of parameters. We apply compression layer-wise, selecting $k_i = \lceil \alpha \cdot d_i \rceil$ coordinates in the i-th layer of size d_i. This approach enables uniform comparison within each group of weights and can be combined with error compensation techniques. For the importance vector w, we solve (1) once per epoch. In practice, we use continuous relaxations: mirror descent on $\Delta_{d_i - 1}$ for ImpK$_s$ and projected gradient descent on $[0, 2]^{d_i}$ for ImpK$_c$.

After obtaining dense w, we form a k-sparse update by keeping the largest $|(w \odot \nabla f(x))_j|$ coordinates and transmitting the reweighted values. With `Error Feedback`, selection uses $\nabla f(x) + \varepsilon$. `SCAM` uses $|(w \odot (\nabla f(x) + \varepsilon))_j|$ for selection and transmits the clean reweighted gradient.

5.2 Convolutional Neural Networks

We evaluate our methods on the *CIFAR-10* classification task using the *ResNet-18* architecture. We compare the proposed ImpK variants against TopK and its error-feedback modification. The batch size is set to 128, with data distributed randomly across batches.

Learning Curves. For learning curves, we train *ResNet-18* on *CIFAR-10* with `CAdamW` (Compressed `AdamW`) — a variant of AdamW [9] that applies the compression operator to the gradient before the optimizer update, $\beta_1 = 0.9$, $\beta_2 = 0.999$, weight decay 0.01, and $\alpha = 1\%$. Learning rates are tuned per method in $[10^{-4}, 2 \cdot 10^{-3}]$. Each run lasts 50 epochs and is averaged over 5 seeds (Fig. 1).

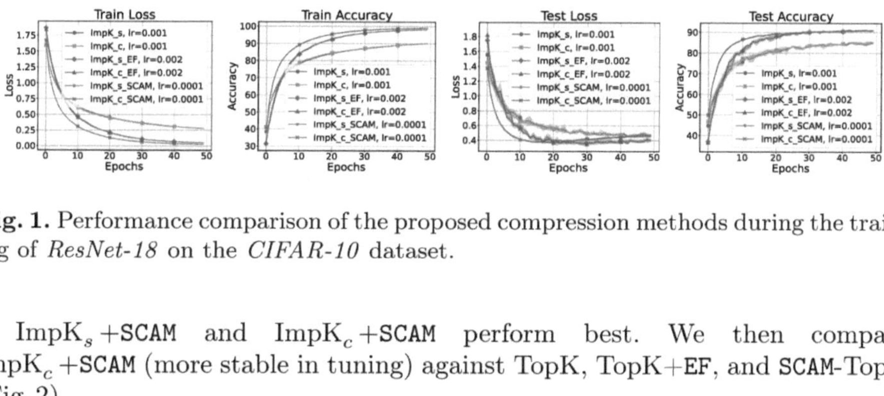

Fig. 1. Performance comparison of the proposed compression methods during the training of *ResNet-18* on the *CIFAR-10* dataset.

ImpK_s +`SCAM` and ImpK_c +`SCAM` perform best. We then compare ImpK_c +`SCAM` (more stable in tuning) against TopK, TopK+`EF`, and `SCAM`-TopK (Fig. 2).

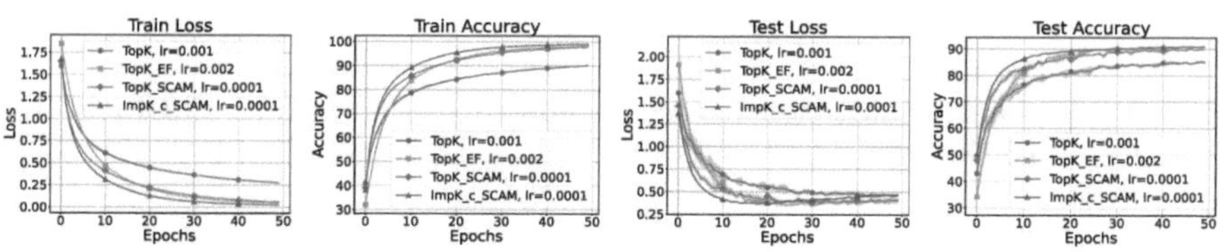

Fig. 2. Performance comparison between proposed method ImpK_c `SCAM` and TopK variations during the training of *ResNet-18* on the *CIFAR-10* dataset.

The plots show that ImpK_c `SCAM` outperforms all TopK-based variants. `SCAM`-TopK is faster early on but trails `EF` after about epoch 20.

The overhead of solving the minimization subproblem once per epoch accounts for approximately 10% of the training time. However, since the importance vector w is computed only once per epoch and reused for all subsequent iterations, this cost is amortized across thousands of communication rounds, making it negligible compared to the reduction in communication time and the gain in convergence speed per epoch.

5.3 Transformer Architecture

We evaluate *GPT-2 Small* (124M, random initialization) on *WikiText2* with batch size 16 and sequence length 1024.

We use `CAdamW` with the same hyperparameters, tune learning rates in $[10^{-4}, 10^{-3}]$, and train for 50 epochs. As in CNN experiments, we first compare ImpK variants and then compare the best one with TopK baselines (Fig. 3).

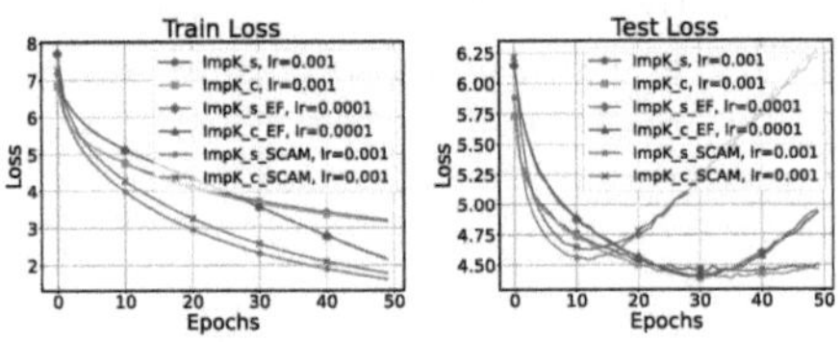

Fig. 3. Performance comparison of the proposed compression methods during the training of *GPT-2* on *WikiText2*.

For the second experiment, we select the best-performing method, ImpK_s `SCAM`, and compare it with methods based on TopK (Fig. 4).

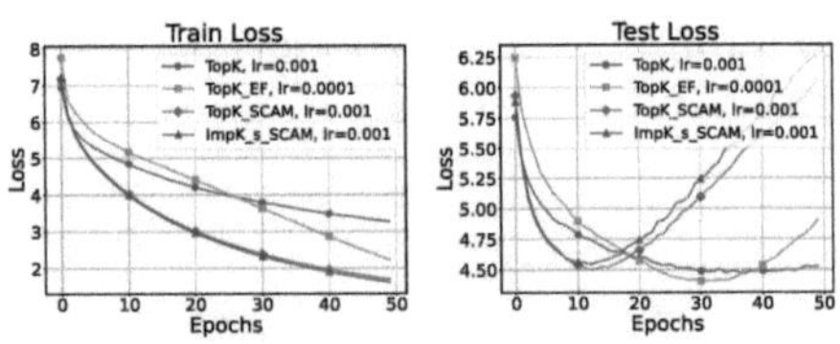

Fig. 4. Performance comparison between proposed method ImpK_s `SCAM` and TopK variations during the training of *GPT-2* on *WikiText2*.

`SCAM` variants perform best; `SCAM` ImpK_s is the strongest method, with a smaller test-loss margin over `SCAM` TopK than in the CNN setting. We observe overfitting on the test set, consistent with the small *WikiText2* size relative to model capacity.

6 Conclusion

We presented ImpK, a new family of biased compression operators that select coordinates based on their importance for minimizing a shared trial function. ImpK accounts for the actual impact of each coordinate on the objective. Our theoretical analysis establishes that ImpK converges linearly to a neighborhood without error compensation. Furthermore, we proposed `SCAM`, a generic mechanism that decouples coordinate selection from transmission. As demonstrated, `SCAM` improves performance not only for ImpK but also for standard compressors like TopK, offering a reliable alternative to traditional `Error Feedback` by eliminating error accumulation in the transmitted message.

Acknowledgments. The work was done in the Laboratory of Federated Learning Problems (Supported by Grant App. No. 2 to Agreement No. 075-03-2024-214).

A Basic Facts and Inequalities

A.1 Combinatorial Identity

For binomial coefficients, the following identity holds:

$$\frac{\binom{d-1}{k-1}}{\binom{d}{k}} = \frac{k}{d}.$$

This follows from the fact that each coordinate $j \in \{1, \dots, d\}$ appears in exactly $\binom{d-1}{k-1}$ subsets of size k.

B Proof of Theorem 1

The `DCGD` method with importance compression performs the update:

$$x^{t+1} = x^t - \gamma \frac{1}{n} \sum_{i=1}^{n} \mathrm{ImpK}_i^t(\nabla f_i(x^t)).$$

For convenience, denote $g_i^t = \mathrm{ImpK}_i^t(\nabla f_i(x^t))$. Under Ω_k, we have $g_i^t = w_i^t \odot \nabla f_i(x^t)$ for some k-sparse vector w_i^t with $\|w_i^t\|_0 = k$. We analyze the progress in terms of the trial function $\hat{f}$.

Since $\hat{f}$ is convex (see Assumption 2), by Jensen's inequality we have:

$$\hat{f}(x^{t+1}) = \hat{f}\left(x^t - \gamma \frac{1}{n} \sum_{i=1}^{n} g_i^t\right) = \hat{f}\left(\frac{1}{n} \sum_{i=1}^{n} (x^t - \gamma g_i^t)\right) \le \frac{1}{n} \sum_{i=1}^{n} \hat{f}(x^t - \gamma g_i^t).$$

Let $\mathcal{C}_S(x) = \sum_{j \in S} x_j e_j$ denote the compressor that keeps only coordinates indexed by $S \subset \{1, \dots, d\}$ with $|S| = k$. Since we instantiate $\Omega = [0, 2]^d$, for any index set S_i the indicator vector $\mathbf{1}_{S_i}$ belongs to Ω_k. Therefore, by optimality of w_i^t we have for any S_i with $|S_i| = k$:

$$\hat{f}(x^t - \gamma g_i^t) \leq \hat{f}(x^t - \gamma \mathcal{C}_{S_i}(\nabla f_i(x^t))).$$

Using L-smoothness of $\hat{f}$:

$$\hat{f}(x^t - \gamma \mathcal{C}_{S_i}(\nabla f_i(x^t))) \leq \hat{f}(x^t) - \gamma \langle \nabla \hat{f}(x^t), \mathcal{C}_{S_i}(\nabla f_i(x^t)) \rangle + \frac{L\gamma^2}{2} \left\| \mathcal{C}_{S_i}(\nabla f_i(x^t)) \right\|_2^2.$$

For each worker i, we sum the inequality over all $\binom{d}{k}$ possible index sets S_i:

$$\binom{d}{k} \hat{f}(x^t - \gamma g_i^t) \leq \binom{d}{k} \hat{f}(x^t) - \gamma \sum_{S_i} \langle \nabla \hat{f}(x^t), \mathcal{C}_{S_i}(\nabla f_i(x^t)) \rangle + \frac{L\gamma^2}{2} \sum_{S_i} \left\| \mathcal{C}_{S_i}(\nabla f_i(x^t)) \right\|_2^2.$$

For the linear term, each coordinate j appears in exactly $\binom{d-1}{k-1}$ sets, hence:

$$\sum_{S_i} \langle \nabla \hat{f}(x^t), \mathcal{C}_{S_i}(\nabla f_i(x^t)) \rangle = \binom{d-1}{k-1} \langle \nabla \hat{f}(x^t), \nabla f_i(x^t) \rangle.$$

Similarly, for the quadratic term:

$$\sum_{S_i} \left\| \mathcal{C}_{S_i}(\nabla f_i(x^t)) \right\|_2^2 = \binom{d-1}{k-1} \left\| \nabla f_i(x^t) \right\|_2^2.$$

Dividing by $\binom{d}{k}$ and using $\frac{\binom{d-1}{k-1}}{\binom{d}{k}} = \frac{k}{d}$:

$$\hat{f}(x^t - \gamma g_i^t) \leq \hat{f}(x^t) - \frac{k}{d} \gamma \langle \nabla \hat{f}(x^t), \nabla f_i(x^t) \rangle + \frac{k}{d} \frac{L\gamma^2}{2} \left\| \nabla f_i(x^t) \right\|_2^2.$$

Combining Jensen's inequality with (B) and averaging over all workers:

$$\hat{f}(x^{t+1}) \leq \hat{f}(x^t) - \frac{k}{d} \gamma \frac{1}{n} \sum_{i=1}^{n} \langle \nabla \hat{f}(x^t), \nabla f_i(x^t) \rangle + \frac{k}{d} \frac{L\gamma^2}{2} \frac{1}{n} \sum_{i=1}^{n} \left\| \nabla f_i(x^t) \right\|_2^2.$$

Note that $\frac{1}{n} \sum_{i=1}^{n} \nabla f_i(x^t) = \nabla f(x^t)$. By Assumption 3:

$$\frac{1}{n} \sum_{i=1}^{n} \left\| \nabla f_i(x^t) \right\|_2^2 \leq \left\| \nabla f(x^t) \right\|_2^2 + \sigma^2.$$

Thus:

$$\hat{f}(x^{t+1}) \leq \hat{f}(x^t) - \frac{k}{d} \gamma \langle \nabla \hat{f}(x^t), \nabla f(x^t) \rangle + \frac{k}{d} \frac{L\gamma^2}{2} \left(\left\| \nabla f(x^t) \right\|_2^2 + \sigma^2 \right).$$

We decompose the inner product as:

$$\begin{aligned}\langle \nabla \hat{f}(x^t), \nabla f(x^t) \rangle &= \langle \nabla \hat{f}(x^t), \nabla \hat{f}(x^t) \rangle + \langle \nabla \hat{f}(x^t), \nabla f(x^t) - \nabla \hat{f}(x^t) \rangle \\ &= \left\| \nabla \hat{f}(x^t) \right\|_2^2 + \langle \nabla \hat{f}(x^t), \nabla f(x^t) - \nabla \hat{f}(x^t) \rangle.\end{aligned}$$

Using Young's inequality $\langle a, b \rangle \geq -\frac{1}{2} \|a\|_2^2 - \frac{1}{2} \|b\|_2^2$:

$$\langle \nabla \hat{f}(x^t), \nabla f(x^t) - \nabla \hat{f}(x^t) \rangle \geq -\frac{1}{2} \left\| \nabla \hat{f}(x^t) \right\|_2^2 - \frac{1}{2} \left\| \nabla f(x^t) - \nabla \hat{f}(x^t) \right\|_2^2.$$

Therefore:

$$\langle \nabla \hat{f}(x^t), \nabla f(x^t) \rangle \geq \frac{1}{2} \left\| \nabla \hat{f}(x^t) \right\|_2^2 - \frac{1}{2} \left\| \nabla f(x^t) - \nabla \hat{f}(x^t) \right\|_2^2.$$

Similarly, for the quadratic term we use the identity $\|a\|_2^2 = \|b\|_2^2 + 2\langle b, a - b \rangle + \|a - b\|_2^2$:

$$\|\nabla f(x^t)\|_2^2 = \left\| \nabla \hat{f}(x^t) \right\|_2^2 + 2\langle \nabla \hat{f}(x^t), \nabla f(x^t) - \nabla \hat{f}(x^t) \rangle + \left\| \nabla f(x^t) - \nabla \hat{f}(x^t) \right\|_2^2.$$

By Lemma 1, $\left\| \nabla f(x^t) - \nabla \hat{f}(x^t) \right\|_2^2 \leq \zeta$. Using Young's inequality on the cross-term:

$$\|\nabla f(x^t)\|_2^2 \leq \left\| \nabla \hat{f}(x^t) \right\|_2^2 + \left\| \nabla \hat{f}(x^t) \right\|_2^2 + \zeta + \zeta = 2 \left\| \nabla \hat{f}(x^t) \right\|_2^2 + 2\zeta.$$

Substituting all bounds:

$$\begin{aligned}\hat{f}(x^{t+1}) &\leq \hat{f}(x^t) - \frac{k}{d}\gamma \left(\frac{1}{2} \left\| \nabla \hat{f}(x^t) \right\|_2^2 - \frac{\zeta}{2} \right) + \frac{k}{d} \frac{L\gamma^2}{2} \left(2 \left\| \nabla \hat{f}(x^t) \right\|_2^2 + 2\zeta + \sigma^2 \right) \\ &= \hat{f}(x^t) - \frac{k}{d}\gamma \left(\frac{1}{2} - L\gamma \right) \left\| \nabla \hat{f}(x^t) \right\|_2^2 + \frac{k\gamma}{2d}\zeta + \frac{k}{d} \frac{L\gamma^2}{2} \left(2\zeta + \sigma^2 \right).\end{aligned}$$

Rearranging:

$$\hat{f}(x^{t+1}) \leq \hat{f}(x^t) - \frac{k}{d}\frac{\gamma}{2}(1 - 2L\gamma) \left\| \nabla \hat{f}(x^t) \right\|_2^2 + \frac{k\gamma}{2d} \left(\zeta + 2L\gamma\zeta + L\gamma\sigma^2 \right).$$

By strong convexity of $\hat{f}$ (Assumption 2), we have $\left\| \nabla \hat{f}(x^t) \right\|_2^2 \geq 2\mu(\hat{f}(x^t) - \hat{f}^*)$. Substituting:

$$\begin{aligned}\hat{f}(x^{t+1}) - \hat{f}^* &\leq (\hat{f}(x^t) - \hat{f}^*) - \frac{k}{d}\frac{\gamma}{2}(1 - 2L\gamma) \cdot 2\mu(\hat{f}(x^t) - \hat{f}^*) \\ &\quad + \frac{k\gamma}{2d} \left(\zeta + 2L\gamma\zeta + L\gamma\sigma^2 \right) \\ &= \left(1 - \frac{k}{d}\mu\gamma(1 - 2L\gamma) \right) (\hat{f}(x^t) - \hat{f}^*) \\ &\quad + \frac{k\gamma}{2d} \left(\zeta + 2L\gamma\zeta + L\gamma\sigma^2 \right).\end{aligned}$$

Setting $\gamma = \frac{1}{4L}$, the one-step recursion becomes:

$$\hat{f}(x^{t+1}) - \hat{f}^* \le \left(1 - \frac{k}{d}\mu \cdot \frac{1}{4L} \cdot \frac{1}{2}\right)(\hat{f}(x^t) - \hat{f}^*) + \frac{k}{8Ld}\left(\zeta + \frac{\zeta}{2} + \frac{\sigma^2}{4}\right)$$
$$= \left(1 - \frac{k\mu}{8Ld}\right)(\hat{f}(x^t) - \hat{f}^*) + \frac{k}{8Ld}\left(\frac{3\zeta}{2} + \frac{\sigma^2}{4}\right).$$

Denote $\rho = 1 - \frac{k\mu}{8Ld}$ and $c = \frac{k}{8Ld}\left(\frac{3\zeta}{2} + \frac{\sigma^2}{4}\right)$. Unrolling the recursion $a_{t+1} \le \rho a_t + c$ gives:

$$\hat{f}(x^T) - \hat{f}^* \le \rho^T(\hat{f}(x^0) - \hat{f}^*) + c \cdot \frac{1-\rho^T}{1-\rho}.$$

Since $1 - \rho = \frac{k\mu}{8Ld}$, we have:

$$\frac{c}{1-\rho} = \frac{\frac{k}{8Ld}\left(\frac{3\zeta}{2} + \frac{\sigma^2}{4}\right)}{\frac{k\mu}{8Ld}} = \frac{1}{\mu}\left(\frac{3\zeta}{2} + \frac{\sigma^2}{4}\right).$$

This yields the final estimate:

$$\hat{f}(x^T) - \hat{f}^* \le \left(1 - \frac{k\mu}{8Ld}\right)^T (\hat{f}(x^0) - \hat{f}^*) + \frac{1}{\mu}\left(\frac{3\zeta}{2} + \frac{\sigma^2}{4}\right).$$

To obtain the horizon-tuned bound from the theorem statement, define $a_t := \hat{f}(x^t) - \hat{f}^*$. From the one-step bound derived above, for any step size $\gamma \le \frac{1}{4L}$, we have $2L\gamma \le \frac{1}{2}$. Thus, the contraction factor satisfies:

$$1 - \frac{k}{d}\mu\gamma(1 - 2L\gamma) \le 1 - \frac{k\mu\gamma}{2d},$$

and the error term is bounded by:

$$\frac{k\gamma}{2d}\left((1 + 2L\gamma)\zeta + L\gamma\sigma^2\right) \le \frac{k\gamma}{2d}\left(\frac{3}{2}\zeta + L\gamma\sigma^2\right).$$

The one-step recursion thus becomes:

$$a_{t+1} \le \left(1 - \frac{k\mu\gamma}{2d}\right)a_t + \frac{k\gamma}{2d}\left(\frac{3}{2}\zeta + L\gamma\sigma^2\right).$$

Unrolling this recursion over T steps and summing the geometric progression, we obtain:

$$a_T \le \left(1 - \frac{k\mu\gamma}{2d}\right)^T a_0 + \sum_{t=0}^{T-1}\left(1 - \frac{k\mu\gamma}{2d}\right)^t \frac{k\gamma}{2d}\left(\frac{3}{2}\zeta + L\gamma\sigma^2\right)$$
$$\le \exp\left(-\frac{k\mu\gamma}{2d}T\right)a_0 + \frac{\frac{k\gamma}{2d}\left(\frac{3}{2}\zeta + L\gamma\sigma^2\right)}{\frac{k\mu\gamma}{2d}}$$
$$= \exp\left(-\frac{k\mu\gamma}{2d}T\right)a_0 + \frac{L\gamma\sigma^2}{\mu} + \frac{3\zeta}{2\mu}.$$

Now we analyze the specific choice of the step size tuned to the horizon T:

$$\gamma = \min\left\{\frac{1}{4L}, \frac{2d}{k\mu T}\log(B)\right\}, \quad \text{where} \quad B = \max\left\{1, \frac{k\mu^2 T a_0}{2Ld\sigma^2}\right\}.$$

We consider two cases depending on which term minimizes γ.

Case 1: $\gamma = \frac{1}{4L}$. This occurs when $\frac{1}{4L} \le \frac{2d}{k\mu T}\log(B)$, which implies $T \le \frac{8Ld}{k\mu}\log(B)$. Substituting $\gamma = \frac{1}{4L}$ into (B) yields:

$$a_T \le \exp\left(-\frac{k\mu}{8Ld}T\right)a_0 + \frac{\sigma^2}{4\mu} + \frac{3\zeta}{2\mu}.$$

Since $T \le \frac{8Ld}{k\mu}\log(B)$ in this case, we can upper-bound the variance term as $\frac{\sigma^2}{4\mu} \le \frac{2Ld\sigma^2}{k\mu^2 T}\log(B)$.

Case 2: $\gamma = \frac{2d}{k\mu T}\log(B)$. Substituting this γ into the first term of (B):

$$\exp\left(-\frac{k\mu\gamma}{2d}T\right)a_0 = \exp(-\log(B))a_0 = \frac{a_0}{B}.$$

By the definition of B, we have $B \ge \frac{k\mu^2 T a_0}{2Ld\sigma^2}$, meaning:

$$\frac{a_0}{B} \le \frac{2Ld\sigma^2}{k\mu^2 T}.$$

For the second term in (B), substituting γ yields:

$$\frac{L\gamma\sigma^2}{\mu} = \frac{2Ld\sigma^2}{k\mu^2 T}\log(B).$$

Summarizing both cases, we consistently bound the error by:

$$a_T \le \exp\left(-\frac{k\mu}{8Ld}T\right)a_0 + \frac{2Ld\sigma^2}{k\mu^2 T}(\log(B)+1) + \frac{3\zeta}{2\mu}.$$

Hiding the explicit logarithmic dependence on T, a_0, and other constants within the standard $\tilde{\mathcal{O}}$ notation, we arrive at the final claim:

$$\hat{f}(x^T) - \hat{f}^* \le \tilde{\mathcal{O}}\left(\exp\left(-\frac{k\mu}{8Ld}T\right)(\hat{f}(x^0) - \hat{f}^*) + \frac{Ld\sigma^2}{k\mu^2 T}\right) + \frac{3\zeta}{2\mu}.$$

This completes the proof.

C Experiments

In this section, we provide supplementary information on the hyperparameters used for the methods employed and technical details.

C.1 Technical Information

The *ResNet-18* experiments were conducted on an NVIDIA GeForce RTX 2080 Ti GPU with 12 GB of video memory; 50 training epochs took approximately 25 minutes. The *GPT-2* experiments on the *WikiText2* dataset were conducted on an NVIDIA A100 GPU with 40 GB of video memory; 50 training epochs took approximately 2 h.

C.2 Finding the Best Hyperparameters

Table 1. Hyperparameters for methods with and without error-feedback ($\alpha = 1\%$)

compressor	epochs	LR (no EF)	LR (with EF)	internal steps	internal LR
TopK	50	0.001	0.002	–	–
ImpK_s	50	0.001	0.002	50	$1 \cdot 10^7$
ImpK_c	50	0.001	0.002	50	$1 \cdot 10^7$

References

1. Aji, A.F., Heafield, K.: Sparse communication for distributed gradient descent (2017). arXiv preprint arXiv:1704.05021
2. Alistarh, D., Grubic, D., Li, J., Tomioka, R., Vojnovic, M.: QSGD: communication-efficient SGD via gradient quantization and encoding. Adv. Neural. Inf. Process. Syst. **30**, 1709–1720 (2017)
3. Beck, A., Teboulle, M.: Mirror descent and nonlinear projected subgradient methods for convex optimization. Oper. Res. Lett. **31**(3), 167–175 (2003). https://doi.org/10.1016/S0167-6377(02)00231-6, https://www.sciencedirect.com/science/article/pii/S0167637702002316
4. Bernstein, J., Wang, Y.X., Azizzadenesheli, K., Anandkumar, A.: signSGD: compressed optimisation for non-convex problems. In: International Conference on Machine Learning, pp. 560–569. PMLR (2018)
5. Beznosikov, A., Horváth, S., Richtárik, P., Safaryan, M.: On biased compression for distributed learning. J. Mach. Learn. Res. **24**(276), 1–50 (2023)
6. Defazio, A., Bottou, L.: On the ineffectiveness of variance reduced optimization for deep learning. In: Wallach, H., Larochelle, H., Beygelzimer, A., d'Alché-Buc, F., Fox, E., Garnett, R. (eds.) Advances in Neural Information Processing Systems. vol. 32, pp. 1753–1763. Curran Associates, Inc. (2019). https://proceedings.neurips.cc/paper_files/paper/2019/file/84d2004bf28a2095230e8e14993d398d-Paper.pdf
7. Horváth, S., Ho, C.Y., Horvath, L., Sahu, A.N., Canini, M., Richtárik, P.: Natural compression for distributed deep learning. In: Mathematical and Scientific Machine Learning, pp. 129–141. PMLR (2022)
8. Karimireddy, S.P., Rebjock, Q., Stich, S.U., Jaggi, M.: Error feedback fixes SignSGD and other gradient compression schemes (2019). https://arxiv.org/abs/1901.09847

9. Loshchilov, I., Hutter, F.: Decoupled weight decay regularization. In: International Conference on Learning Representations (2019). https://openreview.net/forum?id=Bkg6RiCqY7
10. Molodtsov, G., et al.: Bant: byzantine antidote via trial function and trust scores (2025). https://arxiv.org/abs/2505.07614
11. Patel, V., Berahas, A.S.: Gradient descent in the absence of global Lipschitz continuity of the gradients. SIAM J. Math. Data Sci. **6**(3), 602–626 (2024). https://doi.org/10.1137/23M154474X
12. Richtárik, P., Sokolov, I., Fatkhullin, I.: EF21: a new, simpler, theoretically better, and practically faster error feedback. Adv. Neural. Inf. Process. Syst. **34**, 4384–4396 (2021)
13. Stich, S.U., Cordonnier, J.B., Jaggi, M.: Sparsified SGD with memory. Adv. Neural. Inf. Process. Syst. **31**, 4452–4463 (2018)
14. Verbraeken, J., Wolting, M., Katzy, J., Kloppenburg, J., Verbelen, T., Rellermeyer, J.S.: A survey on distributed machine learning. ACM Comput. Surv. **53**(2), 1–33 (2020). https://doi.org/10.1145/3377454
15. Vogels, T., Karimireddy, S.P., Jaggi, M.: PowerSGD: practical low-rank gradient compression for distributed optimization. Adv. Neural. Inf. Process. Syst. **32**, 14236–14245 (2019)

A Finite-Horizon Multistage Spatial Model of Renewable Resource Extraction

Denis Kuzyutin[1,2], Nadezhda Smirnova[2](✉), and Maria Bulgakova[1]

[1] Saint Petersburg State University, Universitetskaya nab. 7/9, 199034 St. Petersburg, Russia
{d.kuzyutin,m.a.bulgakova}@spbu.ru

[2] HSE University, Soyuza Pechatnikov ul. 16, 190008 St. Petersburg, Russia
nvsmirnova@hse.ru

Abstract. We study a finite-horizon multistage model of renewable resource extraction, like a fish war game, when the players may harvest in two areas with different fish reproduction conditions, taking into account the resource migration process. Non-cooperative and cooperative solutions are derived in non-stationary feedback strategies for the game with two asymmetric players. The results of the numerical simulation are provided to detect and clarify the solutions properties.

Keywords: Renewable resource extraction · dynamic game · multistage game · fishery · subgame perfect equilibria · cooperative solution · migration

1 Introduction

Dynamic game-theoretical models provide suitable tools to describe and study the long-term extraction of common renewable resource in a competitive environment. For instance, starting from seminal papers [5,6,12] the dynamic game models are extensively used to explore a worldwide fishery both in cooperative and non-cooperative settings as well as to design fishing policies to ensure sustainable long-term exploitation of fish stocks (see, e.g., [1–4,10,14–16,18,21,26]).

Considering spatial fishery management problems where water basin is divided into several regions with different marine biological conditions, fishery regimes, e.t.c. (see, e.g., [16,17,25]), one need to take into account the fish migration process between the regions. A two-player, two-region dynamic fishery management model with migration where the regulator controls the size of each fishery area was introduced in [14,15]. The paper [4] explored an extended version of this single-criterion model with migration between three marine areas, namely, an open access fishing area with n players, a MPA with limited fishing activity of only one player and a no-take area ("fish bank"). A multicriteria

The research of the third author was supported by RSF, grant No. 24-18-00479, https://rscf.ru/project/24-18-00479/, Saint Petersburg State University.

A. Kazakov et al. (Eds.): MOTOR 2026, LNCS 16687, pp. 294–308, 2026.
https://doi.org/10.1007/978-3-032-31390-4_19

extention of those models that takes into account the players' environmental concern was introduced and explored in [11].

In the paper, we consider a spatial finite-horizon multistage model with two different areas and resource migration assuming that the players harvest in both areas. It is assumed that one of this marine areas (usually, the smaller one) provides more favorable biological or(and) administrative policy conditions for the successful spawning and feeding of commercial fish. The well-known examples of such areas include marine coastal waters like the marine zones around Svalbard island, Lofoten islands e.t.c. (see [17,23,25]), marine protected areas [2,4,13,14]), areas with warm sea currents [24]. The proposed model allows us to describe and study the processes of simultaneous fishery in marine zones with different conditions. In addition, one of the marine areas could be considered as a future marine protected area (see [2,4,11,13]) where fishing is planned to be restricted or completely prohibited starting from some future time T.

The payoff function (we also use terms "objective function") of each player takes into account the stage harvests and the residual fish stocks at time T. Using dynamic programming, we derive the Nash equilibria as well as the cooperative solution in non-stationary feedback strategies.

We also carried out a numerical simulation to clarify the solutions properties and provide sensitivity analysis. In particular, it turns out that:

- the players' asymmetry (of special type adopted in the paper, see also [1,8,20]) is barely perceptible at the first game stages, but obviously appears closer to the last stage;
- throughout the essential part of the game duration, the players' optimal strategies seem to be "almost stationary", i.e. each player catches an approximately constant share of the current fish stock at each fishing area;
- cooperative scenario is always more preferable for the environment than the non-cooperative one.

Another interesting problem we focus on in the paper is to evaluate how the level of the players' asymmetry may affect the solutions properties.

The contributions of this paper are twofold:

- We derive and compare non-cooperative and cooperative solution in non-stationary feedback strategies for finite-stage spatial fishery management problem with asymmetric players and migration, when both players may harvest in two marine zones with different fish reproduction conditions.
- Using numerical simulation, we reveal and clarify the properties of the solutions obtained.

The remainder of the paper is organized as follows. In Sect. 2, we specify the finite-stage game model with two different areas of resource exploitation and migration. In Sects. 3 and 4, we derive a non-cooperative solution (a Nash equilibrium in non-stationary feedback strategies that is known to be subgame perfect) and a cooperative solution, respectively. The results of numerical simulation which shed light on the properties of the solutions obtained are provided in Sect. 5. We briefly conclude in Sect. 6.

2 The Model

Following [4,14,15], consider a discrete time finite-horizon spatial model of renewable resource extraction. One could interpret a resource as a single species commercial fish, and a model as the 2-player "fish war" game with two fishing areas Q and S [11]. We denote by q and s their respective sizes, $q \in (0,1)$, $s \in (0,1)$, $q+s=1$. Let x_t and y_t denote fish stocks at time $t = 0,1,\ldots,T-1,T$ in marine areas Q and S respectively. In the absence of human fishing activity the resource stocks evolution is as follows (see [4–6,14,15]):

$$\begin{cases} x_{t+1} = x_t^{\alpha_x} \cdot \left(\frac{y_t}{x_t}\right)^{(1-s)\beta_x} \\ y_{t+1} = y_t^{\alpha_y} \cdot \left(\frac{x_t}{y_t}\right)^{s\beta_y} \end{cases}, t = 0,\ldots,T, \tag{1}$$

with given initial conditions x_0 and y_0, where $\alpha_x \in (0,1)$ and $\alpha_y \in (0,1)$ are the natural growth parameters in zones Q and S, $\beta_x \in (0,1)$ is a fish migration coefficient from S to Q while $\beta_y \in (0,1)$ is a fish migration rate from Q to S. We adopt the following assumptions on biological parameters (see [4,11]):

$$\alpha_x > \alpha_y, \alpha_x + \beta_x < 1, \alpha_y + \beta_y < 1. \tag{2}$$

The first inequality implies that fish are more successful at spawning in area S (see., e.g., [1,2]). The last two constraints guarantee that the positive steady-state stocks levels in Q and S in corresponding infinite horizon model are equal to 1 when no fishing activity occurs (see [4,11]).

We focus on a two player game, when each player $i \in N = \{1,2\}$ takes fish in both areas Q and S. Let $u_{it} \geq 0$ and $w_{it} \geq 0$ denote the i-th player's admissible catches in period t, $t = 0,1,\ldots,T-1$, in areas Q and S respectively.

$$u_t = u_{1t} + u_{2t} < x_t, \; w_t = w_{1t} + w_{2t} < y_t. \tag{3}$$

Using notations

$$\begin{cases} \widehat{\alpha}_x = \alpha_x - (1-s)\beta_x, \;\; \widehat{\beta}_x = (1-s)\beta_x, \\ \widehat{\alpha}_y = \alpha_y - s\beta_y, \;\; \widehat{\beta}_y = s\beta_y. \end{cases} \tag{4}$$

we derive the fish stock dynamics in the following form

$$\begin{cases} x_{t+1} = (x_t - u_t)^{\widehat{\alpha}_x} \cdot (y_t - w_t)^{\widehat{\beta}_x}, \\ y_{t+1} = (y_t - w_t)^{\widehat{\alpha}_y} \cdot (x_t - u_t)^{\widehat{\beta}_y}, \end{cases}, t = 0,\ldots,T-1. \tag{5}$$

In the paper, we assume the feedback information structure, i. e. $u_i(\cdot) = u_i(t,x_t,y_t)$, $w_i(\cdot) = w_i(t,x_t,y_t)$, $t = 0,\ldots,T-1$, $i = 1,2$.

We assume (see, e. g., [1,4–6,10,12,14]) that the j-th player's payoff function in the original game starting at time $t = 0$ has the following form

$$H_j(\cdot) = \sum_{\tau=0}^{T-1} \delta^\tau \cdot (\ln u_{j\tau} + \ln w_{j\tau}) + \delta^T (K_j^x \ln x_T + K_j^y \ln y_T), \; j = 1,2, \tag{6}$$

where $\delta \in (0,1)$ is a discount factor, while K_j^x and K_j^y are positive parameters which specify the j-th player valuation of the residual resource stocks x_T and y_T after the T-stage fishing program ends. In the paper, we employ the same discount factor for both players, and consider the players' asymmetry as different values of coefficients K_1^x and K_2^x (and / or K_1^y and K_2^y).

Denote by $G^0(x_0, y_0, T)$ the T-stage 2-player game starting at time $t = 0$ with discrete dynamics (5), initial conditions x_0 and y_0, objective functions (6) and feedback information structure.

Each time instant $t = 1, \ldots, T-1$ and a pair of intermediate states x_t and y_t specify a $(T-t)$-stage subgame $G^t(x_t, y_t, T-t)$ starting at time $\tau = t$ with initial conditions x_t, y_t and the payoff functions in this subgame (the so-called "subgame payoff functions")

$$H_j^t(\cdot) = \sum_{\tau=t}^{T-1} \delta^{\tau-t} \cdot (\ln u_{j\tau} + \ln w_{j\tau}) + \delta^{T-t}(K_j^x \ln x_T + K_j^y \ln y_T), \ j = 1, 2. \quad (7)$$

3 Non-cooperative Solution

Following [5–7,22,26] we employ a subgame-perfect equilibrium (SPE) as a non-cooperative solution in a dynamic game.

Definition 1. *A strategy profile $(\overline{u}_j(\cdot), \overline{w}_j(\cdot))_{j=1,2}$ is a Nash equilibrium (NE) in $G^0(x_0, y_0, T)$ if*

$$H_j(u_j, w_j, \overline{u}_{3-j}; \overline{w}_{3-j}) \leq H_j(\overline{u}_j, \overline{w}_j; \overline{u}_{3-j}, \overline{w}_{3-j})$$

for any admissible strategy $(u_j(\cdot), w_j(\cdot))$ of every player $j = 1, 2$.

Definition 2. *A strategy profile $(\overline{u}(\cdot), \overline{w}(\cdot)) = (\overline{u}, \overline{w})$ is a SPE in $G^0(x_0, y_0, T)$ if for each $t = 0, \ldots, T-1$ and intermediate states x_t, y_t the restriction of $(\overline{u}(\cdot), \overline{w}(\cdot))$ in the subgame $G^t(x_t, y_t, T-t)$ forms a NE in that subgame.*

We derive a Nash equilibrium in feedback (non-stationary) strategies (that is known to be SPE) for multistage game $G^0(x_0, y_0, T)$ using the dynamic programming method [7,26]. Denote by $V_j(t, x, y)$ the j-th player's payoff in the subgame $G^t(x, y, T-t)$, $t = 0, \ldots, T-1$, corresponding to the SPE strategy profile in this subgame. Following [7,26], we will call $V_j(t, x, y)$ the player's j value function in the subgame.

Proposition 1. *A finite-horizon multistage game $G^0(x_0, y_0, T)$ with asymmetric players has a unique SPE in non-stationary strategies:*

$$\begin{cases} \overline{u}_{jt} = P_{3-j}^u(t) \cdot \left[P_1^u(t) + P_2^u(t) + \delta P_1^u(t) \cdot P_2^u(t)\right]^{-1} \cdot x_t, \\ \overline{w}_{jt} = P_{3-j}^w(t) \cdot \left[P_1^w(t) + P_2^w(t) + \delta P_1^w(t) \cdot P_2^w(t)\right]^{-1} \cdot y_t \end{cases}, t = 0, \ldots, T-1, \quad (8)$$

j=1, 2; where

$$P_j^u(t) = \frac{A_j(t)-1}{\delta} = A_j(t+1)\widehat{\alpha}_x + B_j(t+1)\widehat{\beta}_y, \quad P_j^w(t) = \frac{B_j(t)-1}{\delta} = A_j(t+1)\widehat{\beta}_x + B_j(t+1)\widehat{\alpha}_y \tag{9}$$

while the coefficients $A_j(t)$, $B_j(t)$ are calculated backwards using the following recurrence formulae

$$\begin{aligned} A_j(t) &= 1 + \delta \cdot (A_j(t+1)\widehat{\alpha}_x + B_j(t+1)\widehat{\beta}_y) \\ B_j(t) &= 1 + \delta \cdot (A_j(t+1)\widehat{\beta}_x + B_j(t+1)\widehat{\alpha}_y) \end{aligned}, t = 0, \dots, T-1; j = 1, 2, \tag{10}$$

and boundary conditions

$$A_j(T) = K_j^x, B_j(T) = K_j^y, j = 1, 2. \tag{11}$$

Proof. Taking the objective finctions (6) structure into account we guess the following functional form of the value functions:

$$V_j(t, x, y) = A_j(t)\ln x + B_j(t)\ln y + D_j(t), \ t = 0, \dots, T-1. \tag{12}$$

Following [7,26], to obtain the appropriate boundary conditions we specify the value functions for the "terminal stage" subgame $G^T(x_T, y_T, 0)$ (i.e. the players' payoffs at time instant T when the states x_T and y_T have been already reached) as follows:

$$V_j(T, x_T, y_T) = K_j^x \ln x_T + K_j^y \ln y_T, \ j = 1, 2. \tag{13}$$

The j-th player's SPE strategy $(\overline{u}_j(\cdot), \overline{w}_j(\cdot))$ has to satisfy the Bellman equation

$$\begin{aligned} V_j(t, x, y) &= \max_{u_j, w_j} \left\{ \ln u_j + \ln w_j + \delta \cdot V_j(t+1, x_{t+1}, y_{t+1}) \right\} = \\ &= \max_{u_j, w_j} \Big\{ \ln u_j + \ln w_j + \delta \cdot \Big[A_j(t+1) \cdot \ln\big((x - u_j - \overline{u}_{3-j}(t, x, y))^{\widehat{\alpha}_x} \cdot \\ &\cdot (y - w_j - \overline{w}_{3-j}(t, x, y))^{\widehat{\beta}_x}\big) + B_j(t+1) \cdot \ln\big((x - u_j - \overline{u}_{3-j}(t, x, y))^{\widehat{\beta}_y} \cdot \\ &\cdot (y - w_j - \overline{w}_{3-j}(t, x, y))^{\widehat{\alpha}_y}\big) + D_j(t+1)\Big] \Big\}, t = 0, 1, \dots, T-1; j = 1, 2, \end{aligned} \tag{14}$$

with boundary conditions (13).

Differentiating the r.h.s. of (14) with respect to u_j and w_j yields the linear (in u_j and w_j) system of the first order conditions. Solving that system we obtain the interior solution (8).

Then we substitute strategies (8) in (14) and extract the coefficients for $\ln x$ and $\ln y$ in both parts of (14) to obtain the recurrence formulae (10).

Note that all the coefficients $A_j(t)$, $B_j(t)$, $P_j^u(t)$, $P_j^w(t)$ are positive for all $t = 0, \dots, T$ due to (9)-(11) and the model assumptions. Hence,

$$\begin{cases} \overline{u}_t = \overline{u}_{1t} + \overline{u}_{2t} = \frac{P_1^u(t) + P_2^u(t)}{P_1^u(t) + P_2^u(t) + \delta P_1^u(t) \cdot P_2^u(t)} \cdot x_t < x_t, \\ \overline{w}_t = \overline{w}_{1t} + \overline{w}_{2t} = \frac{P_1^w(t) + P_2^w(t)}{P_1^w(t) + P_2^w(t) + \delta P_1^w(t) \cdot P_2^w(t)} \cdot y_t < y_t, \end{cases} t = 0, \dots, T-1, \tag{15}$$

and SPE strategies (8) constitute an admissible strategy profile.

Remark 1. The SPE state trajectories $\overline{x}^t$ and $\overline{y}^t$, $t = 0, \ldots, T-1$, could be determined in the following form

$$\begin{cases} \overline{x}_{t+1} = \left[\frac{\delta P_1^u(t) \cdot P_2^u(t)}{P_1^u(t)+P_2^u(t)+\delta P_1^u(t) \cdot P_2^u(t)} \cdot x_t\right]^{\widehat{\alpha}_x} \cdot \left[\frac{\delta P_1^w(t) \cdot P_2^w(t)}{P_1^w(t)+P_2^w(t)+\delta P_1^w(t) \cdot P_2^w(t)} \cdot y_t\right]^{\widehat{\beta}_x}, \\ \overline{y}_{t+1} = \left[\frac{\delta P_1^u(t) \cdot P_2^u(t)}{P_1^u(t)+P_2^u(t)+\delta P_1^u(t) \cdot P_2^u(t)} \cdot x_t\right]^{\widehat{\beta}_y} \cdot \left[\frac{\delta P_1^w(t) \cdot P_2^w(t)}{P_1^w(t)+P_2^w(t)+\delta P_1^w(t) \cdot P_2^w(t)} \cdot y_t\right]^{\widehat{\alpha}_y}. \end{cases} \tag{16}$$

4 Cooperative Solution

Since we allow the players' asymmetry, to seek a cooperative solution we suppose that players at each subgame $G^t(x_t, y_t, T-t)$ jointly choose their strategies to maximize the weighted sum of their payoff functions (7).

Denote by $\mu_j \in (0,1)$ the coefficient of the j-th player's relative importance (when reaching a cooperative agreement), $\mu_1 + \mu_2 = 1$. The maximal weighted sum of the players' payoff functions, which we will call "cooperative total value function" in $G^t(x, y, T-t)$ takes the form

$$\begin{aligned} & V^C(t,x,y) = \mu_1 H_1^t(\cdot) + \mu_2 H_2^t(\cdot) = \\ & = \mu_1 \left[\sum_{\tau=t}^{T-1} \delta^{\tau-t} \cdot (\ln u_{1\tau} + \ln w_{1\tau}) + \delta^{T-t}(K_1^x \ln x_T + K_1^y \ln y_T)\right] + \\ & + \mu_2 \left[\sum_{\tau=t}^{T-1} \delta^{\tau-t} \cdot (\ln u_{2\tau} + \ln w_{2\tau}) + \delta^{T-t}(K_2^x \ln x_T + K_2^y \ln y_T)\right]. \end{aligned} \tag{17}$$

Proposition 2. *A T-stage game $G^0(x_0, y_0, T)$ for each pair (μ_1, μ_2) possesses a unique cooperative solution in non-stationary strategies:*

$$\begin{cases} u_{jt}^C = \frac{\mu_j}{A^C(t)} \cdot x_t, \\ w_{jt}^C = \frac{\mu_j}{B^C(t)} \cdot y_t, \end{cases} , t = 0, \ldots, T-1; j = 1, 2, \tag{18}$$

where coefficients $A^C(t)$, $B^C(t)$ are calculated backwards using the recurrence formulae

$$\begin{cases} A^C(t) = 1 + \delta \cdot (A^C(t+1)\widehat{\alpha}_x + B^C(t+1)\widehat{\beta}_y) \\ B^C(t) = 1 + \delta \cdot (A^C(t+1)\widehat{\beta}_x + B^C(t+1)\widehat{\alpha}_y) \end{cases} , t = 0, \ldots, T-1, \tag{19}$$

and boundary conditions

$$A^C(T) = \mu_1 K_1^x + \mu_2 K_2^x, \; B^C(T) = \mu_1 K_1^y + \mu_2 K_2^y. \tag{20}$$

The proof of Prop. 2 follows the same pattern as the proof of Prop. 1, applying the dynamic programming method to the cooperative value function (17). Obviously, the recurrence formulae (19) are similar to (10), but with different boundary conditions.

Remark 2. The cooperative strategies (18) form an admissible strategy profile meeting (3). Namely, using (20) and (19) one can easily verify that $A^C(t) > 1$, $B^C(t) > 1$ for all $t = 0, \dots, T-1$. Hence, $u_t^C = \frac{x_t}{A^C(t)} < x_t$ and $w_t^C = \frac{y_t}{B^C(t)} < y_t$.

Remark 3. The cooperative solution defined above implies that the payoffs are transferable and the players may redistribute the stage payoffs when implementing the cooperative strategies (18) using an appropriate payoff distribution procedure (see, e.g. [7,19,26]). However, the construction of such payoff distribution procedures as well as considering a cooperation with non-transferable payoffs when the choice of weights μ_j should be based on the players' individual rationality conditions goes far beyond of the scope of the paper.

Proposition 3. *In the* T*-stage game* $G^0(x_0, y_0, T)$ *with symmetric players* ($K_1^x = K_2^x$ *and* $K_1^y = K_2^y$) *the cooperative behavior (18) implies that at each stage* $t = 0, 1, \dots, T-1$ *the players harvest the smaller share of the current stocks* x_t, y_t *than under non-cooperative scenario (8).*

Proof. For the case of symmetric players boundary conditions (11) for non-cooperative solution take the form

$$A_1(T) = A_2(T) = K_1^x, B_1(T) = B_2(T) = K_1^y.$$

It follows from (10) that

$$\begin{cases} A_1(t) = A_2(t),\ B_1(t) = B_2(t), \\ P_1^u(t) = P_2^u(t),\ P_1^w(t) = P_2^w(t), \end{cases} t = 0, \dots, T-1.$$

Taking into account (15), (8), (9), we obtain

$$\begin{cases} \overline{u}_t = \frac{2P_1^u(t)}{2P_1^u(t) + \delta\left(P_1^u(t)\right)^2(t)} \cdot x_t = \frac{2}{1+A_1(t)} \cdot x_t, \\ \overline{w}_t = \frac{2}{1+B_1(t)} \cdot y_t, \end{cases} t = 0, \dots, T-1. \tag{21}$$

Under cooperative scenario boundary conditions (20) for each pair (μ_1, μ_2) take the form

$$A^C(T) = \mu_1 K_1^x + (1-\mu_1)K_2^x = K_1^x = A_j(T),\ B^C(T) = K_1^y = B_j(T),\ j = 1, 2.$$

Then

$$A^C(t) = A_j(t), B^C(t) = B_j(t), t = 0, \dots, T-1 \tag{22}$$

due to (19). Hence,

$$u_t^C = \frac{1}{A_1(t)} \cdot x_t,\ w_t^C = \frac{1}{B_1(t)} \cdot y_t,\ t = 0, \dots, T-1. \tag{23}$$

Comparing (23) and (21) yields

$$\begin{cases} \frac{1}{A_1(t)} < \frac{2}{1+A_1(t)} \\ \frac{1}{B_1(t)} < \frac{2}{1+B_1(t)} \end{cases} \iff \begin{cases} A_1(t) > 1 \\ B_1(t) > 1 \end{cases}.$$

The latter inequalities hold for all $t = 0, \dots, T-1$ due to (22) and Rem. 2.

5 Numerical Simulation

For numerical simulation we first accept the following values of the parameters as a baseline scenario (note that biological parameters are close to those employed in [4,11]): $\alpha_x = 0.6$; $\alpha_y = 0.5$; $\beta_x = 0.1$; $\beta_y = 0.2$; $x_0 = 0.1$; $y_0 = 0.2$; $s = 0.2$; $\delta = 0.95$. We assume in a baseline scenario that the players differently value the residual fish stocks: $K_1^x = 1.5$; $K_1^y = 1$; $K_2^x = 1$; $K_2^y = 0.5$, i. e. $K_j^x > K_j^y$, $j = 1, 2$, and $K_1^x > K_2^x$, $K_1^y > K_2^y$.

First, we run the model for $t = 0, 1, \ldots, T = 10$ to observe the stocks evolution, the stage harvests of each player at both areas, and to compare SPE and cooperative solution. A typical pattern of the fish stocks evolution is presented in Fig. 1. One can notice, that:

- cooperative regime is always more preferable for the environment than the non-cooperative one;
- the stocks in both areas first grow up (because of the small initial stocks – compare with Fig. 7), then retain at almost constant level (this is even more obvious for larger values of T or for larger initial stocks as in Fig. 7), and finally decrease at several last periods but still to the levels which are higher than the initial stocks;
- the higher coefficient μ_1 of the relative importance of the first player, who values the residual stocks more than another player, the higher are the final stocks under cooperative scenario.

A typical patterns of the players' harvest programs are presented in Figs. 2 and 3 for non-cooperative and cooperative regime, respectively. As we can see, the players' catches in a given fishing area increase and almost coincide under SPE scenario at all stages, except for the last few (because the players' asymmetry appears only closer to the final stage). In several last stages the first player, who values the residual stocks more, reduces her catches while the second player does the opposite. Note that the mentioned difference in catching programs disappears when we consider the cooperative mode in Fig. 3.

The graphs of the players' relative catches (the j-th player's harvested share of the current available stock) in Fig. 4 show that:

- the players' SPE strategies are practically stationary, i.e. each player catches a constant share of the current fish stock, throughout the essential part of the game duration;
- one can see that the players' asymmetry does not play a role at the first game stages but obviously appears closer to the last stage;
- the first player, who values the residual stocks more, reduces her relative catch (the harvested share of the available stock) at the last few stages while the second player sharply increases her relative catch.

The last feature mentioned in Fig. 4 for the first player under SPE regime does not take place when we consider the cooperative scenario in Fig. 5. Namely, both players employ almost stationary strategies throughout the most part of

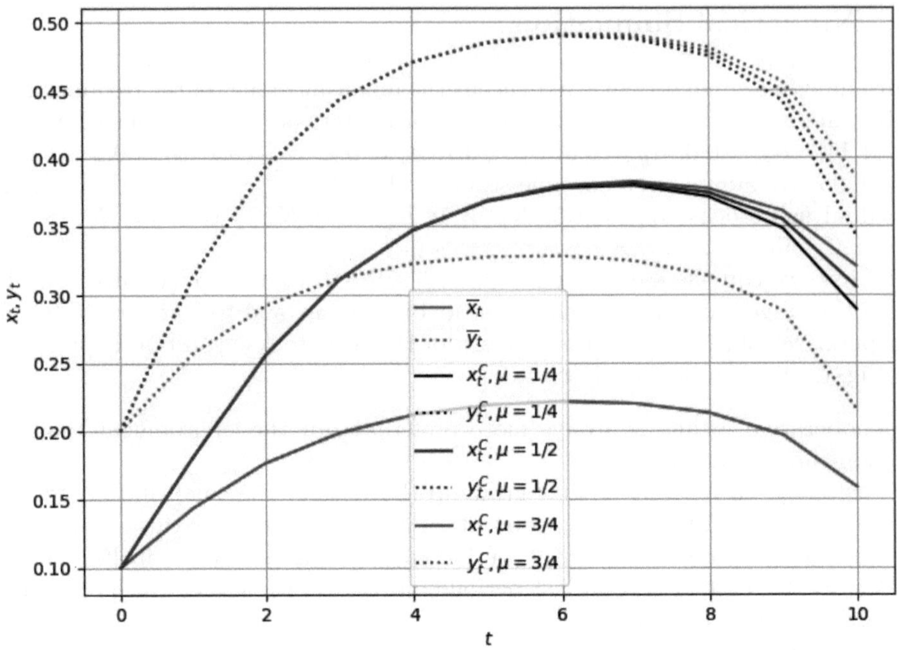

Fig. 1. Stocks evolution (SPE vs Coop), small initial stocks

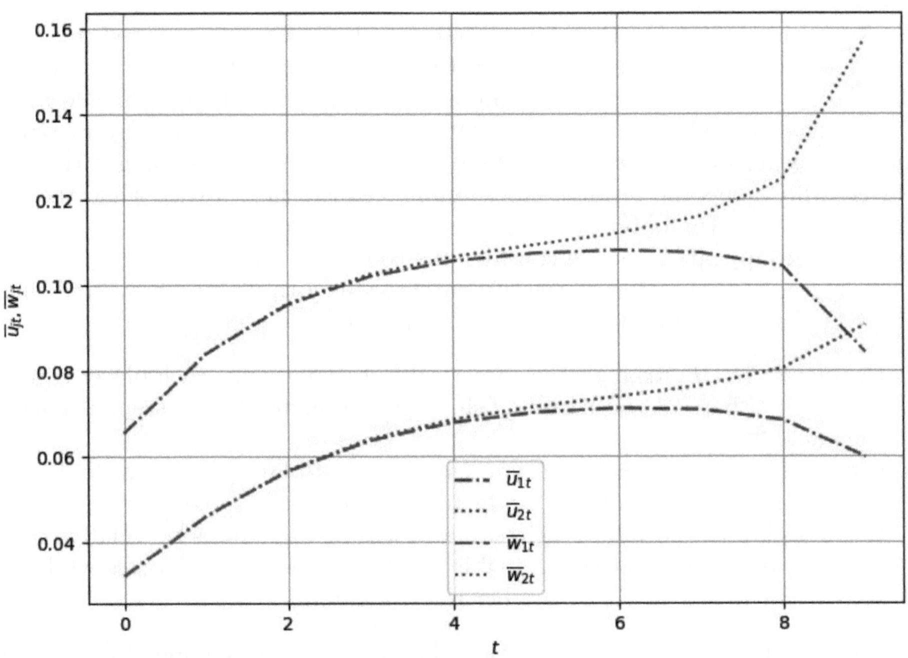

Fig. 2. Non-cooperative strategies

the game duration and then manage to increase their relative catch at the last few stages. Note that this property holds for different values of the relative importance coefficient μ_1.

The distribution of the total stage harvest between the players under non-cooperative scenario is presented in Fig. 6 as the time graph. Each player's share retains at the level approximately equal to 0.5 throughout the most part of the game duration. Then the first player, who values the residual stocks more, needs to reduce her share in favour of the second player. Note that under cooperative regime this drawback disappears and the j-th player's share of the total cooperative harvest is equal μ_j at each stage.

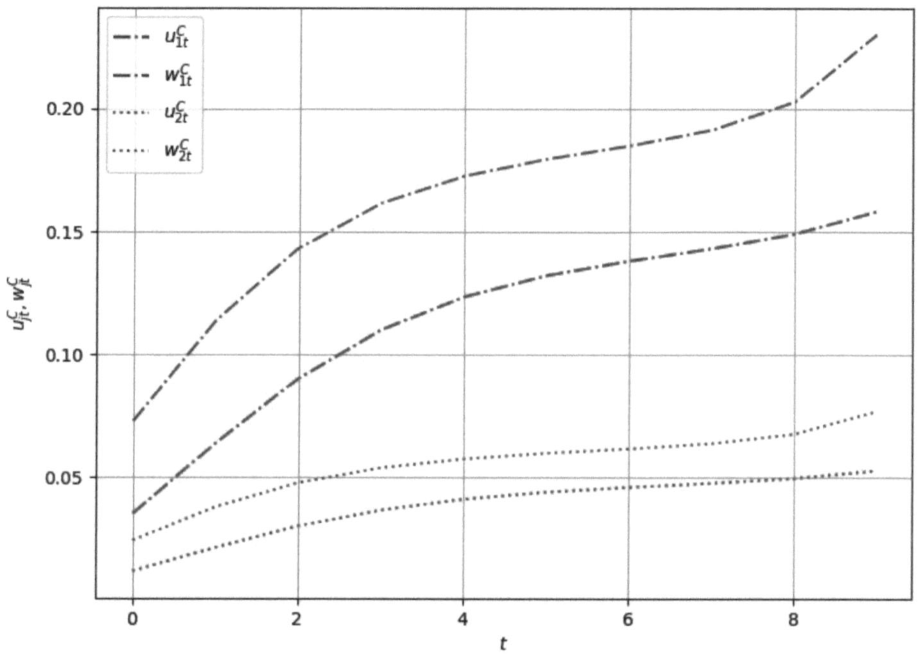

Fig. 3. Cooperative strategies ($\mu_1 = 3/4$)

Now we fix all the parameters values from the baseline scenario except initial stocks and run the model for different values of x_0 and y_0 to verify whether the shape of the time graphs in Fig. 1 crucially depends on the initial stocks. It turns out that initial conditions only affect the behavior of stocks in the first several stages (a typical pattern can be seen in Fig. 7 for higher values of initial stocks and compared with Fig. 1).

Another interesting problem is to evaluate how the level of the players' asymmetry may affect the stocks evolution. The differences $\Delta^x = K_1^x - K_2^x$ and $\Delta^y = K_1^y - K_2^y$ could be considered as two indices of the players' asymmetry (for the baseline scenario $\Delta^x = \Delta^y = 0.5$). Now consider the following vector of

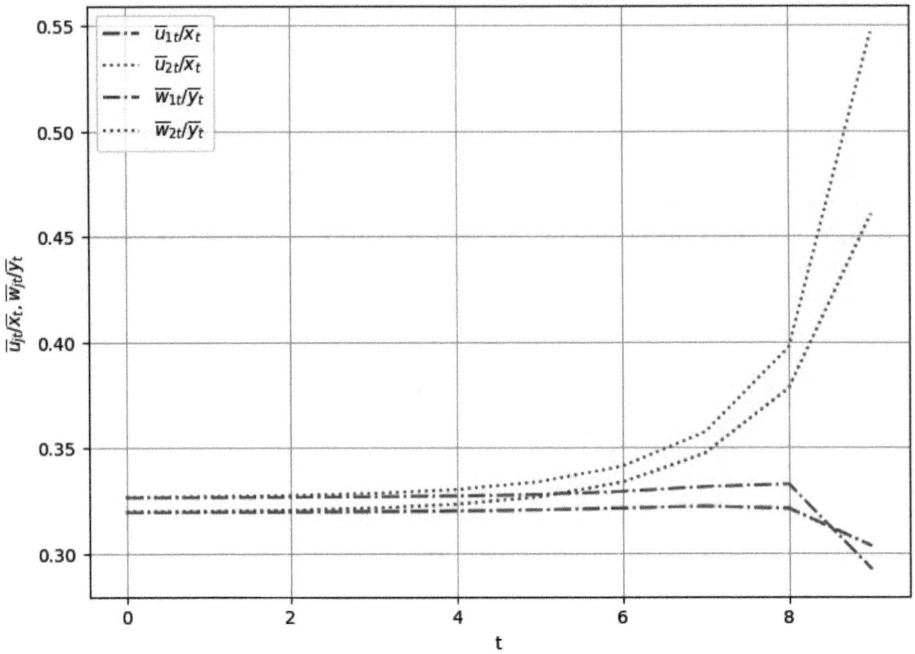

Fig. 4. *SPE* catch as a share of stock

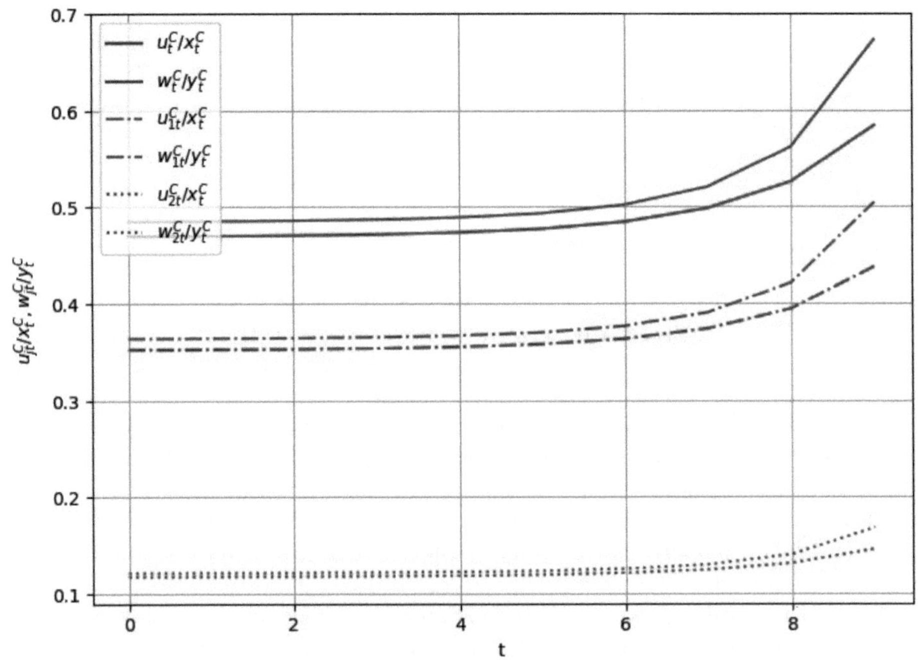

Fig. 5. Cooperative catch as a share of stock ($\mu_1 = 3/4$)

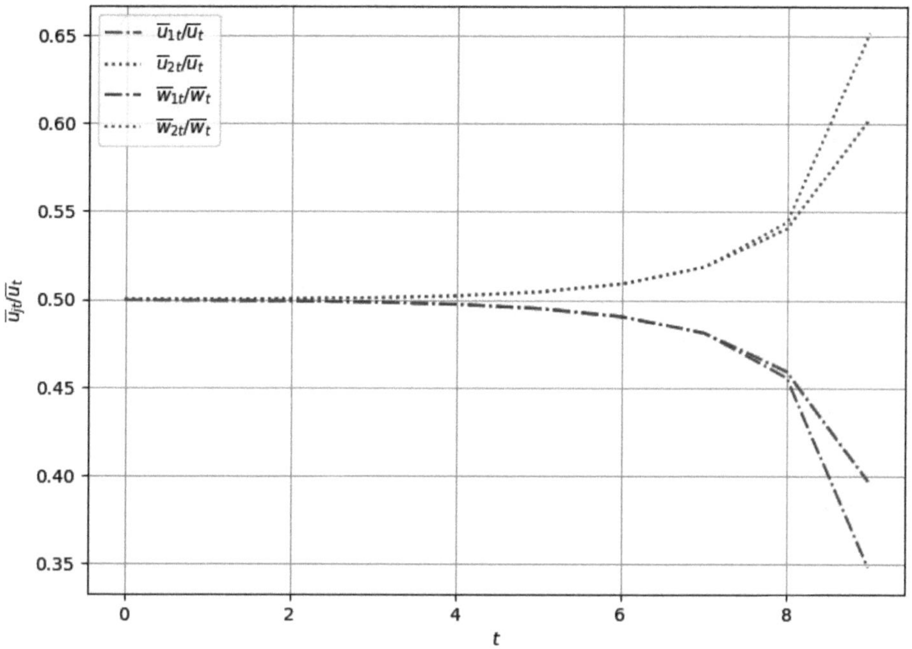

Fig. 6. The j-th player's catch as a share of total SPE harvest

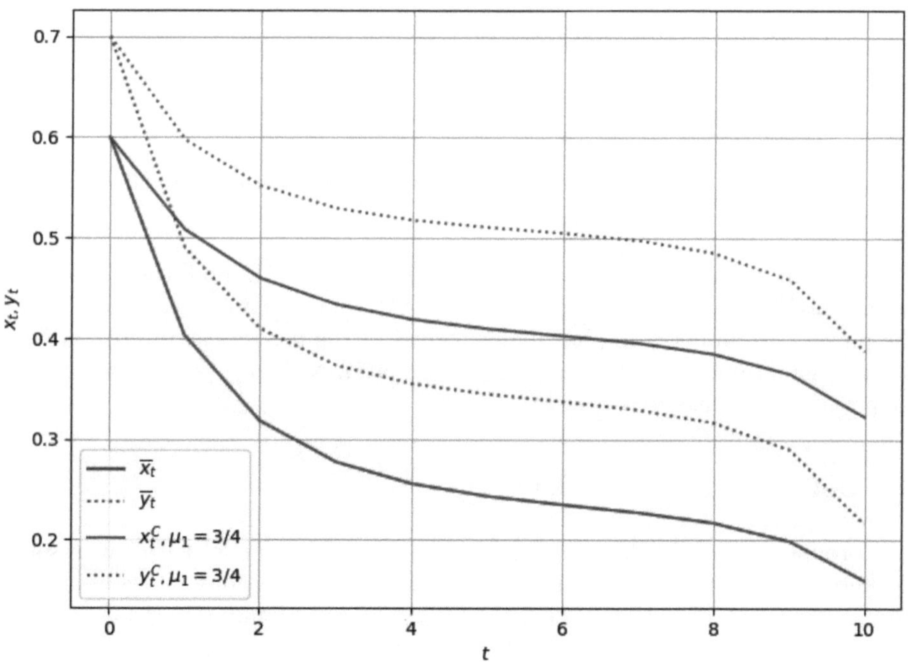

Fig. 7. Stocks evolution, large initial stocks

coefficients $K_1^x = 2$; $K_1^y = 1.25$; $K_2^x = 0.5$; $K_2^y = 0.25$ which differs from that accepted in the baseline scenario and implies the higher values of the asymmetry indices, namely $\Delta^x = 1.5$ and $\Delta^y = 1$ (note that the sum of all four coefficients retains the same). The corresponding fish stocks evolution graphs are presented in Fig. 8.

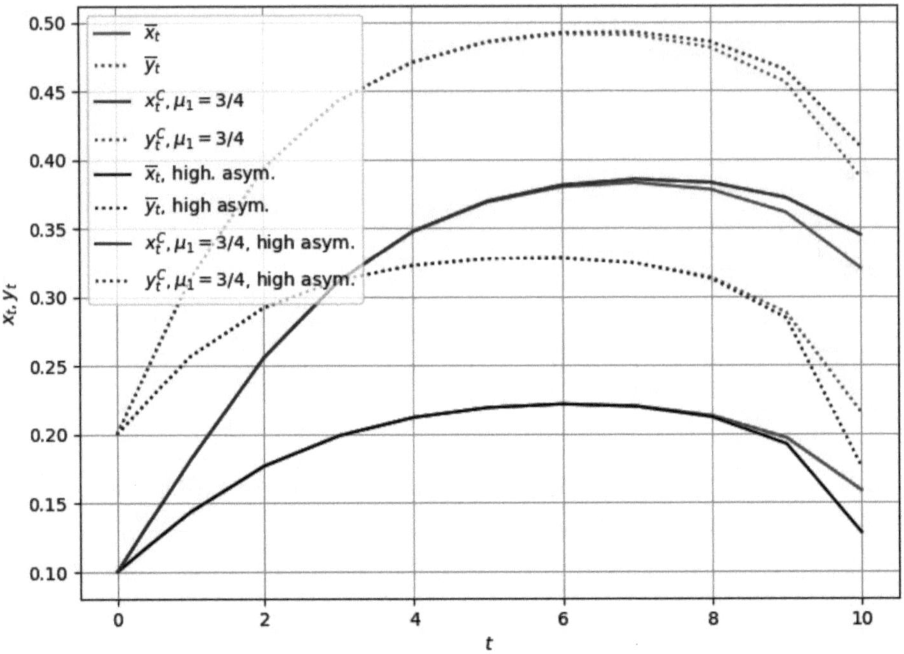

Fig. 8. Baseline scenario vs scenario with higher asymmetry indices

One can notice a valuable difference between SPE and the cooperative regime in Fig. 8 when varying the indices Δ^x and Δ^y: increasing the level of the players' asymmetry implies higher values of the fish stocks in both areas at the last game stages under cooperative scenario, and vice versa – under non-cooperative regime.

6 Concluding Remarks

The proposed spatial fishery management competitive model with two resource exploitation areas with different marine biological conditions allows us to take into account the migration of fish, the finite players' planning horizon as well as the players asymmetry in a sense that players may differently value the residual resource stocks. Although the solutions imply non-stationary feedback strategies, the finite-horizon model retains some advantageous properties of the simpler infinite-horizon model (namely, a cooperative regime always guaranties higher fish stocks, and the players' optimal harvest programs imply that each player catches an approximately constant share of the current fish stock, throughout the most part of the game duration).

Moreover, the model enables us to evaluate the expected levels of the fish stocks in both areas depending on the different biological parameters, length of the planning horizon, different levels of the players' concern about residual stocks, e.t.c. In addition, this model will be useful when considering the fishery management problems with regime shift, for example, the problems which involve an establishment of the marine protected area in the future.

Lastly, it worth noting that the approaches elaborated in the paper could be applied to multistage models of exploitation of renewable resources of different types. For example, multistage models of controversy and mutual influence of various competing religious and ideological movements in a given historical social environment seem to be a promising research direction (see, e.g., [9]).

Acknowledgements. The research of the third author was supported by RSF, grant No. 24-18-00479, https://rscf.ru/project/24-18-00479/, Saint Petersburg State University.

References

1. Breton, M., Keoula, M.Y.: A great fish war model with asymmetric players. Ecol. Econ. **97**, 209–223 (2014). https://doi.org/10.1016/j.ecolecon.2013.11.002
2. Costello, C., Molina, R.: Transboundary marine protected areas. Resource Energy Econ. **65**, 101239 (2021). https://doi.org/10.1016/j.reseneeco.2021.101239
3. Dahmouni, I., Parilina, E.M., Zaccour, G.: Great fish war with moratorium. Math. Biosci. **355**, 108939 (2023). https://doi.org/10.1016/j.mbs.2022.108939
4. Dahmouni, I., Sumaila, R.U.: A dynamic game model for no-take marine reserves. Ecol. Model. **481**, 110360 (2023). https://doi.org/10.1016/j.ecolmodel.2023.110360
5. Fischer, R.D., Mirman, L.J.: Strategic dynamic interaction: fish wars. J. Econ. Dyn. Control **16**(2), 267–287 (1992). https://doi.org/10.1016/0165-1889(92)90034-C
6. Fischer, R.D., Mirman, L.J.: The compleat fish wars: biological and dynamic interactions. J. Environ. Econ. Manag. **30**(1), 34–42 (1996)
7. Haurie, A., Krawczyk, J.B., Zaccour, G.: Games and dynamic games. World Sci. Publishing Company (2012). https://doi.org/10.1142/8442
8. Kuzyutin, D., Skorodumova, Y., Smirnova, N.: A Cooperation Scheme in Multistage Game of Renewable Resource Extraction with Asymmetric Players. In : Pardalos, P., Khachay, M., Mazalov, V. (eds.) Mathematical Optimization Theory and Operations Research, 21st International Conference, MOTOR 2022, vol. 13367, pp. 235–249. Springer (2022)
9. Tantlevskij, I., Kuzyutin, D., Smirnova, N.: A Signed Network Model of the Interaction Between Religious Movements and Authority in Judea. In: Modeling and Simulation of Social-Behavioral Phenomena in Creative Societies, pp. 32–45. Springer Nature Switzerland, Cham. https://doi.org/10.1007/978-3-031-72260-8_3
10. Kuzyutin, D., Smirnova, N.: A dynamic multicriteria game of renewable resource extraction with environmentally concerned players. Econ. Lett. **226**, 111078 (2023). https://doi.org/10.1016/j.econlet.2023.111078

11. Kuzyutin, D., Smirnova, N.: A dynamic bicriteria fish war model with fish bank. Math. Soc. Sci. **140**, 102511 (2026). https://doi.org/10.1016/j.mathsocsci.2026.102511
12. Levhari, D., Mirman, L.J.: The great fish war: an example using a dynamic Cournot-Nash solution. Bell J. Econ. **11**(1), 322–334 (1980)
13. Maestro, M., Pérez-Cayeiro, M.L., Chica-Ruiz, J.A., Reyes, H.: Marine protected areas in the 21st century: Current situation and trends. Ocean Coastal Manage. **171**, 28–36 (2019). https://doi.org/10.1016/j.ocecoaman.2019.01.008
14. Mazalov, V.V., Rettieva, A.N.: Fish wars and cooperation maintenance. Ecol. Model. **221**(12), 1545–1553 (2010). https://doi.org/10.1016/j.ecolmodel.2010.03.011
15. Mazalov, V. V., Rettieva, A. N.: Cooperation maintenance in fishery problems. Fishery Management, 151–198 (2012)
16. Mazalov, V.V., Rettieva, A.N.: Asymmetric dynamic resource management problem with different regimes. Mat. Teor. Igr Pril. **17**(3), 92–110 (2025)
17. Østhagen, A., Jørgensen, A.-K., Moe, A.: The Svalbard Fisheries Protection Zone: How Russia and Norway Manage an Arctic Dispute. Arktika i Sever [Arctic and North] **40**, 183–205 (2020). https://doi.org/10.37482/issn2221-2698.2020.40.183
18. Ougolnitsky, G., Usov, A.: Spatially distributed differential game theoretic model of fisheries. Mathematics **7**, 732 (2019). https://doi.org/10.3390/math7080732
19. Petrosyan, L., Danilov, N.: Stability of the solutions in nonantagonistic differential games with transferable payoffs. Vestnik Leningrad Univ. **1**, 52–59 (1979)
20. Rettieva, A.N.: Dynamic multicriteria games with asymmetric players. J. Global Optim. , 1–17 (2020). https://doi.org/10.1007/s10898-020-00929-5
21. Rettieva, A.N.: Cooperation maintenance in dynamic discrete-time multicriteria games with application to bioresource management problem. J. Comput. Appl. Math. **441**, 115699 (2024). https://doi.org/10.1016/j.cam.2023.115699
22. Selten, R.: Reexamination of the perfectness concept for equilibrium points in extensive games. Int. J. Game Theory **4**, 25–55 (1975). https://doi.org/10.1007/BF01766400
23. Steinsson, S.: The Cod Wars: a re-analysis. Eur. Secur. **25**(2), 256–275 (2016). https://doi.org/10.1080/09662839.2016.1160376
24. U.R. Sumaila: Infinity Fish: Economics and the Future of Fish and Fisheries. Academic Press (2021)
25. Yaragina, N. A.: Biology of Atlantic Cod Reproduction (based on populations in the Barents Sea) [Doct. Diss.] (2006)
26. Yeung, D., Petrosyan, L.: Subgame Consistent Economic Optimization: An Advanced Cooperative Dynamic Game Analysis. Springer, New York (2012)

Last Iterate Convergence of AdaGrad-Norm for Convex Non-smooth Optimization

Margarita Preobrazhenskaia[1], Makar Sidorov[1](✉), Igor Preobrazhenskii[1], and Eduard Gorbunov[2]

[1] Yaroslavl Demidov State University, Sovetskaya str. 14, 150003 Yaroslavl, Russia
makar_sidorov2014@mail.ru

[2] Mohamed Bin Zayed University of Artificial Intelligence, Masdar city 1B, Abu Dhabi, United Arab Emirates

Abstract. We study the convergence of the last iterate (i.e., the $(N+1)$-th iterate) of the AdaGrad method. Although AdaGrad—an adaptive subgradient method—underpins a wide class of algorithms, most existing convergence analyses focus on averaged (or best) iterates. We derive worst-case upper bounds on the suboptimality of the final point and show that, with an optimally tuned stepsize parameter, the last iterate converges at the rate $O(1/N^{1/4})$. We complement this guarantee with matching lower-bound constructions, proving that this rate is tight and that AdaGrad's last-iterate rate is strictly worse than the classical $O(1/N^{1/2})$ rate for its averaged iterate. Technically, our analysis introduces an exponent parameter that captures the growth of the cumulative squared subgradients; combined with the last-iterate inequality of [40], this reduces the problem to bounding a particular series.

Keywords: Convex optimization · Nonsmooth optimization · Subgradient method · Last-iterate convergence · AdaGrad

1 Introduction

First-order optimization methods are popular for solving large-scale optimization problems arising in various applications, including (but not limited to) Machine Learning [1], Signal Processing [14], Image Processing [3], and Optimal Control [2]. The basic example of such a method is a subgradient method [35], and more recent variants including stochastic [28], adaptive [13,21,26], and parameter-free [31] extensions.

The convergence theory of these methods is well developed across a range of settings, including convex non-smooth optimization, which is the main focus of our paper. However, most available results in this area are stated for *averaged* or *best* iterates. By contrast, last-iterate convergence—despite being the default choice in many implementations—has been analyzed only in a small number of

A. Kazakov et al. (Eds.): MOTOR 2026, LNCS 16687, pp. 309–328, 2026.
https://doi.org/10.1007/978-3-032-31390-4_20

works [9,19,22,24,25,34,39,40]. In particular, a tight worst-case characterization for the last iterate of the standard AdaGrad [13,26], which is *one of the most popular first-order adaptive methods*[1], *without stepsize modifications*, has remained elusive.

Contributions. We analyze AdaGrad with the standard scalar stepsizes and a horizon-dependent scaling of the base stepsize. For convex problems with bounded subgradients, we establish a worst-case convergence bound of order $O(1/N^{1/4})$ for the last iterate under an optimal choice of the stepsize parameter (see Theorem 1 and Corollary 2). Our proof takes an asymptotic viewpoint and introduces a key parameter δ that captures the growth exponent of the cumulative squared subgradients over N iterations. Combined with a last-iterate inequality for subgradient methods due to Zamani and Glineur [40], this reduces the analysis to estimating the sum

$$\sum_{k=1}^{N-1} \frac{y_k}{(N-k)(1+\sum_{i=1}^{k} y_i)}$$

for $0 < y_i \leqslant 1$, $1+\sum_{i=1}^{N-1} y_i = N^{2\delta}$, $\delta \in [0, \frac{1}{2}]$. Moreover, we provide matching lower-bound constructions (see Theorem 2), showing that the $O(1/N^{1/4})$ rate *cannot be improved* for the last iterate of standard AdaGrad; in particular, it is strictly slower than $O(1/N^{1/2})$ averaged-iterate rate for AdaGrad [13].

Structure of the Paper. Section 2 contains definitions, assumptions and the problem statement. In Sect. 3, we discuss closely related works. Section 4 contains the formulation and discussion of the main results. Section 5 describes the numerical experiments. Section 6 presents the conclusions and describes further research directions. The detailed proofs of the main results are deferred to Appendices A and B.

2 Preliminaries

Notation. Let $\langle \cdot, \cdot \rangle$ be the standard inner product in $\mathbb{R}^d$, and let $\|\cdot\|$ be the induced Euclidean norm. The subdifferential of a function $f : \mathbb{R}^d \to \mathbb{R}$ at a point x is denoted by $\partial f(x) := \{g \in \mathbb{R}^d : \ \forall y \in \mathbb{R}^d \ f(y) \geqslant f(x) + \langle g, y - x \rangle\}$. We also use $P_X(y)$ to denote the Euclidean projection onto a closed convex set $X \subseteq \mathbb{R}^d$: $P_X(y) := \operatorname{argmin}_{x \in X} \|x - y\|$.

Problem Setup. We consider a convex optimization problem

$$\min_{x \in X} f(x), \tag{1}$$

where $X \subseteq \mathbb{R}^d$ is a closed convex set, $f : X \to \mathbb{R}$ is the objective function. We make the following standard assumptions used in prior works, e.g., see [39,40].

[1] Indeed, AdaGrad serves as a foundation for modern optimal stochastic first-order methods; see, e.g., [20,23,33], and is also a precursor to Adam [21] – the de facto method of choice for training large neural networks.

Assumption 1 (Convexity and Existence of the Minimizer). *Function $f : X \to \mathbb{R}$ is convex and there exists a minimizer $x^* \in \operatorname{argmin}_{x \in X} f(x)$. We assume that the starting point x^1 of the algorithm satisfies for some constant $R > 0$*

$$\|x^1 - x^*\| \leqslant R. \tag{2}$$

Assumption 2 (Bounded subgradients). *Function $f : X \to \mathbb{R}$ has G-bounded subgradients:*

$$\|g\| \leqslant G, \quad \forall g \in \partial f(x), x \in X. \tag{3}$$

Here, $G > 0$ is a fixed constant (Lipschitz constant of f on X).

Considered Method. A standard approach for solving problem (1) is the projected subgradient method, summarized in Algorithm 1.

Algorithm 1 Projected subgradient method with generic stepsizes

---

Require: number of iterations N, sequence of positive stepsizes $\{h_k\}_{1 \leqslant k \leqslant N}$, convex set X, convex function f defined on X, initial iterate $x^1 \in X$

1: **for** $k = 1, \dots, N$ **do**
2: Receive a subgradient $g^k \in \partial f(x^k)$
3: Compute $x^{k+1} = P_X(x^k - h_k g^k)$
4: **end for**

Ensure: last iterate x^{N+1}

AdaGrad[2] can be seen as a special case of Algorithm 1 with

$$h_k = \frac{h}{\sqrt{G^2 + \sum_{i=1}^{k} \|g^i\|^2}} \quad \text{for} \quad k = 1, \dots, N \text{ and some } h > 0. \tag{4}$$

If $h > 0$ is fixed, then the last iterate of AdaGrad is not guaranteed to converge, as we illustrate in the following proposition.

Proposition 1. *There exists problem* (1) *satisfying Assumptions 1 and 2 such that for any $N \geq 1$, $h > 0$ Algorithm 1 with stepsizes defined in* (4) *satisfies:*

$$f(x^{N+1}) - f(x^*) = \frac{Gh}{\sqrt{2}}. \tag{5}$$

[2] The original AdaGrad [13,26] uses coordinate-wise stepsizes, whereas the scalar variant in (4) is often called AdaGrad-Norm [37]. In this paper, we restrict attention to scalar stepsizes and refer to the instance of Algorithm 1 with stepsizes (4) simply as AdaGrad.

Proof. Let $X = \mathbb{R}$ and $f(x) = G|x - x^1|$. Since subgradient g^k can be selected arbitrarily and $\partial f(x^1) = [-G, G]$, consider the following choice: $g^k = 0$ for $k = 1, \dots, N-1$ and $g^N = G$. Then, we have: $x^k = x^1$ for $k = 1, \dots, N$ and

$$x^{N+1} = x^1 - \frac{hG}{\sqrt{G^2 + G^2}} = x^1 - \frac{h}{\sqrt{2}} \Longrightarrow f(x^{N+1}) - f(x^*) = G|x^{N+1} - x^1| = \frac{Gh}{\sqrt{2}}.$$

Proposition 1 suggests that h should either decrease with k (e.g., as in [24]) or depend explicitly on the horizon N. Therefore, we consider the choice $h = R/N^\gamma$ for some $\gamma > 0$. Note that any fixed stepsize h can be written as $h = R/N^\gamma$ for a fixed natural N and suitable $R > 0, \gamma > 0$.

Remark 1 (On G and R). In practice, the constants G and R may be unknown. This, however, is not a major limitation: AdaGrad is known to exhibit robustness with respect to hyperparameter choices; see, e.g., [37]. Since the primary goal of this paper is to obtain a tight characterization of the last-iterate convergence of AdaGrad – rather than to study its robustness – we focus on a version with stepsizes that explicitly depend on G, following prior works [11,13,20,23]. We also note that the constant G in (4) can be replaced by any positive constant $B > 0$. While this substitution does not affect the core arguments of the proof, it leads to more cumbersome expressions in the final bounds. For instance, in (6), the term $GN^{2\delta}$ would be replaced by $B + G(N-1)^{2\delta}$, unnecessarily complicating the presentation without altering the resulting rates or conclusions.

Moreover, our results apply to any choice of h. As discussed above, for any $h > 0$, there exists $\gamma > 0$ such that $h = R/N^\gamma$. Importantly, we do not assume that R is known exactly. If $\gamma \in (0, 1/2]$, we establish both an upper bound (Corollary 2) and a matching lower bound (Theorem 2). In contrast, if $\gamma > 1/2$, our lower bound implies that AdaGrad fails to converge.

3 Related Works

The literature on AdaGrad-type methods and, more broadly, first-order optimization methods is vast. We therefore focus on results most closely related to our setting: convex (possibly non-smooth) minimization of a Lipschitz function via AdaGrad-type methods, and the convergence of the *last iterate* for the first-order methods.

Averaged/Best-Iterate Guarantees for AdaGrad-Type Methods. AdaGrad was introduced in online learning and stochastic optimization [13,26]. The original analyses give regret bounds which, via standard online-to-batch conversion, translate into $\mathcal{O}(1/\sqrt{N})$ guarantees for the best/weighted-average iterate on convex Lipschitz objectives. A large body of work extends these results in different directions (acceleration, parameter-free tuning, and richer preconditioners), see, e.g., [4–8,10,15,16,23,27,30,36]. Since our contribution concerns last-iterate guarantees, we do not survey this line in detail.

Last-Iterate Convergence for (Stochastic) Subgradient Methods. For general convex Lipschitz objectives, [34] proved that projected SGD (and, as a special case, the deterministic projected subgradient method) with stepsizes $h_k \propto 1/\sqrt{k}$ achieves an expected last-iterate accuracy of order $\mathcal{O}(\log(N)/\sqrt{N})$ on bounded domains. [17] showed that this logarithmic factor is unavoidable for this stepsize schedule by constructing matching lower bounds, and also derived high-probability analogues for SGD. To obtain the information-theoretically optimal $\Theta(1/\sqrt{N})$ last-iterate rate, [19] proposed a horizon-dependent stepsize schedule (piecewise-constant, with a "doubling" structure[3]) and proved both in-expectation and high-probability last-iterate bounds under bounded-domain assumptions.

Sharp Deterministic Rates and Impossibility of Anytime Optimal Schedules. [40] developed a refined proof technique for the deterministic projected subgradient method based on tracking the distance to a *time-varying* reference point. Using this approach, they obtained exact worst-case last-iterate bounds (with matching lower bounds including constants) for constant stepsizes and constant step lengths; optimizing the constant stepsize yields a rate of order $\mathcal{O}(\sqrt{\log(N)}/\sqrt{N})$. They further proposed a horizon-dependent "optimal" subgradient method with linearly decaying stepsizes $h_k = R(N+1-k)/G\sqrt{(N+1)^3}$, achieving the minimax rate $GR/\sqrt{N}$ and matching the lower bound of [12]. They also showed that some dependence on the horizon is unavoidable: there is no universal stepsize sequence that achieves the optimal last-iterate bound at *every* iteration. [22] strengthened this impossibility result by proving that any horizon-independent stepsize sequence must incur a polylogarithmic penalty, namely a lower bound of order $\Omega(\log^{1/8}(N)/\sqrt{N})$. Related work also studied other adaptive choices of stepsizes. In particular, [39] analyzed Polyak stepsizes [32]: they proved that the standard Polyak rule $h_k = (f(x^k)-f(x^*))/\|g^k\|^2$ yields a worst-case last-iterate rate $\mathcal{O}(N^{-1/4})$ and constructed instances showing this is tight; they also introduced a modified, horizon-dependent Polyak rule $h_k = (N+1-k)(f(x^k)-f(x^*))/(N+1)\|g^k\|^2$ which recovers the optimal $\mathcal{O}(1/\sqrt{N})$ rate.

From Regret to Last-Iterate and Implications for AdaGrad. A complementary viewpoint is to reduce last-iterate guarantees to online learning regret bounds. [9] established a general regret-to-last-iterate conversion which recovers (and simplifies) the analysis of horizon-dependent linear-decay schedules for subgradient methods, including the optimal schedule of [40]. This framework has recently been used by [24] to obtain last-iterate guarantees in online convex optimization under heavy-tailed feedback for online gradient descent, dual averaging [29,38], and AdaGrad. In particular, their AdaGrad last-iterate bound applies to a *horizon-dependent, linearly decayed* variant of the stepsize (multiplying the

[3] More precisely, [19] propose the following stepsize schedule: $h_k \propto 2^{-i}/\sqrt{N}$, where $k \in (N_i, N_{i+1}]$ and N_i are such that $0 =: N_0 < N_1 < \ldots < N_k = N-1 < N_{k+1} := N$, k is the smallest integer such that $N \cdot 2^{-i} \leq 1$, and $N_i := N - \lceil N \cdot 2^{-i} \rceil$ for $i = 0, 1, \ldots, k$.

standard AdaGrad step by $(N-k)/N)$, whereas our analysis studies the *standard* AdaGrad stepsizes without a linearly decaying stepsize schedule.

Finally, [25] extended and refined the approach of [40] to stochastic gradient methods, providing both in-expectation and high-probability last-iterate bounds under broader assumptions.

4 Main Results

Key Technical Novelty. A key technical ingredient of our analysis is an exponent parameter $\delta \geq 0$ defined as

$$\delta := \frac{1}{2}\log_N\left(1 + \sum_{i=1}^{N-1}\frac{\|g^i\|^2}{G^2}\right) \Longleftrightarrow G^2 + \sum_{i=1}^{N-1}\|g^i\|^2 = G^2 N^{2\delta}. \tag{6}$$

Under Assumption 2, we have $G^2 \leqslant G^2 + \sum_{i=1}^{N-1}\|g^i\|^2 \leqslant G^2 N$, so $\delta \in [0, 1/2]$. Since δ depends on the entire trajectory of the method, it is not known a priori and is used only as an analytical device. This parameterization allows us to derive tight upper bounds for terms that naturally arise in the proof and capture the interaction between the adaptive stepsizes h_k and the subgradient norms $\|g^k\|$.

We now state our main upper bound on $f(x^{N+1}) - f(x^*)$.

Theorem 1. *Let Assumptions 1 and 2 be satisfied. Assume that Algorithm 1 with stepsize* (4) *is run for N iterations and $h = R/N^\gamma$ for some $\gamma \in (0, 1/2]$. Then, the last iterate satisfies*

$$f(x^{N+1}) - f(x^*) \leqslant \frac{GR}{2}\Big(\frac{N^\gamma\sqrt{N^{2\delta}+1}}{2N+1} + \frac{N^{-\gamma}\left(4\log(N^{2\delta})+5\right)\sqrt{N^{2\delta}+1}}{2\max\{N^{2\delta}-1,1\}} + N^{-\gamma-2\delta}\sqrt{N^{2\delta}+1} + N^{-\gamma-\delta}\Big), \tag{7}$$

where constant δ is defined in (6).

Proof. The proof of this theorem is given in Appendix A.

Note that for all natural N,

$$\sqrt{N^{2\delta}+1} \leqslant \sqrt{2}N^\delta, \quad \frac{1}{2N+1} \leqslant \frac{1}{2N}, \quad \frac{1}{\max\{N^{2\delta}-1,1\}} \leqslant 2N^{-2\delta}.$$

Therefore, the following corollary holds.

Corollary 1. *Under the conditions of Theorem 1, the following bound holds:*

$$f(x^{N+1}) - f(x^*) \leqslant C_1 N^{\gamma+\delta-1} + C_2 N^{-\gamma-\delta}\log(N^{2\delta}) + C_3 N^{-\gamma-\delta},$$

where $C_1 = \frac{\sqrt{2}}{4}GR, C_2 = 2\sqrt{2}GR, C_3 = \frac{1}{2}GR(6\sqrt{2}+1)$.

For $0 < \gamma + \delta \leqslant \frac{1}{2}$, the leading term is $C_2 N^{-\gamma-\delta}\log(N^{2\delta})$, whereas for $\frac{1}{2} < \gamma + \delta \leqslant 1$, the dominant term is $C_1 N^{\gamma+\delta-1}$. This yields the following consequence.

Corollary 2. *Under the conditions of Theorem 1, the following bound holds*

$$f(x^{N+1}) - f(x^*) = \begin{cases} O\left(GRN^{-\gamma-\delta}(1+\log(N^{2\delta}))\right), & 0 < \gamma+\delta \leqslant \frac{1}{2}, \\ O\left(GRN^{\gamma+\delta-1}\right), & \frac{1}{2} < \gamma+\delta \leqslant 1. \end{cases}$$

In particular, for any choice of $\gamma \in (0, 1/2]$ there exists problem (1) satisfying Assumptions 1 and 2 such that (6) holds with $\delta = \lfloor \gamma + 3/4 \rfloor / 2$, implying that

$$f(x^{N+1}) - f(x^*) = O\left(\max\left\{\frac{GR}{N^\gamma}, \frac{GR}{N^{\frac{1}{2}-\gamma}}\right\}\right), \tag{8}$$

where we used that $\delta = 0 \implies \log(N^{2\delta}) = 0$ for $\gamma \in (0, 1/4]$. The upper bound above is minimized at $\gamma = 1/4$: in this case, $f(x^{N+1}) - f(x^*) = O(GR/N^{1/4})$.

Even for the best choice of γ, the bound in (8) is worse than the standard $O(GR/N^{1/2})$ guarantee for the averaged iterate of AdaGrad [13]. It is also worse than the $O(GR\sqrt{\log(N)}/N^{1/2})$ bound established for the last iterate of the (non-adaptive) subgradient method with a constant stepsize [40]. These discrepancies naturally raise the question of whether (8) is tight. The next theorem answers this question in the affirmative.

Theorem 2. *For any $\gamma > 0$ and natural N, there exists problem (1) with $d = 1$ for $\gamma \in (0, 1/4]$ and $d = \max\left\{\lceil 16N^{1-2\gamma}\rceil, 1\right\}$ for $\gamma > 1/4$ satisfying Assumptions 1 and 2 such that after N iterations of Algorithm 1 with stepsizes (4) and $h = R/N^\gamma$, $R \geqslant \|x^1 - x^*\|$ the following inequality holds:*

$$f(x^{N+1}) - f(x^*) = \Omega\left(\max\left\{\frac{GR}{N^\gamma}, \min\left\{\frac{GR}{N^{\frac{1}{2}-\gamma}}, GR\right\}\right\}\right). \tag{9}$$

Proof. The proof of this theorem is given in Appendix B.

Combining the upper bound in (8) with the lower bound in (9), we conclude that the optimal worst-case last-iterate rate for AdaGrad with $h = R/N^\gamma$ is $O(GR/N^{1/4})$, attained at $\gamma = 1/4$. We also note that for $\gamma > 1/2$ the lower bound becomes $\Omega(GR)$, which is why we do not consider $\gamma > 1/2$ in Theorem 1.

5 Numerical Experiments

We conduct numerical experiments to assess how tight the bound of Theorem 1 is in practice for the last iterate of AdaGrad. We consider the one-dimensional convex non-smooth objective $f(x) = |x|$, with the initial point $x^1 = 0$. To construct a worst-case instance consistent with the theoretical assumptions, we define the following time-dependent subgradient oracle:

$$g^k = \begin{cases} 0, & \text{if } k < N - m, \\ -1, & \text{if } k > N - m \text{ and } x^k = 0, \\ \text{sign}(x^k), & \text{if } k > N - m \text{ and } x^k \neq 0, \end{cases}$$

where $m = \lceil N^{2\delta} \rceil$. This choice ensures that only the last m iterations contribute nontrivially to the cumulative squared subgradient norm, allowing us to explicitly control the parameter δ. For fixed values of the parameters γ and δ, we run Algorithm 1 for different values of N. For each run, we compute the empirical error $f(x^N) - f(x^*)$ and compare it with the theoretical upper bound from Theorem 1.

In the experiments, we also report the error for the averaged iterate $\bar{x} = \frac{1}{N}\sum_{k=1}^{N} x^k$ of AdaGrad to illustrate its superiority compared to the last iterate. However, the primary goal of the reported experimental results is to highlight the tightness of the bound from Theorem 1 *for the last iterate*, not to benchmark against other convergence measures.

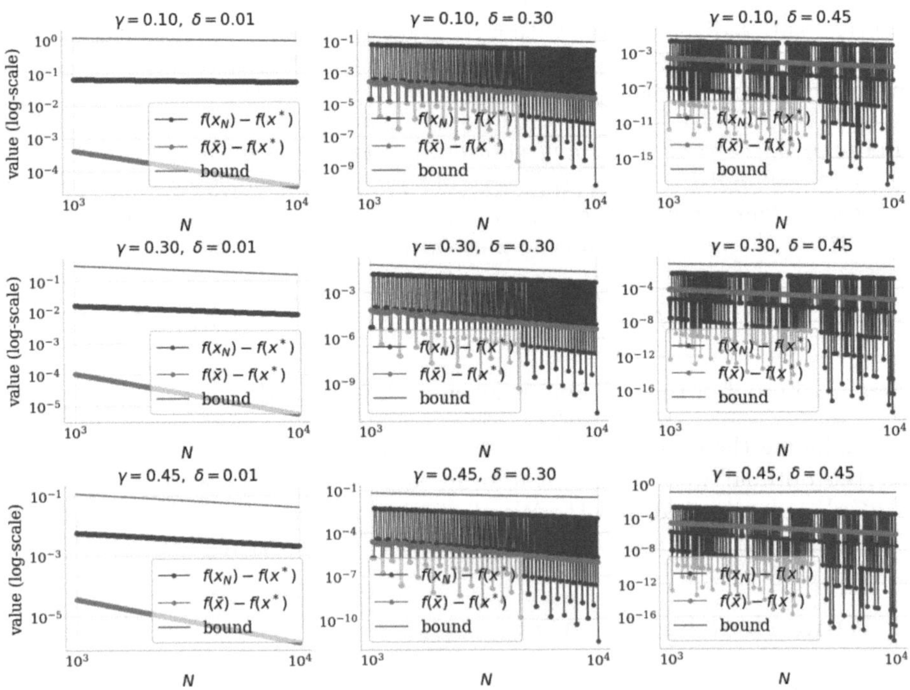

Fig. 1. Empirical last iterate and averaged iterate error and the theoretical bound for AdaGrad with fixed values of δ and different choices of γ. Here, x_N is the last iterate and $\bar{x}$ is the averaged iterate.

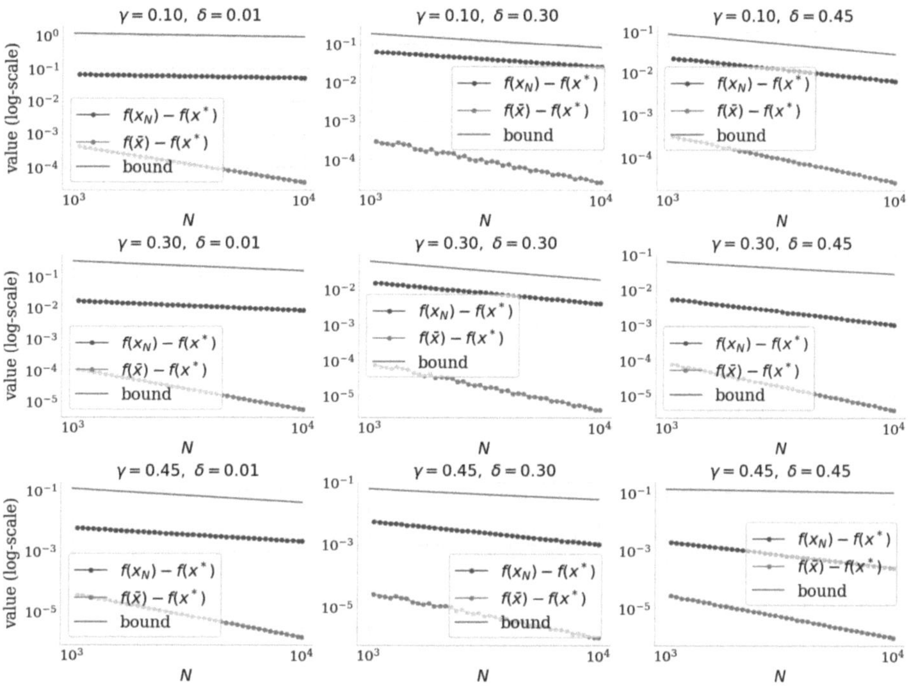

Fig. 2. Empirical last iterate and averaged iterate error and the theoretical bound for AdaGrad with fixed values of δ and different choices of γ with window maximum applied. Here, x_N is the last iterate and $\bar{x}$ is the averaged iterate.

Table 1. Empirical and theoretical slopes for different parameter combinations. Each cell shows the empirical slope k_{emp} (top) and the bound slope k_{bound} (bottom).

δ	γ					
	0.01	0.10	0.20	0.30	0.40	0.49
0.01	-0.0101	-0.1006	-0.2013	-0.3019	-0.4026	-0.4931
	-0.0305	-0.1206	-0.2209	-0.3217	-0.4233	-0.5100
0.10	-0.5692	-0.6606	-0.7620	-0.8635	-0.9650	-1.0564
	-0.1294	-0.2197	-0.3209	-0.4231	-0.5190	-0.5542
0.20	-0.1956	-0.2856	-0.3856	-0.4857	-0.5857	-0.6757
	-0.1913	-0.2824	-0.3850	-0.4831	-0.5250	-0.4347
0.30	-0.3077	-0.3981	-0.4985	-0.5989	-0.6993	-0.7897
	-0.2677	-0.3601	-0.4596	-0.5104	-0.4164	-0.2669
0.40	-0.4135	-0.5045	-0.6057	-0.7069	-0.8081	-0.8991
	-0.3565	-0.4464	-0.5038	-0.4199	-0.2540	-0.1290
0.49	-0.5032	-0.5939	-0.6946	-0.7954	-0.8961	-0.9868
	-0.4388	-0.4991	-0.4392	-0.2740	-0.1302	-0.0261

As illustrated in Fig. 1, the dependence $m = \lceil N^{2\delta} \rceil$ introduces visible discontinuities in the empirical curves. These "jumps" are caused by rounding in the definition of m: the effective value of δ realized in a given run does not exactly coincide with the prescribed one, but only approximates it.

To reduce the visual impact of the discontinuities caused by the rounding of $m = \lceil N^{2\delta} \rceil$, we smooth the empirical curves by applying a windowed maximum: for each consecutive block of iterations we plot the maximum value observed in that block. This operation preserves the worst-case trend of the sequence and allows us to approximate the resulting curve by a linear function and compare the slope in log-log scale. The curves smoothed with the windowed maximum are shown in Figure 2, and the results of this comparison are presented in Table 1.

6 Conclusion

In this paper, we derive the convergence bound $O\left(\max\left\{{}^{GR}/_{N^{\gamma}}, {}^{GR}/_{N^{\frac{1}{2}-\gamma}}\right\}\right)$ for the last iterate of AdaGrad with the stepsize parameter $h = {}^{R}/_{N^{\gamma}}$, $\gamma \in (0, {}^{1}/_{2}]$, in convex optimization with G-bounded subgradients, and we show that this guarantee is tight by providing a matching lower bound. In particular, our results imply that the minimax-optimal convergence rate attained by the last iterate of standard AdaGrad is $O({}^{GR}/_{N^{1/4}})$. Our main technical contribution is the introduction of the parameter δ, which describes the growth of the cumulative squared subgradients. This viewpoint enables tight upper bounds for the sums that naturally arise in the analysis and capture the non-trivial interaction between the adaptive stepsizes h_k and the squared subgradient norms $\|g^k\|^2$. We expect this approach to be useful for analyzing other AdaGrad-like methods.

Our paper also leaves several interesting open problems. In particular, it would be valuable to extend our analysis to the classical coordinate-wise version of AdaGrad and to stochastic settings. Another promising direction is to study last-iterate convergence of AdaGrad for smooth convex objectives. Finally, it would be interesting to investigate whether similar last-iterate phenomena arise for closely related adaptive methods such as RMSProp [18] and Adam [21].

Acknowledgments. The work of Margarita Preobrazhenskaia, Makar Sidorov and Igor Preobrazhenskii was carried out within the framework of a development programme for the Regional Scientific and Educational Mathematical Center of the Yaroslavl State University with financial support from the Ministry of Science and Higher Education of the Russian Federation (Agreement on provision of subsidy from the federal budget No. 075-02-2026-1331).

A Proof of Theorem 1

A.1 Proof Sketch

We sketch the proof of Theorem 1. Our analysis follows an asymptotic viewpoint and builds on the following inequality for the last iterate of subgradient methods, proved by Zamani and Glineur in [40].

Lemma 1. **([40]).** *Let f be a convex function and X be a closed convex set. Let $\hat{x} \in X$, $h_{N+1} > 0$, and*

$$0 \leqslant v_0 \leqslant v_1 \leqslant \ldots \leqslant v_N \leqslant v_{N+1}. \tag{10}$$

Then, if the subgradient method (Algorithm 1) with starting point $x^1 \in X$ generates $\{(x^k, g^k)\}$, then the inequality holds:

$$\sum_{k=1}^{N+1} c_k(f(x^k) - f(\hat{x})) \leqslant \frac{v_0^2}{2}\|x^1 - \hat{x}\|^2 + \frac{1}{2}\sum_{k=1}^{N+1} h_k^2 v_k^2 \|g^k\|^2, \tag{11}$$

$$\text{where} \quad c_k = h_k v_k^2 - (v_k - v_{k-1})\sum_{i=k}^{N+1} h_i v_i, \quad k = 1, \ldots, N+1. \tag{12}$$

We choose the numbers v_k so that all coefficients in (12) vanish except the last one:

$$v_{N+1} = 1, \quad v_N = 1, \quad v_k = \frac{h_{N+1}}{\sum\limits_{i=k+1}^{N+1} h_i v_i} \quad \text{for} \quad k = 0, \ldots, N.$$

Lemma 3 in Appendix A.3 shows that this choice of v_k guarantees $c_{N+1} = h_{N+1}$ and $c_k = 0$ for $k = 1, \ldots, N$. Consequently, for the above v_k, inequality (11) reduces to for chosen v_k, inequality (11) takes the form

$$f(x^{N+1}) - f(\hat{x}) \leqslant \frac{v_0^2}{2h_{N+1}}\|x^1 - \hat{x}\|^2 + \frac{1}{2h_{N+1}}\sum_{k=1}^{N+1} h_k^2 v_k^2 \|g^k\|^2.$$

Next, we separately evaluate the 1-st, N-th, $(N+1)$-th terms on the right-hand side, as well as the sum

$$\sum_{k=1}^{N-1} h_k^2 v_k^2 \|g^k\|^2. \tag{13}$$

Lemma 2 is the key tool for estimating (13). The basic idea is motivated by asymptotic analysis and consists of treating the sum $\sum_{k=1}^{N-1} h_k^2 v_k^2 \|g^k\|^2$ as a quantity of order $N^{2\delta}$. This approach is standard for identifying leading orders of growth. However, in our case, we do not assume that N is large and instead derive estimates that hold for any finite N. Lemma 2 is proved in Appendix A.2.

Appendix A.3 is devoted to technical results. It establishes properties of the sequence v_k, similar to those in [40], and proves Lemma 3, which is used to bound the first and the $(N+1)$-th terms in the sum (13). The final reasoning that completes the proof of Theorem 1 is presented in Appendix A.4.

A.2 Key Lemma

Lemma 2. *Let* y_i, $i = 1, \ldots, N-1$, *be such that* $0 \leqslant y_i \leqslant 1$, *and* $1+\sum_{i=1}^{N-1} y_i = N^{2\delta}$, $\delta \in [0, \frac{1}{2}]$. *Then,*

$$\sum_{k=1}^{N-1} \frac{y_k}{(N-k)(1+\sum_{i=1}^{k} y_i)} \leqslant \frac{4\log(N^{2\delta})+5}{\max\{N^{2\delta}-1, 1\}}. \tag{14}$$

Proof. We introduce the following notation:

$$S_k := 1 + \sum_{i=1}^{k} y_i \quad \text{and} \quad m := \lceil N^{2\delta} \rceil .$$

The latter means that $m-1 < N^{2\delta} \leqslant m$. We consider three possible cases.

Case 1: $m \geqslant 3$. We will show that for $m \geqslant 3$

$$\sum_{k=1}^{N-1} \frac{y_k}{(N-k)S_k} \leqslant \frac{4H_{m-2}+1}{m-1} \leqslant \frac{4\log(N^{2\delta})+5}{N^{2\delta}-1}, \tag{15}$$

where $H_k = 1 + \frac{1}{2} + \cdots + \frac{1}{k}$ is the k-th harmonic number.

We split the indices into "layers" based on the level of partial sums. For $j = 2, \ldots, m-1$, we set

$$\mathcal{I}_j := \{k : j < S_k \leqslant j+1\} \quad \text{and} \quad \mathcal{I}_1 := \{k : 1 \leqslant S_k \leqslant 2\} .$$

At each layer, the following holds: $S_k \geqslant j$ for $k \in \mathcal{I}_j$, and $\sum_{k\in\mathcal{I}_j} y_k \leqslant 2$. Therefore, for $j = 1, \ldots, m-1$

$$\sum_{k\in\mathcal{I}_j} \frac{y_k}{(N-k)S_k} \leqslant \frac{1}{j} \sum_{k\in\mathcal{I}_j} \frac{y_k}{N-k} \leqslant \frac{1}{j} \max_{k\in\mathcal{I}_j} \frac{2}{N-k}. \tag{16}$$

Let $k_j := \max_{k\in\mathcal{I}_j} k$. From $S_{k_j} \leqslant j+1$ and $N^{2\delta} = S_{N-1} \leqslant S_{k_j} + \sum_{i=k_j+1}^{N-1} y_i \leqslant S_{k_j} + N - k_j - 1$, we obtain that for any $k \in \mathcal{I}_j$ and $j = 1, \ldots, m-2$

$$N - k \geqslant N - k_j \geqslant N^{2\delta} + 1 - S_{k_j} \geqslant \lceil N^{2\delta} \rceil - j - 1 = m - j - 1.$$

For $j = m-1$, we have $k_{m-1} := \max_{k\in\mathcal{I}_{m-1}} k = N-1$, i.e., for $k \in \mathcal{I}_{m-1}$ we have $N - k \geq N - k_{m-1} = 1$. Substituting into (16) and summing over layers, we obtain

$$\sum_{k=1}^{N-1} \frac{y_k}{(N-k)S_k} \leqslant \sum_{j=1}^{m-2} \frac{2}{j(m-j-1)} + \frac{1}{m-1} =$$

$$\frac{2}{m-1} \sum_{j=1}^{m-2} \left(\frac{1}{j} + \frac{1}{m-j-1}\right) + \frac{1}{m-1} = \frac{1}{m-1}\left(4\sum_{j=1}^{m-2}\frac{1}{j}+1\right) = \frac{4H_{m-2}+1}{m-1},$$

which yields the first inequality in (15). Next, we use $H_{m-2} < \log(m-2)+1 \leqslant \log\left(N^{2\delta}-1\right)+1$, as well as $m-1 \geqslant N^{2\delta}-1$, yielding the second part of (15).

Case 2: $m = 2$. In this case, we have that $\mathcal{I}_1 = \{1, \dots, N-1\}$ and

$$\sum_{k=1}^{N-1} \frac{y_k}{(N-k)S_k} = \sum_{k \in \mathcal{I}_1} \frac{y_k}{(N-k)S_k} \overset{(16)}{\leqslant} 2 \leqslant \frac{4\log(N^{2\delta}) + 5}{\max\{N^{2\delta} - 1, 1\}},$$

since $\max\{N^{2\delta} - 1, 1\} = 1$ for $m = 2$.

Case 3: $m = 1$. If $m = 1$, then $\delta = 0$ and $y_i = 0$ for all $i = 1, \dots, N-1$. In this case, (14) is satisfied since the left-hand side of (14) equals 0, while the right-hand side of (14) equals 5.

A.3 Technical Results

Lemma 3. *Let*

$$v_{N+1} = 1, \quad v_k = \frac{h_{N+1}}{\sum\limits_{i=k+1}^{N+1} h_i v_i} \quad \textit{for} \quad k = 0, \dots, N. \tag{17}$$

Then $v_N = 1$, *and for* c_k *defined by equality (12), the following is true:*

$$c_{N+1} = h_{N+1}, \quad c_k = 0 \quad \textit{for} \quad k = 1, \dots, N. \tag{18}$$

Proof. The expression for v_k, according to (17), for $k = N$:

$$v_N = \frac{h_{N+1}}{h_{N+1} v_{N+1}} = 1.$$

The expression for c_k, according to (12), for $k = N+1$:

$$c_{N+1} = h_{N+1} v_{N+1}^2 - (v_{N+1} - v_N) h_{N+1} v_{N+1} = v_N h_{N+1} v_{N+1} = h_{N+1}.$$

The expression for c_k, according to (12), with $k = 1, \dots, N$:

$$c_k = h_k v_k^2 - (v_k - v_{k-1}) \sum_{i=k}^{N+1} h_i v_i =$$

$$h_k v_k^2 - v_k (v_k h_k + \sum_{i=k+1}^{N+1} h_i v_i) + v_{k-1} \sum_{i=k}^{N+1} h_i v_i =$$

$$- v_k \sum_{i=k+1}^{N+1} h_i v_i + v_{k-1} \sum_{i=k}^{N+1} h_i v_i = h_{N+1} - h_{N+1} = 0.$$

Remark 2. If $h_i > 0$, then v_k, defined by the equalities (17), are positive and increasing, i.e., (10) holds.

Remark 3. For v_k chosen according to (17), inequality (11) takes the form

$$f(x^{N+1}) - f(\hat{x}) \leqslant \frac{v_0^2}{2h_{N+1}}\|x^1 - \hat{x}\|^2 + \frac{1}{2h_{N+1}} \sum_{k=1}^{N+1} h_k^2 v_k^2 \|g^k\|^2. \tag{19}$$

Lemma 4. *The coefficients v_k, defined in (17), have the following properties.*

1. $\dfrac{1}{v_k} = \dfrac{1}{v_{k+1}} + \dfrac{h_{k+1}v_{k+1}}{h_{N+1}}$ *for* $k = 0, \dots, N-1$.
2. $\dfrac{1}{v_k^2} = 1 + \dfrac{2}{h_{N+1}} \sum_{i=k+1}^{N} h_i + \dfrac{1}{h_{N+1}^2} \sum_{i=k+1}^{N} h_i^2 v_i^2$ *for* $k = 0, \dots, N-1$.

Proof. 1. This part follows from $\dfrac{1}{v_k} = \dfrac{1}{h_{N+1}} \sum_{i=k+1}^{N+1} h_i v_i = \dfrac{1}{h_{N+1}} \sum_{i=k+2}^{N+1} h_i v_i + \dfrac{h_{k+1}v_{k+1}}{h_{N+1}} = \dfrac{1}{v_{k+1}} + \dfrac{h_{k+1}v_{k+1}}{h_{N+1}}$ for $k = 0, \dots, N-1$.

2. From part 1 of this lemma we have $\dfrac{1}{v_k^2} = \left(\dfrac{1}{v_{k+1}} + \dfrac{h_{k+1}v_{k+1}}{h_{N+1}}\right)^2 = \dfrac{1}{v_{k+1}^2} + \dfrac{2h_{k+1}}{h_{N+1}} + \dfrac{h_{k+1}^2 v_{k+1}^2}{h_{N+1}^2} = \dfrac{1}{v_{N+1}^2} + \dfrac{2}{h_{N+1}} \sum_{i=k+1}^{N} h_i + \dfrac{1}{h_{N+1}^2} \sum_{i=k+1}^{N} h_i^2 v_i^2 = 1 + \dfrac{2}{h_{N+1}} \sum_{i=k+1}^{N} h_i + \dfrac{1}{h_{N+1}^2} \sum_{i=k+1}^{N} h_i^2 v_i^2.$

Corollary 3. *For $k = 0, \dots, N-1$, the inequality*

$$v_k^2 \leqslant \frac{1}{2(N-k)+1}. \tag{20}$$

Proof. Using part 2 of Lemma 4 and the fact that $h_j \geqslant h_{N+1}$ for $j = 1, \dots, N+1$, we get

$$\frac{1}{v_k^2} \geqslant 1 + \frac{2}{h_{N+1}} \sum_{i=k+1}^{N} h_i \geqslant 1 + \frac{2}{h_{N+1}} \sum_{i=k+1}^{N} h_{N+1} = 1 + 2(N-k)$$

for $k = 0, \dots, N-1$. Then we have (20).

Remark 4. Note that this is the first time in the proof, when we used some assumptions on sequence h_j (besides $h_j > 0$), i.e., that $h_j \geqslant h_{N+1}$ for $j = 1, \dots, N+1$. Since h_{N+1} is not used to generate x^{N+1}, we can select $h_{N+1} := h_N$ to satisfy all the conditions needed for the proof.

A.4 Proof of Theorem 1

From (19), using $v_{N+1} = v_N = 1$ and (3), we obtain

$$f(x^{N+1}) - f(\hat{x}) \leqslant \frac{R^2 v_0^2}{2h_{N+1}} + \frac{1}{2h_{N+1}} \sum_{k=1}^{N-1} h_k^2 v_k^2 \|g^k\|^2 + \frac{G^2 h_N^2}{2h_{N+1}} + \frac{G^2}{2} h_{N+1}. \quad (21)$$

Due to (4), (6), and Remark 4 ($h_{N+1} := h_N$), we have

$$\frac{RN^{-\gamma}}{G\sqrt{N^{2\delta}+1}} \leqslant h_N \leqslant \frac{R}{G} N^{-\gamma-\delta}, \quad (22)$$

$$\frac{RN^{-\gamma}}{G\sqrt{N^{2\delta}+1}} \leqslant h_{N+1} \leqslant \frac{R}{G} N^{-\gamma-\delta}. \quad (23)$$

To prove (7), we upper bound each term from the right-hand side of inequality (21) separately.

1. Using Corollary 3 for $k = 0$ and inequality (22), we obtain that $v_0^2 \leqslant \frac{1}{2N+1}$ and the first term is bounded as

$$\frac{R^2 v_0^2}{2h_{N+1}} \leqslant \frac{R^2}{2h_{N+1}(2N+1)} \leqslant \frac{R^2}{2(2N+1)} \frac{G\sqrt{N^{2\delta}+1}}{RN^{-\gamma}}$$
$$= \frac{RG}{2} \frac{N^{\gamma}\sqrt{N^{2\delta}+1}}{2N+1}. \quad (24)$$

2. Now we estimate the sum $\frac{1}{2h_{N+1}} \sum_{k=1}^{N-1} h_k^2 v_k^2 \|g^k\|^2$. Using Corollary 3 and formula (4), we have $v_k^2 \leqslant \frac{1}{2(N-k)+1}$ and obtain

$$\frac{1}{2h_{N+1}} \sum_{k=1}^{N-1} h_k^2 v_k^2 \|g^k\|^2 \leqslant \frac{1}{2h_{N+1}} \sum_{k=1}^{N-1} \frac{h_k^2 \|g^k\|^2}{2(N-k)+1} \leqslant \frac{1}{4h_{N+1}} \sum_{k=1}^{N-1} \frac{h_k^2 \|g^k\|^2}{N-k} =$$
$$\frac{h^2}{4h_{N+1}} \sum_{k=1}^{N-1} \frac{\|g^k\|^2}{(N-k)\left(G^2 + \sum_{i=1}^{k} \|g^i\|^2\right)}.$$

The multiplier in front of the sum is upper-bounded as:

$$\frac{h^2}{4h_{N+1}} \overset{(23)}{\leqslant} \frac{R^2 N^{-2\gamma}}{4} \frac{G\sqrt{N^{2\delta}+1}}{RN^{-\gamma}} = \frac{RG}{4} N^{-\gamma} \sqrt{N^{2\delta}+1}.$$

To estimate the sum $\sum_{k=1}^{N-1} \frac{\|g^k\|^2}{(N-k)(G^2+\sum_{i=1}^{k}\|g^i\|^2)}$, we apply Lemma 2 for

$y_k = \dfrac{\|g^k\|^2}{G^2}$, $k = 1, \ldots, N-1$. Then

$$\sum_{k=1}^{N-1} \frac{\|g^k\|^2}{(N-k)\left(G^2 + \sum\limits_{i=1}^{k} \|g^i\|^2\right)} = \sum_{k=1}^{N-1} \frac{\|g^k\|^2/G^2}{(N-k)\left(1 + \sum\limits_{i=1}^{k} \|g^i\|^2/G^2\right)}$$
$$\leqslant \frac{4\log(N^{2\delta}) + 5}{\max\{N^{2\delta} - 1, 1\}}.$$

Hence,

$$\frac{1}{2h_{N+1}} \sum_{k=1}^{N-1} h_k^2 v_k^2 \|g^k\|^2 \leqslant \frac{RG}{4} N^{-\gamma} \frac{\sqrt{N^{2\delta}+1}}{\max\{N^{2\delta}-1, 1\}} (4\log(N^{2\delta}) + 5). \quad (25)$$

3. Taking into account inequalities (22) and (23), for the 3-rd term in the right-hand side of (21) we obtain

$$\frac{G^2 h_N^2}{2h_{N+1}} \leqslant \frac{G^2}{2} \frac{R^2}{G^2} N^{-2\gamma-2\delta} \frac{G\sqrt{N^{2\delta}+1}}{RN^{-\gamma}} = \frac{RG}{2} N^{-\gamma-2\delta} \sqrt{N^{2\delta}+1}. \quad (26)$$

4. Given inequality (23), for the last term in the right-hand side of (21) we have

$$\frac{G^2}{2} h_{N+1} \leqslant \frac{G^2}{2} \frac{R}{G} N^{-\gamma-\delta} = \frac{RG}{2} N^{-\gamma-\delta}. \quad (27)$$

Summing inequalities (24)–(27) and taking $\hat{x} = x^*$, we obtain (7).

B Proof of Theorem 2

Recall from Corollary 2 that

$$f(x^{N+1}) - f(x^*) = O\left(\max\left\{\frac{GR}{N^\gamma}, \frac{GR}{N^{\frac{1}{2}-\gamma}}\right\}\right),$$

We prove that this upper bound is tight by considering two possible situations.

Case 1: $\gamma \in (0, \frac{1}{4}]$. Let $X = \mathbb{R}$ and consider the problem

$$\min_{x\in\mathbb{R}} f(x), \quad f(x) = G|x - x_1|,$$

for given $x^1 \in \mathbb{R}$. Then $x^* = x^1$, and Assumptions 1 and 2 hold. Since g^k may be chosen arbitrarily from $\partial f(x^k)$ and $\partial f(x^1) = [-G, G]$, we choose

$$g^k = 0 \quad \text{for } k = 1, \ldots, N-1, \quad g^N = G.$$

Then $x^k = x^1$ for $k = 1, \ldots, N$, and therefore

$$G^2 + \sum_{i=1}^{N} \|g^i\|^2 = G^2 + G^2 = 2G^2 \Longrightarrow h_N = \frac{h}{\sqrt{2}G}.$$

The last iterate satisfies

$$x^{N+1} = x^1 - h_N G = x^1 - \frac{h}{\sqrt{2}}.$$

Since $h = R/N^\gamma$ for some $R > 0$, we obtain

$$f(x^{N+1}) - f(x^*) = G|x^{N+1} - x^1| = \frac{Gh}{\sqrt{2}} = \frac{GR}{\sqrt{2}N^\gamma} = \Omega\left(\frac{GR}{N^\gamma}\right).$$

Case 2: $\gamma > \frac{1}{4}$. Let $X = \mathbb{R}^d$ and define

$$f(x) = G\|x\|_\infty = G \max_{1 \leq i \leq d} |x_i|. \tag{28}$$

Then $x^* = 0$, and Assumptions 1 and 2 hold. Without loss of generality, we choose the initial point

$$x^1 = \left(\frac{R}{\sqrt{d}}, \ldots, \frac{R}{\sqrt{d}}\right)^\top, \quad \|x^1\| = R.$$

Indeed, for arbitrary initialization x^1, one can always change the coordinates in such a way that x^1 is defined as written above. Note that ∂ f(x) $=$ $G \operatorname{conv}\{e_i \operatorname{sign}(x_i) : i \in I(x)\}$, where e_i is the i-th basis vector and $I(x) = \{i : |x_i| = \|x\|_\infty\}$. At each iteration, we choose

$$g^k = Ge_{i_k} \operatorname{sign}(x_{i_k}^k) \quad \text{for some} \quad i_k \in I(x^k).$$

In this case, only the i_k-th coordinate is updated in iterations k and $\|g^k\|_1 = \|g^k\| = G$ for all $k = 1, \ldots, N$. After N iterations,

$$x^{N+1} = x^1 - \sum_{k=1}^{N} h_k g^k,$$

hence, the total ℓ_1-distance "traveled" by the method is bounded as

$$\|x^{N+1} - x^1\|_1 \leqslant \sum_{k=1}^{N} h_k \|g^k\|_1 = G \sum_{k=1}^{N} h_k.$$

Using

$$h_k = \frac{h}{\sqrt{G^2 + \sum_{i=1}^{k} \|g^i\|^2}} = \frac{h}{G\sqrt{1+k}},$$

we obtain

$$\sum_{k=1}^{N} h_k = \frac{h}{G}\sum_{k=1}^{N}\frac{1}{\sqrt{1+k}} \leqslant \frac{h}{G}\int_1^{N+1}\frac{dx}{\sqrt{x}} \leqslant \frac{2h}{G}\sqrt{N}.$$

Therefore, $\|x^{N+1} - x^1\|_1 \leqslant 2h\sqrt{N} = 2RN^{\frac{1}{2}-\gamma}$. Since $\|x^{N+1} - x^1\|_1 = \sum_{i=1}^{d}|x_i^{N+1} - x_i^1|$, there exists $i \in \{1,\ldots,d\}$ such that

$$|x_i^{N+1} - x_i^1| \leqslant \frac{2RN^{\frac{1}{2}-\gamma}}{d}.$$

Choosing $d = \max\left\{\lceil 16N^{1-2\gamma}\rceil, 1\right\}$, we ensure that

$$|x_i^{N+1} - x_i^1| \leqslant \frac{2RN^{\frac{1}{2}-\gamma}}{d} \leqslant \frac{R}{2\sqrt{d}}.$$

Hence,

$$x_i^{N+1} \geqslant x_i^1 - |x_i^{N+1} - x_i^1| \geqslant \frac{R}{\sqrt{d}} - \frac{R}{2\sqrt{d}} = \frac{R}{2\sqrt{d}}.$$

Therefore, taking into account that $d = \max\left\{\lceil 16N^{1-2\gamma}\rceil, 1\right\}$, we get

$$\begin{aligned} f(x^{N+1}) - f(x^*) = G\|x^{N+1}\|_\infty \geqslant \frac{GR}{2\sqrt{d}} &= \frac{GR}{2\sqrt{\max\left\{\lceil 16\,N^{1--2\gamma}\rceil, 1\right\}}} \\ &= \Omega\left(\min\left\{\frac{GR}{N^{\frac{1}{2}-\gamma}}, GR\right\}\right). \end{aligned}$$

Final bound. Combining Cases 1 and 2, we conclude that for any $\gamma > 0$,

$$f(x^{N+1}) - f(x^*) = \Omega\left(\max\left\{\frac{GR}{N^\gamma}, \min\left\{\frac{GR}{N^{\frac{1}{2}-\gamma}}, GR\right\}\right\}\right).$$

References

1. Bottou, L., Curtis, F.E., Nocedal, J.: Optimization methods for large-scale machine learning. SIAM Rev. **60**(2), 223–311 (2018)
2. Bryson, A.E., Ho, Y.C.: Applied Optimal Control: Optimization, Estimation, and Control. Advanced Textbooks in Control and Signal Processing, Taylor & Francis, New York, NY, USA (1975)
3. Chambolle, A., Pock, T.: A first-order primal-dual algorithm for convex problems with applications to imaging. J. Math. Imaging Vision **40**(1), 120–145 (2011)
4. Cutkosky, A.: Adaptive online learning with varying norms. arXiv preprint arXiv:2002.03963 (2020)
5. Cutkosky, A.: Better full-matrix regret via parameter-free online learning. Adv. Neural. Inf. Process. Syst. **33**, 8836–8846 (2020)
6. Cutkosky, A.: Parameter-free, dynamic, and strongly-adaptive online learning. In: International Conference on Machine Learning, pp. 2250–2259. PMLR (2020)

7. Cutkosky, A., Orabona, F.: Black-box reductions for parameter-free online learning in banach spaces. In: Conference On Learning Theory, pp. 1493–1529. PMLR (2018)
8. Cutkosky, A., Sarlos, T.: Matrix-free preconditioning in online learning. In: International Conference on Machine Learning, pp. 1455–1464. PMLR (2019)
9. Defazio, A., Cutkosky, A., Mehta, H., Mishchenko, K.: Optimal linear decay learning rate schedules and further refinements. arXiv preprint arXiv:2310.07831 (2023)
10. Defazio, A., Jelassi, S.: A momentumized, adaptive, dual averaged gradient method. J. Mach. Learn. Res. **23**(144), 1–34 (2022)
11. Defazio, A., Mishchenko, K.: Learning-rate-free learning by d-adaptation. In: International conference on machine learning, pp. 7449–7479. PMLR (2023)
12. Drori, Y., Teboulle, M.: An optimal variant of Kelley's cutting-plane method. Math. Program. **160**(1), 321–351 (2016)
13. Duchi, J., Hazan, E., Singer, Y.: Adaptive subgradient methods for online learning and stochastic optimization. J. Mach. Learn. Res. **12**, 2121–2159 (2011)
14. Eldar, Y.C., Kutyniok, G.: Compressed sensing: theory and applications. Cambridge university press (2012)
15. Feinberg, V., Chen, X., Sun, Y.J., Anil, R., Hazan, E.: Sketchy: memory-efficient adaptive regularization with frequent directions. Adv. Neural. Inf. Process. Syst. **36**, 75911–75924 (2023)
16. Gupta, Vineet an Koren, T., Singer, Y.: A unified approach to adaptive regularization in online and stochastic optimization. arXiv preprint arXiv:1706.06569 (2017)
17. Harvey, N.J., Liaw, C., Plan, Y., Randhawa, S.: Tight analyses for non-smooth stochastic gradient descent. In: Conference on Learning Theory, pp. 1579–1613. PMLR (2019)
18. Hinton, G., Srivastava, N., Swersky, K.: Neural networks for machine learning lecture 6A overview of mini-batch gradient descent. Cited on **14**(8), 2 (2012)
19. Jain, P., Nagaraj, D.M., Netrapalli, P.: Making the last iterate of SGD information theoretically optimal. SIAM J. Optim. **31**(2), 1108–1130 (2021)
20. Kavis, A., Levy, K.Y., Bach, F., Cevher, V.: Unixgrad: a universal, adaptive algorithm with optimal guarantees for constrained optimization. Adv. Neural Inf. Process. Syst. **32** (2019)
21. Kingma, D.P., Ba, J.: Adam: A method for stochastic optimization. In: International Conference on Learning Representations (2015)
22. Kornowski, G., Shamir, O.: Gradient descent's last iterate is often (slightly) suboptimal. In: OPT 2025: Optimization for Machine Learning (2025)
23. Levy, K.Y., Yurtsever, A., Cevher, V.: Online adaptive methods, universality and acceleration. Adv. Neural Inf. Process. Syst. **31** (2018)
24. Liu, Z.: Online convex optimization with heavy tails: old algorithms, new regrets, and applications. arXiv preprint arXiv:2508.07473 (2025)
25. Liu, Z., Zhou, Z.: Revisiting the last-iterate convergence of stochastic gradient methods. arXiv preprint arXiv:2312.08531 (2023)
26. McMahan, H.B., Streeter, M.: Adaptive bound optimization for online convex optimization. arXiv preprint arXiv:1002.4908 (2010)
27. Mohri, M., Yang, S.: Accelerating online convex optimization via adaptive prediction. In: Gretton, A., Robert, C.C. (eds.) Proceedings of the 19th International Conference on Artificial Intelligence and Statistics. Proceedings of Machine Learning Research, vol. 51, pp. 848–856. PMLR, Cadiz, Spain (2016). https://proceedings.mlr.press/v51/mohri16.html

28. Nemirovski, A., Juditsky, A., Lan, G., Shapiro, A.: Robust stochastic approximation approach to stochastic programming. SIAM J. Optim. **19**(4), 1574–1609 (2009)
29. Nesterov, Y.: Primal-dual subgradient methods for convex problems. Math. Program. **120**(1), 221–259 (2009)
30. Orabona, F., Pál, D.: Scale-free algorithms for online linear optimization. In: International Conference on Algorithmic Learning Theory, pp. 287–301. Springer (2015)
31. Orabona, F., Pál, D.: Training deep networks without learning rates through coin betting. In: Advances in Neural Information Processing Systems (2016)
32. Polyak, B.T.: Minimization of unsmooth functionals. USSR Comput. Math. Math. Phys. **9**(3), 14–29 (1969)
33. Rodomanov, A., Jiang, X., Stich, S.: Universality of adagrad stepsizes for stochastic optimization: inexact oracle, acceleration and variance reduction. Adv. Neural. Inf. Process. Syst. **37**, 26770–26813 (2024)
34. Shamir, O., Zhang, T.: Stochastic gradient descent for non-smooth optimization: Convergence results and optimal averaging schemes. In: International Conference on Machine Learning, pp. 71–79. PMLR (2013)
35. Shor, N.Z.: Minimization Methods for Non-Differentiable Functions. Springer, Berlin, Heidelberg (1985)
36. Wan, Y., Wei, N., Zhang, L.: Efficient adaptive online learning via frequent directions. In: Proceedings of the Twenty-Seventh International Joint Conference on Artificial Intelligence, IJCAI-18, pp. 2748–2754. International Joint Conferences on Artificial Intelligence Organization (2018). https://doi.org/10.24963/ijcai.2018/381
37. Ward, R., Wu, X., Bottou, L.: Adagrad stepsizes: sharp convergence over nonconvex landscapes. J. Mach. Learn. Res. **21**(219), 1–30 (2020)
38. Xiao, L.: Dual averaging method for regularized stochastic learning and online optimization. Adv. Neural Inf. Process. Syst. **22** (2009)
39. Zamani, M., Glineur, F.: Exact convergence rate of the subgradient method by using Polyak step size. arXiv preprint arXiv:2407.15195 (2024)
40. Zamani, M., Glineur, F.: Exact convergence rate of the last iterate in subgradient methods. SIAM J. Optim. **35**(3), 2182–2201 (2025)

Optimal Control

Zonal Control Synthesis with Memory for Rod Heating Process Under Partial Observability

Kamil R. Aida-Zade[1], Samir Z. Guliyev[2(✉)], and Seymur E. Mirzabayov[3]

[1] Azerbaijan University of Architecture and Construction, 11 A. Sultanov Str, AZ1073 Baku, Azerbaijan

[2] Azerbaijan State Oil and Industry University, Baku, Azadliq Avenue 34, AZ1010 Baku, Azerbaijan

azcopal@gmail.com

[3] French-Azerbaijani University (UFAZ), 183 Nizami St, AZ1000 Baku, Azerbaijan

seymour.mirzabayov@ufaz.az

Abstract. This paper proposes a numerical approach for designing feedback controllers in distributed parameter systems, using a rod heating process as a case study. In this setup, the control inputs are the heating power levels of spatially localized heat sources along the rod. The controller uses feedback based on temperature readings at selected sensor locations and employs a zonal control strategy extended to include memory. In the memory-based zonal approach, the control law at time t depends on both the current observed state and a previous state from time $t - \Delta$, , partitioning the augmented state space (current plus recent past) into discrete zones each associated with a fixed control value. Unlike continuously varying feedback laws, this zonal-with-memory strategy updates the control only when the combined state trajectory crosses a zone boundary, resulting in fewer adjustments. This discrete update policy reduces actuator wear and enhances robustness against uncertainties in the initial rod temperature and ambient conditions. The control design problem is transformed into a finite-dimensional optimization over the set of zonal control values. We derive explicit formulas for the gradient of the objective function with respect to these control parameters, enabling efficient gradient-based numerical optimization. A conjugate gradient projection method is used to iteratively tune the zone control values.

Keywords: Zonal Feedback Control · Memory-Dependent Control · Distributed Parameter Systems · Heat Conduction Process · Gradient-Based Optimization · Lumped Sources

1 Introduction

Advances in industrial process control over recent decades have led to the development of sophisticated control technologies and sensing devices [1, 2]. Modern instrumentation provides real-time measurement and computation capabilities, allowing feedback controllers initially developed for lumped-parameter systems to be deployed in a wide range of engineering applications. In parallel, significant effort has been devoted to extending

A. Kazakov et al. (Eds.): MOTOR 2026, LNCS 16687, pp. 331–342, 2026.
https://doi.org/10.1007/978-3-032-31390-4_21

control methods to distributed parameter systems (DPS), which are governed by partial differential equations and have infinitely many degrees of freedom. DPS arise in numerous fields, including heat and mass transfer, chemical processes, structural mechanics, fluid dynamics, and others. Foundational work by Bryson and Ho [3], Lions [4], Lurie [5], Fursikov [6], etc., provided the theoretical underpinnings of DPS control. Despite this progress, designing effective feedback controllers for DPS remains challenging due to the systems' infinite-dimensional nature and the lack of a general design framework.

This work presents a strategy for DPS feedback control based on a special class of control actions termed "zonal controls". In a zonal control paradigm, the space of observable system states is partitioned into a set of predefined regions (zones), and the control input remains constant when the system state lies within any given zone. Unlike conventional feedback laws where the control continuously varies with the state, the zonal approach only permits changes in control values when the system's state trajectory crosses from one zone into another. This structure is especially suitable for systems with measurements available at discrete times. Here, we extend the zonal control paradigm by incorporating memory into the feedback law. Specifically, each control action will be a function of both the current observed state and a recent past state, introducing a dependence on the system's history. This memory-based extension effectively enlarges the state space considered in the controller – from the instantaneous measured state to an augmented state that includes a past (previous) measurement. Incorporating memory improves the robustness of the control system by preventing unnecessary control switching due to noise or momentary deviations. The idea of using not only the current state but also information from a previous moment of time in the synthesis of zonal controls was previously considered in [7]. In that work, the memory-based zonal principle was formulated for control laws depending on the current and preceding states of a lumped-parameter object. The present paper builds on this methodological idea but extends it to a substantially different class of systems, namely distributed-parameter thermal processes observed only through a finite number of temperature sensors. In contrast to the earlier formulation, the current study incorporates spatially localized heat sources, uncertainty in the initial temperature distribution and ambient temperature, a finite-dimensional optimization formulation for the augmented current–previous observation space, and adjoint-based gradient formulas adapted to the rod-heating problem. Thus, the contribution of the present paper is not the abstract idea of memory in zonal control itself, but its development and realization for a partially observable distributed-parameter heating process.

The following sections formulate the rod-heating feedback control problem, detail the memory-dependent zonal control strategy, and derive an explicit expression for the gradient of the objective function, which facilitates efficient gradient-based optimization of the zonal parameters, and present results of numerical experiments.

2 Problem Statement

To demonstrate the proposed approach, we consider a feedback control problem for a one-dimensional thermal process (rod heating). The goal is to regulate the temperature distribution along a rod of length ℓ by adjusting heat input from localized sources. Let

$u(x, t)$ denote the temperature at position $x \in [0, \ell]$ and time $t \geq 0$. The rod's temperature dynamics are governed by the heat conduction equation with convective losses and point heat sources. In particular, $u(x, t)$ satisfies the parabolic partial differential equation:

$$u_t(x,t) = au_{xx}(x,t) + \mu_0[\theta - u(x,t)] + \sum\nolimits_{i=1}^{N} v_i(t)\delta(x - \overline{x}_i),$$
$$\text{for } 0 < x < \ell \text{ and } 0 < t \leq T. \tag{1}$$

subject to boundary conditions:

$$\eta\, u_x(0,t) = \mu_1[\theta - u(0,t)], \quad \eta\, u_x(\ell,t) = \mu_2[u(\ell,t) - \theta]. \tag{2}$$

Here, a is the thermal diffusivity of the rod material; η the thermal conductivity coefficient; μ_1 and μ_2 are heat transfer coefficients at the rod's left and right ends, respectively; $\delta(\cdot)$ the Dirac delta distribution function; $v_i(t)$, $t \in [0, T]$, the control defining the heat power of the source located at $\overline{x}_i$, $i = 1, 2, \ldots, N$; θ the ambient temperature; and μ_0 the effective heat transfer coefficient. The time horizon T may be a fixed constant or, in some cases, treated as a variable to be optimized (with an upper bound $T_{\max}$).

The rod's initial temperature distribution is not known exactly; but it is known to belong to a family of functions characterized by a parameter vector $\gamma \in \Gamma \subset \mathbb{R}^r$:

$$u(x,0) = \varphi(x,\gamma), \quad x \in [0,\ell], \tag{3}$$

where $\varphi(x, \gamma)$ is a known function, Γ denotes the set of admissible parameter values and $\rho_\Gamma(\gamma)$ a probability density function on Γ describing our uncertainty in the initial condition. Similarly, the ambient temperature θ is uncertain, with respect to which it is assumed that $\theta \in \Theta \subset \mathbb{R}$ with known probability density $\rho_\Theta(\theta)$. These probability densities satisfy normalization conditions:

$$\rho_\Gamma(\gamma) \geq 0, \quad \rho_\Theta(\theta) \geq 0, \int\limits_\Gamma \rho_\Gamma(\gamma)\mathrm{d}_\gamma = 1, \int\limits_\Theta \rho_\Theta(\theta)\mathrm{d}_\theta = 1.$$

In particular cases, the initial rod temperature and/or ambient temperature might be effectively constant, which is common in long-running processes where the system reaches a steady regime. We note that formulating the problem with unspecified initial conditions (treated in a probabilistic sense) is aligned with the concept of a system without a perfectly specified initial state, as discussed in the works of Tikhonov [8].

Let $\mathrm{v}(t) = (v_1(t), v_2(t), \ldots, v_N(t))^*$ denote the control vector (with $*$ designating the transposition sign). We seek the control in the class of piecewise continuous functions on $[0, T]$. To satisfy physical and technological constraints, the control values have to be bounded:

$$\mathrm{v}(t) \in V = \left\{v_i : v_i^{\min} \leq v_i \leq v_i^{\max}, i = 1, 2, \ldots, N\right\}, t \in [0, T], \tag{4}$$

where $v_i^{\min}$ and $v_i^{\max}$, $i = 1, 2, \ldots, N$, are given quantities.

Under standard coercivity/boundedness assumptions on the coefficients, for each fixed $\gamma \in \Gamma$ and $\theta \in \Theta$, for each admissible control vector $\mathrm{v}(\cdot)$ with piecewise continuous components, and for initial state $\varphi(\cdot, \gamma) \in L^2(0, \ell)$, problem (1)–(4) admits a unique weak solution $u(x, t)$ such that, for each $t > 0$, $u(\cdot, t)$ is continuous on $[0, \ell]$ and is piecewise smooth with respect to x (in particular, C^2 on each subinterval between neighboring source locations).

We formalize performance using an objective functional that aggregates the final tracking error (deviation of $u(x, T)$ from $\tilde{u}(x)$) and accounts for the uncertainties in γ and θ. One convenient choice is to minimize the expected (density-weighted) squared terminal tracking error, plus a regularization term on the control inputs. Specifically, we define:

$$\mathcal{G}(\mathrm{v}) = \int\limits_{\Theta}\int\limits_{\Gamma} \mathcal{F}(u, T; \mathrm{v}, \theta, \gamma)\rho_\Gamma(\gamma)\rho_\Theta(\theta)\mathrm{d}_\gamma \mathrm{d}_\theta + \varepsilon \sum\nolimits_{i=1}^{N} \int_0^T v_i^2(t)\, \mathrm{d}_t, \tag{5}$$

$$\mathcal{F}(u, T; \mathrm{v}, \theta, \gamma) = \int\limits_0^\ell \omega(x)\left[u(x, T; \mathrm{v}, \theta, \gamma) - \tilde{u}(x)\right]^2 \mathrm{d}_x. \tag{6}$$

where $\omega(x) > 0$ is a weighting function, $u(x, t; \mathrm{v}, \theta, \gamma)$ denotes the solution of (1)–(3) at time t under the control input v and parameters (γ, θ). The expectation is taken over $\gamma \in \Gamma$ and $\theta \in \Theta$ with densities ρ_Γ and ρ_Θ, reflecting that control performance is assessed on average with respect to uncertainty in the initial condition (via γ) and in the ambient temperature (via θ). The second term in $\mathcal{G}(\cdot)$ is a regularization term with the regularization parameter $\varepsilon > 0$.

Assume that the system state is observed only at M sensor locations $\hat{x}_1, \hat{x}_2, \ldots, \hat{x}_M$ along the rod and only at time instants $\tau_k \in [0, T]$, $k = 0, 1, 2, \ldots$, where $\tau_0 = 0$ and $\tau_k = \tau_{k-1} + \Delta$, with $\Delta > 0$. At each such instant τ_k, the rod temperature is measured at the sensor locations, yielding the M-dimensional vector $\xi(\tau_k) = (u(\hat{x}_1, \tau_k), u(\hat{x}_2, \tau_k), \ldots, u(\hat{x}_M, \tau_k))^* \in \mathbb{R}^M$. In contrast to memoryless zonal-control formulations [9, 10], and following the general memory-based zonal-control idea introduced in [7], here the feedback control law is formulated as

$$v_i(t) = \Phi_i(\xi(\tau_k), \xi(\tau_{k-1})),\ t \in \left[\tau_k, \tau_{k+1}\right),\ i = 1, 2, \ldots, N,\ k = 0, 1, 2, \ldots, \tag{7}$$

for a mapping $\Phi_i : \mathbb{R}^{2M} \to \left[v_i^{\min}, v_i^{\max}\right]$, where $\xi(\tau_{k-1})$ is the previously observed state vector and each $\Phi_i(\cdot, \cdot)$ maps the current and last measured states to a permissible control value. The inclusion of the delayed state in (7) is a key modification: it means the control decision at the current time is influenced by the previous state of the system, not just the current state. In summary, our control design problem is to determine a feedback mapping

$$\Phi : \left(\xi_{\text{curr}}, \xi_{\text{prev}}\right) \mapsto (v_1, v_2, \ldots, v_N)$$

such that the resulting closed-loop system (1)–(4), (7) drives the temperature distribution of the rod at the final time T to a desired target temperature $\tilde{u}(x)$.

3 Memory-Based Zonal Feedback Control Strategy

In the feedback design considered here, the control vector is computed from measurements of the system state taken at prescribed spatial points. Rather than using a feedback law that changes continuously with these measurements, we employ a zonal design: the state space is partitioned into predefined regions, and the control input is assigned a constant value on each region. We assume that the temperatures recorded at the observation points lie within a known interval $[u_{\min}, u_{\max}]$. We partition this interval into L subintervals by choosing threshold levels: $u_{\min} = \sigma_0 < \sigma_1 < \cdots < \sigma_L = u_{\max}$. These threshold values divide the possible measurement range into L intervals for each sensor.

In a memoryless zonal control [9], one would define zones in the L-dimensional state space as the Cartesian product of each sensor's interval. That is, a typical zone in the observable state space would be specified by a multi-index $\mathrm{i} = (i_1, i_2, \ldots, i_L)$, where each index $i_j \in \{1, 2, \ldots, L\}$ corresponds to the interval number associated with the j-th sensor. The corresponding zone $Z_{\mathrm{i}} \subset \mathbb{R}^M$ would be defined as

$$Z_{\mathrm{i}} = \left\{\xi \in \mathbb{R}^M \,|\, \xi_1 \in \left[\sigma_{i_1-1}, \sigma_{i_1}\right), \xi_2 \in \left[\sigma_{i_2-1}, \sigma_{i_2}\right), \ldots, \xi_M \in \left[\sigma_{i_M-1}, \sigma_{i_M}\right)\right\}.$$

Thus, the observable state vector $\xi = (\xi_1, \xi_2, \ldots, \xi_M)^*$ belongs to the zone Z_{i} if and only if each of its components lies in the corresponding temperature interval assigned to that zone.

In the memory-based formulation, the controller operates not on the instantaneous measurement alone, but on an augmented state vector that includes both the current and previous (past) observations. At the sampling instant τ_k, we therefore consider the pair $(\xi(\tau_k), \xi(\tau_{k-1}))$ as a single point in the space $\mathbb{R}^{2M}$. To construct zones in this $2M$-dimensional augmented state space, we combine the partitions of the current and previous sensor readings. A memory-dependent zone is indexed by a pair of multi-indices (i, j), where $\mathrm{i} = (i_1, i_2, \ldots, i_M)$, $\mathrm{j} = (j_1, j_2, \ldots, j_M)$, and each $i_s, j_s \in \{1, 2, \ldots, L\}$, $s = 1, 2, \ldots, M$. The corresponding zone $Z_{\mathrm{i,j}} \subset \mathbb{R}^{2M}$ is defined as

$$\begin{aligned} Z_{\mathrm{i,j}} = &\left\{\left(\xi_{\mathrm{curr}}, \xi_{\mathrm{prev}}\right) \in \mathbb{R}^{2M} \,|\, \xi_{\mathrm{curr},s} \in \left[\sigma_{i_s-1}, \sigma_{i_s}\right), \xi_{\mathrm{prev},s} \in \left[\sigma_{j_s-1}, \sigma_{j_s}\right)\right\}, \\ &s = 1, 2, \ldots, M. \end{aligned} \tag{8}$$

Thus, a state belongs to the zone $Z_{\mathrm{i,j}}$ if, for each sensor s, the current measurement lies in the i_s-th temperature interval and the previous measurement lies in the j_s-th temperature interval. Each zone $Z_{\mathrm{i,j}}$ is an $2M$-dimensional hyper-rectangle (a Cartesian product of intervals) representing a combination of the current sensor readings falling into interval i_s at sensor s and the previous readings falling into interval j_s at the same sensor. The total number of such combined zones is L^{2M} if we allow all possible combinations. In practice, not all theoretical combinations may occur or be relevant, depending on how the system evolves and the relationship between current and previous measurements.

In the proposed feedback strategy, we restrict the functional form of $\Phi(\cdot, \cdot)$ in (7) to be piecewise constant over partitioned regions of the augmented state space. So, a fixed control vector is assigned to each zone in the augmented state space. In the case of N heat sources, this control vector has N components. Specifically, for each memory-dependent zone $Z_{\mathrm{i,j}}$, we introduce a constant control vector

$$\mathrm{v}_{\mathrm{i,j}} = \left(v_{1;\mathrm{i,j}}, v_{2;\mathrm{i,j}}, \ldots, v_{N;\mathrm{i,j}}\right)^*,$$

where each component satisfies the bounds (4):

$$v_s^{\min} \le v_{s;\mathrm{i,j}} \le v_s^{\max}, \quad s = 1, 2, \ldots, N.$$

The collection of all vectors $\mathrm{v}_{\mathrm{i,j}}$ over all zones $Z_{\mathrm{i,j}}$ constitutes the set of design parameters of the controller. These parameters are determined by solving the finite-dimensional optimization problem associated with the zonal feedback synthesis. Using these zone-dependent control values, the memory-based feedback law is defined as follows. At each observation time τ_k, the control applied to the s-th heat source is

$$v_s(t) = v_{s;\mathrm{i,j}} = \text{const}, \ \ \text{for}\, t \in \left[\tau_k, \tau_{k+1}\right), \ \ \text{if}\, (\xi(\tau_k), \xi(\tau_{k-1})) \in Z_{\mathrm{i,j}}, s = 1, 2, \ldots, N. \tag{9}$$

In other words, at the moment τ_k the controller determines which memory-dependent zone $Z_{\mathrm{i,j}}$ contains the pair consisting of the current measurement $\xi(\tau_k)$ and the previous measurement $\xi(\tau_{k-1})$, and then applies the corresponding pre-assigned control vector $\mathrm{v}_{\mathrm{i,j}}$. As long as the augmented state $(\xi(\tau_k), \xi(\tau_{k-1}))$ remains in the same zone $Z_{\mathrm{i,j}}$ for successive observations times τ_k, the control values do not change. The control is updated only when the augmented state enters a different zone, that is, at the next sampling instant τ_{k+1}, if $(\xi(\tau_{k+1}), \xi(\tau_k)) \in Z_{\mathrm{p,q}} \neq Z_{\mathrm{i,j}}$, then the control switches to the new zonal value $\mathrm{v}_{\mathrm{p,q}}$. This piecewise-constant feedback law in the augmented (current, past) state space generalizes the standard memoryless zonal control [9, 10], which is recovered when only the current-state index i is used and the previous-state index j is ignored.

Note that the memory-dependent hysteresis smooths the control actions and prevents rapid switching caused by transient disturbances or noise. In other words, the feedback law exhibits a form of persistence: a sustained change in the measured state is required to produce a new control response. As a result, robustness to measurement noise is improved and unnecessary actuator cycling is reduced, leading to better stability and longer hardware lifetime. From a design perspective, the introduction of memory into the zonal structure also simplifies the control synthesis problem. Instead of searching for an arbitrary nonlinear mapping $\Phi(\xi(t), \xi(t-\Delta))$, we only need to determine a finite set of zone-dependent control values $\left\{\mathrm{v}_{\mathrm{i,j}}\right\}$. By exploiting the piecewise-constant structure of the feedback law in the augmented state space, the original feedback synthesis problem is therefore reduced to a finite-dimensional constrained optimization problem for the decision vector $\mathrm{v} \in \mathbb{R}^{N \cdot L^{2M}}$. In summary, the feedback synthesis problem is reduced to determining a finite set of constant control values associated with the memory-dependent zones.

4 Zonal Controls Optimization

The zonal feedback control problem under consideration consists in determining the constant control parameters $\mathrm{v}_{\mathrm{i,j}}$ for each zone that minimize the objective function $\mathcal{G}(\cdot)$. To efficiently solve the resulting finite-dimensional optimization problem, we use projected gradient method [11–13], which is well suited to the simple form of the admissible control set V. This requires explicit formulas for the gradient of the objective functional $\mathcal{G}(\cdot)$ with respect to the components of the control vector v. These formulas are derived via the method of variations, by analyzing how $\mathcal{G}(\cdot)$ changes under small perturbations of the control values. The corresponding result is stated in the following theorem.

Theorem 1. The components of the gradient of the functional in the problem (1)–(3), (5)–(7), in the space of zonal controls (8) and (9), for an arbitrary control v, are determined by the formula:

$$\frac{\partial \mathcal{G}(\mathrm{v})}{\partial v_{s;\mathrm{i,j}}} = \int_{\Theta}\int_{\Gamma} \frac{\partial \mathcal{F}(u, T; \mathrm{v}, \theta, \gamma)}{\partial v_{s;\mathrm{i,j}}} \rho_{\Gamma}(\gamma)\rho_{\Theta}(\theta)\mathrm{d}\gamma\mathrm{d}\theta + 2\varepsilon v_{s;\mathrm{i,j}}, \tag{10}$$

$$\begin{aligned} \frac{\partial \mathcal{F}(u,T;\mathrm{v},\theta,\gamma)}{\partial v_{s;\mathrm{i,j}}} &= \sum_{k:(\xi(\tau_k),\xi(\tau_{k-1}))\in Z_{\mathrm{i,j}}} \int_{\tau_k}^{\tau_{k+1}} \psi(\overline{x}_s, t; \mathrm{v}, \theta, \gamma)\mathrm{d}t, \\ & s = 1, 2, \ldots, N, \quad k \geq 0. \end{aligned} \tag{11}$$

Here, the adjoint function $\psi(x, t) = \psi(x, t; \mathrm{v}, \theta, \gamma)$ is twice continuously differentiable with respect to $x \in (0, \ell)$ and continuously differentiable with respect to t, except at the time instants when the observed state of the process crosses the zone boundaries, at which it remains continuous and satisfies the following boundary-value problem:

$$a\psi_{xx}(x, t) + \psi_t(x, t) - \mu_0\psi(x, t) = 0, (x, t) \in (0, \ell) \times [0, T), \tag{12}$$

$$\eta\psi_x(0, t) + \mu_1\psi(0, t) = 0, \eta\psi_x(\ell, t) - \mu_2\psi(\ell, t) = 0, t \in [0, T], \tag{13}$$

$$\psi(x, T) = -2\omega(x)[u(x, T) - \tilde{u}(x)], \quad x \in [0, \ell]. \tag{14}$$

In general, the sum in (11) may consist of several disjoint time intervals, since the system trajectory can enter and leave the same zone multiple times. Therefore, the integral is taken over the union of all such intervals. This gradient formula follows from a first-order variational analysis of the objective functional. To derive it, we perturb zonal control values $v_{\mathrm{i,j}}$, construct the associated Lagrangian that incorporates the PDE constraint via adjoint variables, and enforce that the first-order variation of $\mathcal{G}(\cdot)$ vanishes at the optimum. By choosing the adjoint equation and its terminal and boundary conditions appropriately, all undesired first-order terms cancel, leaving the expression above as the derivative of $\mathcal{G}(\cdot)$.

Theorem 2. For feedback law defined by (8) and (9), if the termination time $T = T(\gamma, \theta)$ is also subject to optimization, then the derivative of $\mathcal{G}(\cdot)$ with respect to variable time T is given by

$$\frac{\partial \mathcal{G}(\mathrm{v})}{\partial T} = \int_{\Theta}\int_{\Gamma} \frac{\partial \mathcal{F}(u, T; \mathrm{v}, \theta, \gamma)}{\partial T} \rho_{\Gamma}(\gamma)\rho_{\Theta}(\theta)\mathrm{d}\gamma\mathrm{d}\theta, \tag{15}$$

$$\begin{aligned} \frac{\partial \mathcal{F}(u,T;\mathrm{v},\theta,\gamma)}{\partial T} &= -\int_{0}^{\ell} \psi(x, T; \mathrm{v}, \theta, \gamma)\Big[a\frac{\partial^2 u(x,T)}{\partial x^2} + \mu_0[\theta - u(x, T)]\Big]\mathrm{d}x \\ & -\textstyle\sum_{s=1}^{N} \psi(\overline{x}_i, T; \mathrm{v}, \theta, \gamma) \cdot v_{s;T}, \end{aligned} \tag{16}$$

where $\psi(x, t; \mathrm{v}, \theta, \gamma)$ solves the adjoint problem (12)–(14), and $v_{s;T}$ is one of the constant values $v_{s;\mathrm{i,j}}$ the control input v_s assumes according to the last taken measurement of the state $(\xi(T), \xi(T - \Delta))$. A proof can be constructed by following the scheme used in [3] for problems with variable termination time.

For practical implementation of the proposed zonal control synthesis procedure, several key issues related to feedback organization must be taken into account.

Remark 1 – Reachability of Zones
Only those memory-dependent zones $Z_{i,j}$ that can actually be visited by the system should be included in the optimization. If the augmented state $(\xi(t), \xi(t-\Delta))$ never enters a particular zone under any admissible conditions, then the control value associated with that zone has no effect on the system and can be omitted. Therefore, the sampling instants and the temperature partition thresholds must be chosen so that every zone that can influence the system's behavior is detectable in the measurement data. This prevents the inclusion of redundant parameters that cannot be identified from observations.

Remark 2 – Choice of the Number of Zones
The partition of the temperature range, determined by the number of intervals L and the number of sensors M, has a critical influence on performance. A finer partition (larger L or more sensors) provides greater flexibility, since the controller can react differently to a larger number of distinct operating regimes. This usually leads to a smaller value of the objective functional $\mathcal{G}(\cdot)$, because the controller can better compensate for deviations. However, an excessive number of zones makes the controller overly sensitive, increases switching frequency, and may reduce robustness. On the other hand, too few zones lead to a coarse feedback law that may be unable to correct deviations effectively, possibly degrading performance or even causing instability. A practical design strategy is therefore recurrent: start with a moderate number of zones, solve the optimization problem, and then examine the resulting control values. If neighbouring zones yield nearly identical optimal controls, they can be merged. If a single zone covers a wide range and produces a large tracking error, it should be subdivided. This refinement process is repeated until the objective value $\mathcal{G}(\cdot)$ stabilizes.

Remark 3 – Computational Complexity
The number of optimization variables grows rapidly with the dimension of the augmented state. In the memory-based case, zones lie in $\mathbb{R}^{2M}$, and their number is on the order of L^{2M}. Consequently, the total number of decision variables grows exponentially with the number of sensors M and polynomially with the number of intervals L. Each optimization step requires repeated solutions of the forward heat equation and the adjoint equation, possibly for many samples of the uncertain parameters γ and θ. This makes the computation expensive. However, the structure of the objective functional as an expectation over independent scenarios allows for efficient parallelization. Forward simulations, adjoint solves, and gradient evaluations for different parameter samples and different zones can be distributed across multiple processors or threads. In our implementation, this parallel strategy is used to handle different parts of the expectation and gradient summation concurrently, which significantly mitigates the computational burden imposed by the memory-augmented state space. Standard techniques from parallel scientific computing (e.g., Dongarra et al. and Ridgway et al. [14, 15]) are applied to keep the overall computation tractable.

5 Numerical Simulations

To evaluate the effectiveness of the proposed zonal feedback control strategy, a series of numerical simulations has been carried out for the rod-heating model introduced earlier. The objectives of these simulations were to assess how accurately the controller drives the final temperature distribution $u(x, T)$ toward the prescribed target profile $\tilde{u}(x)$ under different configurations. The simulations were performed across a range of controller configurations and discretization settings, including different numbers of temperature intervals (zones), sampling schedules $\{\tau_k\}$ (hence the delay Δ), sensor/source layouts, and uncertainty realizations of (θ, γ). All simulations were performed using a custom software framework that includes a numerical solver for the forward heat-conduction problem (1)–(3), a solver for the adjoint problem (12)–(14) integrated backward in time from $t = T$, and routines for evaluating the objective functional $\mathcal{G}(\cdot)$ and its gradient $\nabla\mathcal{G}(\cdot)$ using the formulas listed earlier.

Let us consider the control problem for the rod heating process described by system (1)–(3), where the following parameter values were selected for the model: $a = 0.78$, $\mu_0 = \mu_1 = \mu_2 = 5$, length of the rod $\ell = 1$. For numerical simulations, the Dirac delta function $\delta(x)$ was approximated using narrow peaked functions, such as the Gaussian and the rectangular pulse,

$$\delta_\in(x) = \frac{1}{\in\sqrt{\pi}} e^{-x^2/\in^2}; \quad \delta_\in(x) = \begin{cases} 1/\in \text{ if } |x| <\in /2, \\ 0 \quad \text{otherwise,} \end{cases}$$

respectively, where $\in << 1$. As the initial state $u(x, 0)$, the function

$$\varphi(x, \gamma) = e^\gamma x^2 (x-1)^2$$

was chosen (arbitrarily), where the parameter γ took values from the finite discrete set $\Gamma = \{1, 2, 3, 4, 5\}$. The ambient temperature θ was taken from the finite discrete set $\Theta = \{25, 26, 27, 28, 29, 30\}$. Under this configuration, the objective functional (5), (6) takes the following form:

$$\mathcal{G}(\mathrm{v}) = \frac{1}{30} \sum_{i=1}^{6} \sum_{j=1}^{5} \int_0 \omega(x) \big[u\big(x, T; \mathrm{v}, \theta_i, \gamma_j\big) - \tilde{u}(x)\big]^2 dx + \varepsilon \mathrm{v}^2_{\mathbb{R}^z},$$

with z being the dimension of the control vector. The state of the system at fixed points along the rod was observed continuously over the time interval [0, 20], i.e., $T = 20$ time units, at discrete time instances τ_k, $k = 0, 1, 2, \ldots$. As the weighting function $\omega(x)$ in the objective functional, the expression $e^{-10(x-0.5)^2}$ was used. This function assigns greater weight to points located near the center of the rod and lesser weight to those near its ends. The regularization parameter ε in $\mathcal{G}(\mathrm{v})$ was varied in the range from 10^{-2} to 10^{-8} during the optimization process. The desired temperature distribution $\tilde{u}(x)$ was taken constant equal to 20.

The values of all control inputs $v_i(t)$, $i = 1, 2, \ldots, N$, for $t \in [0, T]$, were constrained to the range $\left[v^{\min}, v^{\max}\right] = [-25, 25]$. Preliminary calculations showed that the temperature within the rod varied approximately in the range $[u_{\min}, u_{\max}] \approx [0, 35]$.

This interval was divided into a finite number of subintervals using the points $\sigma = \{0, 5, 10, 15, 20, 25, 30, 35\}$, which, based on the temperature readings at specific points along the rod, collectively formed the zones in the state space. For all experiments reported below, the optimization was initialized with a random control vector whose components were drawn independently from the uniform distribution on $\left[v^{\min}, v^{\max}\right]$ using a pseudorandom number generator.

In the first experiment, we placed 4 heat sources at the positions $\overline{x} = (0.2, 0.4, 0.6, 0.8)$ and a single sensor at the center of the rod, i.e., at the point $\hat{x} = 0.5$. The number of zones was 7, resulting in a control vector of dimension 28. As a result of solving the optimization problem, we obtained the average value of the objective functional $\mathcal{G}(\mathrm{v}) \approx 1.19551$. Figure 1 shows the temperature profiles in the rod at the final time T for various values of the parameters γ and θ (depicted as colored dots), along with the desired temperature distribution (shown as a thick black solid line).

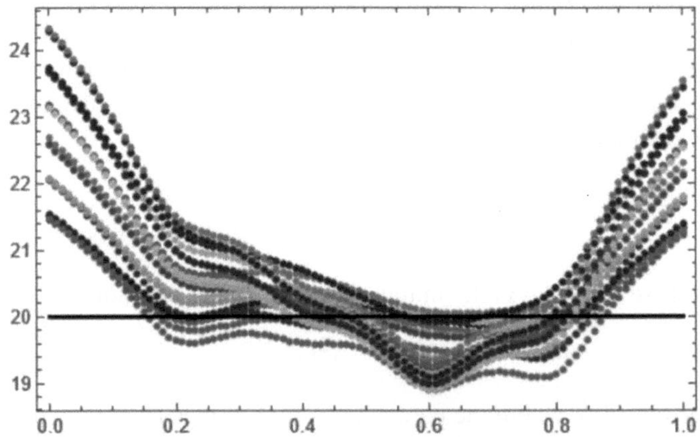

Fig. 1. State trajectories at the final moment along with the desired temperature profile in the 1st experiment.

In the second experiment, we placed 5 heat sources at the positions $\overline{x} = (0.1, 0.3, 0.5, 0.7, 0.9)$ and a single sensor at the point $\hat{x} = 0.5$. The number of zones was 7, resulting in a control vector of dimension 35. As a result of solving the optimization problem, we obtained the average value of the objective functional $\mathcal{G}(\mathrm{v}) \approx 0.430597$. Figure 2 displays the temperature profiles in the rod at the final time T for various values of the parameters γ and θ (represented by colored dots), as well as the desired temperature distribution (depicted by a thick black solid line).

In the third experiment, we placed 5 heat sources at the positions $\overline{x} = (0.1, 0.3, 0.5, 0.7, 0.9)$ and 2 sensors at the points $\hat{x} = (0.35, 0.65)$. The number of zones was 49, resulting in a control vector of size 245. As a result of solving the optimization problem, we obtained the average value of the objective functional $\mathcal{G}(\mathrm{v}) \approx 0.421654$. Repositioning these 2 sensors at the points $\hat{x} = (0.2, 0.8)$ resulted in the average value of the objective functional $\mathcal{G}(\mathrm{v}) \approx 0.425644$. Increasing the number of partitions of the interval $[u_{\min}, u_{\max}]$ in the neighborhood of the desired temperature profile $\tilde{u}(x)$, namely, taking the partitions $\sigma = \{0, 5, 10, 15, 18, 20, 22, 25, 35\}$, and thus increasing the number of zones, allowed us to decrease the average value of the objective functional $\mathcal{G}(\mathrm{v})$ to ≈ 0.386821.

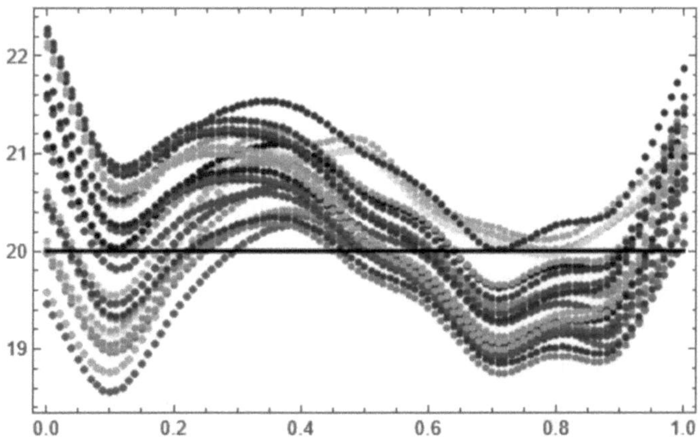

Fig. 2. State trajectories at the final moment along with the desired temperature profile in the 2nd experiment.

6 Conclusion

A feedback control strategy for distributed-parameter systems has been developed by combining zonal control with memory-dependent feedback. In this framework, the observable state space is augmented by delayed measurements and partitioned into a finite set of zones. Each zone is assigned a constant control value, and control updates occur only when the augmented state moves to a new zone. This formulation converts the original infinite-dimensional feedback synthesis problem into a finite-dimensional optimization problem over zonal control parameters. The main advantages of the proposed approach are as follows:

Discrete updates**:** Control signals change only when zone boundaries are crossed, significantly reducing switching frequency and actuator wear.
Hysteresis through memory**:** The delayed state introduces persistence, preventing rapid switching due to noise or transient fluctuations and improving stability.
Efficient optimization**:** Adjoint-based gradient formulas allow fast and reliable tuning of zone control values.
Scalability**:** The structure of the problem supports parallel computation, making large numbers of zones computationally manageable.

The proposed zonal control framework with memory is not limited to the rod-heating problem. It is applicable to a wide class of distributed-parameter systems in which measurements are available at discrete locations and times and where excessive actuator switching is undesirable. By partitioning the augmented state space (including previous measurements) into zones and optimizing the corresponding zone-based control values, one can obtain a feedback law that is both effective, in the sense of accurately achieving control objectives, and robust, in the sense of tolerating uncertainty, modeling errors, and measurement noise.

References

1. Levine, W.S.: The Control Systems Handbook, 2nd edn. CRC Press, Boca Raton, Florida (2011)
2. Franklin, G.F., Powell, J.D., Emami-Naeini, A.: Feedback Control of Dynamic Systems, 8th edn. Pearson, Harlow, UK (2019)
3. Bryson, A.E., Yu-Chi, H.: Applied Optimal Control: Optimization, Estimation and Control. 1st edn. CRC Press, New-York (1975)
4. Lions, J.L.: Optimal control of systems governed by partial differential equations. Springer-Verlag, Berlin (1971)
5. Lurie, K.A.: Applied Optimal Control Theory of Distributed Systems. Springer-Verlag, New York (1993). https://doi.org/10.1007/978-1-4757-9262-1
6. Fursikov, A.V.: Optimal Control of Distributed Systems: Theory and Applications. American Mathematical Society, Providence, Rhode Island (1999)
7. Aida-zade, K.R., Guliyev, S.Z.: Feedback control approach based on classes of zonal control actions. In: Ronzhin, A., Truong, X.-T., Meshcheryakov, R. (eds.) Interactive Collaborative Robotics. ICR 2025. LNCS, vol. 16304, pp. 227–242. Springer, Cham (2026). https://doi.org/10.1007/978-3-032-11903-2_18
8. Tikhonov, A.N., Samarskii, A.A.: Equations of Mathematical Physics. Dover Publications, New York (2011)
9. Guliyev, S.Z.: Synthesis of zonal controls for a problem of heating with delay under nonseparated boundary conditions. Cybern. Syst. Anal. **54**, 110–121 (2018)
10. Aida-Zade, K.R., Kuliev, S.Z.: Hydraulic resistance coefficient identification in pipelines. Autom. Remote. Control. **77**(7), 1225–1239 (2016). https://doi.org/10.1134/S0005117916070092
11. Calamai, P.H., More, J.J.: Projected gradient methods for linearly constrained problems. Math. Program. **39**, 93–116 (1986)
12. Pengjie, L. Jinbao, J., Xianzhen, J.: A New Conjugate Gradient Projection Method for Convex-Constrained Nonlinear Equations. Wiley, Complexity, Hoboken (2020)
13. Nocedal, J., Wright, S.: Numerical Optimization. Springer-Verlag, New-York (2006). https://doi.org/10.1007/978-0-387-40065-5
14. Dongarra, J., et al.: Sourcebook of Parallel Computing. Morgan Kaufmann, San Francisco (2003)
15. Ridgway, S., Clark, T., Bagheri, B.: Scientific Parallel Computing. Princeton University Press, Princeton, New Jersey (2005)

Feedback Control of Boundary Conditions and Measurement Point Movements During Rod Heating

Kamil R. Aida-Zade[1,2] and Vugar A. Hashimov[1,3](✉)

[1] Institute of Mathematics of the Ministry of Science and Education of Republic of Azerbaijan, 68, B. Vahabzadeh, Baku AZ1141, Azerbaijan
kamil_aydzade@rambler.ru

[2] Azerbaijan University of Architecture and Construction, 11, Ayna Sultanova, Baku AZ1073, Azerbaijan

[3] National Aviation Academy of Azerbaijan, 30, Mardakan Ave., Baku AZ1045, Azerbaijan
vugarhashimov@gmail.com
https://www.isi.az, https://www.azmiu.edu.az, https://naa.edu.az

Abstract. Using the example of boundary controls of the rod heating process, the problem of synthesizing control boundary actions in systems with distributed parameters is investigated. Feedback is achieved by measuring the state of the process using moving remote measuring devices (sensors). Linear dependencies of the current values of control actions on the boundary conditions and on the movement of measuring devices on the measured values of the process state are proposed. In the problems, the constant feedback parameters, that is, the coefficients involved in these dependencies, are synthesized. To determine the optimal values of the feedback parameters, formulas for the gradient of the objective functional were obtained, which make it possible to use effective numerical methods of first-order optimization. The results of numerical experiments are presented.

Keywords: rod heating · control of synthesis · moving measuring point · non-local conditions · gradient projection method

1 Introduction

In this article, the problem of optimal synthesis of control actions on the state of a process by acting the boundary conditions and the movements of the devices measuring the state of the process (sensors) is investigated in relation to systems with distributed parameters. As an example, the control of the rod heating process is considered using controlled heat sources placed near the boundaries of the rod and influencing the boundary values. The values of control actions are assigned depending on the measured values of the state (temperature) of the rod by moving remote measuring devices (sensors). It is necessary to determine

A. Kazakov et al. (Eds.): MOTOR 2026, LNCS 16687, pp. 343–357, 2026.
https://doi.org/10.1007/978-3-032-31390-4_22

the optimal values of the velocity of movement of measurement points and the dependencies of the synthesized control actions on the measured values of the state, at which the minimum value of the specified functional that determines the quality of control of the heating process will be achieved.

The problems of synthesis of control of objects with distributed parameters, described by partial differential equations [1–4], in contrast to systems with lumped parameters [5–9], have been studied much less. A significant difference between distributed parameter objects is the fundamental impossibility of quickly and accurately measuring the state of both the object itself and its surrounding environment at all points, both at the initial and current moments of time during the control process. For control problems of objects with distributed parameters, the problem of optimizing the placement of control points on the controlled object has been practically unstudied, both in the case where state measurements can be carried out continuously in time and if measurements are carried out at specified discrete moments.

In this work, using the example of synthesis of boundary controls for the process of heating a rod with feedback, the problem of 1) controlling the movement of a given number of measuring devices is solved; 2) optimizing the parameters of the linear dependence of control on the current values of the measured states. The problem under consideration is reduced to a parametric problem of optimal control of a process described by a parabolic differential control with non-local boundary conditions including non-separated intermediate states.

To solve the obtained problem, it is proposed to use first-order numerical optimization methods. In this regard, formulas for the gradient of the objective functional of the problem were obtained for the feedback parameters involved in the dependence of control actions on the boundary conditions and on the movement of remote measuring devices. The results of numerical experiments obtained by solving test problems are presented and analyzed.

2 Problem Statement

Let us consider the problem of controlling the heating process of a rod of length l by supplying heat to its boundaries [10]:

$$u_t(x,t) = a^2 u_{xx}(x,t) - \lambda_0 \left[u(x,t) - \theta\right], \quad x \in (0,l), \quad t \in (0,T], \tag{1}$$

$$u_x(0,t) = \lambda_1 \left(u(0,t) - \vartheta_1(t)\right), \quad t \in (0,T], \tag{2}$$

$$u_x(l,t) = -\lambda_2 \left(u(l,t) - \vartheta_2(t)\right), \quad t \in (0,T]. \tag{3}$$

Here $u(x,t)$ is the temperature of the rod at point x at time t; θ is the temperature of the external ambient; the given value T determines the duration of the heating process; $\vartheta_1(t)$ and $\vartheta_2(t)$ are the temperatures of the sources, which are controls from the class of piecewise continuous functions and satisfy the technological constraints:

$$\underline{\vartheta_1} \le \vartheta_1(t) \le \overline{\vartheta_1}, \quad \underline{\vartheta_2} \le \vartheta_2(t) \le \overline{\vartheta_2}, \quad t \in [0,T]. \tag{4}$$

The values of the coefficients a^2, λ_0, λ_1, λ_2 and the parameters $\underline{\vartheta_1}$, $\overline{\vartheta_1}$, $\underline{\vartheta_2}$, $\overline{\vartheta_2}$ are given.

It is assumed that the initial temperature of the heated rod is constant along its length, but is not exactly specified, and the set of its possible values is defined: Φ

$$u(x,0) = \varphi(t) = \varphi = \text{const}, \quad x \in [0,l], \tag{5}$$

with the density function $\rho_{\Phi}(\varphi) \geq 0$. The set of possible initial conditions can be finite $\Phi = \{\varphi_1, \varphi_2, \ldots, \varphi_{N_\Phi}\}$, with a given probability of taking the i-th value

$$p_i^{\varphi} = P(\varphi = \varphi_i) \in [0,1], \quad i = 1,2,\ldots,N_\Phi, \quad \sum_{i=1}^{N_\Phi} p_i^{\varphi} = 1.$$

The ambient temperature $\theta(t) = \theta = \text{const}$ can also be specified not exactly, but determined by a set Θ of possible values:

$$\theta \in \Theta, \quad t \in [0,T], \tag{6}$$

with a given density function $\rho_\Theta(\theta)$. It is possible that the set of values of external temperature is discrete: $\Theta = \{\theta_1, \theta_2, \ldots, \theta_{N_\Theta}\}$, with given probability values

$$p_j^{\theta} = P(\theta = \Theta_j) \in [0,1], \quad j = 1,2,\ldots,N_\Theta, \quad \sum_{j=1}^{N_\Theta} p_j^{\theta} = 1.$$

It is required to determine the control action $\vartheta = \vartheta(t) = (\vartheta_1(t), \vartheta_2(t))$, at which the given functional reaches its minimum, which determines the average value of the standard deviation of the state at the final moment of time from the given desired temperature distribution along the rod $U(x)$, $x \in [0,l]$ over all permissible external temperatures θ and initial conditions φ:

$$J(\vartheta) = \int_\Theta \int_\Phi I(\vartheta;\varphi,\theta)\,\rho_\Phi(\varphi)\,\rho_\Theta(\theta)\,d\varphi d\theta, \tag{7}$$

$$I(\vartheta;\varphi,\theta) = \int_0^l \mu(x)\left[u(x,T;\vartheta,\varphi,\theta) - U(x)\right]^2 dx + \varepsilon\|\vartheta(t) - \hat{\vartheta}\|^2_{L_2[0,T]}. \tag{8}$$

Here $u(x,T;\vartheta,\varphi,\theta)$ is the solution of the initial boundary value problems (1)–(3), (5) under an admissible initial condition $u(x,0) = \varphi$ and under an admissible ambient temperature θ; ε, $\hat{\vartheta}$ are the regularization parameters of the functional; $\mu(x) \geq 0$, $x \in [0,l]$ is the given weight function.

If the set of initial conditions Φ and external temperature values Θ are finite, then instead of (7) the functional will be used:

$$J(\vartheta) = \sum_{i=1}^{N_\Phi}\sum_{j=1}^{N_\Theta} I(\vartheta;\varphi_i,\theta_j)\,p_i^{\varphi} p_j^{\theta}. \tag{9}$$

Let there be N moving sensors for measuring the temperature value at the points of the rod to assign the current value of the boundary control temperature. The movements of the sensors are controlled and described by Cauchy problems with respect to nonlinear differential equations:

$$\dot{\xi}_i(t) = f_i(\xi_i(t), \omega_i(t), t), \quad t \in (0, T], \quad i = 1, 2, \dots, N, \tag{10}$$

$$\xi_i(0) = \xi_i^0 \in [0, l], \quad i = 1, 2, \dots, N. \tag{11}$$

Here $\xi_i(t)$ is the position of the ith temperature sensor at the time $t \in [0, T]$; the given functions $f_i(\cdot, \cdot, t)$, $i = 1, 2, \dots, N$ are continuously differentiable with respect to the first two arguments and piecewise continuous with respect to the third argument; $\omega_i(t)$, $i = 1, 2, \dots, N$ are the optimized control actions on the sensor movements; ξ_i^0, $i = 1, 2, \dots, N$ are the given initial positions of the sensors on the rod.

Control actions on sensor movements must satisfy technological constraints:

$$\underline{\omega_i} \le \omega_i(t) \le \overline{\omega_i}, \quad t \in [0, T], \quad i = 1, 2, \dots, N. \tag{12}$$

Each of the sensors during the heating process must remotely take measurements on the entire rod or on its specified parts:

$$0 < \underline{\xi_i} \le \xi_i(t) \le \overline{\xi_i} < l, \quad t \in [0, T], \quad i = 1, 2, \dots, N. \tag{13}$$

Here $\underline{\xi_i}$, $\overline{\xi_i}$ are the given values and, as a rule, the parts of the rod on which different sensors move should not intersect:

$$\left[\underline{\xi_i}, \overline{\xi_i}\right] \cap \left[\underline{\xi_j}, \overline{\xi_j}\right] = \emptyset, \quad \text{at} \quad i \ne j, \quad i, j = 1, 2, \dots, N.$$

In conditions (12), (13) the values $\underline{\omega_i}$, $\overline{\omega_i}$, $\underline{\xi_i}$, $\overline{\xi_i}$ are given.

Let us denote by

$$\hat{u}_i(t) = u(\xi_i(t), t), \quad t \in [0, T], \quad i = 1, 2, \dots, N. \tag{14}$$

continuously measured in time by the ith sensor temperature at point $\xi_i(t)$ at time t, $i = 1, 2, \dots, N$. In the case of measurements at discrete moments of time $\tau_s \in [0, T]$, $s = 1, 2, \dots, N_t$, $\tau_0 = 0$ we denote by:

$$u_i^s = u(\xi_i(\tau_s), \tau_s), \quad s = 1, 2, \dots, N_t, \quad i = 1, 2, \dots, N, \tag{15}$$

the value of the temperature measured by the i-th sensor at the time τ_s at the point $\xi_i(\tau_s)$.

The values of the controlled boundary effects $\vartheta_1(t)$ and $\vartheta_2(t)$ in the control process will be determined based on the results of current measurements in the form of their linear dependence on the state of the process at the controlled points. In the case of continuous feedback (14), we define these dependencies using the formulas:

$$\vartheta_1(t; \mathbf{k}) = \sum_{i=1}^{N} k_i^1 \left[u(\xi_i(t), t) - U(\xi_i(t))\right], \quad t \in [0, T], \tag{16}$$

$$\vartheta_2(t;\mathbf{k}) = \sum_{i=1}^{N} k_i^2 \left[u(\xi_i(t),t) - U(\xi_i(t))\right], \quad t \in [0,T], \tag{17}$$

where the feedback parameters k_i^1 and k_i^2 are the gain factors, $i = 1, 2, \dots, N$.

Similarly, for the dependence of control actions on the movement of sensors on the current temperature values measured by the sensors at the measurement points, we use the formula

$$\omega_i(t;\mathbf{k}) = k_i^3 \left[u(\xi_i(t),t) - U(\xi_i(t))\right], \quad t \in [0,T], \quad i = 1, 2, \dots, N, \tag{18}$$

where k_i^3, $i = 1, 2, \dots, N$ are gain factors.

In (16), (17), (18) and further the following notations are used:

$$k^1 = \left(k_1^1, k_2^1, \dots, k_N^1\right), \quad k^2 = \left(k_1^2, k_2^2, \dots, k_N^2\right), \quad k^3 = \left(k_1^3, k_2^3, \dots, k_N^3\right),$$

$$\mathbf{k} = \left(k^1, k^2, k^3\right) \in \mathbb{R}^{3N}.$$

It is clear that functions (16), (17), (18) are continuous in $t \in [0,T]$ and determine the control actions on the process of heating the rod described by problem (1)–(13).

Let us substitute formulas (16) and (17) into the boundary conditions (2) and (3), respectively.

$$u_x(0,t) = \lambda_1 \left(u(0,t) - \sum_{i=1}^{N} k_i^1 \left[u(\xi_i(t),t) - U(\xi_i(t))\right] \right), \quad t \in (0,T], \tag{19}$$

$$u_x(l,t) = -\lambda_2 \left(u(l,t) - \sum_{i=1}^{N} k_i^2 \left[u(\xi_i(t),t) - U(\xi_i(t))\right] \right), \quad t \in (0,T]. \tag{20}$$

we obtain non-local (loaded) boundary conditions with non-separated intermediate conditions. Substituting formula (18) into the differential Eqs. (10), we obtain:

$$\dot{\xi}_i(t) = f_i\left(\xi_i(t), k_i^3\left[u(\xi_i(t),t) - U(\xi_i(t))\right], t\right), \tag{21}$$
$$t \in (0,T], \quad i = 1, 2, \dots, N.$$

The differential equations (19)–(21) depend on the synthesized constant feedback parameters $\mathbf{k}$.

It is easy to show that the solution of the boundary value problem (1), (19), (20), (5) for arbitrarily given parameters $\mathbf{k}$ is continuously differentiable with respect to t and twice continuously differentiable with respect to $x \in (0,l)$.

In the case of discrete-time measurements (15) of the state at the measurement points $\xi_i(t)$, $i = 1, 2, \dots, N$, the control actions on the boundary conditions and on the movements of the sensors, taking into account the feedback, are defined as follows:

$$\vartheta_1(t;\mathbf{k}) = \sum_{i=1}^{N} k_i^1 \left[u(\xi_i(\tau_s),\tau_s) - U(\xi_i(\tau_s))\right], \tag{22}$$

$$t \in (\tau_s, \tau_{s+1}], \quad s = 0, 1, \ldots, N_t,$$

$$\vartheta_2(t; \mathbf{k}) = \sum_{i=1}^{N} k_i^2 \left[u\left(\xi_i(\tau_s), \tau_s\right) - U\left(\xi_i(\tau_s)\right)\right], \tag{23}$$

$$t \in (\tau_s, \tau_{s+1}], \quad s = 0, 1, \ldots, N_t,$$

$$\omega_i(t; \mathbf{k}) = k_i^3 \left[u\left(\xi_i(\tau_s), \tau_s\right) - U\left(\xi_i(\tau_s)\right)\right], \tag{24}$$

$$t \in [\tau_s, \tau_{s+1}), \quad s = 0, 1, \ldots, N_t, \quad i = 1, 2, \ldots, N.$$

In this case, the synthesized control is a piece-wise constant function, and its parameters k^1, k^2, k^3 have the same meaning as in (16), (17), (18).

The boundary conditions (2), (3) and differential Eq. (10) in the case of control (22), (23) will take the form:

$$u_x(0, t) = \lambda_1 \left(u(0, t) - \sum_{i=1}^{N} k_i^1 \left[u\left(\xi_i(\tau_s), \tau_s\right) - U\left(\xi_i(\tau_s)\right)\right] \right), \tag{25}$$

$$t \in (\tau_s, \tau_{s+1}], \quad s = 0, 1, \ldots, N_t,$$

$$u_x(l, t) = -\lambda_2 \left(u(l, t) - \sum_{i=1}^{N} k_i^2 \left[u\left(\xi_i(\tau_s), \tau_s\right) - U\left(\xi_i(\tau_s)\right)\right] \right), \tag{26}$$

$$t \in (\tau_s, \tau_{s+1}], \quad s = 0, 1, \ldots, N_t,$$

$$\dot{\xi}_i(t) = f_i\left(\xi_i(t), k_i^3 \left[u\left(\xi_i(\tau_s), \tau_s\right) - U\left(\xi_i(\tau_s)\right)\right], t\right), \tag{27}$$

$$t \in (\tau_s, \tau_{s+1}], \quad s = 0, 1, \ldots, N_t, \quad i = 1, 2, \ldots, N.$$

It can be proved that the solution of the initial boundary value problem under consideration $u(x, t)$ under the boundary control actions of the form (25), (26) is twice continuously differentiable with respect to $x \in (0, l)$ and continuously differentiable with respect to t except for the time moments $t = \tau_s$, $s = 1, 2, \ldots, N_t$, at which the solution is continuous.

In this case, the mathematical description of the process (1)

$$u_t(x, t) = a^2 u_{xx}(x, t) - \lambda_0 \left[u(x, t) - \theta\right], \tag{28}$$

$$x \in (0, l), \quad t \in (\tau_s, \tau_{s+1}], \quad s = 0, 1, \ldots, N_t,$$

will be supplemented by the condition of continuity of its state at the moments of measurement time τ_s:

$$u(x, \tau_s) = u\left(x, \tau_s^-\right) = u\left(x, \tau_s^+\right), \quad s = 1, 2, \ldots, N_t, \tag{29}$$

where

$$\tau_s^- = \tau_s - 0, \quad \tau_s^+ = \tau_s + 0.$$

In the case of continuous measurement (14) of the heating process, the minimized functional (7), (8) is written as:

$$J(\mathbf{k}) = \int_{\Theta}\int_{\Phi} I(\mathbf{k};\varphi,\theta)\,\rho_{\Phi}(\varphi)\,\rho_{\Theta}(\theta)\,d\varphi d\theta, \tag{30}$$

$$I(\mathbf{k};\varphi,\theta) = \int_0^l \mu(x)\left[u(x,T;\mathbf{k},\varphi,\theta) - U(x)\right]^2 dx + \varepsilon\|\mathbf{k} - \hat{\mathbf{k}}\|^2_{\mathbb{R}^{3N}}. \tag{31}$$

Here $u(x,T;\mathbf{k},\varphi,\theta)$ is the solution of the initial boundary value problem (1), (19), (20), (5) for arbitrarily given admissible values of the feedback parameters $\mathbf{k}$, the initial condition $u(x,0) = \varphi$ and the ambient temperature θ; $\varepsilon \geq 0$, $\hat{\mathbf{k}} \in \mathbb{R}^{3N}$ are the regularization parameters of the functional.

From dependencies (16), (17) and (18) it is clear that the control $\vartheta_1(t;\mathbf{k})$, $\vartheta_2(t;\mathbf{k})$, $\omega_i(t;\mathbf{k})$ nonlinearly depends on the synthesized parameters, and the most significant is the nonlinearity along the vector $\xi(t) \in \mathbb{R}^N$. Therefore, the functional (30) represents a complex dependence on the optimized parameters, and it is not possible to conduct any research on its convexity. This implies the possibility of multi-extremality of the functional in the space of synthesized control parameters.

The issues of existence and uniqueness of solutions of loaded differential equations with both ordinary and partial derivatives were investigated in such works as [11,12]. In the works [13,14], numerical methods were investigated for solving both Cauchy problems and initial boundary value problems with local and non-local boundary conditions.

The resulting problem (1), (4)–(6), (11), (12), (19)–(21), (30), (31) can be classified as a class of parametric problems of optimal control of a system with distributed parameters [3,6,7]. The main features of the problem are: 1) the nonlinear participation of the optimized parameters in the problem, which indicates the non-convexity of the control problem as a whole, and, consequently, the possible multi-extremality of the functional; 2) the presence of an non-separated boundary conditions, in which the states of the process at the internal points (control) of the rod participate, which causes additional difficulties in solving the corresponding boundary value problems for given values of the control parameters.

3 Obtaining Formulas for the Numerical Solution of the Problem

Let's introduce some transformations and the following notation. The sensor positions must satisfy conditions (13), which we write as follows:

$$g_i(t;\mathbf{k}) = \left|g_i^0(t;\mathbf{k})\right| - \frac{\overline{\xi_i} - \underline{\xi_i}}{2} \leq 0, \quad t \in [0,T], \quad i = 1,2,\ldots,N. \tag{32}$$

$$g_i^0\left(t;\mathbf{k}\right)=\xi_i\left(t\right)-\frac{\overline{\xi_i}+\underline{\xi_i}}{2},$$

Let us assume that the range of possible temperatures on the rod during the heating process is known for all permissible control values

$$\underline{u}\leq u\left(x,t\right)\leq\overline{u},\quad x\in\left[0,l\right],\quad t\in\left[0,T\right],$$

and the desired temperature at the points of the rod satisfies the condition:

$$\underline{U}\leq U\left(x\right)\leq\overline{U},\quad x\in\left[0,l\right],$$

$\underline{U}$, $\overline{U}$ are given.

Then, taking into account these conditions and dependencies (16), (17) and (18), we obtain the following linear constraints on the synthesized feedback parameters:

$$\begin{cases}\underline{\vartheta_1}\leq\sum_{i=1}^N k_i^1[\overline{u}-\overline{U}]\leq\overline{\vartheta_1}, & \underline{\vartheta_1}\leq\sum_{i=1}^N k_i^1[\overline{u}-\underline{U}]\leq\overline{\vartheta_1},\\ \underline{\vartheta_1}\leq\sum_{i=1}^N k_i^1[\underline{u}-\overline{U}]\leq\overline{\vartheta_1}, & \underline{\vartheta_1}\leq\sum_{i=1}^N k_i^1[\underline{u}-\underline{U}]\leq\overline{\vartheta_1},\\ \underline{\vartheta_2}\leq\sum_{i=1}^N k_i^2[\overline{u}-\overline{U}]\leq\overline{\vartheta_2}, & \underline{\vartheta_2}\leq\sum_{i=1}^N k_i^2[\overline{u}-\underline{U}]\leq\overline{\vartheta_2},\\ \underline{\vartheta_2}\leq\sum_{i=1}^N k_i^2[\underline{u}-\overline{U}]\leq\overline{\vartheta_2}, & \underline{\vartheta_2}\leq\sum_{i=1}^N k_i^2[\underline{u}-\underline{U}]\leq\overline{\vartheta_2},\\ \underline{\omega_i}\leq k_i^3[\overline{u}-\overline{U}]\leq\overline{\omega_i}, \quad \underline{\omega_i}\leq k_i^3[\overline{u}-\underline{U}]\leq\overline{\omega_i}, & i=1,2,\ldots,N,\\ \underline{\omega_i}\leq k_i^3[\underline{u}-\overline{U}]\leq\overline{\omega_i}, \quad \underline{\omega_i}\leq k_i^3[\underline{u}-\underline{U}]\leq\overline{\omega_i}, & i=1,2,\ldots,N.\end{cases} \tag{33}$$

To take into account phase constraints (32), we use the external penalty method, reducing the problem to the sequential minimization of the following penalty functional:

$$J_{\mathcal{R}}\left(\mathbf{k}\right)=\int\limits_{\Theta}\int\limits_{\Phi}I_{\mathcal{R}}\left(\mathbf{k};\varphi,\theta\right)\rho_{\Phi}\left(\varphi\right)\rho_{\Theta}\left(\theta\right)d\varphi d\theta, \tag{34}$$

$$I_{\mathcal{R}}\left(\mathbf{k};\varphi,\theta\right)=\int\limits_0^l\mu\left(x\right)\left[u\left(x,T;\mathbf{k},\varphi,\theta\right)-U\left(x\right)\right]^2dx+\varepsilon\|\mathbf{k}-\hat{\mathbf{k}}\|_{\mathbb{R}^{3N}}^2 \tag{35}$$

$$+\mathcal{R}\int\limits_0^T\left[g_i^+\left(t;\mathbf{k}\right)\right]^2dt,$$

where $\mathcal{R}\geq 0$ is a positive penalty coefficient tending to $+\infty$. The notation $g_i^+\left(t;\mathbf{k}\right)$ means that $g_i^+\left(t;\mathbf{k}\right)=g_i\left(t;\mathbf{k}\right)$ if $g_i\left(t;\mathbf{k}\right)>0$ and $g_i\left(t;\mathbf{k}\right)=0$ if $g_i\left(t;\mathbf{k}\right)\leq 0$, $i=1,2,\ldots,N$.

Next, we will use the function $\operatorname{sgn}\left(\cdot\right)$, which is equal to -1 for negative argument values, zero for an argument equal to 0, and 1 for positive argument values.

To minimize the penalty functional (34), (35), taking into account the linearity of the constraints (33), we use the gradient projection method [3]:

$$\mathbf{k}^{n+1} = \mathcal{P}_{(33)}\left[\mathbf{k}^n - \alpha^n \operatorname{grad}_{\mathbf{k}} J_{\mathcal{R}}\left(\mathbf{k}^n\right)\right], \tag{36}$$

$$\alpha^n = \arg\min_{\alpha \geq 0} J_{\mathcal{R}}\left(\mathcal{P}_{(33)}\left[\mathbf{k}^n - \alpha \operatorname{grad}_{\mathbf{k}} J_{\mathcal{R}}\left(\mathbf{k}^n\right)\right]\right), \quad n = 0, 1, \ldots .$$

Here α^n is the one-dimensional minimization step, $\mathbf{k}^0$ is an arbitrary initial search point from $\mathbb{R}^{3N}$; $\mathcal{P}_{(33)}\left[\cdot\right]$ is the operator of projecting an arbitrary point $\mathbf{k} \in \mathbb{R}^{3N}$ onto the feasible region defined by the constraints (33). Given the linearity of the constraints (33), the operator $\mathcal{P}_{(33)}\left[\cdot\right]$ can be easily constructed constructively [3].

In the implementation of the procedure (36), an important role is played by the analytical formulas for the components of the gradient of the objective functional given in the accompanying theorem.

Theorem 1. *Under the conditions on the functions and parameters involved in the problem (1), (4)–(6), (11), (12), (19)–(21), (30), (31), the penalty functional (34), (35) for a given value of the penalty coefficient $\mathcal{R}$ is differentiable with respect to the feedback parameters $\mathbf{k}$, and the components of its gradient are determined by the formulas:*

$$\frac{\partial J_{\mathcal{R}}\left(\mathbf{k}\right)}{\partial k_i^1} = \int\limits_{\Theta}\int\limits_{\Phi}\left\{-\lambda_1 a^2 \int\limits_0^T \psi\left(0,t\right) u\left(\xi_i\left(t\right),t\right) - U\left(\xi_i\left(t\right)\right) dt\right\} \tag{37}$$

$$\times\, \rho_{\Phi}\left(\varphi\right) \rho_{\Theta}\left(\theta\right) d\varphi d\theta,$$

$$\frac{\partial J_{\mathcal{R}}\left(\mathbf{k}\right)}{\partial k_i^2} = \int\limits_{\Theta}\int\limits_{\Phi}\left\{-\lambda_2 a^2 \int\limits_0^T \psi\left(l,t\right) u\left(\xi_i\left(t\right),t\right) - U\left(\xi_i\left(t\right)\right) dt\right\} \tag{38}$$

$$\times\, \rho_{\Phi}\left(\varphi\right) \rho_{\Theta}\left(\theta\right) d\varphi d\theta,$$

$$\frac{\partial J_{\mathcal{R}}\left(\mathbf{k}\right)}{\partial k_i^3} = \int\limits_{\Theta}\int\limits_{\Phi}\left\{\int\limits_0^T \varphi_i\left(t\right) \frac{\partial f_i\left(\xi_i\left(t\right),\omega_i\left(t\right),t\right)}{\partial \omega_i} u\left(\xi_i\left(t\right),t\right) - U\left(\xi_i\left(t\right)\right) dt\right\} \tag{39}$$

$$\times\, \rho_{\Phi}\left(\varphi\right) \rho_{\Theta}\left(\theta\right) d\varphi d\theta,$$

where $i = 1, 2, \ldots, N$. The function $\psi\left(x,t\right) = \psi\left(x,t;\mathbf{k},\varphi,\theta\right)$, continuously differentiable with respect to $t \in \left[0,T\right]$ and twice continuously differentiable $x \in \left(0,l\right)$ except for the points $\xi_i\left(t\right) \in \left(0,l\right)$, $t \in \left(0,T\right)$, $i = 1, 2, \ldots, N$, is a solution to the following initial-boundary value problem:

$$\psi_t\left(x,t\right) = -a^2 \psi_{xx}\left(x,t\right) + \lambda_0 \psi\left(x,t\right), \tag{40}$$

$$x \in \left(\xi_i\left(t\right), \xi_{i+1}\left(t\right)\right), \quad i = 1, 2 \ldots, N, \quad t \in \left[0,T\right),$$

$$\psi(x,T) = -2\mu(x)(u(x,T) - U(x)), \quad x \in [0,l], \tag{41}$$

$$\psi_x(0,t) = \lambda_1 \psi(0,t), \quad t \in [0,T), \tag{42}$$

$$\psi_x(l,t) = -\lambda_2 \psi(l,t), \quad t \in [0,T). \tag{43}$$

At the measurement points $\xi_i(t) \in (0,l)$, $i = 1,2,\ldots,N$ the following conditions are met:

$$\psi_x\left(\xi_i\left(t^+\right),t^+\right) - \psi_x\left(\xi_i\left(t^-\right),t^-\right) \tag{44}$$

$$= -\left\{\lambda_1 a^2 k_i^1 \psi(0,t) + \lambda_2 a^2 k_i^2 \psi(l,t) + k_i^3 \frac{\partial f_i(\xi_i(t),\omega_i(t),t)}{\partial \omega_i} \varphi_i(t)\right\},$$

$$i = 1,2,\ldots,N,$$

$$\psi\left(\xi_i\left(t^+\right),t^+\right) = \psi\left(\xi_i\left(t^-\right),t^-\right), \quad i = 1,2,\ldots,N, \tag{45}$$

continuously differentiable functions $\varphi_i(t)$ are solutions of Cauchy problems

$$\dot{\varphi}_i(t) = -\frac{\partial f_i(\xi_i(t),\omega_i(t),t)}{\partial \xi_i} \varphi_i(t) \tag{46}$$

$$-\left\{\lambda_1 a^2 k_i^1 \psi(0,t) + \lambda_2 a^2 k_i^2 \psi(l,t) + k_i^3 \frac{\partial f_i(\xi_i(t),\omega_i(t),t)}{\partial \omega_i}\right\}$$

$$\times \left[u_x(\xi_i(t),t) - U_x(\xi_i(t))\right] + 2\mathcal{R} sgn\left(g_i^0(t;\mathbf{k})\right) g_i^+(t;\mathbf{k}),$$

$$t \in [0,T), \quad i = 1,2,\ldots,N,$$

$$\varphi_i(T) = 0, \quad i = 1,2,\ldots,N. \tag{47}$$

To prove the theorem, one can use the methodology of works [3,15,16], based on the assessment of the increment of the objective functional due to the increment of the optimized parameters.

Theorem 2. *For the problem of synthesis of control of the process (15), (28), (29) of heating a rod with time discrete feedback of the form (22)–(24), the formulas for the components of the gradient of the functional $J_{\mathcal{R}}(\mathbf{k})$ for the current values of the synthesized parameters $\mathbf{k} \in \mathbb{R}^{3N}$ have the form:*

$$\frac{\partial J_{\mathcal{R}}(\mathbf{k})}{\partial k_i^1} = \int_{\Theta}\int_{\Phi}\left\{-\lambda_1 a^2 \sum_{s=0}^{N_t} \int_{\tau_s}^{\tau_{s+1}} \psi(0,t)\, u(\xi_i(\tau_s),\tau_s) - U(\xi_i(\tau_s))\, dt\right\} \tag{48}$$

$$\times \rho_{\Phi}(\varphi)\, \rho_{\Theta}(\theta)\, d\varphi d\theta,$$

$$\frac{\partial J_{\mathcal{R}}(\mathbf{k})}{\partial k_i^2} = \int_{\Theta}\int_{\Phi}\left\{-\lambda_2 a^2 \sum_{s=0}^{N_t} \int_{\tau_s}^{\tau_{s+1}} \psi(l,t)\, u(\xi_i(\tau_s),\tau_s) - U(\xi_i(\tau_s))\, dt\right\} \tag{49}$$

$$\times \rho_{\Phi}(\varphi)\, \rho_{\Theta}(\theta)\, d\varphi d\theta,$$

$$\frac{\partial J_{\mathcal{R}}(\mathbf{k})}{\partial k_i^3} = \int\limits_{\Theta}\int\limits_{\Phi}\Big\{\sum_{s=0}^{N_t}\int\limits_{\tau_s}^{\tau_{s+1}}\varphi_i(t)\frac{\partial f_i(\xi_i(t),\omega_i(t),t)}{\partial \omega_i}u(\xi_i(\tau_s),\tau_s)-U(\xi_i(\tau_s))\,dt\Big\} \tag{50}$$
$$\times\rho_\Phi(\varphi)\rho_\Theta(\theta)\,d\varphi d\theta,$$

and the function $\psi(x,t)=\psi(x,t;\mathbf{k},\varphi,\theta)$, which is continuously differentiable with respect to $t\in(0,T)$ except for the time moments τ_s, $s=1,2,\dots,N_t$ and twice continuously differentiable with respect to $x\in(0,l)$ except for the points $\xi_i(t)\in(0,l)$, $t\in(0,T)$, $i=1,2,\dots,N$, is a solution to the following adjoint initial boundary value problem:

$$\psi_t(x,t)=-a^2\psi_{xx}(x,t)+\lambda_0\psi(x,t), \tag{51}$$
$$x\in(\xi_i(t),\xi_{i+1}(t)),\quad i=1,2\dots,N,\quad t\in[\tau_s,\tau_{s+1}),\quad s=1,2,\dots,N_t,$$
$$\psi(x,T)=-2\mu(x)(u(x,T)-U(x)),\quad x\in[0,l], \tag{52}$$
$$\psi_x(0,t)=\lambda_1\psi(0,t),\quad t\in[\tau_s,\tau_{s+1}),\quad s=1,2,\dots,N_t, \tag{53}$$
$$\psi_x(l,t)=-\lambda_2\psi(l,t),\quad t\in[\tau_s,\tau_{s+1}),\quad s=1,2,\dots,N_t. \tag{54}$$

At the moments of time τ_s at the measurement points $\xi_i(\tau_s)\in(0,l)$, $s=1,2,\dots,N_t$, $i=1,2,\dots,N$ the following conditions are met:

$$\psi_x(\xi_i(\tau_s^+),\tau_s^+)-\psi_x(\xi_i(\tau_s^-),\tau_s^-) \tag{55}$$
$$=-\int\limits_{\tau_s}^{\tau_{s+1}}\lambda_1 a^2k_i^1\psi(0,t)+\lambda_2a^2k_i^2\psi(l,t)+k_i^3\frac{\partial f_i(\xi_i(t),\omega_i(t),t)}{\partial\omega_i}\varphi_i(t)\,dt,$$
$$s=1,2,\dots,N_t,\quad i=1,2,\dots,N,$$
$$\psi(\xi_i(\tau_s^+),\tau_s^+)=\psi(\xi_i(\tau_s^-),\tau_s^-),\quad s=1,2,\dots,N_t,\quad i=1,2,\dots,N, \tag{56}$$

continuously differentiable functions $\varphi_i(t)$ are solutions of Cauchy problems

$$\dot\varphi_i(t)=-\frac{\partial f_i(\xi_i(t),\omega_i(t),t)}{\partial\xi_i}\varphi_i(t) \tag{57}$$
$$-\int\limits_{\tau_s}^{\tau_{s+1}}\lambda_1a^2k_i^1\psi(0,t)+\lambda_2a^2k_i^2\psi(l,t)+k_i^3\frac{\partial f_i(\xi_i(t),\omega_i(t),t)}{\partial\omega_i}dt$$
$$\times[u_x(\xi_i(\tau_s),\tau_s)-U(\xi_i(\tau_s))]+2\mathcal{R}sgn\left(g_i^0(t;\mathbf{k})\right)g_i^+(t;\mathbf{k}),$$
$$t\in[\tau_s,\tau_{s+1}),\quad s=1,2,\dots,N_t,$$
$$\varphi_i(T)=0,\quad i=1,2,\dots,N. \tag{58}$$

The solution of the adjoint initial boundary value problem in (48)–(58) the function $\psi(x,t)=\psi(x,t;\mathbf{k},\varphi,\theta)$ satisfies the following conditions: 1) it is continuously differentiable with respect to t and twice continuously differentiable with respect to x for $t\in(\tau_s,\tau_{s+1})$ and $x\in(\xi_i(\tau_s),\xi_{i+1}(\tau_s))$; 2) for $t=\tau_s$ it is continuous in x for $x\in(\xi_i(\tau_s),\xi_{i+1}(\tau_s))$; 3) for $x=\xi_i(t)$ it is continuous in t, and its first derivatives in x, $\psi(\xi_i(t),t)$, have a finite discontinuity; 4) the points $(\xi_i(\tau_s),\tau_s)$ themselves are points of continuity in x and discontinuous in t.

4 Numerical Experiments

Here are the results of a test problem solving for controlling a furnace heating process with continuous feedback. The following parameter values were used:

$$a^2 = 1, \quad \lambda_0 = 0.001, \quad \lambda_1 = 1000, \quad \lambda_2 = 1200,$$

$$l = 1, \quad T = 1, \quad \underline{\vartheta_1} = 0, \quad \overline{\vartheta_1} = 60, \quad \underline{\vartheta_2} = 0, \quad \overline{\vartheta_2} = 50,$$

$$N_\Phi = N_\Theta = 4, \quad \tilde{\Phi} = \{0.11; 0.12; 0.13; 0.14\}, \quad \tilde{\Theta} = \{3.1; 3.2; 3.3; 3.4\},$$

$$p_i^\varphi = p_j^\theta = 0.25, \quad i = 1, 2, \ldots, 4, \quad j = 1, 2, \ldots, 4.$$

$$U(x) = 10, \quad x \in [0; 1], \quad \underline{U} = 0.95, \quad \overline{U} = 10.05, \quad \underline{u} = 0, \quad \overline{u} = 11,$$

$$-2.5 \le \omega_i(t) \le 2.5, \quad t \in [0; 1], \quad i = 1, 2, 3, \quad N = 3,$$

$$\xi_1^0 = 0.22, \quad \dot{\xi}_1(t) = \omega_1(t), \quad 0.05 \le \xi_1(t) \le 0.95,$$

$$\xi_2^0 = 0.45, \quad \dot{\xi}_2(t) = \omega_2(t), \quad 0.06 \le \xi_2(t) \le 0.95,$$

$$\xi_3^0 = 0.83, \quad \dot{\xi}_3(t) = \omega_3(t), \quad 0.05 \le \xi_3(t) \le 0.95.$$

To solve the direct (1), (19), (20), (5) and adjoint (40)–(45) loaded boundary value problems, the approach proposed in [13,14] was used, using a modified scheme of the Crank-Nicholson method with steps in the spatial variable $h_x = 0.01$ and in the time variable $h_t = 0.01$ [13].

To solve the direct (21), (11) and adjoint (46), (47) Cauchy problems, a modified scheme of the Euler method with a time step $h_t = 0.01$ was used.

The penalty function was minimized three times with the initial value of the penalty coefficient $\mathcal{R}^1 = 1$, and then $\mathcal{R}^{m+1} = 10\mathcal{R}^m$, $m = 1, 2, 3$, and the value of the parameter $\varepsilon = 0.1$ was decreased by 10 times.

Table 1 presents the results of solving the problem of synthesizing the parameters $\mathbf{k} = (k^1, k^2, k^3)$ with continuous feedback, obtained for different initial values $\mathbf{k}^0 = (k^1, k^2, k^3)^0$ for the iterative procedure (36), where the values of the feedback parameters $\mathbf{k}$ are given row by row in the following order: $(k_1^1, k_2^1, k_3^1, k_1^2, k_2^2, k_3^2, k_1^3, k_2^3, k_3^3)$.

Figure 1 shows the graphs of the function $\tilde{J}(t; \mathbf{k}^*)$ indicating the behavior of the heating process, with the obtained optimal feedback parameters for $t \in [0;\ 1]$ (solid lines). The dotted line shows the graph of the function $\tilde{J}\left(t; \tilde{\mathbf{k}}^*\right)$ obtained by solving the original problem under the assumption that the measurement points do not move, and their locations are optimized. In this case, the feedback parameters form the vector $\tilde{\mathbf{k}} = (k^1, k^2, k^3, \xi)$. From these graphs it is show that the quality of control of the rod heating process when the measuring points are moving is significantly better compared to the case when the stationary locations of the measuring devices are optimized [15].

Figure 2 shows the graphs of the trajectories of the measuring sensors $\xi_i(t)$ and the corresponding controls $\omega_i(t)$ for the initial value of the feedback parameter vector $\mathbf{k}^0$ (dashed lines) and for the synthesized optimal vector $\mathbf{k}^*$ (solid lines).

Table 1. Results of solving the problem for different initial points $\mathbf{k}^0$

	$\mathbf{k}^0$			$J_{\mathcal{R}}(\mathbf{k}^0)$	$\mathbf{k}^*$			$J_{\mathcal{R}}(\mathbf{k}^*)$
1	-0.0819	-0.0374	-0.0311	3.7532	-0.3374	-0.5048	-0.5429	0.0001
	-0.0334	0.0875	-0.0121		-0.5224	-0.3679	-0.3262	
	-0.0038	-0.0136	-0.0018		0.2817	0.2141	0.7510	
2	-0.0246	-0.0432	-0.0346	4.2512	-0.1785	-0.2953	-0.5034	0.0002
	0.0408	0.0569	-0.0320		-0.4494	-0.5265	-0.5233	
	-0.0135	-0.0405	-0.0543		-0.3562	-0.5653	-0.7156	
3	-0.0256	-0.0435	-0.0497	3.2928	-0.4727	-0.2780	-0.3572	0.0001
	0.0272	-0.0218	-0.0245		-0.4361	-0.2457	-0.4437	
	-0.0451	-0.0242	-0.0378		-0.2365	-0.4836	-0.2864	

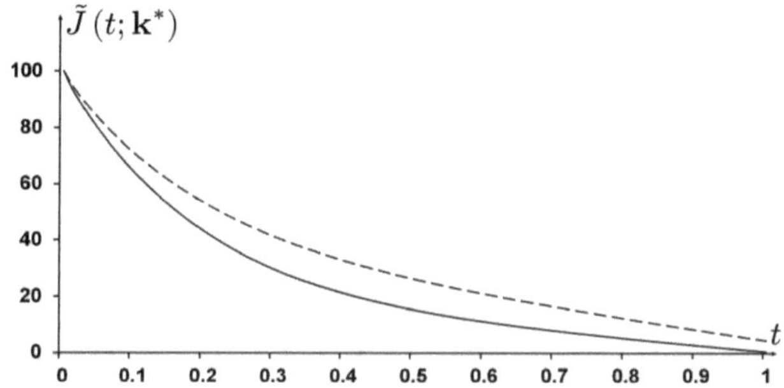

Fig. 1. Graphs of the function $\tilde{J}(t;\mathbf{k}^*)$ obtained with moving measurement points (——) and with their optimal stationary location (- - -).

Figure 3 shows the graphs of the $\vartheta(t;\mathbf{k}) = (\vartheta_1(t;\mathbf{k}), \vartheta_2(t;\mathbf{k}))$ boundary control functions for the initial values of the parameter vector $\mathbf{k}^0$ and for the synthesized optimal vector of feedback parameters $\mathbf{k}^*$.

Figure 4 shows the results of controlling the heating process (solution of the initial boundary value problem (1), (19), (20), (5) with control parameters $\mathbf{k}^*$, initial conditions $\varphi = 0.11$, at an external temperature $\theta = 3.1$ and in the presence of noise (errors) in temperature measurement. Taking into account the errors, the measurement values were determined by the formula

$$\breve{u}(\xi_i(t), t) = u(\xi_i(t), t)(1 + e(t)\chi), \quad i = 1, 2, \ldots, N,$$

where $e(t)$ for each t is a random variable uniformly distributed on the interval $[-1; 1]$; $\chi \geq 0$ is a value determined by the noise level: $\chi = 0$ means that there is no noise; $\chi = 0.01$ means that the absolute value of the noise does not exceed the measured value by more than 1%. Let us introduce the function $\Delta u(t) = \max_{x\in[0,l]} |u(x,t) - U(x)|$, which determines at the moment of time t the maximum deviation along the length of the rod of the achieved temperature from the desired one. From the graphs provided it is clear that with an noise

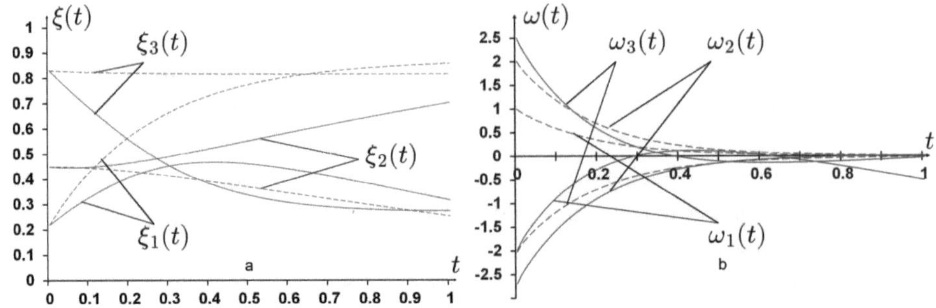

Fig. 2. Graphs of the trajectories of the measurement sensors (a), the corresponding controls (b) for the initial value of the feedback parameter vector $\mathbf{k}^0$ (dashed lines) and for the synthesized optimal vector $\mathbf{k}^*$ (solid lines).

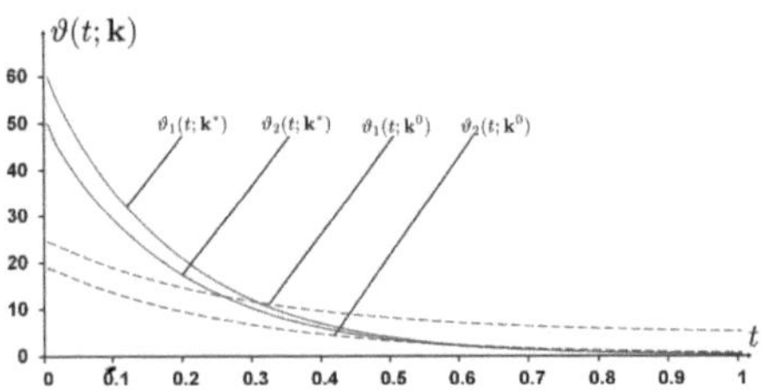

Fig. 3. Graphs of boundary controls for the initial parameter vector $\mathbf{k}^0$ (dashed lines) and for the synthesized optimal parameter vector $\mathbf{k}^*$ (solid lines).

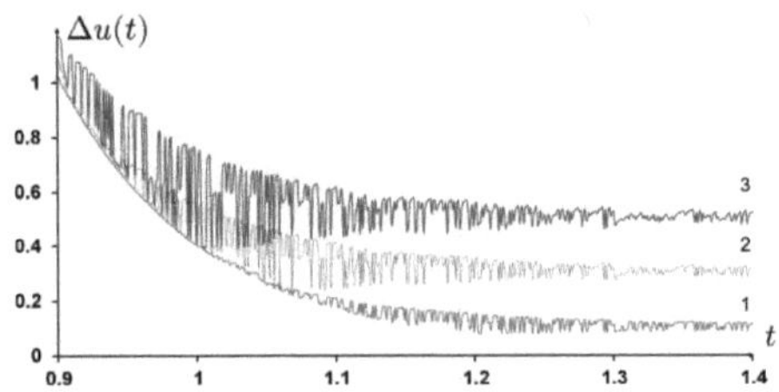

Fig. 4. Graphs of the function $\Delta u\,(t)$ for noise levels in measurements at control points of 1%, 3%, 5% (respectively lines 1, 2, 3) for $t \geq 0.9$.

level of 5%, the quality of control of the heating process deteriorates, namely, the deviation from the desired value $U\,(x)$ increases significantly.

Conclusion

In this paper, using the example of feedback control of the rod heating process by means of boundary actions, a mathematical formulation and approach to solving the problem of synthesizing optimal control together with optimization of the movement of measurement points are proposed. The research methodology and approach to obtaining calculation formulas and constructing a numerical scheme for solving the problem proposed in the work can be used in many problems when constructing automatic control and regulation systems for other technological processes and technical objects. The proposed approach can easily be extended to processes described by other types and kinds of partial differential equations and initial boundary conditions of higher phase space dimension.

The proposed approach to feedback control of systems with distributed parameters, taking into account the current state of computing and technical automation tools, is quite feasible in practice.

References

1. Utkin, V.I.: Sliding Modes in Optimization and Control Problems. Nauka, Moscow (1981). (in russian)
2. The Control Handbook, Levine, W.S., Ed., Boca Raton: CRC Press, New York, pp. 895–908. IEEE Press (1996)
3. Vasil'ev, F.P.: Metody optimizatsii. Faktorial Press, Moscow (2002). (in russian)
4. Polyak, B.T., Shcherbakov, P.S.: Robust Stability and Control. Nauka, Moscow (2002). (in russian)
5. Ray, W.H.: Advanced Process Control. McGraw-Hill, New York (1981)
6. Butkovskii, A.G.: Control Methods for Systems with Distributed Parameters. Nauka, Moscow (1984). (in russian)
7. Egorov, A.I.: Fundamentals of Control Theory. Fizmatlit, Moscow (2004). (in russian)
8. Sergienko, I.V., Deineka, V.S.: Optimal Control of Distributed Systems with Conjugation Conditions. Kluwer Acad. Publ, New York (2005)
9. Aida-zade, K.R.: An Approach to Synthesis of Focused Control in Distributed Systems. Autom. Control Comput. Sci. Allenton Press, INC. **39**(3), 16–22 (2005)
10. Tikhonov, A.N.: Samarskii, A.A.: Equations of Mathematical Physics, Nauka, Moscow (1966)
11. Alikhanov, A.A., Berezgov, A.M., Shkhanukov-Lafishev, M.X.: Boundary value problems for certain classes of loaded differential equations and solving them by finite difference methods. Comput. Math. and Math. Phys., 48, 1581–1590 (2008). https://doi.org/10.1134/S096554250809008X
12. Aida-zade, K.R.: An approach for solving nonlinearly loaded problems for linear ordinary differential equations. Proc. Inst. Math. Mech. **44**(2), 338–350 (2018)
13. Abdullaev, V.M., Aida-Zade, K.R.: Numerical method of solution to loaded nonlocal boundary value problems for ordinary differential equations. Comput. Math. Math. Phys. **54**(7), 1096–1109 (2014). https://doi.org/10.1134/S0965542514070021
14. Abdullayev, V.M., Aida-zade, K.R.: Approach to the numerical solution of optimal control problems for loaded differential equations with nonlocal conditions. Comput. Math. Math. Phys. **59**(5), 696–707 (2019). https://doi.org/10.1134/S0965542519050026
15. Aida-zade, K.R., Hashimov, V.A.: Optimization of Measurement Points Positioning in a Border Control Synthesis Problem for the Process of Heating a Rod. Autom Remote Control **79**, 1643–1660 (2018). https://doi.org/10.1134/S0005117918090096
16. Hashimov, V.A.: Synthesis of boundary conditions control and state measurement points motion for fast damping of string vibrations, Proc. Inst. Math. Mech. **51**(2), 294–317 (2025). https://doi.org/10.30546/2409-4994.2025.51.2.4110
17. Samarskii, A.A.: Theory of Difference Schemes. Nauka, Moscow (1989). (in russian)

A Priori Estimation of the Perfomance Time for Linear Discrete-Time Systems with Bounded Control Based on the Eigenset Theory

Danis N. Ibragimov(✉) and Sofya R. Guseva

Moscow Aviation Institute (National Research University), Volokolamskoe shosse, 4, Moscow, Russia 125993
rikk.dan@gmail.com

Abstract. A linear discrete-time system with geometric constraints on the control is considered. It is assumed that the system matrix is non-degenerate and diagonalizable via a similarity transformation, and the set of admissible control values is convex and compact. For the given system, the time-optimization problem is studied. In particular, it is required to construct a priori estimates of the minimum time as a function of the initial state and system parameters in the form of unimprovable inequalities. The solution is based on the theory of eigensets of a linear transformation. For given eigensets of the system matrix, an optimality criterion for the minimum time estimates is formulated and proved. A numerical algorithm for constructing optimal estimates is developed, which in the general case reduces to using the dichotomy method. Numerical calculations are carried out demonstrating the efficiency and accuracy of the developed method.

Keywords: Linear Discrete-Time System · Time-Optimization Problem · A Priori Estimates of the Optimal Criterion Function · Eigenset

1 Introduction

The time-optimization control problem is one of the classical problems of optimal control with high practical value. This statement is significant when achieving a given state in the shortest possible time is necessary. For example, it is in damping high-rise structures, normalizing patients' medical parameters, or stabilizing the orbit of spacecraft. The time-optimization problem was first considered in detail in monographs on mathematical control theory [1,2]. For continuous-time linear systems, it was shown that optimal control is piecewise-constant and can be found from the maximum principle. For discrete-time systems, the extremum

Supported by grant from the Russian Science Foundation № 25-21-00487.

A. Kazakov et al. (Eds.): MOTOR 2026, LNCS 16687, pp. 358–372, 2026.
https://doi.org/10.1007/978-3-032-31390-4_23

of the time-optimization problem is irregular for almost all initial states. As a result, the conditions of the discrete maximum principle degenerate, making it impossible to compute the optimal control.

Most papers on the time-optimization problem in discrete time are devoted to the study of special cases. They are often limited to two-dimensional and three-dimensional systems with linear control constraints [3–10]. However, a geometric approach to solving the discrete time-optimization problem, based on the method of reachable and null-controllable sets, has recently begun to actively develop. Methods exist that involve explicitly constructing these sets [11,12], their polyhedral approximation [13], and obtaining their description in terms of support functions [14] or support points [15]. A significant drawback of these methods is their computational complexity. This is due to the fact that the number of steps required to transfer the system to the origin is calculated by a linear search. That is, for all values of the maximum number of steps, it is proposed to sequentially solve the optimal control problem with a constraint on the terminal state. In the case where the minimum time is calculated or estimated from above, the time-optimization problem can be reduced to the terminal stabilization problem. Classical numerical methods can be used to solve it [16,17]. Therefore, the development of the algorithm for a priori calculation and estimation of the minimum time for discrete-time systems is relevant.

The approach based on approximating the set of admissible control values by eigensets of the system matrix has demonstrated high efficiency and ease of use. An eigenset is a set whose image is congruent to the preimage for a given linear transformation [18,19]. For two-dimensional systems, the a priori estimation problem was completely solved using circles and rectangles as the only possible eigensets on the plane [20]. In particular, a numerical algorithm was developed that allows the construction of unimprovable minimum time estimates [21]. For systems of arbitrary dimensions [22], the eigenset method is described in detail, and a priori estimates for the minimum number of steps are constructed.

The paper develops the algorithm that allows one to refine the minimum time estimates [22]. In particular, necessary and sufficient conditions are formulated under which these estimates cannot be improved. The problem of multidimensional optimization with respect to parameters defining the set of eigensets of the system matrix can be reduced to finding the root of a monotone function of a scalar argument. This means that the complexity of this algorithm is independent of the system's dimension, unlike that of the time-optimization methods in [11,13,14].

2 Problem Statement

We consider a linear stationary system with discrete time and geometric constraints on the control $(A,\mathcal{U})$:

$$\begin{aligned} &x(k+1) = Ax(k) + u(k), \\ &x(0) = x_0,\ \ u(k) \in \mathcal{U},\ \ k \in \mathbb{N} \cup \{0\}, \end{aligned} \tag{1}$$

where $x(k) \in \mathbb{R}^n$ is the state vector, $u(k) \in \mathbb{R}^n$ is the control action at the k-th step, $A \in \mathbb{R}^{n\times n}$ is the system matrix, $\mathcal{U} \subset \mathbb{R}^n$ is the set of admissible control values. It is assumed that $\mathcal{U}$ is a convex compact set, $0 \in \operatorname{int}\mathcal{U}$, $\det A \neq 0$.

For system (1), the time-optimization problem is studied, i.e., the problem of transferring the system from a given initial state x_0 to the origin in a minimum number of steps $N_{\min}$:

$$N_{\min} = \min\{N \in \mathbb{N} \cup \{0\} \colon \exists u(0), \ldots, u(N-1) \in \mathcal{U} \colon x(N) = 0\}.$$

In particular, the problem of a priori estimation of the value $N_{\min}$ as a function of A, x_0 and $\mathcal{U}$ without constructing an optimal process is discussed:

$$\underline{N_{\min}}(x_0; A, \mathcal{U}) \leqslant N_{\min} \leqslant \overline{N_{\min}}(x_0; A, \mathcal{U}). \tag{2}$$

3 Eigensets Method

The approach to constructing estimates (2) based on the method of eigensets [22] is known. An eigenset of a linear transformation defined by a matrix $A \in \mathbb{R}^{n\times n}$ is a compact set $\mathcal{U} \subset \mathbb{R}^n$ satisfying condition $A\mathcal{U} = \alpha\mathcal{U}$ for some $\alpha \in \mathbb{R}$. If $\mathcal{U}$ is convex and $0 \in \operatorname{int}\mathcal{U}$, then $\mathcal{U}$ is called a nontrivial eigenset. The criterion for the existence of nontrivial eigensets is known.

Theorem 1 ([22, Theorem 1]). *Let $A \in \mathbb{R}^{n\times n}$ have a nontrivial eigenset $\mathcal{U} \subset \mathbb{R}^n$ corresponding to the eigenvalue α.*

Then either $A = O$ and every compact set is eigenset, $\alpha = 0$, or $\alpha \neq 0$ and the following conditions hold:

1. $\det A \neq 0$;
2. *A is diagonalizable, i.e., it has n linearly independent eigenvectors;*
3. *all eigenvalues of A are equal in absolute value to $|\alpha|$;*
4. *$\mathcal{U} \subset \mathbb{R}^n$ is a nontrivial eigenset of A if and only if $S^{-1}\mathcal{U}$ is a nontrivial eigenset of Λ, where*

$$S^{-1}AS = \alpha\Lambda = \alpha \operatorname{diag}\Bigg(\underbrace{1, \ldots, 1}_{n_1}, \underbrace{-1, \ldots, -1}_{n_2}, A_{\varphi_1}, \ldots A_{\varphi_{n_3}}\Bigg), \tag{3}$$

$$A_\varphi = \begin{pmatrix} \cos\varphi & \sin\varphi \\ -\sin\varphi & \cos\varphi \end{pmatrix}, \quad \varphi \in [0; 2\pi), \ n_1 + n_2 + 2n_3 = n.$$

Example 1. Let us consider the matrices

$$A = S\begin{pmatrix} -2 & 0 & 0 \\ 0 & \sqrt{2} & \sqrt{2} \\ 0 & -\sqrt{2} & \sqrt{2} \end{pmatrix} S^{-1}, \quad S = \begin{pmatrix} 1 & 0 & 1 \\ 0 & 1 & 1 \\ 1 & 1 & 0 \end{pmatrix}.$$

The matrix A has eigenvalues $\lambda_1 = 2$ and $\lambda_{2,3} = 2e^{\pm i\frac{\pi}{4}}$, $|\lambda_1| = |\lambda_2| = |\lambda_3| = 2$. Then, by Theorem 1, there exists a nontrivial eigenset $\mathcal{U}$ of the linear mapping A corresponding to the eigenvalue 2. For geometric reasons, the auxiliary set $S^{-1}\mathcal{U}$ must be symmetric with respect to the 1st coordinate and invariant to rotation by $\frac{\pi}{4}$ in the plane of the 2nd and 3rd coordinates. For example, we can choose the intersection of two cones:

$$S^{-1}\mathcal{U} = \left\{u \in \mathbb{R}^3 \colon (1 - |u_1|)^2 \geqslant u_2^2 + u_3^2\right\}.$$

As shown in Fig. 1, the equality $A\mathcal{U} = 2\mathcal{U}$ holds.

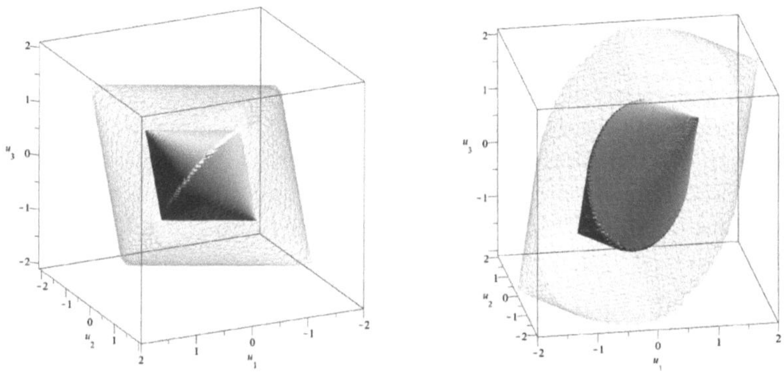

Fig. 1. The set $\mathcal{U}$ (solid rendering) and $A\mathcal{U}$ (outline rendering) from Example 1.

For $x \in \mathbb{R}^n$ and a convex $\mathcal{X} \subset \mathbb{R}^n$, $0 \in \operatorname{int} \mathcal{X}$, let $\mu(x, \mathcal{X})$ denote the Minkowski functional [23, §15, Ch. III]:

$$\mu(x, \mathcal{X}) = \inf\{t > 0 \colon x \in t\mathcal{X}\}.$$

Let us introduce the function $\hat{N}(\cdot, \cdot, \mathcal{X}) \colon \mathbb{R}^n \times (0; +\infty) \to [0; +\infty]$

$$\hat{N}(x, \alpha, \mathcal{X}) = \begin{cases} \mu(x, \mathcal{X}), & \alpha = 1, \\ -\dfrac{\ln(1 - (\alpha - 1)\mu(x, \mathcal{X}))}{\ln \alpha}, & \alpha \neq 1. \end{cases} \tag{4}$$

Also, for an arbitrary $\gamma \in \mathbb{R}$, $\lceil \gamma \rceil$ denotes the minimal integer not less than γ:

$$\lceil \gamma \rceil = \min\{k \in \mathbb{Z} \colon \gamma \leqslant k\}.$$

Then the following estimate of $N_{\min}$ is valid.

Theorem 2 (**[22, Theorem 3]**). *Let A be a diagonalizable matrix in system* (1) *and* $\det A \neq 0$, *i.e., there exists a non-singular $S \in \mathbb{R}^{n \times n}$, $\alpha_1, \ldots, \alpha_m > 0$, $\Lambda_1 \in \mathbb{R}^{n_1 \times n_1}, \ldots, \Lambda_m \in \mathbb{R}^{n_m \times n_m}$ of form* (3) *such that*

$$S^{-1}AS = \Lambda = \operatorname{diag}(\alpha_1 \Lambda_1, \ldots, \alpha_m \Lambda_m). \tag{5}$$

Then for any $x_0 \in \mathbb{R}^n$ *the following estimates are valid:*

$$\left[\max_{i=\overline{1,m}} \hat{N}(y_{0,i}, \alpha_i, -\overline{\mathcal{U}}_i)\right] \leqslant N_{\min} \leqslant \left[\max_{i=\overline{1,m}} \hat{N}(y_{0,i}, \alpha_i, -\underline{\mathcal{U}}_i)\right],$$

where $S^{-1}x_0 = \left(y_{0,1}^{\mathrm{T}}, \ldots, y_{0,m}^{\mathrm{T}}\right)^{\mathrm{T}}$, *for all* $i = \overline{1,m}$, *it is true that* $y_{0,i} \in \mathbb{R}^{n_i}$, $\overline{\mathcal{U}}_i$, $\underline{\mathcal{U}}_i$ *are nontrivial eigensets of* Λ_i *corresponding to the eigenvalue 1 and satisfying the inclusion*

$$\underline{\mathcal{U}}_1 \times \ldots \times \underline{\mathcal{U}}_m \subset S^{-1}\mathcal{U} \subset \overline{\mathcal{U}}_1 \times \ldots \times \overline{\mathcal{U}}_m.$$

The function $\hat{N}(y_{0,i}, \alpha_i, -\mathcal{U}_i)$ can take infinite values. From the point of view of Theorem 2, this means that the time-optimization problem for a given initial state x_0 is not solvable or the corresponding upper bound in (2) turns out to be trivial.

Theorem 2 does not specify how exactly sets $\overline{\mathcal{U}}_1, \underline{\mathcal{U}}_1, \ldots, \overline{\mathcal{U}}_m, \underline{\mathcal{U}}_m$ should be chosen. The form and structure of each such set can be determined by Theorem 1. However, even if we fix for each $i = \overline{1,m}$ a nontrivial eigenset $\mathcal{U}_i \subset \mathbb{R}^{n_i}$ of the mapping Λ_i and take $\overline{\mathcal{U}}_i = \beta_i\mathcal{U}_i$, $\underline{\mathcal{U}}_i = \gamma_i\mathcal{U}_i$ for some $\beta_i, \gamma_i > 0$, then the problem of choosing numbers $\beta_1^*, \ldots, \beta_m^*, \gamma_1^*, \ldots, \gamma_m^*$ satisfying the following conditions still remains:

$$\max_{i=\overline{1,m}} \hat{N}(y_{0,i}, \alpha_i, -\beta_i^*\mathcal{U}_i) = \max_{\substack{\beta \in \mathbb{R}^m \\ S^{-1}\mathcal{U} \subset \beta_1\mathcal{U}_1 \times \ldots \times \beta_m\mathcal{U}_m}} \max_{i=\overline{1,m}} \hat{N}(y_{0,i}, \alpha_i, -\beta_i\mathcal{U}_i), \tag{6}$$

$$\max_{i=\overline{1,m}} \hat{N}(y_{0,i}, \alpha_i, -\gamma_i^*\mathcal{U}_i) = \min_{\substack{\gamma \in \mathbb{R}^m \\ \gamma_1\mathcal{U}_1 \times \ldots \times \gamma_m\mathcal{U}_m \subset S^{-1}\mathcal{U}}} \max_{i=\overline{1,m}} \hat{N}(y_{0,i}, \alpha_i, -\gamma_i\mathcal{U}_i). \tag{7}$$

Hence, it is necessary to determine a set of parameters $\beta_1^*, \ldots, \beta_m^*, \gamma_1^*, \ldots, \gamma_m^*$ that would guarantee an optimal solution to the problem of estimating $N_{\min}$ by Theorem 2.

4 Refinement of a Priori Estimates

For each $i = \overline{1,m}$, we fix a convex compact set $\mathcal{U}_i \subset \mathbb{R}^{n_i}$, $0 \in \operatorname{int}\mathcal{U}_i$, a number $\alpha_i > 0$ and a vector $y_{0,i} \in \mathbb{R}^{n_i}$.

$$f_i(\gamma_i) = \hat{N}(y_{0,i}, \alpha_i, -\gamma_i\mathcal{U}_i). \tag{8}$$

We can eliminate the infinite values of $\hat{N}$ in definition (8) by defining the function f_i on $(o_i; +\infty)$, where

$$o_i = \max\{0, (\alpha_i - 1)\mu(y_{0,i}, -\mathcal{U}_i)\}. \tag{9}$$

Let us formulate this fact and the other properties of the function f_i in the form of the following lemma.

Lemma 1. *Let the function* $f_i\colon (o_i; +\infty) \to [0; +\infty)$ *be defined according to* (8). *Then either* f_i *is continuously decreasing and* $y_{0,i} \neq 0$, *or* $f_i(\gamma_i) \equiv 0$ *and* $y_{0,i} = 0$.

Proof. According to the definition of the Minkowski functional, $\mu(y_{0,i}, -\gamma_i\mathcal{U}_i) = 0$ if and only if $y_{0,i} = 0$. In this case, it follows from (4) and (8) that

$$f_i(\gamma_i) = \hat{N}(0, \alpha_i, -\gamma_i\mathcal{U}_i) = \begin{cases} \mu(0, -\gamma_i\mathcal{U}_i), & \alpha_i = 1, \\ -\dfrac{\ln(1-(\alpha_i-1)\mu(0,-\gamma_i\mathcal{U}_i))}{\ln\alpha_i}, & \alpha_i \neq 1, \end{cases} = 0.$$

If $y_{0,i} \neq 0$, then from the definition of the Minkowski functional it follows that

$$\mu(y_{0,i}, -\gamma_i\mathcal{U}_i) = \frac{1}{\gamma_i}\inf\{t\gamma_i > 0\colon y_{0,i} \in -t\gamma_i\mathcal{U}_i\} = \frac{1}{\gamma_i}\mu(y_{0,i}, -\mathcal{U}_i).$$

Then

$$f_i(\gamma_i) = \begin{cases} \dfrac{1}{\gamma_i}\mu(y_{0,i}, -\mathcal{U}_i), & \alpha_i = 1, \\ -\dfrac{\ln\left(1-(\alpha_i-1)\frac{1}{\gamma_i}\mu(y_{0,i},-\mathcal{U}_i)\right)}{\ln\alpha_i}, & \alpha_i \neq 1. \end{cases} \tag{10}$$

It follows that f_i takes finite values for $\gamma_i \in (o_i; +\infty)$ and decreases continuously, since it is a composition of continuous and monotonic functions.

Lemma 1 is proved.

If $y_{0,i} = 0$, the corresponding subvector and the parameters β_i, γ_i can be ignored in problems (6) and (7), since any admissible values will be optimal due to Lemma 1. Therefore, without loss of generality, we assume that $y_{0,i} \neq 0$ for all $i = \overline{1,m}$.

We first consider problem (6), which is simpler.

Theorem 3. *The solution to problem* (6) *is*

$$\beta_i^* = \max_{(y_{0,1}^{\mathrm{T}},\ldots,y_{0,m}^{\mathrm{T}})^{\mathrm{T}} \in S^{-1}\mathcal{U}} \mu(y_{0,i}, \mathcal{U}_i).$$

Proof. Taking into account Lemma 1, the following equalities are true:

$$\max_{\substack{\beta\in\mathbb{R}^m \\ S^{-1}\mathcal{U}\subset\beta_1\mathcal{U}_1\times\ldots\times\beta_m\mathcal{U}_m}} \max_{i=\overline{1,m}} \hat{N}(y_{0,i}, \alpha_i, -\beta_i\mathcal{U}_i) = \max_{i=\overline{1,m}} \max_{\substack{\beta\in\mathbb{R}^m \\ S^{-1}\mathcal{U}\subset\beta_1\mathcal{U}_1\times\ldots\times\beta_m\mathcal{U}_m}} f_i(\beta_i) =$$

$$= \max_{i=\overline{1,m}} f_i\left(\min_{\substack{\beta\in\mathbb{R}^m \\ S^{-1}\mathcal{U}\subset\beta_1\mathcal{U}_1\times\ldots\times\beta_m\mathcal{U}_m}} \beta_i\right).$$

The inclusion $S^{-1}\mathcal{U} \subset \beta_1\mathcal{U}_1\times\ldots\times\beta_m\mathcal{U}_m$ is true if and only if for each $i = \overline{1,m}$ the projection of $S^{-1}\mathcal{U}$ onto the corresponding n_i-dimensional subspace is embedded in $\beta_i\mathcal{U}_i$. Therefore, the following equality holds:

$$\beta_i^* = \max_{(y_1^{\mathrm{T}},\ldots,y_m^{\mathrm{T}})^{\mathrm{T}} \in S^{-1}\mathcal{U}} \mu(y_i, \mathcal{U}_i).$$

Indeed, let $(y_1^{\mathrm{T}}, \ldots, y_m^{\mathrm{T}})^{\mathrm{T}} \in S^{-1}\mathcal{U}$. Then, by the definition of the Minkowski functional, $y_i \in \beta_i^* \mathcal{U}_i$ and, consequently, $y \in \beta_1^* \mathcal{U}_1 \times \ldots \times \beta_m^* \mathcal{U}_m$. If we choose any $\beta_i < \beta_i^*$, then there exists $y \in S^{-1}\mathcal{U}$ for which $y_i \notin \beta_i \mathcal{U}_i$ and, as a consequence, $y \notin \beta_1 \mathcal{U}_1 \times \ldots \times \beta_m \mathcal{U}_m$.

Theorem 3 is proved.

An approach similar to Theorem 3 is inapplicable to problem (7), since (7) is a minimax problem, unlike (6). Moreover, this problem may have no solution at all in the case of unstable systems. This happens if the time-optimization problem becomes unsolvable for x_0, when the set $S^{-1}\mathcal{U}$ is reduced to $\gamma_1 \mathcal{U}_1 \times \ldots \times \gamma_m \mathcal{U}_m$. Let us define the set of all admissible values of $\gamma \in \mathbb{R}^m$ in problem (7):

$$\Gamma = \{\gamma \in (o_1; +\infty) \times \ldots \times (o_m; +\infty) \colon \gamma_1 \mathcal{U}_1 \times \ldots \times \gamma_m \mathcal{U}_m \subset S^{-1}\mathcal{U}\}.$$

Due to the convexity of $S^{-1}\mathcal{U}$, we note an important property of Γ:

$$\tilde{\gamma} \in \Gamma \Rightarrow (o_1; \tilde{\gamma}_1] \times \ldots \times (o_m; \tilde{\gamma}_m] \subset \Gamma. \tag{11}$$

Lemma 1 implies the strict monotonicity of f_i, which allows us to define its inverse function $\bar{\gamma}_i \colon (0; +\infty) \to (o_i; +\infty)$. Due to representations (4) and (8), it is true that

$$\bar{\gamma}_i(c) = \begin{cases} \dfrac{1}{c}\mu(y_{0,i}, -\mathcal{U}_i), & \alpha_i = 1, \\ \dfrac{\alpha_i - 1}{1 - \alpha_i^{-c}}\mu(y_{0,i}, -\mathcal{U}_i), & \alpha_i \neq 1. \end{cases}$$

Also we can define a vector function $\bar{\gamma}(c) = (\bar{\gamma}_1(c), \ldots, \bar{\gamma}_m(c))^{\mathrm{T}}$.

Theorem 4. *Let $\Gamma \neq \varnothing$, for all $i = \overline{1, m}$ it is true that $y_{0,i} \neq 0$. Then*

1. *there exists a unique $c^* > 0$ such that $\bar{\gamma}(c^*) \in \Gamma \setminus \operatorname{int} \Gamma$;*
2. *$\gamma^* = \bar{\gamma}(c^*)$ is a solution to problem (7);*
3. *$[c^*]$ is an optimal upper bound on the minimum time for a given set of nontrivial eigensets $\mathcal{U}_1, \ldots, \mathcal{U}_m$.*

Proof. We carry out the proof in four stages.

1. By assumption, there exists $\tilde{\gamma} \in \Gamma$. We denote by $\tilde{\Gamma} = \{\gamma \in \Gamma \colon \gamma_i \geqslant \tilde{\gamma}_i,\ i = \overline{1, m}\}$. Since Lemma 1, $f_1, \ldots, f_m$ are monotonically decreasing, then

$$\min_{\gamma \in \Gamma} \max_{i = \overline{1, m}} f_i(\gamma_i) = \min_{\gamma \in \tilde{\Gamma}} \max_{i = \overline{1, m}} f_i(\gamma_i).$$

Moreover, the set $\tilde{\Gamma}$ is compact in $\mathbb{R}^m$, and therefore, by the Weierstrass theorem [24, Ch. VII, §2, Sec. 2] there exists $\gamma^1 \in \tilde{\Gamma}$ for which it is true that

$$\min_{\gamma \in \tilde{\Gamma}} \max_{i = \overline{1, m}} f_i(\gamma_i) = \max_{i = \overline{1, m}} f_i(\gamma_i^1) = c^*.$$

Taking into account definition (8) and Theorem 2, the value $[c^*]$ is a solution to problem (7), i.e., it is an optimal estimate of the minimum time for a given set of nontrivial eigensets $\mathcal{U}_1, \ldots, \mathcal{U}_m$.

2. Let us take into account that from representation (10) it follows that

$$\lim_{\gamma_i \to o_i + 0} f_i(\gamma_i) = +\infty.$$

Due to Lemma 1, the function f_i is continuous, then by the intermediate value theorem [24, Ch. III, §2, Corollary of Theorem 2] for all $i = \overline{1,m}$ there exist $\gamma_i^* \in (o_i; \gamma_i^1]$ such that

$$c^* = f_1(\gamma_1^*) = \ldots = f_m(\gamma_m^*).$$

Moreover, $\gamma^* \in \Gamma$ due to (11). By the definition of the inverse function, we also obtain that $\gamma^* = \bar{\gamma}(c^*)$ is the minimum point in problem (7).

3. Suppose that $\gamma^* \in \operatorname{int} \Gamma$. Then there exists a neighborhood $\mathcal{O}_\varepsilon(\gamma^*) \subset \Gamma$ and a point $\gamma^\varepsilon \in \mathcal{O}_\varepsilon(\gamma^*)$ satisfying the condition $\gamma_i^\varepsilon > \gamma_i^*$, $i = \overline{1,m}$. Since the functions f_i decrease, by Lemma 1 we obtain

$$\max_{i=\overline{1,m}} f_i(\gamma_i^\varepsilon) < \max_{i=\overline{1,m}} f_i(\gamma_i^*) = \min_{\gamma \in \Gamma} \max_{i=\overline{1,m}} f_i(\gamma_i).$$

This leads to a contradiction. Therefore, $\gamma^* \in \Gamma \setminus \operatorname{int} \Gamma$.

4. Now we consider some $c_1 \neq c^*$. If $c_1 < c^*$, then, due to c^* is a solution to problem (7), it is true that $\bar{\gamma}(c_1) \notin \Gamma$. If $c_1 > c^*$, then, by monotonicity, we get $\bar{\gamma}_i(c_1) < \bar{\gamma}_i(c^*) = \gamma^*$, $i = \overline{1,m}$.

$$\bar{\gamma}(c_1) \in (o_1; \gamma_1^*) \times \ldots \times (o_m; \gamma_m^*) \subset \operatorname{int} \Gamma.$$

Finally, according to Step 4, there exists a unique point of the intersection of the graph of the function $\bar{\gamma}$ and set $\Gamma \setminus \operatorname{int} \Gamma$. By Step 3, this point can be calculated as $\gamma^* = \bar{\gamma}(c^*)$. According to Step 2, γ^* represents a minimum point in problem (7), and according to Step 1, c^* is the minimum value of the criterion, i.e., $[c^*]$ is the optimal upper bound for the minimum time. Theorem 4 is proved.

Theorem 4 does not claim that every solution to problem (7) can be represented as $\bar{\gamma}(c^*)$, since (7) may have more than one solution in the general case. However, among the solutions of this optimization problem, there is a point representable as $\gamma^* = \bar{\gamma}(c^*)$. Since the function $\bar{\gamma}$ depends on a scalar argument, this solution can be calculated numerically using the dichotomy method.

5 Algorithm for a Priori Estimation of Minimum Time

Using Theorems 3 and 4 requires the ability to verify the condition $\gamma \in \Gamma$, which is equivalent to the inclusion $\mathcal{U}_1 \times \ldots \times \mathcal{U}_m \subset S^{-1}\mathcal{U}$. For convex sets, such an inclusion is usually formulated in terms of the Minkowski functional and the support function, which are also used in describing sets. The support function [23, §4, Ch. I] of a convex and compact set $\mathcal{X} \subset \mathbb{R}^n$ at $p \in \mathbb{R}^n$ is

$$\mathcal{S}(p, \mathcal{X}) = \max_{x \in \mathcal{X}} \langle p, x \rangle.$$

From the definition of the support function and the Minkowski functional it follows that for two arbitrary convex compacts $\mathcal{X}, \mathcal{U} \subset \mathbb{R}^n$ containing the origin, the inclusion $\mathcal{X} \subset \mathcal{U}$ holds if and only if

$$\text{for all } x \in \mathbb{R}^n \text{ the inequality } \mu(x,\mathcal{X}) \geqslant \mu(x,\mathcal{U}) \text{ holds;} \tag{12}$$

$$\text{for all } p \in \mathbb{R}^n \text{ the inequality } \mathcal{S}(p,\mathcal{X}) \leqslant \mathcal{S}(p,\mathcal{U}) \text{ holds.} \tag{13}$$

Conditions (12) and (13) can be reduced to equivalent optimization problems:

$$\max_{x \in \mathbb{R}^n} (\mu(x,\mathcal{U}) - \mu(x,\mathcal{X})) \leqslant 0; \tag{14}$$

$$\max_{p \in \mathbb{R}^n} (\mathcal{S}(p,\mathcal{X}) - \mathcal{S}(p,\mathcal{U})) \leqslant 0; \tag{15}$$

$$\max_{x \in \mathbb{R}^n \setminus \{0\}} \frac{\mu(x,\mathcal{U})}{\mu(x,\mathcal{X})} \leqslant 1; \tag{16}$$

$$\max_{p \in \mathbb{R}^n \setminus \{0\}} \frac{\mathcal{S}(p,\mathcal{X})}{\mathcal{S}(p,\mathcal{U})} \leqslant 1. \tag{17}$$

Due to Minkowski's theorem [23, Corollary 18.5.1] and the fact that for all extreme points of the set $\mathcal{X}$ the Minkowski functional is equal to 1, we can obtain a condition equivalent to (14) and (16):

$$\max_{x \in \operatorname{Ext} \mathcal{X}} \mu(x,\mathcal{U}) \leqslant 1, \tag{18}$$

where $\operatorname{Ext} \mathcal{X}$ denotes the set of extreme points of $\mathcal{X}$ [23, §18, Ch. IV]. In the case where $\mathcal{X} = \operatorname{conv}\{x^1, \ldots, x^N\}$, i.e., $\mathcal{X}$ is a polyhedron, condition (18) reduces to enumeration of vertices:

$$\max_{j=\overline{1,N}} \mu\left(x^j,\mathcal{U}\right) \leqslant 1. \tag{19}$$

A problem similar to (18) can also be formulated using support functions, reducing the set of admissible values of p to the set of extreme points of the polar $\mathcal{U}$ [23, §14, Ch. III]. However, this approach is not considered here due to the complexity of constructing a polar set. We can consider the case when $\mathcal{U} = \bigcap\limits_{k=1}^{M} \{x \in \mathbb{R}^n \colon \langle p^k, x\rangle \leqslant 1\}$ for some $p^1, \ldots, p^M \in \mathbb{R}^n \setminus \{0\}$, i.e., $\mathcal{U}$ is a polyhedron. Then problems (15) and (17) are reduced to enumerating normals to faces of $\mathcal{U}$:

$$\max_{k=\overline{1,M}} \mathcal{S}(p^k,\mathcal{X}) \leqslant 1. \tag{20}$$

Finally, for a polyhedron $\mathcal{X}$ and a polyhedron $\mathcal{U}$, conditions (19) and (20) are reduced to the inequality

$$\max_{j=\overline{1,N}} \max_{k=\overline{1,M}} \langle p^k, x^j\rangle \leqslant 1. \tag{21}$$

We note that for $\mathcal{U} = \mathcal{U}_1 \times \ldots \times \mathcal{U}_m$, from the definition of the Minkowski functional and the support function it follows that

$$\mathcal{S}(p,\mathcal{U}) = \sum_{i=1}^{m} \mathcal{S}(p_i,\mathcal{U}_i), \; p = \left(p_1^{\mathrm{T}}, \ldots, p_m^{\mathrm{T}}\right)^{\mathrm{T}},$$

$$\mu(x,\mathcal{U}) = \max_{i=\overline{1,m}} \mu(x_i,\mathcal{U}_i),\ x = \left(x_1^{\mathrm{T}},\ldots,x_m^{\mathrm{T}}\right)^{\mathrm{T}}.$$

Let us represent the solution to the problem of a priori minimum time estimation as an algorithm. The algorithm receives parameters of system (1) as input.

Algorithm 1. For given $A \in \mathbb{R}^{n\times n}$, $\mathcal{U} \subset \mathbb{R}^n$, $x_0 \in \mathbb{R}^n$:

1. construct the spectral decomposition of the matrix A of form (5), calculate $y_0 \triangleq S^{-1}x_0$, $S^{-1}\mathcal{U}$;
2. for each $i = \overline{1,m}$, fix a nontrivial eigenset $\mathcal{U}_i$ of the mapping Λ_i corresponding to the eigenvalue 1;
3. calculate $\beta_1^*,\ldots,\beta_m^*$ according to Theorem 3;
4. calculate the lower bound for the minimum time using Theorem 2:

$$N_{\min}^{down} \triangleq \max_{i=\overline{1,m}} \left\lceil \hat{N}(y_{0,i},\alpha_i,-\beta_i^*\mathcal{U}_i)\right\rceil;$$

5. if $N_{\min}^{down} = \infty$, terminate the algorithm, since the time-optimization problem for a given x_0 is unsolvable;
6. if necessary, fix new nontrivial eigensets $\mathcal{U}_1,\ldots,\mathcal{U}_m$ and calculate the values $\beta_1^*,\ldots,\beta_m^*$ according to Theorem 3;
7. calculate the values $o_1,\ldots,o_m$ according to (9);
8. if for any $i = \overline{1,m}$ the inequality $\beta_i^* < o_i$ holds, return the upper bound for the minimum time $N_{\min}^{up} = \infty$ and terminate the algorithm;
9. set $\underline{c}_0 = \max_{i=\overline{1,m}} f_i(\beta_i^*)$, $\overline{c}_0 = \infty$, $k = 0$;
10. if $\lceil\overline{c}_k\rceil = \lceil\underline{c}_k\rceil$, return the upper bound for the minimum time $N_{\min}^{up} = \lceil\overline{c}_k\rceil$ and terminate the algorithm;
11. calculate

$$c_{k+1} = \begin{cases} \dfrac{\overline{c}_k + \underline{c}_k}{2}, & \overline{c}_k < \infty,\\ 2\underline{c}_k, & \overline{c}_k = \infty;\end{cases}$$

12. if $\bar{\gamma}(c_{k+1}) \in \Gamma$, put $\overline{c}_{k+1} = c_{k+1}$, $\underline{c}_{k+1} = \underline{c}_k$, otherwise put $\overline{c}_{k+1} = \overline{c}_k$, $\underline{c}_{k+1} = c_{k+1}$;
13. increase k by 1 and go to step 10.

For Algorithm 1 to work, it is necessary to construct the Minkowski functionals of the sets $\mathcal{U}_1,\ldots,\mathcal{U}_m$, since they are used in definitions (4) and (8). At step 12, the condition $\gamma \in \Gamma$ is checked by solving one of optimization problems (14)–(21). The algorithm can also return the estimates $N_{\min}^{down} = \infty$ or $N_{\min}^{up} = \infty$ only in the case of an unstable system, which follows from definition (8). For stable systems, the numbers $N_{\min}^{down}$ and $N_{\min}^{up}$ will always have finite values.

A significant limitation when using Algorithm 1 is the diagonalizability of the matrix A. This follows from the fact that the underlying Theorem 2 relies on the apparatus of eigensets and Theorem 1. In the future, this limitation can be overcome by replacing the condition $A\mathcal{U} = \alpha\mathcal{U}$ in the definition of an eigenset with the pair of inclusions $\alpha\mathcal{U} \subset A\mathcal{U} \subset \beta\mathcal{U}$, which can be satisfied even for a non-diagonalizable matrix. Nevertheless, such a set $\mathcal{U}$ will no longer satisfy Theorem 1 and will require new conditions for its construction.

6 Numerical Experiments

Let us consider the problem of computing $N_{\min}$ for two systems.

Example 2. Let the following parameter values be selected in system (1) with $n = 3$:

$$A = \begin{pmatrix} -0.1930 & 0.6369 & 0.1814 \\ -0.1814 & -1.0114 & 0.6369 \\ -0.8183 & -0.8183 & -0.3745 \end{pmatrix}, \quad x_0 = \begin{pmatrix} 150 \\ -130 \\ -340 \end{pmatrix},$$

$$\mathcal{U} = \mathrm{conv}\left\{ \begin{pmatrix} 1 \\ 1 \\ 1 \end{pmatrix}, \begin{pmatrix} -1 \\ 1 \\ 1 \end{pmatrix}, \begin{pmatrix} 1 \\ -1 \\ 1 \end{pmatrix}, \begin{pmatrix} 1 \\ 1 \\ -1 \end{pmatrix}, \begin{pmatrix} -1 \\ -1 \\ 1 \end{pmatrix}, \begin{pmatrix} -1 \\ 1 \\ -1 \end{pmatrix}, \begin{pmatrix} 1 \\ -1 \\ -1 \end{pmatrix} \begin{pmatrix} -1 \\ -1 \\ -1 \end{pmatrix} \right\}.$$

Let us sequentially execute all steps of Algorithm 1.

1. The following spectral decomposition of the matrix A is true:

$$A = S\,\mathrm{diag}\,(0.83\Lambda_1, 0.9\Lambda_2)\, S^{-1},$$

$$S = \begin{pmatrix} 1 & -1 & 0 \\ -1 & 0 & 1 \\ 0 & 1 & 1 \end{pmatrix}, \quad \Lambda_1 = (-1), \quad \Lambda_2 = \begin{pmatrix} \cos 2 & \sin 2 \\ -\sin 2 & \cos 2 \end{pmatrix}.$$

Matrix A has eigenvalues $\lambda_1 = -0.83$, $\lambda_{2,3} = 0.9\mathrm{e}^{\pm 2\mathrm{i}}$. And in terms of Theorem 3, the following equalities are true: $m = 2$, $n_1 = 1$, $n_2 = 2$, $\alpha_1 = 0.83$, $\alpha_2 = 0.9$. The following equalities also hold:

$$y_0 = \begin{pmatrix} y_{0,1} \\ y_{0,2} \end{pmatrix}, \quad y_{0,1} = -30, \quad y_{0,2} = \begin{pmatrix} -180 \\ -160 \end{pmatrix},$$

$$S^{-1}\mathcal{U} = \mathrm{conv}\left\{ \begin{pmatrix} 0.5 \\ -0.5 \\ 1.5 \end{pmatrix}, \begin{pmatrix} -0.5 \\ 0.5 \\ -1.5 \end{pmatrix}, \begin{pmatrix} -0.5 \\ -1.5 \\ 0.5 \end{pmatrix}, \begin{pmatrix} 0.5 \\ 1.5 \\ -0.5 \end{pmatrix}, \right.$$
$$\left. \begin{pmatrix} 0.5 \\ -0.5 \\ -0.5 \end{pmatrix}, \begin{pmatrix} -0.5 \\ 0.5 \\ 0.5 \end{pmatrix}, \begin{pmatrix} 1.5 \\ 0.5 \\ 0.5 \end{pmatrix}, \begin{pmatrix} -1.5 \\ -0.5 \\ -0.5 \end{pmatrix} \right\}.$$

2. According to the definition, the only nontrivial eigensets for Λ_1 and Λ_2 are a symmetric segment and a circle, respectively. Therefore, we choose

$$\mathcal{U}_1 = [-1; 1], \ \mathcal{U}_2 = \left\{ \begin{pmatrix} u_2 \\ u_3 \end{pmatrix} \in \mathbb{R}^2 : u_2^2 + u_3^2 \leqslant 1 \right\}.$$

The following expressions are also true for Minkowski functionals and support functionals:

$$\mu(u_1,\mathcal{U}_1) = |u_1|,\ \mu\left((u_2\ u_3)^{\mathrm{T}},\mathcal{U}_2\right) = \sqrt{u_2^2+u_3^2},$$

$$\mathcal{S}(p_1,\mathcal{U}_1) = |p_1|,\ \mathcal{S}\left((p_2\ p_3)^{\mathrm{T}},\mathcal{U}_2\right) = \sqrt{p_2^2+p_3^2},$$

3. According to Theorem 3, the values $\beta_1^* = 1.5$, $\beta_2^* = 1.5811$ are obtained.

4. According to (4) and (8), the lower bound for the minimum time is as follows:

$$N_{\min}^{down} = \max\left\{\left\lceil\frac{\ln\left(1-\frac{(\alpha_1-1)|y_{0,1}|}{\beta_1^*}\right)}{-\ln\alpha_1}\right\rceil, \left\lceil\frac{\ln\left(1-\frac{(\alpha_2-1)\sqrt{y_{0,2}^2+y_{0,3}^2}}{\beta_2^*}\right)}{-\ln\alpha_2}\right\rceil\right\} = 27.$$

Steps 5–6 of Algorithm 1 are skipped.

7–8. According to (9), $o_1 = o_2 = 0$.

9. The initial values are $\underline{c}_0 = 26.4516$, $\overline{c}_0 = \infty$.

10–13. Since $S^{-1}\mathcal{U}$ is a polyhedron, condition at step 12 is equivalent to problem 20. The results of calculations are presented in Table 1.

Table 1. Refining the upper bound on the minimum time by Algorithm 1 in Example 2.

k	0	1	2	3	4	5	6
$\underline{c}_k$	26.4516	26.4516	26.4516	33.0645	33.0645	33.0645	33.0645
$\overline{c}_k$	∞	52.9032	39.6774	39.6774	36.3709	34.7177	33.8911

As follows from Table 1 and Theorem 4, $N_{\min}^{upper} = 34$. Hence, estimate (2) takes the form

$$27 \leqslant N_{\min} \leqslant 34.$$

Example 3. Let the following parameter values be selected in system (1) with $n = 4$:

$$A = \begin{pmatrix} 0.2177 & 0.2261 & 0.9584 & -1.1845 \\ -0.2261 & 0.3577 & -0.3661 & 0.9584 \\ 0.9584 & 0.3661 & 0.3577 & -0.3661 \\ 1.1845 & 0.9584 & -0.2261 & 0.2177 \end{pmatrix},\ x_0 = \begin{pmatrix} 127 \\ -54 \\ 85 \\ 92 \end{pmatrix},$$

$$\mathcal{U} = \left\{u \in \mathbb{R}^4 \colon u_1^2 + u_2^2 + u_3^2 + u_4^2 \leqslant 1\right\}.$$

Let us sequentially execute all steps of Algorithm 1.

1. The following spectral decomposition of the matrix A is true:

$$A = S\,\mathrm{diag}(0.9\Lambda_1, 0.95\Lambda_2)S^{-1},$$

$$S = \begin{pmatrix} 1 & -1 & 0 & 1 \\ -1 & 0 & 1 & -1 \\ 0 & 1 & 1 & 1 \\ 1 & 1 & 1 & 0 \end{pmatrix}, \ \Lambda_1 = \begin{pmatrix} \cos 2 & -\sin 2 \\ \sin 2 & \cos 2 \end{pmatrix}, \ \Lambda_2 = \begin{pmatrix} 1 & 0 \\ 0 & 1 \end{pmatrix}.$$

Matrix A has eigenvalues $\lambda_{1,2} = 0.9\mathrm{e}^{\pm 2\mathrm{i}}$, $\lambda_{3,4} = 0.95$. And in terms of Theorem 3, the following equalities are true: $m = 2$, $n_1 = 2$, $n_2 = 2$, $\alpha_1 = 0.9$, $\alpha_2 = 0.95$. The following equalities also hold:

$$y_0 = \begin{pmatrix} y_{0,1} \\ y_{0,2} \end{pmatrix}, \ y_{0,1} = \begin{pmatrix} 57.4 \\ -19.2 \end{pmatrix}, \ y_{0,2} = \begin{pmatrix} 53.8 \\ 50.4 \end{pmatrix},$$

$$S^{-1}\mathcal{U} = \left\{ u \in \mathbb{R}^4 \colon 3u_1^2 - 4u_1u_2 + 3u_2^2 + 3u_3^2 + 4u_3u_4 + 3u_4^2 \leqslant 1 \right\}.$$

2. A nontrivial eigenset for Λ_1 is the circle centered at 0, and for Λ_2, it is an arbitrary convex compact body. Therefore, we choose the unit circle as $\mathcal{U}_1$, and the section of $S^{-1}\mathcal{U}$ by the plane of the 3rd and 4th coordinates as $\mathcal{U}_2$:

$$\mathcal{U}_1 = \left\{ \begin{pmatrix} u_1 \\ u_2 \end{pmatrix} \in \mathbb{R}^2 \colon u_1^2 + u_2^2 \leqslant 1 \right\}, \mathcal{U}_2 = \left\{ \begin{pmatrix} u_3 \\ u_4 \end{pmatrix} \in \mathbb{R}^2 \colon 3u_3^2 + 4u_3u_4 + 3u_4^2 \leqslant 1 \right\}.$$

The following expressions are also true for Minkowski functionals and support functionals:

$$\mu\left((u_1\ u_2)^{\mathrm{T}}, \mathcal{U}_1\right) = \sqrt{u_1^2 + u_2^2}, \ \mu\left((u_3\ u_4)^{\mathrm{T}}, \mathcal{U}_2\right) = \sqrt{3u_3^2 + 4u_3u_4 3u_4^2}.$$

3. According to Theorem 3, the values $\beta_1^* = 1$, $\beta_2^* = 1$ are obtained.

4. According to (4) and (8), the upper bound for the minimum time is as follows:

$$N_{\min}^{down} = \max\left\{ \left\lceil \hat{N}(y_{0,1}, \alpha_1, -\beta_1^*\mathcal{U}_1) \right\rceil, \left\lceil \hat{N}(y_{0,2}, \alpha_2, -\beta_2^*\mathcal{U}_2) \right\rceil \right\} =$$
$$= \max\left\{ \lceil 18.5401 \rceil, \lceil 43.3466 \rceil \right\} = 44.$$

Steps 5–6 of Algorithm 1 are skipped.

7–8. According to (9), $o_1 = o_2 = 0$.

9. The initial values are $\underline{c}_0 = 43.3466$, $\overline{c}_0 = \infty$. To check inclusion $\gamma \in \Gamma$ at step 12, problem (18) is solved. In Example 3, it is reduced to the following form:

$$\max_{\substack{u_1^2+u_2^2=\gamma_1^2,\\ 3u_3^2+4u_3u_4 3u_4^2=\gamma_2^2}} \left(3u_1^2 - 4u_1u_2 + 3u_2^2 + 3u_3^2 + 4u_3u_4 + 3u_4^2\right) \leqslant 1.$$

10–13. The results of calculations are presented in Table 2.

As can be seen from Table 2, the minimum time is calculated exactly: $N_{\min} = 44$. However, for an arbitrary initial state, Algorithm 1 only constructs estimates of form (2). Note that in the Maple 2018 software environment, the calculations took less than 1 s, while the exact methods from [13, 14] usually take several minutes of computational time in the case of $n = 4$.

Table 2. Refining the upper bound on the minimum time by Algorithm 1 in Example 3.

k	0	1	2	3	4	5	6	7	8
$\underline{c}_k$	43.3466	43.3466	43.3466	43.3466	43.3466	43.3466	43.3466	43.3466	43.3466
$\overline{c}_k$	∞	86.6832	65.0199	54.1832	48.7649	46.0557	44.7012	44.0239	43.6852

7 Conclusion

This paper continues and completes the ideas from [20–22]. It develops the algorithm for calculating upper and lower bounds for the number of steps required to bring a discrete-time linear system to the origin. Importantly, the algorithm does not require the construction of any control processes, and the resulting estimates are a priori and can be obtained in a relatively short computational time compared to solving optimal control problems. This fact is demonstrated in Examples 2 and 3. Another feature of the algorithm is that the resulting minimum time estimates are unimprovable, as proved in Theorems 3 and 4. It is important to note the disadvantages of the algorithm: it is applicable only to systems with a diagonalizable matrix, and its use involves constructing a spectral decomposition of the matrix, which is not stable in general.

References

1. Pontryagin, L.S.: Mathematical Theory of Optimal Processes. Routledge, London (1987). https://doi.org/10.1201/9780203749319
2. Boltyanskii, V.G.: Optimal Control of Discrete Systems. I.P.S.T, USA (1979)
3. Blanchini, F.: Polyhedral set constrained control for discrete-time systems with unknown additive disturbances. IFAC Proc. Vol. **24**(8), 95–100 (1991). https://doi.org/10.1016/s1474-6670(17)54151-8
4. Blanchini, F., Ukovich, W.: Linear programming approach to the control of discrete-time periodic systems with uncertain inputs. J. Optim. Theory Appl. **78**(3), 523–539 (1993). https://doi.org/10.1007/bf00939880
5. Keerthi, S., Gilbert, E.: Computation of minimum-time feedback control laws for discrete-time systems with state-control constraints. IEEE Trans. Autom. Control **32**(5), 432–435 (1987). https://doi.org/10.1109/tac.1987.1104625
6. Abdelhak, A., Rachik, M.: The linear quadratic minimum-time problem for a class of discrete systems. J. Math. Programm. Oper. Res. **59**(4), 575–587 (2010). https://doi.org/10.1080/02331930801954672
7. Amato, F., Cosentino, C., Tommasi, G.D., Pironti, A., Romano, M.: Input-output finite-time stabilization of linear time-varying discrete-time systems. IEEE Trans. Autom. Control **67**(9), 4438–4450 (2022). https://doi.org/10.1109/TAC.2022.3161374
8. Chen, D., Bako, L., Lecoeuche, S.: The minimum-time problem for discrete-time linear systems: a non-smooth optimization approach. In: 2012 Proceedings of the IEEE International Conference on Control Applications, pp. 196–201 (2012). https://doi.org/10.1109/CCA.2012.6402693

9. Lee, J., Haddad, W.M.: Fixed time stability and optimal stabilisation of discrete autonomous systems. Int. J. Control **96**(9), 2341–2355 (2022). https://doi.org/10.1080/00207179.2022.2092557
10. Yang, H., Xia, Y., Geng, Q.: Stabilization on null controllable region. In: Analysis and Synthesis of Delta Operator Systems with Actuator Saturation. Studies in Systems, Decision and Control 2019, vol. 193. pp. 39–65. Springer, Singapore (2019). https://doi.org/10.1007/978-981-13-3660-7_3
11. Sirotin, A.N., Sypalo, K.I.: On solving the 0-controllability problem for linear discrete systems with scalar control of bounded total magnitude. J. Comput. Syst. Sci. Int. **64**(4), 515–543 (2025). https://doi.org/10.1134/S1064230725700479
12. Ibragimov, D.N.: On the optimal speed problem for the class of linear autonomous infinite-dimensional discrete-time systems with bounded control and degenerate operator. Autom. Remote. Control. **80**(3), 393–412 (2019). https://doi.org/10.1134/S0005117919030019
13. Ibragimov, D.N., Novozhilin, N.M., Portseva, E.Y.: On sufficient optimality conditions for a guaranteed control in the speed problem for a linear time-varying discrete-time system with bounded control. Autom. Remote. Control. **82**(12), 2076–2096 (2021). https://doi.org/10.1134/S000511792112002X
14. Bortakovskii, A.S.: Applying support functions to solve a problem of fast-acting discrete linear systems with a convex set of controls. J. Comput. Syst. Sci. Int. **64**(2), 203–220 (2025). https://doi.org/10.1134/S1064230725700157
15. Podgornaya, V.M.: On the suboptimal solution of the speed-in-action problem for a linear discrete system in the case of asymmetric control constraints. Modell. Data Anal. **14**(3), 63–86 (2024). https://doi.org/10.17759/mda.2024140304
16. Propoi, A.I.: Elements of the Theory of Optimal Discrete Processes. Nauka, Russia (1973). (in Russian)
17. Krotov, V.F., Gurman, V.I.: Methods and Problems of Optimal Control. Nauka, Russia (1973). (in Russian)
18. Murota, K.: Eigensets and power products of a bimatroid. Adv. Math. **80**(1), 78–91 (1990). https://doi.org/10.1016/0001-8708(90)90015-F
19. Johnson, C.R.: Eigenset generalizations of the eigenvalue concept. J. Res. Natl. Bur. Stand. **82**(2), 133–136 (1977). https://doi.org/10.6028/jres.082.013
20. Ibragimov, D.N., Guseva, S.R.: A priori estimates of the objective function in the speed-in-action problem for a linear two-dimensional discrete-time system. In: Khachay, M., Kochetov, Y., Eremeev, A., Khamisov, O., Mazalov, V., Pardalos, P. (eds.) MOTOR 2023. LNCS, vol. 13930, pp. 378–393. Springer, Cham (2023). https://doi.org/10.1007/978-3-031-35305-5_26
21. Ibragimov, D.N.: On the method for refining a priori estimates of the objective function in the speed-in-action problem for a linear discrete-time system. In: Eremeev, A., Khachay, M., Kochetov, Y., Mazalov, V., Pardalos, P. (eds.) MOTOR 2024. CCIS, vol. 2239, pp. 199–213. Springer, Cham (2024). https://doi.org/10.1007/978-3-031-73365-9_14
22. Ibragimov, D.N., Guseva, S.R.: A priori estimation of the minimal stabilization time for linear discrete-time systems with bounded control based on the apparatus of eigensets. Modell. Data Anal. **15**(1), 110–132 (2025). https://doi.org/10.17759/mda.2025150106
23. Rockafellar R.T.: Convex Analysis. Princeton University Press, USA (1970). https://doi.org/10.1515/9781400873173
24. Zorich, V.A.: Mathematical Analysis I. Springer Berlin, Heidelberg (2015). https://doi.org/10.1007/978-3-662-48792-1

Operations Research

Comparison of Path Generation Techniques for Optical Network Optimization

Egor Grishin[1(✉)], Nadezda Lutovinova[2,3], and Sergey Usmanov[3]

[1] Skoltech, Moscow, Russia
mkgrgor@gmail.com
[2] V.A. Trapeznikov Institute of Control Sciences of RAS, Moscow, Russia
[3] Cloud Quantum Technologies LLC, Moscow, Russia

Abstract. Routing and spectrum allocation (RSA) involves simultaneous path selection and contiguous frequency slot assignment, subject to spectrum continuity and non-overlapping constraints. Under these constraints, the RSA problem is known to be NP-hard. It is typically modeled using either node-link or path-link integer linear programming formulations. The node-link model used in this work selects directed arcs for each demand and assigns a left spectrum slot to the selected route. In contrast, path-link formulations rely on predefined candidate routes, leading to more compact models whose performance largely depends on the candidate path set. In this work, we compare several path-generation heuristics against a node-link reference formulation for solving the RSA problem.

Keywords: Integer Linear Programming (ILP) · Routing and Spectrum Allocation (RSA) · Elastic Optical Networks (EON) · Path Generation

1 Introduction

Traditional optical networks based on Wavelength Division Multiplexing (WDM) map traffic demands to fixed-grid wavelengths through the Routing and Wavelength Assignment (RWA) problem [1]. However, to accommodate the heterogeneous and increasing bandwidth demands of modern applications, flex-grid Elastic Optical Networks (EONs) have emerged as a flexible-grid optical network architecture. EON architectures replace the fixed-frequency grid with flexible, fine-grained frequency slots, enabling more efficient, distance-adaptive spectrum utilization [2,3]. This change turns the traditional RWA problem into the Routing and Spectrum Allocation (RSA) problem. In RSA, establishing a connection requires simultaneous path selection and the allocation of a continuous frequency slot block. This process is subject to spectrum continuity constraints (the same slots must be used on all links of a path), spectrum contiguity constraints (the

A. Kazakov et al. (Eds.): MOTOR 2026, LNCS 16687, pp. 375–389, 2026.
https://doi.org/10.1007/978-3-032-31390-4_24

allocated slots must be adjacent in the frequency domain), and non-overlapping constraints [4,5].

Due to the complexity of these constraints, the RSA problem is known to be NP-hard [3]. Exact mathematical modeling of RSA is typically approached using Integer Linear Programming (ILP), with node-link and path-link formulations being two common modeling families [6,7]. These formulation families and related variants have already been studied in the RSA literature. For example, Wang et al. classify RSA ILP models into path/slot, path/channel, node/slot, and node/channel categories and compare improved bounds and formulations [8]. Zheng et al. further study improved ILP models and heuristics for routing and resource allocation in optical networks, including the effect of the number of candidate paths in path-based models [9]. Channel-based RSA formulations have also been studied for slice protection and QoS-aware settings [10]. Thus, this paper does not claim a new RSA formulation. Instead, it evaluates how the choice of candidate paths affects a path-link formulation when compared with a node-link reference under the same spectrum-width objective.

To keep RSA solvable at practical instance sizes, heuristic frameworks are used, broadly classified under Route-First (RF) and Spectrum-First (SF) methods [11]. Topology plays an important role in performance of the methods used for RSA. For example, in [12], a framework for analyzing the effectiveness of RSA algorithms was proposed, where it was shown through the use of conflict graphs that neither the SF nor the RF methods were more effective for elastic rings than the other. This was a contradiction to the numerical results obtained earlier, where the SF method was always more effective than the RF method. An extensive study of existing real-world topologies was provided in [13].

Candidate-path generation is also a mature topic. The standard k-shortest paths method is widely used and is based on classical shortest-path enumeration [14]. Disjoint-path construction is related to survivable-network routing and diverse-path algorithms [15]. Penalty-based ideas also appear in alternative-route generation, for example in the alternative route graph approach of Bader et al. [16]. More advanced diversity-aware methods for optical networks are proposed by Mori and Shigeno [17]; they study integer programming-based selection, clustering-based selection, and link-sharing suppression, in addition to the standard k-shortest paths method. Dynamic generation methods, including column-generation approaches, have also been studied for RSA variants [18]. Our generators should therefore be interpreted as lightweight baselines and not as new route-generation algorithms.

In this work, we study penalty-based heuristics for RSA. We compare them with the k-shortest paths method and the clustering-based method from [17]. We also compare path-link and node-link formulations under the same spectrum-width objective. The formulations share the same spectrum-assignment and non-overlap semantics; however, the path-link model is exact only when its candidate path set contains a globally optimal route combination. Our results show that node-link models provide a useful reference for small and medium-size instances and help to evaluate route-selection heuristics.

2 Problem Statement

There are two widely used mathematical formulations for the RSA problem [6]: the node-link formulation and the path-link formulation.

The node-link formulation considered here works on the physical network arcs directly. For each demand, it selects a route as a set of directed arcs and assigns one contiguous spectrum interval to that route. The model therefore does not require a precomputed candidate-path set.

The path-link formulation assumes the same fixed network topology, but first generates a finite set of candidate paths for each demand. The optimization model then chooses one candidate path per demand and assigns spectrum intervals. Common objectives include minimizing the total number of frequency slots [3,5,12] or minimizing unserved bandwidth when resources are limited [4]. We consider the following ILP problem statement based on the $DR-AOV$ formulation from [7].

Notation. Let $G=(V,E)$ denote the directed graph representing the optical fiber network. By $\delta^-(v)$ and $\delta^+(v)$ we denote the sets of incoming and outgoing arcs of node v, respectively. The available spectrum is given as a set of frequency slots $\{0,1,\ldots,F-1\}$. Let D denote the set of demands, where each demand corresponds to a connection request. Each demand $d\in D$ is characterized by a tuple $(s(d),t(d),w(d),n(d))$, where $s(d)$ and $t(d)$ are the source and target nodes, respectively, $w(d)\in\mathbb{Z}_+$ is the number of contiguous frequency slots requested by the connection, and $n(d)\in\{1,2,...,|D|\}$ is a unique demand index used to order demand pairs.

The RSA problem consists of establishing a valid lightpath for each demand such that no two lightpaths overlap in the spectrum domain. Specifically, each demand $d\in D$ must be assigned a route in G from $s(d)$ to $t(d)$ and a half-open spectrum interval $[l_d,\ l_d+w(d))$ of $w(d)$ consecutive slots within $[0,F)$. If the paths assigned to demands d and d' share at least one physical arc, their respective spectrum intervals must be strictly disjoint. The objective is to minimize the spectrum width, i.e., the right endpoint $\max\limits_{d\in D}\big(l_d+w(d)\big)$ of the used spectrum interval.

ILP Node-Link Formulation. We introduce a set of problem variables:

- $y_{de}\in\{0,1\}$ equals 1 if directed arc $e\in E$ is used by demand d, and 0 otherwise.
- $l_d\in[0,F-w(d)]\subset\mathbb{Z}$ is the leftmost slot index assigned to demand d;
- $o_{dd'}\in\{0,1\}$ is a global ordering variable for each unordered pair of demands $d,\ d'$ with $n(d)<n(d')$:

$$o_{dd'}=\begin{cases}1, & \text{if demand } d \text{ is placed before } d',\\ 0, & \text{if demand } d' \text{ is placed before } d.\end{cases}$$

Equivalently, when the two demands must be separated, $o_{dd'} = 1$ corresponds to $l_d + w(d) \le l_{d'}$, whereas $o_{dd'} = 0$ corresponds to $l_{d'} + w(d') \le l_d$. The variable $o_{dd'}$ is not arc-specific: if two demands share several arcs, the same relative spectrum order is enforced on all of them.
- $c \in [\max_{d \in D} w(d), F] \subset \mathbb{Z}$ is the spectrum envelope variable;

The objective is to minimize the spectrum width:

$$\min c \tag{1}$$

The constraints are modeled as follows.

1. *Flow conservation:* selected arcs for a demand must form a feasible flow

$$\sum_{e \in \delta^-(v)} y_{de} - \sum_{e \in \delta^+(v)} y_{de} = \begin{cases} -1, & \text{if } v = s(d), \\ 1, & \text{if } v = t(d), \\ 0, & \text{otherwise,} \end{cases} \qquad \forall d \in D,\ \forall v \in V. \tag{2}$$

2. *Degree constraints:* ensure correctness of node degrees for each demand

$$\sum_{e \in \delta^-(s(d))} y_{de} = 0, \qquad \sum_{e \in \delta^+(s(d))} y_{de} = 1, \quad \forall d \in D, \tag{3}$$

$$\sum_{e \in \delta^+(t(d))} y_{de} = 0, \qquad \sum_{e \in \delta^-(t(d))} y_{de} = 1, \quad \forall d \in D, \tag{4}$$

$$\sum_{e \in \delta^-(v)} y_{de} \le 1, \qquad \sum_{e \in \delta^+(v)} y_{de} \le 1, \qquad \forall d \in D,\ \forall v \in V \setminus \{s(d), t(d)\}. \tag{5}$$

3. *Envelope constraint:* the envelope variable c upper-bounds the right endpoint of every allocated spectrum interval:

$$l_d + w(d) \le c, \qquad \forall d \in D. \tag{6}$$

Therefore, minimizing c is equivalent to minimizing the overall spectrum envelope.

4. *Non-overlap on shared arcs:*
$\forall d, d' \in D : n(d) < n(d'),\ \forall e \in E :$

$$l_d + w(d) \le l_{d'} + F(1 - o_{dd'}) + F(2 - y_{de} - y_{d'e}), \tag{7}$$

$$l_{d'} + w(d') \le l_d + F\, o_{dd'} + F(2 - y_{de} - y_{d'e}), \tag{8}$$

The term $F(2 - y_{de} - y_{d'e})$ acts as an arc-usage gate:
- If $y_{de} = y_{d'e} = 1$, then both demands use arc e, the gate equals 0, and the ordering variable $o_{dd'}$ becomes active.
- If at least one of y_{de} or $y_{d'e}$ is 0, then the gate is at least F, which is large enough to relax both inequalities, making them non-binding on arc e.

When the gate is active, the binary variable $o_{dd'}$ selects exactly one of the two possible non-overlap orders:

- If $o_{dd'} = 1$, then (7) reduces to

$$l_d + w(d) \leq l_{d'},$$

while (8) is relaxed by the term $F\,o_{dd'} = F$. Thus, demand d is placed before demand d' on arc e.

- If $o_{dd'} = 0$, then (8) reduces to

$$l_{d'} + w(d') \leq l_d,$$

while (7) is relaxed by the term $F(1 - o_{dd'}) = F$. Thus, demand d' is placed before demand d on arc e.

The formulation contains $O(|D||E| + |D|^2)$ variables and $O(|D||V| + |D|^2|E|)$ constraints.

ILP Path-Link Formulation. We introduce additional notations for path-link formulation:

- $P(d)$ is a precomputed finite set of candidate paths for each demand d;
- $E(p) \subseteq E$ is a set of directed arcs used by $p \in P(d)$;
- C is the set of potentially conflicting demand pairs:

$$C = \{(d, d') \in D \times D : n(d) < n(d'), \exists p \in P(d), \exists q \in P(d') \text{ s.t. } E(p) \cap E(q) \neq \emptyset\}$$

- $\Gamma_d(e)$ is the set of all paths of demand d that use arc e:

$$\Gamma_d(e) = \{p \in P(d) : e \in E(p)\}$$

- $Q_{dd'}(p)$ is the set of all candidate paths of demand d' that conflict with candidate path p of demand d by sharing at least one arc:

$$Q_{dd'}(p) = \{q \in P(d') : E(p) \cap E(q) \neq \emptyset\}$$

$Q_{dd'}(p)$ can be equivalently expressed as $Q_{dd'}(p) = \bigcup_{e \in E(p)} \Gamma_{d'}(e)$.

We introduce the following decision variables:

- $x_{dp} \in \{0, 1\}$ equals 1 if candidate path $p \in P(d)$ is selected for demand d, and 0 otherwise;
- $l_d \in [0, F - w(d)] \subset \mathbb{Z}$ denotes the leftmost slot index assigned to demand d;
- $s_{dd'} \in \{0, 1\}$ for each $(d, d') \in C$ indicates whether the selected paths of demands d and d' share at least one arc;
- $o_{dd'} \in \{0, 1\}$ for each $(d, d') \in C$ encodes the spectrum ordering of the two demands whenever $s_{dd'} = 1$: $o_{dd'} = 1$ means that d is placed before d', while $o_{dd'} = 0$ means that d' is placed before d;

- $c \in [\max_{d\in D} w(d), F] \subset \mathbb{Z}$ is the spectrum envelope variable;

The objective is to minimize the spectrum envelope:

$$\min\ c. \tag{9}$$

The constraints are modeled as follows.

1. *Path-selection:* ensure that each demand is assigned exactly one candidate route:
$$\sum_{p\in P(d)} x_{dp} = 1, \quad \forall d \in D. \tag{10}$$
2. *Envelope constraint:* the envelope variable c upper-bounds the right endpoint of every allocated spectrum interval:
$$l_d + w(d) \leq c, \quad \forall d \in D. \tag{11}$$
3. *Correct identification of conflicting pairs:* model correctness of the binary variable $s_{dd'}$, which is used to indicate whether the selected paths of demands d and d' share at least one directed arc:
$$s_{dd'} \geq \sum_{p\in\Gamma_d(e)} x_{dp} + \sum_{q\in\Gamma_{d'}(e)} x_{d'q} - 1, \quad \forall (d,d') \in C,$$
$$\forall e \in E \text{ such that } \Gamma_d(e) \neq \emptyset,\ \Gamma_{d'}(e) \neq \emptyset. \tag{12}$$
$$s_{dd'} \leq 1 - x_{dp} + \sum_{q\in Q_{dd'}(p)} x_{d'q}, \quad \forall (d,d') \in C,\ \forall p \in P(d). \tag{13}$$
Together, constraints (12)–(13) enforce the logical equivalence:
$$s_{dd'} = 1 \iff \text{the selected paths of } d \text{ and } d' \text{ share at least one arc.}$$
If the chosen path of demand d uses arc e and the chosen path of demand d' also uses arc e, then both sums in (12) are equal to 1, and (12) forces $s_{dd'} = 1$. $s_{dd'}$ is prevented from taking value 1 unless the chosen paths actually conflict:
   - If path p is not selected for demand d, then (13) is inactive because of the term $1 - x_{dp}$.
   - If path p is selected, then the constraint reduces to $s_{dd'} \leq \sum_{q\in Q_{dd'}(p)} x_{d'q}$, which means that $s_{dd'}$ can be equal to 1 only if demand d' selects a path that shares at least one arc with p.

4. *Correctness of the ordering variables:* The binary variable $o_{dd'}$ determines the relative order of the spectrum intervals of demands d and d', but only when the selected paths share an arc:
$$o_{dd'} \leq s_{dd'}, \quad \forall (d,d') \in C. \tag{14}$$
Hence, if the selected paths do not share any arc, then $s_{dd'} = 0$ and (14) forces $o_{dd'} = 0$, so the ordering variable becomes irrelevant.

5. *Spectrum non-overlap:*
   $\forall(d, d') \in C$:

$$l_d + w(d) \leq l_{d'} + F\big(1 - o_{dd'}\big) + F\big(1 - s_{dd'}\big), \quad (15)$$

$$l_{d'} + w(d') \leq l_d + F\, o_{dd'} + F\big(1 - s_{dd'}\big). \quad (16)$$

   The term $F(1 - s_{dd'})$ acts as a conflict gate:
   - If $s_{dd'} = 0$, then both inequalities are relaxed by F and become non-binding, which is correct because demands d and d' do not need spectrum separation when their selected paths are arc-disjoint.
   - If $s_{dd'} = 1$, then the gate disappears and (15)–(16) enforce one of the two admissible spectrum orders.

   The term involving $o_{dd'}$ selects which order is active:
   - If $o_{dd'} = 1$, then (15) reduces to $l_d + w(d) \leq l_{d'}$, while (16) is relaxed by the term $F\, o_{dd'} = F$. Thus, demand d is placed before demand d'.
   - If $o_{dd'} = 0$, then (16) reduces to $l_{d'} + w(d') \leq l_d$, while (15) is relaxed by the term $F(1 - o_{dd'}) = F$. In this case, demand d' is placed before demand d.

   Therefore, when $s_{dd'} = 1$, the two inequalities enforce disjoint spectrum intervals and use $o_{dd'}$ to choose the relative order.

Let $k_{\max}$ be the maximum number of candidate paths allowed per demand, so that $|P(d)| \leq k_{\max}$ for all $d \in D$. In the worst case, when $|C| = \binom{|D|}{2}$ (i.e., every pair of demands is potentially conflicting), the formulation contains $O(|D|k_{\max} + |D|^2)$ variables and $O(|D|^2(|E| + k_{\max}))$ constraints (Table 1).

Table 1. Asymptotic size of the ILP formulations.

Formulation	Variables	Constraints
Node-link	$O(\lvert D\rvert\lvert E\rvert + \lvert D\rvert^2)$	$O(\lvert D\rvert\lvert V\rvert + \lvert D\rvert^2\lvert E\rvert)$
Path-link	$O(\lvert D\rvert k_{\max} + \lvert D\rvert^2)$	$O(\lvert D\rvert^2(\lvert E\rvert + k_{\max}))$

3 Path Generators

The formulation of the path-link model assumes a finite set of candidate routes connecting each source-destination pair. Each of the proposed generators yields at most $k_{\max}$ simple candidate routes connecting a given source-destination pair.

- **Shortest Paths.** Candidate routes connecting a given source-destination pair are generated by Yen's algorithm [14] on the unweighted version of the directed graph model of the network. The first route connecting a given source-destination pair is the shortest simple route connecting the source and the destination. Each of the additional routes connecting the source-destination pair is generated by the classical spur-node deviation procedure. In the absence of edge weights, the shortest path computations in Yen's algorithm reduce to a breadth-first search procedure, which implies that the routes generated by the algorithm are ordered in nondecreasing order of their lengths, i.e., their hop counts.

- **Arc-Disjoint Shortest Paths.** Paths are constructed sequentially through repeated use of a breadth-first search. After a path is found, all arcs on that path are removed from the graph before continuing to the next path. As a result, the next path is the shortest path in the updated directed graph. This generator produces a greedy list of pairwise arc-disjoint shortest paths when such paths exist. As arc reuse is not allowed, this generator typically returns fewer than $k_{\max}$ paths on sparse graphs.
- **Path clustering.** This method was proposed in [17]. For each pair of simple paths p and p' between two vertices, a similarity measure (introduced in [19]) is defined. Since edge weights are not considered in our setting, we use the unweighted version of the measure:

$$\rho_{pp'} = \frac{|E(p) \cap E(p')|}{\min\left(|E(p)|, |E(p')|\right)}.$$

  Based on these pairwise similarities, the candidate path pool is partitioned into $k_{\max}$ clusters. One representative path is then selected from each cluster.
- **Penalty-based shortest paths.** This generator uses a weight α_e for each directed arc $e \in E$, initialized as

$$\alpha_e = 1, \quad \forall e \in E.$$

  At each iteration, a shortest path is computed using Dijkstra's algorithm [20] according to the current arc weights. If the selected path is P, the arc weights are updated as follows:
  - **Additive penalty rule:** $\alpha_e \leftarrow \alpha_e + p$, for all $e \in E(P)$,
  - **Multiplicative penalty rule:** $\alpha_e \leftarrow p \cdot \alpha_e$, for all $e \in E(P)$,

  where p is a given penalty factor. As a result, previously used arcs are not forgotten and remain available for use in the next path. The multiplicative rule increases the penalty for reuse much faster than the additive penalty rule. As a consequence, it generates more diverse paths and tends to produce arc-disjoint paths when alternative routes are available.

In the experiments, $k_{\max} = 8$. The clustering generator first builds a pool of 50 shortest paths and then uses average-linkage clustering. The additive and multiplicative penalized generators use penalty parameters $p = 1$ and $p = 10$, respectively; duplicate paths are skipped, so a generator may return fewer than $k_{\max}$ paths.

4 Results

We evaluate one node-link formulation and five path-link candidate generators: shortest paths, shortest disjoint paths, path clustering, and additive and multiplicative penalized shortest paths. The benchmark contains 480 common instances over five standard optical network graphs (ARPANET, EON, NSFNET, UKNet [21], and Spanish Telefonica [4]) and demand counts $|D| \in \{10, 20, 30, 40, 50, 60, 80, 150\}$. All experiments were run on a PC with an AMD R7 7840HS CPU and 32 GB RAM using the Google OR-Tools CP-SAT solver with the default thread setting. In the path-link experiments we used $k_{\max} = 8$ candidate paths per demand. The per-instance time limits were demand-dependent: node-link used 120, 240, 480, 600, 800, 1200, 1200, and 3600 s for $|D| = 10, 20, 30, 40, 50, 60, 80, 150$, respectively, while path-link used 300, 600, 480, 1200, 800, 1200, 2400, and 3600 s (Fig. 1, Table 2).

Table 2. Agreement with the node-link optimum on the 424 instances where the node-link model proved optimal.

Method	Matches node-link optimum (%)	Worse than node-link optimum (%)	Unsolved (%)	Mean gap (%)
node-link reference	100.00	0.00	0.00	0.00
shortest	93.87	6.13	0.00	1.56
disjoint	82.55	17.45	0.00	4.63
clustered	95.99	4.01	0.00	1.00
pen. additive	97.88	2.12	0.00	0.46
pen. multiplicative	97.88	2.12	0.00	0.46

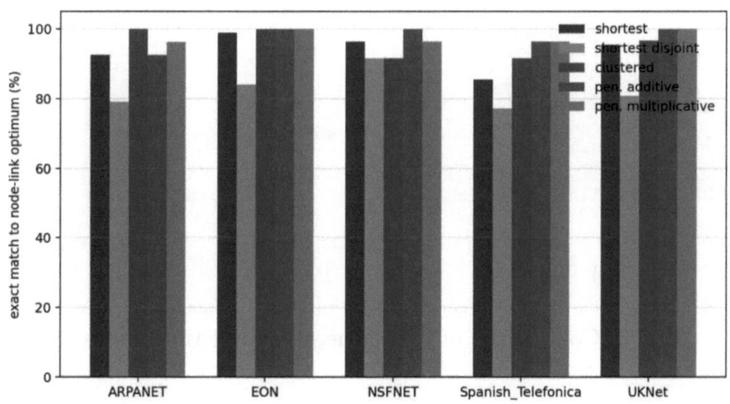

Fig. 1. Exact-match rate to the node-link optimum by graph on the subset where the node-link model proved optimal.

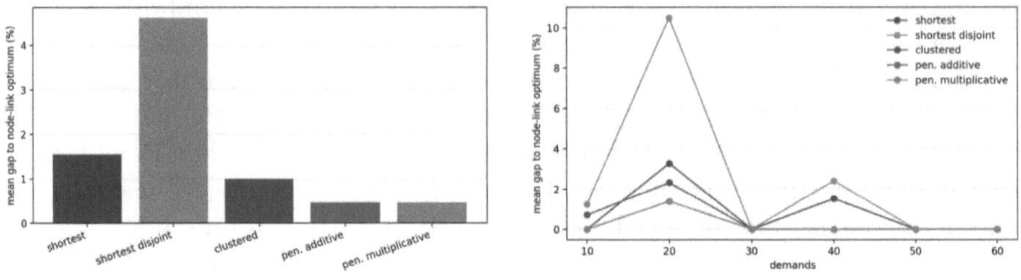

(a) (b)

Fig. 2. Gap to the node-link optimum on the subset where the node-link model proved optimal: overall performance (left) and scaling with demand count (right).

Optimality comparisons are reported only on the 424 instances where the node-link model proved optimal, so that the exact formulation can serve as a reliable reference. This subset contains all 140 instances with $|D| = 10$, all 140 with $|D| = 20$, 131 of the 140 with $|D| = 40$, 8 of the 10 instances with $|D| = 30$, 3 of the 10 instances with $|D| = 50$, 2 of the 10 instances with $|D| = 60$, and none with $|D| = 80$ or $|D| = 150$. Therefore, the quality analysis below should be interpreted as an exact-reference comparison on small and medium-sized cases. On that subset, the additive and multiplicative penalized generators show the best agreement with the node-link optimum, matching it in 97.88% of instances with a mean gap of 0.46%. The clustered generator follows closely with a 95.99% match rate and 1.00% mean gap, the shortest-path generator remains competitive at 93.87% and 1.56%, and the shortest-disjoint generator is clearly less accurate at 82.55% and 4.63%.

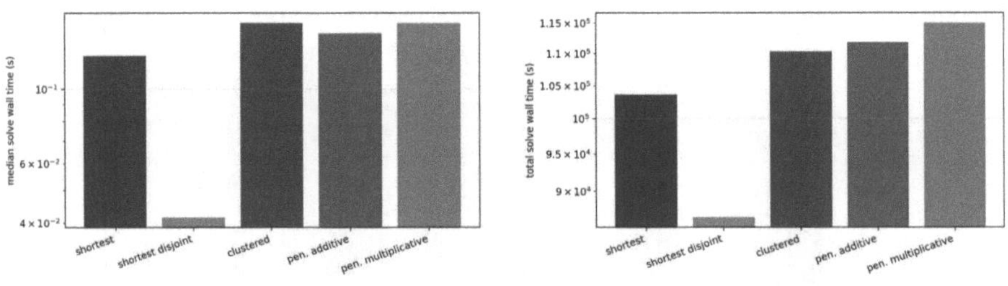

(a) (b)

Fig. 3. Comparison of path-generator solve-time metrics.

Figure 2a and Fig. 2b show that the penalized generators remain closest to the node-link reference across the filtered benchmark. The clustered generator is also consistently accurate, while the shortest-disjoint strategy deteriorates much more noticeably as demand count increases. Some medium-size points are difficult despite having fewer demands because hardness is also driven by spectrum tightness and bottleneck links: when many candidate routes must reuse the same critical arcs, small changes in path diversity can increase the spectrum width. This confirms that richer candidate-path sets improve spectral efficiency, even though they lead to larger path-link models.

The solve-time comparison between path generators is presented in Fig. 3a, Fig. 3b, and Fig. 4a. All path-link variants remain fast on typical instances: the median runtime is 0.04 s for the shortest-disjoint generator, 0.13 s for shortest paths, 0.15 s for penalized additive, and about 0.16 s for clustered and penalized multiplicative. Total runtime is dominated by a relatively small number of difficult instances, but the overall ranking is similar: shortest-disjoint is the fastest, shortest is next, and clustered and penalized generators spend more time exploring larger and more diverse candidate sets.

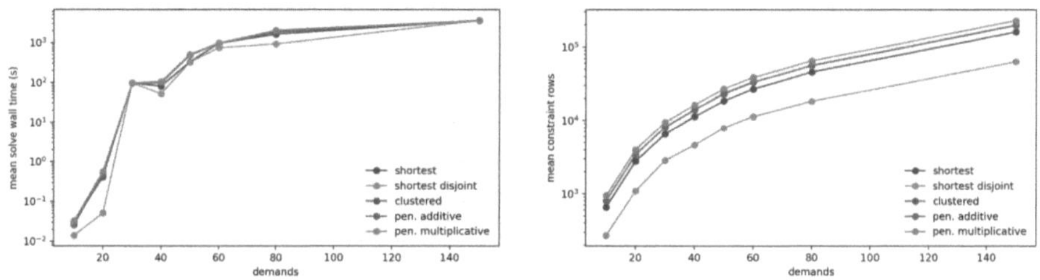

(a) (b)

Fig. 4. Scaling with demand count for path generators: solve-time performance (left) and constraint-matrix size (right).

This behavior is further explained by Fig. 4b, Fig. 5a, and Fig. 5b. For all path-link formulations, the number of candidate-path variables and pairwise conflict checks grows rapidly with $|D|$. The constraint counts differ because the formulation adds lower-bound sharing constraints for arcs that can be used by both demands, upper-bound sharing constraints for candidate paths of a conflicting demand pair, and two non-overlap constraints for each potentially conflicting demand pair. The shortest-disjoint strategy controls this growth most aggressively because it often returns fewer than $k_{\max}$ usable paths, whereas shortest, clustered, and penalized generators keep candidate-path counts close to the requested limit and therefore induce substantially larger path-pair structures.

To illustrate the role of the candidate-path limit, we additionally ran a $k = 1, \ldots, 15$ sweep on the UKNet instance with $|D| = 40$ and number of slots equal to 60. Figure 6 shows that increasing k can sharply reduce spectrum usage when it introduces useful route diversity: the penalized generators reach the best observed spectrum width of 6 slots by $k = 5$–6, while clustering reaches it at $k = 9$ and the plain shortest-path generator only at $k = 15$. At the same time, the number of conflict pairs grows quickly and then saturates once most candidate

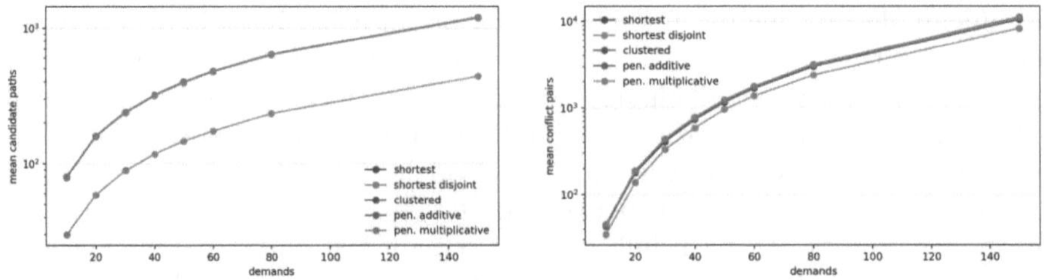

(a) (b)

Fig. 5. Problem-size scaling with demand count for path generators: candidate-path variables (left) and conflict pairs (right).

paths already share the same bottleneck arcs. This supports using $k_{\max} = 8$ in the main benchmark as a practical compromise between quality and model size.

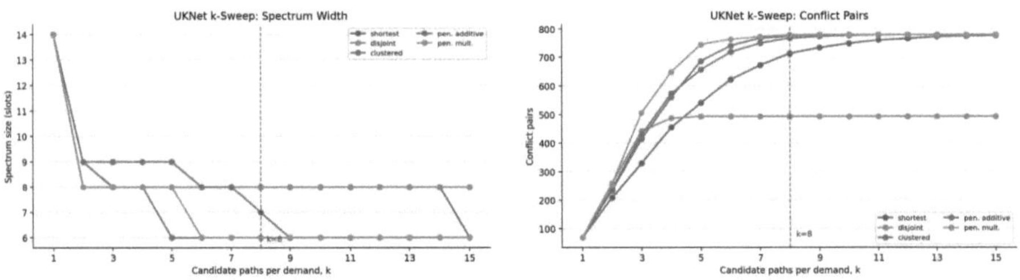

(a) (b)

Fig. 6. Diagnostic UKNet k-sweep: spectrum width (left) and conflict-pair count (right) for `UKNet_test_s1_40_60`.

Finally, performance across graph topologies is compared in Fig. 7b and Fig. 8. The graph-level speed ranking is broadly stable across the tested topologies, with shortest-disjoint usually remaining the fastest. In contrast, solution quality is strongly topology-dependent even on the node-link-optimal subset. The shortest-disjoint generator performs reasonably on sparse graphs but degrades substantially on denser meshes, whereas clustered and penalized generators remain much closer to the node-link reference.

Memory usage is summarized in Fig. 7a. Among the path generators, the shortest-path variant has the smallest mean peak Resident Set Size (RSS) at about 4.93 GB, while the multiplicative-penalty variant reaches the largest value at about 7.15 GB.

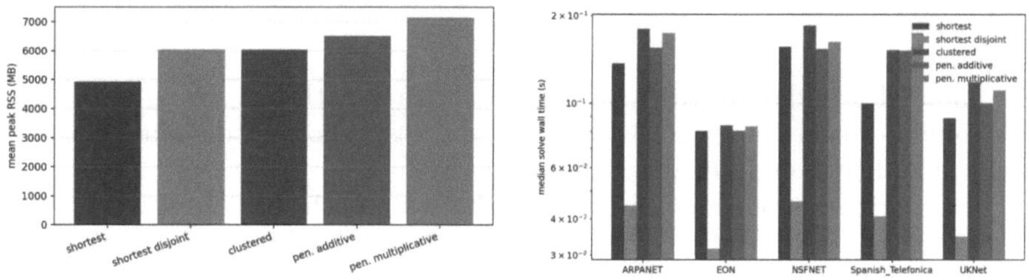

(a) (b)

Fig. 7. Path-generator resource usage and performance: memory footprint (left) and solve time by graph (right).

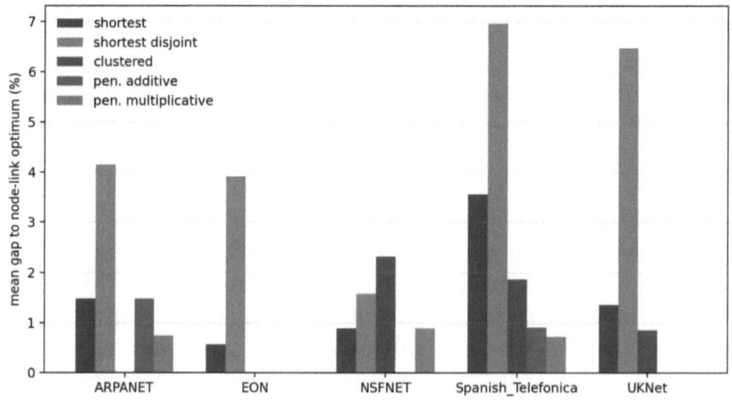

Fig. 8. Mean gap to the node-link optimum by topology on the subset where the node-link model proved optimal.

Table 3. Full-benchmark summary of path generation strategies over all common instances.

Method	Solved (%)	Optimal (%)	Mean gap to best (%)	Median time (s)	Mean RSS (MB)
shortest	100.00	90.62	1.47	0.13	4931
disjoint	100.00	92.29	4.11	0.04	6049
clustered	99.79	89.79	1.03	0.16	6045
penalized additive	99.79	89.17	0.56	0.15	6515
penalized multiplicative	100.00	88.75	0.62	0.16	7151

5 Conclusion

We compared a node-link RSA formulation with a path-link formulation using five candidate-path generation strategies: shortest, shortest disjoint, clustered, penalized additive, and penalized multiplicative.

The experiments show a clear trade-off between path-set richness, solve time, memory usage, and spectral efficiency. For solution-quality analysis, the exact comparison must be restricted to the 424 instances where the node-link formulation proved optimal. On that subset, the two penalized generators perform best, each matching the node-link optimum in 97.88% of cases with a mean gap of 0.46%. The clustered generator is the next-best compromise, while shortest-disjoint is consistently the least accurate. For the broader benchmark, where node-link proves optimality mainly up to $|D| = 40$ and only in a few $|D| = 50$ and $|D| = 60$ cases, the path generators should therefore be compared primarily on their own speed, scaling behavior, and memory footprint. The full-benchmark path-generator summary is presented in Tab. 3. In this table, "Optimal" denotes the percentage of all common instances for which the corresponding path-link run proved optimality. The "gap to best" is computed for each instance relative to the smallest spectrum width found by any compared method on that instance, and then averaged over the benchmark.

Overall, the shortest-disjoint generator is attractive when speed is the main priority, but its quality loss is substantial. The penalized generators provide the smallest full-benchmark gap to the best observed spectrum width, with the additive rule slightly ahead of the multiplicative rule in this aggregate metric. The clustered generator offers a strong intermediate option with lower memory usage than the penalized variants. Future work will focus on broader topology sets, tuning penalty parameters to graph and demand characteristics, and developing adaptive or dynamic candidate-generation schemes, including column-generation approaches, that preserve quality while reducing path-link model size.

References

1. Zang, H., Jue, J.P., Mukherjee, B.: A review of routing and wavelength assignment approaches for wavelength-routed optical WDM networks. Optical Netw. Mag. **1**(1), 47–60 (2000)
2. Chatterjee, B.C., Sarma, N., Oki, E.: Routing and spectrum allocation in elastic optical networks: a tutorial. IEEE Commun. Surv. Tutorials **17**(3), 1776–1800 (2015)
3. Christodoulopoulos, K., Tomkos, I., Varvarigos, E.A.: Elastic bandwidth allocation in flexible OFDM-based optical networks. J. Lightwave Technol. **29**(9), 1354–1366 (2011)
4. Velasco, L., Klinkowski, M., Ruiz, M., Comellas, J.: Modeling the routing and spectrum allocation problem for flexgrid optical networks. Photon Netw. Commun. **24**(3), 177–186 (2012)
5. Klinkowski, M., Walkowiak, K.: Routing and spectrum assignment in spectrum sliced elastic optical path network. IEEE Commun. Lett. **15**(8), 884–886 (2011)

6. Pióro, M., Medhi, D.: Routing, Flow, and Capacity Design in Communication and Computer Networks. Morgan Kaufmann, California (2004)
7. Bertero, F., Bianchetti, M., Marenco, J.: Integer programming models for the routing and spectrum allocation problem. TOP **26**(3), 465–488 (2018). https://doi.org/10.1007/s11750-018-0483-6
8. Wang, J., Shigeno, M., Wu, Q.: ILP models and improved methods for the problem of routing and spectrum allocation. Opt. Switch. Netw. **45**, 100675 (2022)
9. Zheng, W., et al.: Improved ILP models and heuristics for solving routing and resource allocation problems in optical networks. J. Lightwave Technol. **43**(7), 3016–3033 (2025)
10. Assis, K.D.R.: Channel-based RSA approaches for QOS protection of slices over elastic optical networks. IEEE Access **10**, 20714–20726 (2022)
11. Abkenar, F.S., Rahbar, A.G.: Study and analysis of RSA and RMSA algorithms in elastic optical networks. Opt. Switch. Netw. **23**, 11–39 (2017)
12. Wu, H., et al.: Analysis framework of RSA algorithms in elastic optical rings. J. Lightwave Technol. **37**(13), 3110–3121 (2019)
13. Matzner, R., et al.: Topology bench: systematic graph-based benchmarking for core optical networks. J. Optical Commun. Network. **17**(1), 7–27 (2025)
14. Yen, J.Y.: Finding the k shortest loopless paths in a network. Manage. Sci. **17**(11), 712–716 (1971)
15. Bhandari, R.: Survivable Networks: Algorithms for Diverse Routing. Springer, New York (1999)
16. Klinkowski, M., Walkowiak, K.: Routing and spectrum assignment in spectrum sliced elastic optical path network. IEEE Commun. Lett. **15**(8), 884–886 (2011)
17. Mori, Y., Shigeno, M.: Selection of candidate paths based on diversity in resource allocation problems on optical networks. Opt. Switch. Netw. **58**, 100826 (2025)
18. Goscien, R., Walkowiak, K.: A column generation technique for routing and spectrum allocation in cloud-ready survivable elastic optical networks. Int. J. Appl. Math. Comput. Sci. **27**(3), 591–603 (2017)
19. Chondrogiannis, T., Bouros, P., Gamper, J., Leser, U., Blumenthal, D.B.: Finding k-shortest paths with limited overlap. VLDB J. **29**(5), 1023–1047 (2020)
20. Dijkstra, E.W.: A note on two problems in connexion with graphs. Numer. Math. **1**, 269–271 (1959)
21. Baroni, S., Bayvel, P.: Wavelength requirements in arbitrarily connected wavelength-routed optical networks. J. Lightwave Technol. **15**(2), 242–251 (1997)

Finding Equilibrium in Networks with Flow Losses

Oleg V. Khamisov[1(✉)], Ilya M. Minarchenko[1], and Oleg O. Khamisov[2]

[1] Matrosov Institute for System Dynamics and Control Theory, Irkutsk, Russia
khamisov@isem.irk.ru

[2] Skolkovo Institute of Science and Technology, Moscow, Russia

Abstract. The problem of finding equilibrium in oligopolistic network problem is considered. The main attention is devoted to the taking into account the network line losses. A potential-like approach is suggested and results on a test example is given.

Keywords: Network equilibrium · Potential game · Transmission losses

1 Introduction

The network equilibrium problems are of great interest both from theoretical and practical points of view [1]. One of the main problem in this topic is the network problem with transmission capacity constraints [2], which has serious influence on existence and finding network equilibrium. Problem of finding equilibrium in network with transmission capacity constraints is one of the key problems in multiple industrial branches. One of the most common examples is power flow problem in power systems. Similar problems are formulated for hydraulic systems [3]. In this type of problems multiple producers form oligopolistic market, with each producers' goal being profit maximization. There exists a multitude approaches to solve such type of problems: reduction of the game problem to convex optimization or application of machine learning approaches for finding Nash equilibrium and [4]. However, the problem complexity increases significantly with addition of transmission losses. As a result, the considered problem becomes non-potential leading to a significant increase in the computational complexity. While, it is possible in some particular cases to formulate corresponding optimization problem, i.e. optimal power flow in power systems such type of problems is highly non-convex.

In our paper we concentrate the investigation on networks with lines losses. Such problems arise again in equilibrium model of power energy systems. Line Losses bring specific features in the problems, however, from our point of view they are not so restrictive as transmission capacities. We suggest an approach based on an extension of the theory of potential games.

A. Kazakov et al. (Eds.): MOTOR 2026, LNCS 16687, pp. 390–397, 2026.
https://doi.org/10.1007/978-3-032-31390-4_25

2 The Model

The mathematical structure of the model is given by a graph with n nodes. Topology of the graph is described by an $n \times n$ adjacency matrix A, elements a_{ij} of A have standard definition: $a_{ij} = 1$ if node i is connected to node j and $a_{ij} = 0$ otherwise. The set J_s is the set of product supply nodes, the set J_c is the set of consumption (demand) nodes. Sets J_s and J_c can have nonempty intersection. Amount of production volume at node $j \in J_s$ is denoted by x_j, $\overline{x}_j$ is the production capacity ($\overline{x}_j > 0,\ j \in J_s,\ \overline{x}_j = 0,\ j \notin J_s$), $c_j\colon R \to R$ is the cost function. The demand function d_j at node $j \in J_c$ is linear,

$$d_j = \eta_j - \nu_j p, \tag{1}$$

$\eta_j > 0,\ \nu_j > 0$ are given constants, p is the price. For $j \notin J_c$ we assume $\eta_j = \nu_j = 0$.

Network balance relations between production, transmission and consumption are presented by the following system of linear equations

$$x_j + \sum_{i=1}^{n} a_{ij}(1 - \delta_{ij})y_{ij} - \sum_{i=1}^{n} a_{ji}y_{ji} = d_j,\ j = 1, \ldots, n, \tag{2}$$

where y_{ij} is the product flow from node i to node j, $\delta_{ij} \in (0, 1)$ is the loss transmission factor for line $i - j$. Transmission capacities of lines are given by non-negative constants $\overline{y}_{ij}$, $i, j = 1, \ldots, n$,

$$y_{ij} \in [0, \overline{y}_{ij}],\ i, j = 1, \ldots, n. \tag{3}$$

If there is no line between node i and node j, then $\overline{y}_{ij} = 0$.

Assume that production and consumption values x_j and d_j are given. Then system (2) under constraints (3) may be inconsistent or, if consistent, may have non unique solution in y_{ij}. In the latter case the purpose of the network is to provide transmission flows with minimal costs, i.e. the flows are determined as a solution of the problem consisting in minimization of the total transmission costs

$$\sum_{i=1}^{n}\sum_{j=1}^{n} \mu_{ij} y_{ij} \to \min_{y_{ij}} \tag{4}$$

under constraints (2)–(3) with fixed $x_j,\ d_j,\ j = 1, \ldots, n$ and with the line transmission costs $\mu_{ij} > 0$. If problem (2)–(4) is inconsistent then the previously given values x_j and d_j are unacceptable.

Producers behave themselves as participants of an oligopilistic market, i.e. each producer maximizes its own profit that leads to the following system of optimization problems:

$$p x_j - c_j(x_j) \to \max_{x_j}, \tag{5}$$

$$x_j \in [0, \overline{x}_j],\ j \in J_s. \tag{6}$$

The price p can be determined from Eqs. (1)–(2), connecting producers, consumers and network. We can say that producers implicitly determine the price p through Eqs. (1)–(2).

The aim of the model consists in finding such production amounts x_j and price value p that maximize profits of the producers and provide minimal transmission costs in problem (1)–(4).

3 Solution Methodology

The main difficulty in interconnection of problems (5)–(6) and (1)–(4) is caused by network transmission losses. In order to understand this difficulty better let us assume that there are no losses in the network, i.e. $\delta_{ij} = 0, i, j = 1, \ldots, n$. Then from (2) and (1) we have

$$\sum_{j\in J_s} x_j = \sum_{j\in J_c} d_j = \sum_{j\in J_c} (\eta_j + \nu_j p) = \sum_{j\in J_c} \eta_j + \left(\sum_{j\in J_c} \nu_j\right) p = \eta - \nu p, \tag{7}$$

where $\eta = \sum\limits_{j\in J_c} \eta_j, \nu = \sum\limits_{j\in J_c}^{n} \nu_j$. Hence, we can define the inverse demand function

$$p = \frac{1}{\nu}\left(\eta - \sum_{j\in J_s} x_j\right) = r - w \sum_{j\in J_s} x_j, \tag{8}$$

where $r = \frac{\eta}{\nu}$, $w = \frac{1}{\nu}$. In this case system of problems (5)–(6) together with equality (8) is closed or self-sufficient in the following sense. The system of connected optimization problems has now the following form

$$F_j(x_j, x_{-j}) = (r - w \sum_{i\in J_s\setminus\{j\}} x_i - w x_j) x_j - c_j(x_j) \to \max_{x_j}, \tag{9}$$

$$x_j \in [0, \overline{x}_j],\ j \in J_s. \tag{10}$$

Here x_{-j} is the standard notation: $x_{-j} = (x_i : i \in J_s \setminus \{j\})^\top$. The objective function F_j of each producer j is maximized with respect to its own strategy x_j still being dependent on the strategies of other producers presented in the vector x_{-j}. The solution of the optimization problems system (9)–(10) is understood in the sense of Nash-equilibrium [5]. Since the inverse demand function (8) is linear the equilibrium problem (9)–(10) is potential [6,7]. This property is proved in [8]. Therefore, solution of the optimization problems system (9)–(10) can be reduced to the following optimization problem [9]

$$\Pi(x) = -\frac{w}{2} x^\top H x + r \sum_{j\in J_s} x_j - \sum_{j\in J_s}^{n} c_j(x_j) \to \max, \tag{11}$$

$$x \in X = \prod_{j \in J_s} [0, \overline{x}_j], \tag{12}$$

where diagonal elements of the matrix H are equal to 2 and other elements are equal to 1. Matrix H is positive definite and if cost functions $c_j,\ j \in J_s$ are convex and continuous then function Π is continuous concave and problem (11)–(12) is a convex optimization problem. Since $X \neq \varnothing$ problem (11)–(12) always has a solution x^*. Therefore, finding equilibrium production x^*, equilibrium price p^* and equilibrium demand d^* is reduced to the following steps:

Step 1. *Solve optimization problem* (11)–(12) *in order to find the equilibrium production* $x_j^*,\ j \in J_s$.
Step 2. *Determine the equilibrium price:* $p^* = \eta - \nu \sum\limits_{j \in J_s} x_j^*$.
Step 3. *Determine the node demand:* $d_j^* = \eta_j - \nu_j p,\ j \in J_c$.

Now we can switch to the distribution problem (1)–(4) with $x_j = x_j^*,\ j \in J_s$ and $d_j = d_j^*,\ j \in J_c$. Since $\sum\limits_{j \in J_s} x_j^* = \sum\limits_{j \in J_c} d_j^*$, system (2) (with $\delta_{ij} = 0$) is always solvable. If we assume that the corresponding distribution problem is solvable, then finding solution of the network equilibrium problem is reduced to solving two optimization problems: convex optimization problem (11)–(12) and linear optimization problem (1)–(4).

Now let us return to the initial problem with $\delta_{ij} > 0$. Summarizing all Eqs. in (2) we obtain the following equation

$$\sum_{j \in J_s} x_j - \sum_{i=1}^{n} \sum_{j=1}^{n} a_{ij} \delta_{ij} y_{ij} = \sum_{j \in J_c} d_j. \tag{13}$$

In the general case it can be assumed that total losses and total production satisfy the relation

$$\sum_{i=1}^{n} \sum_{j=1}^{n} a_{ij} \delta_{ij} y_{ij} = \Delta \sum_{j \in J_s} x_j \tag{14}$$

for some $\Delta \in (0, 1)$. Then in the same way as in (7) we have

$$(1 - \Delta) \sum_{j \in J_s} x_j = \eta - \nu p \tag{15}$$

and

$$p = \frac{\eta}{\nu} - \frac{(1 - \Delta)}{\nu} \sum_{j \in J_s} x_j. \tag{16}$$

We can try to repeat the above approach in the following way. Fix $\Delta = \hat{\Delta}$ and determine the demand function

$$p = r - \hat{w} \sum_{j \in J_g} x_j \tag{17}$$

with r as above and $\hat{w} = \frac{1-\hat{\Delta}}{\nu}$. Solve problem (11)–(12) for $w = \hat{w}$ and determine the corresponding $\hat{x}^*$, $\hat{p}^*$ and $\hat{d}^*$. Solve the distribution problem (1)–(4) for $x_j = \hat{x}_j^*$, $j \in J_s$ and $d_j = \hat{d}_j^*$, $j \in J_c$. If the latter problem is solvable then stop. If the problem is inconsistent then we have to try another value of Δ.

The source of the inconsistency is the following: either we overestimate or underestimate the optimal value of total losses. In order to overcome this difficulty we substitute equality constraints (2) by inequality ones

$$x_j + \sum_{i=1}^{n} a_{ij}(1-\delta_{ij})y_{ij} - \sum_{i=1}^{n} a_{ji}y_{ji} \leqslant d_j,\ j = 1,\dots,n. \tag{18}$$

If problem (4), (1), (18), (3) is still inconsistent then the current value $\hat{\Delta}$ is to big, due to the overestimated losses production values $\hat{x}_j^*$, $j \in J_s$ are in total greater than the real demand. In this case the factor $\hat{\Delta}$ must be decreased. If problem (4), (1), (18), (3) is consistent, but some inequalities in (18) are fulfilled as strict inequalities then the losses are underestimated and total production is less that the real demand. In this case the factor $\hat{\Delta}$ must be increased. In order to avoid an oscillating process the steps of increment or decrement must be monotonously decreasing. We suggest to set next increment-decrement step values equal to one half of the previous one.

We also have to take into account the following condition. Solution of problem (4), (1), (18), (3) allows opposite flows along the same line, which is forbidden in some models, for example, in competitive energy market models. To cancel the opposite flows the binary variables ω_{ij} are introduced in the standard way together with the constraints [10]

$$y_{ij} \leqslant w_{ij}\overline{y}_{ij},\ i,j = 1,\dots,n, \tag{19}$$

$$\omega_{ij} + \omega_{ji} \leqslant 1,\ i,j = 1,\dots,n. \tag{20}$$

Now we can describe the main result of our paper: algorithm for finding the equilibrium solution.

<u>Step 0.</u> *Set bounding values* $\underline{\Delta}_0$ *and* $\overline{\Delta}_0$ *and* $k = 0$.
<u>Step 1.</u> *Calculate* $\Delta_k = \frac{\underline{\Delta}_k + \overline{\Delta}_k}{2}$.
<u>Step 2.</u> *Set* $w_k = 1 - \Delta_k$.
<u>Step 3.</u> *Solve optimization problem*

$$\Pi(x) = -\frac{w_k}{2}x^\top Hx + r\sum_{j\in J_s} x_j - \sum_{j\in J_s}^{n} c_j(x_j) \to \max, \tag{21}$$

$$x \in X = \prod_{j\in J_s}[0,\overline{x}_j]. \tag{22}$$

Let x^k *be a solution.*

Step 4. *Calculate the total supply* $x_T^k = \sum_{j \in J_s} x_j^k$.
Step 5. *Calculate the price* $p_k = \eta - \nu x_T^k$.
Step 6. *Calculate the nodal demand* $d_j^k = \eta_j - \nu_j p_k,\ j \in J_c$.
Step 7. *Set* $x_j^k = 0$ *for* $j \notin J_s$ *and* $d_j^k = 0$ *for* $j \notin J_c$.
Step 8. *Solve mixed linear boolean problem*

$$\sum_{i=1}^{n}\sum_{j=1}^{n} \mu_{ij} y_{ij} \to \min, \tag{23}$$

$$x_j^k + \sum_{i=1}^{n} a_{ij}(1-\delta_{ij}) y_{ij} - \sum_{i=1}^{n} a_{ji} y_{ji} \leqslant d_j^k,\ j = 1, \ldots, n, \tag{24}$$

$$y_{ij} \leqslant w_{ij} \overline{y}_{ij},\ i, j = 1, \ldots, n, \tag{25}$$

$$\omega_{ij} + \omega_{ji} \leqslant 1,\ i, j = 1, \ldots, n, \tag{26}$$

$$w_{ij} \in \{0, 1\},\ i, j = 1, \ldots, n, \tag{27}$$

$$0 \leqslant y_{ij} \leqslant \overline{y}_{ij},\ i, j = 1, \ldots, n. \tag{28}$$

Step 9. *If problem* (23)–(28) *is inconsistent then set* $\overline{\Delta}_{k+1} = \Delta_k$, $\underline{\Delta}_{k+1} = \underline{\Delta}_k$ *and goto Step 11. Otherwise goto Step 10.*
Step 10. *Let* (y^k, ω^k) *be a solution of problem* (23)–(28). *If there exist inequality constraints in* (24) *that are satisfied as the strict inequalities then set* $\overline{\Delta}_{k+1} = \overline{\Delta}_k$, $\underline{\Delta}_{k+1} = \Delta_k$ *and goto Step 12. Otherwise goto Step 12.*
Step 11. *If* $\overline{\Delta}_{k+1} - \underline{\Delta}_{k+1} \leqslant \varepsilon$ *then stop: no equilibrium solution is determined. Otherwise set* $k = k + 1$ *and goto Step 1.*
Step 12. *Stop:* (x^k, d^k, p_k, y^k) *is an equilibrium solution.*

4 The Example

In this example the network consists of $n = 4$ nodes, the adjacency matrix

$$A = \begin{pmatrix} 0 & 0 & 1 & 1 \\ 0 & 0 & 1 & 1 \\ 1 & 1 & 0 & 1 \\ 1 & 1 & 1 & 0 \end{pmatrix},$$

the set of product supply nodes $J_s = \{1, 3\}$, production capacities $\overline{x}_1 = 500$, $\overline{x}_3 = 500$, cost functions $c_1(x_1) = 0.01x_1^2 + 0.5x_1$, $c_3(x_3) = 0.02x_3^2 + 0.3x_3$, the set of consumption nodes $J_c = \{1, 2, 3, 4\}$, inverse demand functions $d_1 = 100 - 0.1p$, $d_2 = 200 - 0.2p$, $d_3 = 400 - 0.15p$, $d_4 = 300 - 0.05p$, transmission losses $\delta_{ij} = 0.02\ \forall i, j$, transmission capacities $\overline{y}_{ij} = 200\ \forall i, j$, line transmission costs $\mu_{ij} = 1\ \forall i, j$.

Table 1. Results of computations by iterations.

k	$\underline{\Delta}_k$	$\overline{\Delta}_k$	Δ_k	x_1^k	x_3^k	p_k	DPS	Dev
0	0	0.5	0.25	444.24	438.59	675.74	infes	–
1	0	0.25	0.125	380.79	376.65	674.49	infes	–
2	0	0.125	0.0625	355.41	351.79	673.99	infes	–
3	0	0.0625	0.03125	343.94	340.56	673.77	infes	–
4	0	0.03125	0.015625	338.48	335.21	673.66	fes	0

Starting values were chosen as $\underline{\Delta}_0 = 0$, $\overline{\Delta}_0 = 0.5$ because it is unlikely that the network losses can be higher that 50% of the total supply. Results of application of the algorithm from the previous section is given in Table 1.

In Table 1 abbreviation DPS means Distribution Problem Status—solution status of problem (23)–(28), infes corresponds to infeasible, fes corresponds to feasible. Dev is the maximum deviation from equality in constraints (24). Equilibrium solution was found in 5 iterations. Equilibrium supply is given by x_1^4 and x_3^4, equilibrium price p_4, equilibrium demand $d^4 = (32.63, 65.27, 298.95, 266.32)^\top$, nonzero equilibrium flows $y_{13}^4 = 120.01, y_{14}^4 = 185.84, y_{34}^4 = 153.88, y_{42}^4 = 66.60$.

5 Conclusion

The suggested algorithm is of practical importance. If opposite flows are allowed then the mixed boolean problem at the Step 8 can be reduced just to linear optimization problem. However, in this case it is necessary to remember that solution of this problem is in general non-unique. Another extension of the suggested approach is the following. In the case of potential game it is not necessary to assume that cost functions c_j are convex, they may be nonconvex or discontinuous. If so then the corresponding problem (21)–(22) loses its convexity property, but still can be used in finding equilibrium solutions.

Acknowledgments. The research was supported by RSF (project No 24-41-03004).

Disclosure of Interests. The authors have no competing interests to declare that are relevant to the content of this article.

References

1. Nagurney, A.: Networks Economics. A Variational Inequality Approach. 2nd edn. Springer New York, (1999)
2. Liu, Y., Wu, F.F.: Impacts of network constraints on electricity market equilibrium. IEEE Trans. Power Syst. **22**(1), 126–135 (2007)
3. Battiti, R., Kvasov, D.E., Sergeyev, Y.D. (eds.): LION 2017. LNCS, vol. 10556. Springer, Cham (2017). https://doi.org/10.1007/978-3-319-69404-7

4. Meshcheryakov, A.A., Khamisov, O.O.: Search of the Nash equilibrium for a system of two interconnected raw materials and products markets. J. Min. Inst. EDN AULTP (2025)
5. Owen, G.: Game Theory. 2nd edn. Academic Press (1982)
6. Monderer, D., Shapley, L.S.: Potential games. Games Econ. Behav. **14**(1), 124–143 (1996)
7. Potential Game Theory. LNCS, Springer, Cham (2016). https://doi.org/10.1007/978-3-319-30869-2_5
8. Slade, M.: What does an oligopoly maximize? J. Ind. Econ. XLI **I**(1), 45–61 (1994)
9. Muu, L.D., Nguyen, V.H., Quy, N.V.: On Nash-Cournot oligopolistic market equilibrium models with concave cost function. J. Glob. Optim. **41**(3), 351–364 (2008)
10. Balas, E.: Solving (CGLP) k on the LP Simplex Tableau. In: Disjunctive Programming, pp. 107–119. Springer, Cham (2018). https://doi.org/10.1007/978-3-030-00148-3_9

GPU-Parallel Alternatives for Traffic Assignment

Pavel Pishchev[1] and Demyan Yarmoshik[2(✉)]

[1] Moscow Institute of Physics and Technology, Moscow, Russia
pishchev.pm@phystech.edu

[2] Moscow Independent Research Institute of Artificial Intelligence, Moscow, Russia
yarmoshik.d@miriai.org

Abstract. This paper compares several well-known algorithms for solving the traffic assignment problem and introduces a new method based on sequential quadratic programming (SQP). The proposed approach is designed with GPU-parallelization in mind, distinguishing it from traditional algorithms commonly used in this domain. Experimental results show that the method achieves performance comparable to existing techniques in a single-threaded environment while also demonstrating acceleration from GPU-parallelism. These findings highlight the advantages of the proposed algorithm for GPU-parallel computing environments while also identifying scenarios where classical shortest-path–based approaches remain more effective.

Keywords: Multicommodity flows · Primal-dual methods · Path-based methods · Sequential quadratic programming

1 Introduction

1.1 Traffic Assignment

The traffic assignment problem is a core model of equilibrium behavior in transportation networks. Given origin-destination demand, link travel-time functions, and a road graph, we seek a distribution of path flows produced by selfish route choices. At equilibrium, all used routes for each OD pair have minimal and equal travel cost, while unused routes are not cheaper; this corresponds to Wardrop's first principle [3,17]. In static settings, this condition is often represented by the Beckmann formulation, i.e., a convex multicommodity flow problem (4) [1].

Solving traffic assignment yields equilibrium link loads and travel times, which are then used to estimate congestion, test infrastructure upgrades, and compare policy scenarios [3,17]. Modern machine-learning approaches can successfully perform short-term (∼1 h) traffic forecasting [14,20], but unfortunately, they are not yet applicable for long-term traffic modeling due to lack of training data.

A. Kazakov et al. (Eds.): MOTOR 2026, LNCS 16687, pp. 398–412, 2026.
https://doi.org/10.1007/978-3-032-31390-4_26

1.2 Motivation and Contribution

Traffic assignment remains highly relevant because both model fidelity and data scale keep growing. Open transportation benchmarks and recent city-scale datasets include increasingly detailed road graphs and demand matrices, making each equilibrium solve expensive [18,21]. In practical studies this cost multiplies, since assignment is repeatedly called inside calibration and scenario-analysis pipelines [3].

At the same time, available hardware has shifted toward massively parallel architectures. This creates a gap between mature shortest-path-based algorithms and newer linear-algebra-driven formulations that can exploit GPU execution more directly [22,23]. Bridging this gap, while preserving solution quality on realistic networks, is the main motivation of this work.

Nonlinear multicommodity flow algorithms can be categorized into two groups by the approach they use for handling flow conservation constraints. The first, dominating, approach consists of iteratively reevaluating link costs and redistributing a portion of flow along shortest paths for each origin-destination pair. In turn, shortest path calculation for each commodity can be performed sequentially on CPU using Dijkstra's algorithm [3], or can be parallelized on GPU using Bellman-Ford or similar algorithms [12].

The second approach is to formulate flow conservation constraints in Laplacian/Kirchhoff form (6) and either apply structural optimization, usually interior-point (IP) solvers, that can potentially use matrix inversion subroutines, or matrix-free primal-dual methods for convex optimization with affine constraints. Note, that this dichotomy is quite vague, because, on the one hand structural solvers can use matrix-free linear algebra subroutines, and, on the other hand, primal-dual methods can utilize information about individual matrix elements, e.g. in preconditioning. Nevertheless, most Primal-Dual algorithms rely on simple matrix-vector operation and therefore perform best when executed on GPU [23], while IP solvers can benefit from CPU or GPU linear algebraic backend depending on problem properties [7,11]. We summarize the situation in Table 1.

Table 1. Classification of nonlinear multicommodity flow solvers by algorithmic approach to flow conservation constraints

	CPU	GPU
Shortest-path-based	Dijkstra [3]	Bellman-Ford [12]
Laplacian-based	IP (this work)	Primal-dual [22,23], IP (this work)

Laplacian-based formulation of multicommodity flow problems is gaining popularity as it allows to replace path-finding-based flow distribution algorithmic subroutines with pure matrix-vector operations, facilitating simple implementation of solvers with assistance of modern tensor frameworks, which provides a desirable feature of out-of-the-box support for GPU-parallelism [22,23].

To the best of our knowledge, no IP-based algorithms for traffic assignment problem were proposed in the literature, while primal-dual algorithms are ineffective on medium and large-scale instances [22]. We close this gap by providing an IP-based SQP algorithm for traffic assignment problem. We also conduct numerical experiments comparing best path-based, primal-dual and IP algorithms on CPU and GPU. We find out that SQP, path-based and NFW algorithms are competitive in single-threaded CPU setup, while SQP outperforms other algorithms on GPU. In short, our contributions are as follows:

Contribution

- New Laplacian-based SQP algorithm that outperforms other traffic-assignment algorithms on GPU;
- Numerical comparison of SOTA methods for traffic assignment on CPU and GPU.

1.3 Paper Organization

In Sect. 1 we motivate the problem and summarize our main contributions. In Sect. 2 we formalize the traffic assignment problem in both path-based and Laplacian forms. In Sect. 3 we review existing path-based and Laplacian-based first-order methods used as baselines. In Sect. 4 we present our SQP method for the Laplacian formulation. In Sect. 5 we report numerical results in CPU and GPU setups. In Sect. 6 we conclude and discuss the main findings.

2 Problem Statement

In this work we consider the Beckmann model for equilibrium traffic flow assignment [1] which is essential for urban planning and road infrastructure optimization [4,17,21]. Consider the properties that a travel time–flow function should satisfy, i.e., the functional relationship between travel time and traffic volume on a road segment. Even in the absence of traffic, travel incurs a nonzero time cost corresponding to free-flow conditions, which is primarily determined by the segment length and the maximum permitted speed. Furthermore, there exists a threshold level of traffic volume beyond which additional flow cannot be accommodated within the nominal capacity of the facility. It is also evident that travel time is a non-decreasing function of traffic volume and approaches its maximum as flow approaches this threshold. In the proposed model, travel time is represented by a monotonic function of flow; however, strict upper bounds on traffic volume are not imposed. Usually the Bureau of Public Roads (BPR) cost function is used [19], which has the following time-flow relationship:

$$\tau_e(f_e) = \bar{\tau}_e \left(1 + \rho \left(\frac{f_e}{\bar{f}_e} \right)^{1/\mu} \right), \tag{1}$$

where $\bar{f}_e$ has the meaning of the maximum flow on the segment, $\bar{\tau}_e$ is the travel time along a road segment in the absence of traffic, ρ controls how strongly congestion affects travel time, μ controls how sharply travel time increases as congestion grows, and f_e is the flow on road e. The model parameters are determined empirically and differ for each segment.

The traffic assignment problem is then reduced to minimizing the following potential function:

$$\Psi(f) = \sum_{e \in \mathcal{E}} \int_0^{f_e} \tau_e(z)dz = \left[\sigma_e(f_e) = \int_0^{f_e} \tau_e(z)dz\right] = \sum_{e \in \mathcal{E}} \sigma_e(f_e), \tag{2}$$

where $\mathcal{E}$ is the set of links in the graph.

Traffic flows cannot be arbitrary; they must correspond, in some sense, to the movement of people. Let O be the set of origin vertices and D the set of destination vertices. Then we have the set of origin–destination pairs $OD = \{(i,j) \mid i \in O,\ j \in D\}$. For each pair (i,j) in this set a demand d_{ij} is specified, representing the amount of flow from the origin i to the destination j. One can also define the set of all paths from i to j denoted by $\mathcal{P}_{ij}$.

The constraints can be formalized in several ways. The first approach is to define the set of feasible path flow distributions:

$$X = \left\{ x \mid d_{ij} = \sum_{p \in \mathcal{P}_{ij}} x_p \ \forall (i,j) \in OD;\ x_p \geq 0\ \forall p \in \cup_{(i,j) \in OD} \mathcal{P}_{ij} \right\}. \tag{3}$$

Let us introduce the incidence matrix $\Theta = \{\theta_{ep}\}_{e \in \mathcal{E}, p \in \mathcal{P}}$, where $\theta_{ep} = 1$ if link e belongs to path p and $\theta_{ep} = 0$ otherwise. Then $f = \Theta x$, and the traffic assignment problem can be written as

$$\Psi(f) = \sum_{e \in \mathcal{E}} \sigma_e(f_e) \longrightarrow \min_{f = \Theta x, x \in X} \tag{4}$$

The constraints can also be described directly in terms of flows, without using paths. Let us introduce the matrix $F = \{f_{ei}\}_{e \in \mathcal{E}, i \in O}$. The constraints on F can be obtained using the incidence matrix A of the graph representing the transportation network. The product AF contains information about how much flow remains at each vertex j for source i

$$\Delta_{ij} = \sum_{e \in \mathcal{E}_{in}(j)} f_{ei} - \sum_{e \in \mathcal{E}_{out}(j)} f_{ei}.$$

However, this quantity can also be expressed in terms of the demands

$$\begin{cases} \Delta_{ij} = -\sum_{k \neq i} d_{ik}, & j = i \\ \Delta_{ij} = 0, & j \notin D\,. \\ \Delta_{ij} = d_{ij}, & j \in D \end{cases} \tag{5}$$

Using this notation, the problem can be written as follows

$$\Psi(f) = \sum_{e \in \mathcal{E}} \sigma_e(f_e) \longrightarrow \min_{AF=\Delta, F \geq 0}. \tag{6}$$

3 Existing Algorithms

3.1 Finding Flows Along Shortest Paths

For several of the algorithms, we need to be able to solve the following problem: for certain fixed link weights τ, define link flows f such that for each source-target pair, only the shortest paths are used and all demands are satisfied. From a mathematical perspective, the problem can be formulated as follows:

$$\arg\min_{f \in \{\Theta x : x \in X\}} \langle \tau, f \rangle. \tag{7}$$

where X and Θ have same meaning as in problem (4).

For any algorithm for finding the shortest paths while preserving the predecessor of a vertex, any weights vector W and any demand matrix C, this problem is solved by Algorithm 1:

Algorithm 1. Flows on shortest paths

```
procedure FLOWS_ON_SHORTEST(G, C, τ)
    f := ARRAY(num_edges(G))
    for s in Sources(G) do
        sps := SHORTEST_PATHS(G, τ, s, Targets(G))
        for t in Targerts(G) do
            sp := sps[t]
            for e do in EDGES(sp)
                f[e] += C[s][t]
            end for
        end for
    end for
    return f
end procedure
```

As we can see, we need a shortest-path algorithm that returns the predecessor of each vertex and that can handle multiple objectives. In a single-threaded scenario, Dijkstra's algorithm with Fibonacci heap can be chosen as the shortest-path function, as it has the best known complexity of $\Theta(|E| + |V| \log |V|)$ [10]. The algorithm can be efficiently parallelized across sources, since shortest paths for different sources can be computed independently. However, the algorithm for a single source cannot be further parallelized effectively, because in graphs typical for our problem each vertex has only a small number of neighbors.

3.2 Frank-Wolfe Algorithm

The Frank-Wolfe algorithm is based on the linearization of the objective function [9]. Since the gradient of the objective function in our problem is the time (cost) of moving along links, the Frank-Wolfe algorithm shifts the same part of the flow from all paths to the shortest paths.

Algorithm 2. Frank-Wolfe with linesearch

```
procedure FRANK-WOLFE(G, C, n_iters)
    f^0 := FLOWS_ON_SHORTEST(G, C, τ̄)
    for k = 0 to n_iters do
        τ^k := τ(f^k)
        s^k := FLOWS_ON_SHORTEST(G, C, τ^k)
        γ := arg min_{γ∈[0,1]} Ψ(γf^k + (1 − γ)s^k)
        f^{k+1} := γf^k + (1 − γ)s^k
    end for
    return f
end procedure
```

Each iteration of Algorithm 2 is dominated by shortest path finding algorithm. For Dijkstra's algorithm complexity is $\Theta(|E| + |V| \log |V|)$, where E is the set of links and V is the set of vertices. Since we need to calculate the shortest paths for all sources, the total complexity of one iteration is $\Theta(|S|(|E| + |V| \log |V|))$, where S is the set of sources.

3.3 N-Conjugate Frank-Wolfe (NFW)

The Frank-Wolfe algorithm can be improved by updating the flows using not only the last step direction but also several previous steps [13]. The coefficients of the convex combination are chosen such that the current step direction is Hessian-conjugate at the current point to the directions of previous steps, assuming that the previous steps were Hessian-conjugate at the current point:

$$s^k = \alpha_0 s_k^{FW} + \sum_{i=1}^{N_{curr}} \alpha_i s^{k-i} \tag{8}$$

$$A_{k-m} \equiv d^{k-m} H^k d_k^{FW}$$
$$B_{k-m} \equiv d^{k-m} H^k d^{k-m}$$

$$\beta_m = \frac{-A_{k-m}}{B_{k-m}(1-\gamma_{k-m})} + \frac{\gamma_{k-m}}{1-\gamma_{k-m}} \sum_{m=n+1}^{N_{curr}} \beta_m \tag{9}$$

$$\alpha_m \equiv \beta_m \alpha_0$$

$$\alpha_0 = \frac{1}{1 + \sum_{i=1}^{N_{curr}} \beta_i}$$

Generally speaking, they won't be perfectly conjugate, but assuming the Hessian doesn't change significantly over a single step, the error won't be large. At the same time, this assumption greatly simplifies the calculation of the coefficients.

Algorithm 3. N-conjugate Frank-Wolfe with linesearch

procedure N-CONJUGATE FRANK-WOLFE(G, C, n_iters, N, γ_{max})
  $f^0 :=$ FLOWS_ON_SHORTEST(G, C, $\bar{\tau}$)
  $N_{curr} = 1$ ▷ number of step direction to consider
  for $k = 0$ **to** n_iters **do**
    $\tau^k := \tau(f^k)$
    $s_k^{FW} :=$ FLOWS_ON_SHORTEST(G, C, τ^k)
    Calculate $\{\alpha_i\}_{i=0}^{N_{curr}}$ using (9)
    $s^k := \alpha_0 s_k^{FW} + \sum_{i=1}^{N_{curr}} \alpha_i s^{k-i}$
    $d^k := s^k - f^k$
    $\gamma := \arg\min_{\gamma \in [0,1]} \Psi(f + \gamma s^k)$
    $f^{k+1} := f^k + \gamma s^k$
    $N_{curr} = \min(N_{curr} + 1, N)$
    $N_{curr} := 1$ **if** $\gamma > \gamma_{max}$
  end for
  return f
end procedure

Compared to the previous algorithm, operations with complexity $\Theta(N|E|)$ have been added. The best convergence is typically achieved for small N, so the complexity of one iteration remains $\Theta(|S|(|E| + |V| \log |V|))$.

3.4 Gradient Projection

The most time-consuming step in the Frank-Wolfe method is calculating the shortest paths, so it can be useful to store them and reuse them. This is the idea behind path-based algorithms. The essence of path-based algorithms is to store several active paths for each source-sink pair and, at each iteration, find the shortest path, add it to the set of active paths, and balance it. A path is removed from the set of active paths when the flow along the path becomes zero. The difference in path-based algorithms lies in the balancing step. In the gradient projection algorithm, flows are shifted from all paths to the shortest one, and the size of the transferred flow is determined by one iteration of Newton's method [6].

At each iteration, we traverse all origin–destination (OD) pairs (in total, $\Theta(|S|^2)$) and reassign the flows. Let the number of active paths for any OD pair be bounded by K (in practice, this number is small, typically around 1–3), and let the maximum path length between an origin and a destination be L. Then, the complexity of flow reassignment for a single OD pair is $\Theta(KL)$, yielding

Algorithm 4. Gradient projection algorithm (Sketch)

```
procedure GRAD-PROJ(G, C, n_iters)
    For any OD pair find the shortest paths and add them to the active set.
    Assign all flows to the shortest paths.
    Compute the link weights.
    for k = 1 to n_iters do
        For any OD pair find the shortest paths
        Add them to the active set if they are not already included.
        Reassign part of the flow from all paths to the shortest ones
        If there is a path with zero flow, then remove it from the set of active paths
    end for
    return f
end procedure
```

an overall complexity of $\Theta(|S|^2KL + |S|(|E| + |V|\log|V|))$. As a rule, the term $|S|^2KL$ makes a significant contribution to the performance of the algorithm.

Although we are not aware of studies devoted to analyzing the convergence rate of the Algorithm 4, in practice it requires fewer operations than Algorithms 2 and 3.

3.5 Chambolle-Pock Algorithm

By reformulating the problem (6), we can move away from finding the shortest paths to matrix multiplications. In this formulation, we can solve the problem using primal-dual methods, such as the Chambolle-Pock algorithm [5]. The most difficult step is finding the value of the proximal operator for the functions σ_e. Number of variables in this subproblem can be reduced by expressing this value in terms of the values of the proximal operator for σ_e^* – the convex conjugate of σ_e. The proximal operator for the conjugate function is found using Newton's method independently for each variable. The final algorithm is given as Algorithm 5.

Algorithm 5. Chambolle-Pock algorithm

procedure CHAMBOLLE-POCK(G, A, b, n_iters, γ, ν, θ)

$f^0 :=$ ARRAY(num_edges(G), size($Sources(G)$)) $\hat{f}^0 := f^0$

$y^0 :=$ ARRAY(num_nodes(G), size($Sources(G)$))

$z^0 :=$ ARRAY(num_edges(G))

for $k = 0$ **to** n_iters **do**

$y^{k+1} := y^k + \gamma(A\hat{f}^k - b)$

$z_e^{k+1} := \arg\min_z(\sigma_e^*(z) + \frac{1}{2\gamma}||z - z_e^k - \gamma\hat{f}_e^k||^2)\ \forall e$

$f^{k+1} = \max(0, f^k - \nu(A^T y^{k+1} + z^{k+1}))$

$\hat{f}^{k+1} := f^{k+1} + \theta(f^{k+1} - f^k)$

end for

return f

end procedure

Most expensive operation in iteration of Algorithm 5 is matrix multiplications $A\hat{f}^k$ and $A^T y^{k+1}$ with complexity $\Theta(|E||V||S|)$, which is much worse than previous algorithms. But this algorithm has smaller hidden constants in $\Theta(\cdot)$ notation and has no branches, so in practice one iteration requires significantly less time. Furthermore, matrix multiplication can be parallelized much more efficiently than finding shortest paths.

In terms of the number of iterations, the situation is the opposite: due to the fact that the Chambolle-Pock algorithm uses a locally optimal direction, the required number of iterations is much higher than in the previous algorithms considered.

4 Sequential Quadratic Programming

Algorithm 6, like the Chambolle-Pock algorithm, is based on a matrix reformulation. At each iteration of the algorithm, the gradient g and Hessian H of the objective function are found, and the problem is replaced with a quadratic approximation:

$$\frac{1}{2}d^T H(x_k)d + g^T(x_k)d \to \min_d \quad s.t. \quad A(x_k + d) = b, \;\; x_k + d \geq 0 \qquad (10)$$

There are many very effective solvers for quadratic problems. For our specific problem, we will use the CLARABEL [11] and CuCLARABEL [7] solvers. These solvers are based on the interior point method (replacing inequality constraints with barrier functions) and use sparse matrices. The most difficult part of these algorithms is solving a system of linear equations, which can be parallelized.

Let's find the gradient and Hessian for our problem. First, our objective function is decomposed into a sum of functions $\Psi(f) = \sum_{e\in E} \sigma_e(\sum_{s\in S} f_{es})$, each of these functions depends on its own set of variables, so the gradient and Hessian can be split into $|E|$ blocks that can be considered separately. Since $\nabla_{es}\sigma_e(f_e) = \tau_e(f_e)$, the linear approximation of σ_e can be written as $\sum_{s\in S} \tau_e(f_e)\Delta f_{es} = \tau_e(f_e)\sum_{s\in S} \Delta f_{es} = \tau_e(f_e)\Delta f_e$. Similarly, handle the Hessian: $\sum_{i,j\in S} \tau_e^{'}(f_e)\Delta f_{ei}\Delta f_{ej} = \tau_e^{'}(f_e)\sum_{i\in S}\Delta f_{ei}\sum_{j\in S}\Delta f_{ej} = \tau_e^{'}(f_e)(\Delta f_e)^2$.

Since the objective function depends only on Δf_e, and not on individual Δf_{ei}, we can introduce them as separate variables and add the constraints $\hat{d}_e = \sum_{i\in S} d_{ei}$. This will make the matrix sparser and speed up CLARABEL on large networks.

5 Numerical Results

5.1 Experimental Setup

For algorithms such as Frank–Wolfe and gradient projection, which remain within the feasible region at every iteration, the relative gap is typically used as the convergence metric. However, the Chambolle–Pock and SQP algorithms may leave the feasible region, making the relative gap unapplicable.

Algorithm 6. Sequential Quadratic Programming

procedure SEQ-QUAD(G, A, b, n_iters)
  $f^0 :=$ ARRAY(num_edges(G), size($Sources(G)$))
  for $k = 0$ **to** n_iters **do**
    $g := \tau(f^k)$
    $h := \tau^{'}(f^k)$
    $d := \arg\min_{d,\hat{d}}(\frac{1}{2}h^T\hat{d}^2 + g^T\hat{d})$ ▷ Using CLARABEL
      $s.t.\ \ Ad = -Af^k + b,\ d \geq -f^k,\ \hat{d}_e = \sum_{i \in S} d_{ei}$
    $f^{k+1} := f^k + d$
  end for
  return f
end procedure

In addition, for our problem, there are no benchmark values for flows or for the objective function. The model only approximates real road networks, and the observed real flows may not correspond to the minimum of the objective function.

Therefore, the remaining approach is to compare the group of algorithms against the best-performing algorithm. To do this, we run all considered algorithms for a large number of iterations. We then select the best algorithm according to the value of the objective function, excluding Chambolle–Pock (as will be shown later, it fails to converge on many networks), and evaluate the Euclidean distance to the resulting flows as well as the difference from the best objective function value. We do not employ a stopping criterion. If needed, one may use the norm of the difference between the flows at the current and previous iterations. For always-feasible algorithms such as NFW and gradient projection a more trustworthy duality gap convergence metric is available [2]. However, Laplacian-based algorithm can violate flow conservation constraints in approximate solutions, thus on convergence plots we show distance to a highly accurate approximate solution computed by running the best method for each network with increased iterations budget.

The following hyperparameters were chosen for the algorithms:

- **N-conjugate Frank-Wolfe:** $N = 3$, $\gamma_{max} = 0.999$.
- **Chambolle-Pock:** $\gamma = \frac{0.99}{16||A||}$, $\nu = \frac{0.99 \cdot 16}{||A||}$, $\theta = 1$.
- **Sequential Quadratic Programming:** default settings for the CLARABEL solver, except for disabling data equilibration and iterative refinement.

We compare the algorithms on six networks from [18]: Anaheim, Sioux-Falls, Terrassa-Asymmetric, Eastern-Massachusetts, Berlin-Tiergarten, and Berlin-Mitte-Center.

5.2 Single-Threaded Setup

The results of the algorithms running in single-threaded mode. For the experiments, an Intel Core i7-12650H processor was used. Algorithm 1 for finding shortest-paths flows was implemented using Dijkstra's algorithm from high-performance `graph-tool` library [16] for `SHORTEST_PATHS` subroutine, while the flow accumulation loop was implemented using the Python JIT-compiler `numba` [15]. Results are shown in Figs. 1 and 2.

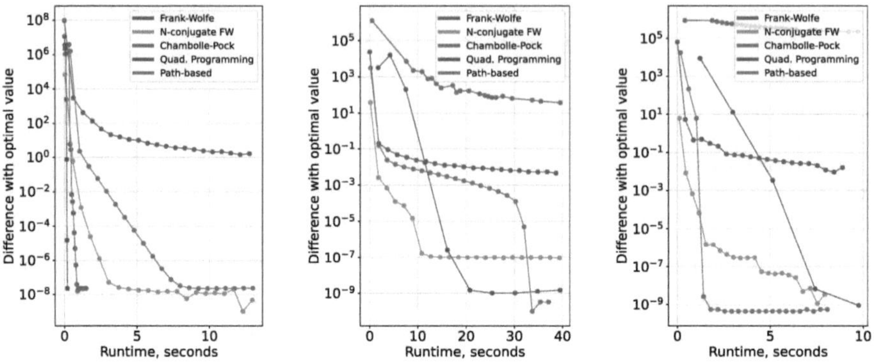

(a) (b) (c)

Fig. 1. Comparison of difference with optimal value for Sioux-Falls (a), Anaheim (b), Berlin-Mitte-Center (c) in single-threaded setup for all methods

As can be seen from the plots, the SQP algorithm outperforms the others on the Sioux Falls, Anaheim, and Terrassa-Asymmetric networks. Its poorer performance in the remaining cases may be related to the presence of links whose travel time does not depend on the flow. This issue can be partially mitigated by inserting one iteration of the Frank–Wolfe algorithm between the inner iterations (which is much cheaper than an inner SQP iteration); however, the convergence rate of the inner iterations still remains too slow.

5.3 GPU Setup

The results of the best shortest-path-based algorithm (using CPU), Chambolle-Pock and SQP (both CPU and GPU) are shown in Figs. 3 and 4. For the experiments, an Intel Core i7-12650H processor and an NVIDIA RTX 4050 Mobile graphics card were used.

While [12] reports acceleration of shortest-paths flow distribution (all-or-nothing assignment) for road graphs on GPU, their implementation was used in a proprietary software Saturn, and, to the best of our knowledge, there is no open-source implementations for all-or-nothing flow assignment available. Note, that an effective implementation must be quite involved, because high-level packages do not provide optimized procedures suitable for aggregating flows from the predecessors map obtained after shortest paths calculation. Even without this issue, we observed that only calculating shortest paths without flow aggregation

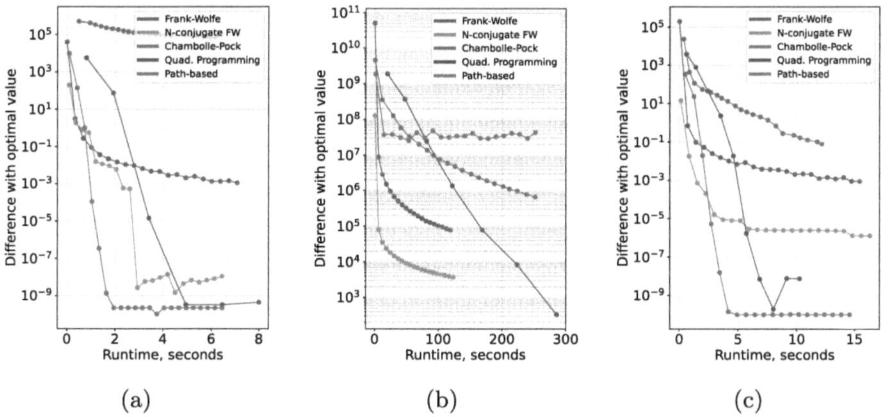

Fig. 2. Comparison of difference with optimal value for Berlin-Tiergarten (a), Terrassa-Asymmetric (b), Eastern-Massachusetts (c) in single-threaded setup for all methods

using NVIDIA's **RAPIDS cuGraph** library with shortest paths algorithm from [8] is approximately 50+ times slower on all tested networks than complete all-or-nothing assignment via **graph-tool** plus **numba** on CPU.

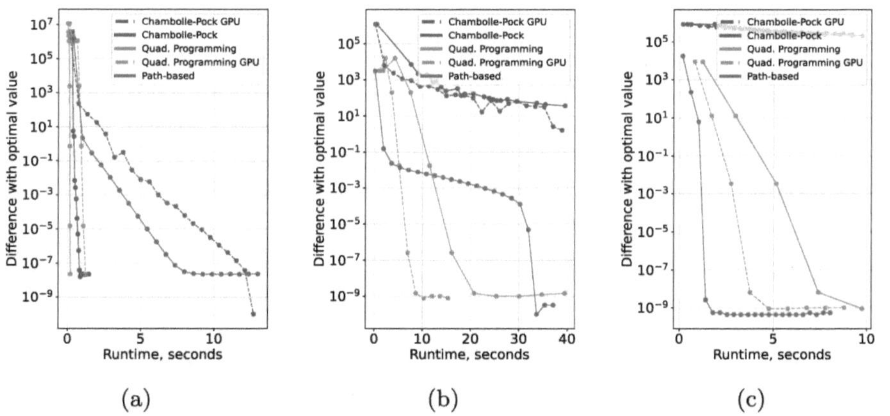

Fig. 3. Comparison of difference with optimal value for Sioux-Falls (a), Anaheim (b), Berlin-Mitte-Center (c) in single-threaded setup for sequential quadratic programming, Chambolle-Pock (single-threaded and GPU) and best shortest-path-based algorithm

As can be seen from the plots, for medium-sized networks and larger, substantial speedups can be achieved using a GPU. The poorer performance on small networks may be due to the need to transfer memory to and from the GPU in the case of Sequential quadratic programming, as well as the inefficiency of parallelizing the computation of the proximal operator in the Chambolle–Pock algorithm.

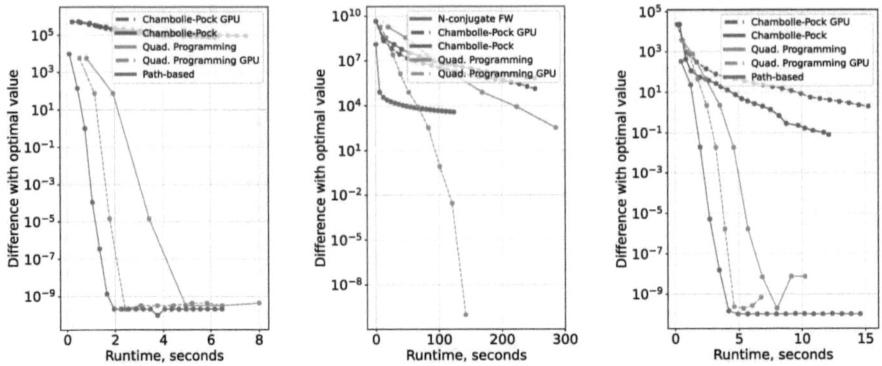

(a) (b) (c)

Fig. 4. Comparison of difference with optimal value for Berlin-Tiergarten (a), Terrassa-Asymmetric (b), Eastern-Massachusetts (c) in single-threaded setup for sequential quadratic programming, Chambolle-Pock (single-threaded and GPU) and best single-threaded method

Table 2 presents the speedups of the algorithms on a GPU (defined as the ratio of execution time in single-threaded mode to execution time on the GPU). It can be observed that, for networks that are not too small (such as Sioux-Falls), the use of sequential quadratic programming allows for significant acceleration through parallelization.

In contrast, the Chambolle–Pock algorithm is strongly limited by the computational cost of evaluating the proximal operator. It is also worth noting that as the network size increases, parallelization becomes more effective; for example, on the largest network, Terrassa-asymmetric, a threefold speedup was achieved.

Table 2. Acceleration for sequential quadratic programming and Chambolle-Pock algorithm for all networks

Name	Acceleration (SQP)	Acceleration (Chambolle-Pock)
Sioux-Falls	0.21	0.05
Anaheim	2.68	0.88
Berlin-Mitte-Center	1.92	0.65
Berlin-Tiergarten	1.83	0.78
Terrassa-Assymetric	3.29	1.48
Eastern-Massachusets	1.52	0.35

6 Conclusion

In this paper, well-known algorithms for solving the traffic assignment problem were compared, and a new method based on Sequential quadratic programming was proposed. The proposed algorithm differs from existing approaches in that it can be efficiently parallelized. It demonstrates a comparable performance in the single-threaded setting and superior performance on GPU for networks of various sizes. While this demonstrates a potential of Laplacian-based approach for solving multicommidity flow problems, in our view, most practical algorithms for today are still path-based and NFW algorithms, with single-source shortest paths subproblems being parallelized on CPU when computing flows on shortest paths between all OD-pairs.

Acknowledgments. The research was supported by the Russian Science Foundation (project No. 21-71-30005-P), https://rscf.ru/en/project/21-71-30005/.

References

1. Beckmann, M., McGuire, C.B., Winsten, C.B.: Studies in the economics of transportation. Technical report (1956)
2. Boyce, D., Ralevic-Dekic, B., Bar-Gera, H.: Convergence of traffic assignments: how much is enough? J. Transp. Eng. **130**(1), 49–55 (2004)
3. Boyles, S.D., Lownes, N.E., Unnikrishnan, A.: Transportation network analysis. Volume I: static and dynamic traffic assignment (2020)
4. Camargo, P.: AequilibraE: a free QGIS add-on for transportation modeling. Foss4g North America (2015)
5. Chambolle, A., Pock, T.: A first-order primal-dual algorithm for convex problems with applications to imaging. J. Math. Imaging Vision **40**(1), 120–145 (2011)
6. Chen, A., Jayakrishnan, R.: A path-based gradient projection algorithm: effects of equilibration with a restricted path set under two flow update policies (1998)
7. Chen, Y., Tse, D., Nobel, P., Goulart, P., Boyd, S.: CuClarabel: GPU acceleration for a conic optimization solver (2024). https://arxiv.org/abs/2412.19027
8. Davidson, A., Baxter, S., Garland, M., Owens, J.D.: Work-efficient parallel GPU methods for single-source shortest paths. In: IEEE 28th International Parallel and Distributed Processing Symposium, pp. 349–359. IEEE (2014)
9. Frank, M., Wolfe, P.: An algorithm for quadratic programming. Naval Res. Logist. Q. **3**(1–2), 95–110 (1956)
10. Fredman, M.L., Tarjan, R.E.: Fibonacci heaps and their uses in improved network optimization algorithms. J. ACM (JACM) **34**(3), 596–615 (1987)
11. Goulart, P.J., Chen, Y.: Clarabel: an interior-point solver for conic programs with quadratic objectives (2024)
12. Heywood, P., et al.: A data-parallel many-source shortest-path algorithm to accelerate macroscopic transport network assignment. Transp. Res. Part C: Emerg. Technol. **104**, 332–347 (2019)
13. Ignashin, I.N., Yarmoshik, D.V.: Modifications of the Frank-Wolfe algorithm in the problem of finding an equilibrium distribution of traffic flows. arXiv preprint arXiv:2403.04863 (2024)

14. Jiang, W., Luo, J.: Graph neural network for traffic forecasting: a survey. Expert Syst. Appl. **207**, 117921 (2022)
15. Lam, S.K., Pitrou, A., Seibert, S.: Numba: a LLVM-based python JIT compiler. In: Proceedings of the Second Workshop on the LLVM Compiler Infrastructure in HPC, LLVM 2015. Association for Computing Machinery, New York (2015). https://doi.org/10.1145/2833157.2833162
16. Peixoto, T.P.: The graph-tool python library (2014). https://doi.org/10.6084/m9.figshare.1164194.v13, https://figshare.com/articles/software/graph_tool/1164194
17. Sheffi, Y.: Urban Transportation Networks, vol. 6. Prentice-Hall, Englewood Cliffs (1985)
18. Transportation Networks for Research Core Team: Transportation networks for research (2024). https://github.com/bstabler/TransportationNetworks. Accessed 29 Feb 2024
19. US Bureau of Public Roads: Trac Assignment Manual. Department of Commerce, Urban Planning Division, Washington D.C. (1964)
20. Velikonivtsev, F., Platonov, O., Bazhenov, G., Prokhorenkova, L.: Fine-grained urban traffic forecasting on metropolis-scale road networks. arXiv preprint arXiv:2510.02278 (2025)
21. Xu, X., Zheng, Z., Hu, Z., Feng, K., Ma, W.: A unified dataset for the city-scale traffic assignment model in 20 US cities. Sci. data **11**(1), 325 (2024)
22. Yarmoshik, D., Persiianov, M.: On the application of saddle-point methods for combined equilibrium transportation models. In: Eremeev, A., Khachay, M., Kochetov, Y., Mazalov, V., Pardalos, P. (eds.) MOTOR 2024. LNCS, vol. 14766, pp. 432–448. Springer,Cham (2024). https://doi.org/10.1007/978-3-031-62792-7_29
23. Zhang, F., Boyd, S.: Solving large multicommodity network flow problems on GPUs. arXiv preprint arXiv:2501.17996 (2025)

ML and MILP Models in Production Planning Under Uncertainty

Varvara A. Rasskazova(✉)

Moscow Aviation Institute (National Research University), Moscow, Russia
varvara.rasskazova@mail.ru

Abstract. The paper investigates the problem of planning production processes at the stage of raw material allocation. For a given set of requirements (production orders), the execution volumes and acceptable limits for various impurities are specified. Similarly, the volumes of each raw material resource are given. However, the levels of impurities in raw material composition are determined only approximately, which significantly complicates the planning process due to obvious risks associated with achieving solution quality indicators.

To solve this problem, a mixed-integer linear programming model (MILP) is proposed as the primary planning tool. This model also implements an approach to form maximal feasible subsystems of constraints based on the penalty function method. At the stage of creating the functional space for the model, we use regression analysis methods, such as linear regression, random forest, and gradient boosting, to predict the qualitative characteristics of resources. A computational experiment is conducted on both simulated and real production data from an industrial enterprise in the non-ferrous metals industry. Another direction of the computational experiment is the comparative analysis of the efficiency of open-source libraries PuLP and HiGHS in Python, for solving optimization problems within the framework of the proposed MILP model.

Keywords: Production Planning · Resource Allocation · Uncertainty · Mixed-Integer Linear Programming (MILP) · Machine Learning (ML)

1 Introduction

The paper investigates a problem of distribution-type planning [5,15]. Such problems are fundamental to operations research and arise in various industrial contexts where limited resources must be allocated efficiently. A common feature of such problems is the necessity to allocate a limited set of resources among a given set of requirements (orders), subject to various quantitative and qualitative constraints [4,16].

One of the most universal and formally well-founded approaches to solving distribution-type planning problems is Mixed-Integer Linear Programming (MILP) [8,21]. This methodological framework provides a rigorous mathematical foundation for decision-making in complex resource allocation scenarios. This

A. Kazakov et al. (Eds.): MOTOR 2026, LNCS 16687, pp. 413–423, 2026.
https://doi.org/10.1007/978-3-032-31390-4_27

method allows for the consideration of both continuous variables, which describe the volumes of resources to be allocated, and discrete variables, associated with the selection or utilization of individual resources [3,22]. However, an increase in the problem's dimensionality leads to significant computational costs, making the study of the efficiency of various software libraries (solvers) a relevant task [2,13].

Another characteristic of the problem under study is that the actual production process exhibits a high degree of uncertainty in the initial data: the chemical composition of the available raw materials is subject to noise, and its online measurement involves substantial time and resource expenditures [7,11]. This uncertainty significantly complicates the planning process and can lead to suboptimal or infeasible solutions if not properly addressed. A promising direction for improving the robustness and accuracy of planning is the application of predictive analytics and machine learning methods [20]. Constructing regression and classification models capable of estimating key raw material quality parameters based on available indirect features allows for a reduction in the uncertainty level of the input data [12,19]. This approach makes it possible to formulate more precise conditions for decision-making regarding batch blending, which ultimately contributes to minimizing production risks and increasing the yield of compliant products [10].

Consequently, the aim of this paper is to investigate the potential of applying ML methods for forecasting input raw material quality, as well as to construct a MILP model for solving the resource allocation problem with impurity constraints. Furthermore, a comparative analysis of contemporary solvers within a computational experiment is of particular interest. The scientific significance of this direction lies in the integration of machine learning-based uncertainty reduction with exact optimization methods, providing a comprehensive framework for production planning under real-world conditions.

2 The Problem of Raw Material Quality Forecasting

Various classes of machine learning models can be applied to solve the problem of forecasting impurity content in the charge [6,14]. Accurate quality forecasting is a critical prerequisite for effective resource allocation, as it directly impacts the feasibility and optimality of subsequent planning decisions. The choice of a specific method depends on the nature of the problem (regression or classification), the expected complexity of the relationship between the features and the target variable, as well as the requirements for the interpretability of the results. In the context of this research, an application case of primary aluminum production is considered. The features may include:

- historical data from chemical analysis (indicators of silicon (Si) and iron (Fe) content in the melt from previous periods),
- technological identifiers (potroom number, electrolytic cell number), which indirectly characterize the equipment type and its condition,

- temporal characteristics (date and time of the last measurement, duration of the smelting process),
- process parameters (metal mass, temperature regimes).

The dataset used in this study comprises 15000 historical records from an industrial enterprise collected over one month. Each record includes continuous (metal mass, temperature, duration) and categorical (potroom ID, cell ID, melt ID, shift, etc.) features. The dataset was split into training (70%), validation (15%), and test (15%) sets using stratified sampling based on time to preserve temporal dependencies.

The following features were considered as target variables for forecasting in this paper:

- continuous indicators – the expected content of a specific impurity (e.g., Fe) in the melt,
- quality class – categorical labels derived from impurity ranges (e.g., grade A with Fe $< 0.1\%$, grade B with $0.1\% \leqslant$ Fe $< 0.2\%$ and grade C with Fe $\geqslant 0.2\%$). However, this paper focuses on the regression task, and the classification results are planned for future work.

The construction and implementation of regression models enable a transition from reactive quality management (based on already completed analyses) to proactive management, where an assessment of raw material quality is available in near real-time. This approach creates a foundation for more accurate and informed subsequent technological decision-making, including for optimal blending problems, which are addressed at the next level of production planning.

For the forecasting task itself, three model types were selected: Random Forest, Linear Regression, and Gradient Boosting. A comparison of their performance was conducted using the classic metrics: MAE (Mean Absolute Error), MSE (Mean Squared Error), RMSE (Root Mean Squared Error), and R^2 (coefficient of determination).

Hyperparameter tuning was performed using cross-validation on the training set with a random grid search. For Random Forest: number of trees $\in [100, 300]$, maximal depth $\in [10, 30]$, minimal samples split $\in [2, 8]$. For Gradient Boosting: learning rate $\in [0.01, 0.2]$, number of trees $\in [100, 500]$, maximal depth $\in [3, 8]$. Linear regression used no hyperparameters. Early stopping (patience) was equal to 20 and applied for boosting models on validation loss to prevent overfitting. No significant overfitting was observed. The difference between training and test R^2 was within 1.5% for all models.

Based on the metrics of the first two learning models for forecasting iron, it can be concluded that linear regression better tracks the variation in iron impurity, implying a linear relationship. In the case of silicon, the situation is reversed: the variation in the impurity percentage exhibits a non-linear character. Consequently, for forecasting silicon, a Gradient Boosting model was additionally considered, which demonstrated a better coefficient of determination (albeit with a trade-off in terms of a higher mean absolute error).

Table 1. Comparison of ML metrics for forecasting impurity content in resources.

Metric	Linear Regression		Random Forest		Gradient Boosting
	Fe	Si	Fe	Si	Si
MAE	0.026189	0.008915	0.026796	0.007711	0.007899
MSE	0.005742	0.000249	0.006630	0.000220	-
RMSE	0.075778	0.015764	0.081427	0.014844	-
R^2	0.989071	0.981456	0.987380	0.983558	0.984329

From the results presented in Table 1, it follows that for solving the problem of raw material quality forecasting, the most rational approach from an implementation perspective is the application of a linear regression model for iron and a gradient boosting model for silicon.

Note, that the performance of linear regression for iron forecasting is explained by domain. The iron content in production process under consideration is strongly correlated with a single dominant factor (the iron content in the input bauxite ore), which itself follows an approximately linear trend over time. In contrast, silicon content depends on multiple interacting factors (temperature, reaction kinetics, flux composition), resulting in nonlinear behavior that tree-based models capture better.

3 The Resource Allocation Problem

The construction of a MILP model is a key stage in solving a distribution-type planning problem. At this stage, the problem is formalized in the form of a rigorous mathematical model, the subsequent solution of which is implemented by means of standard optimization algorithms [17,18]. This modeling approach enables the formal representation of complex real-world constraints and objectives in a mathematically tractable form.

Let us introduce a set of requirements (orders) I, each element of which represents a vector $\left(\chi_i^{si}, \chi_i^{fe}, M_i\right)^T$, where χ_i^{si} is the maximum allowable silicon impurity content, χ_i^{fe} is the maximum allowable iron impurity content, and M_i is the execution volume of the i-th requirement.

We also define a set J of resources available for use. Each element of the set J represents a vector $\left(m_j, \xi_j^{si}, \xi_j^{fe}\right)^T$, where m_j is the volume of the j-th resource, ξ_j^{si} is the silicon impurity content in the j-th resource, and ξ_j^{fe} is the iron impurity content in the j-th resource. Note that the values of the parameters ξ_j^{si} and ξ_j^{fe} are determined at the preceding stage of raw material quality forecasting.

We introduce variables δ_{ij} characterizing the use of the j-th resource to fulfill the i-th requirement. If the j-th resource is used to fulfill the i-th requirement, then $\delta_{ij} = 1$, and $\delta_{ij} = 0$ otherwise.

Using described variables, we construct the set of feasible strategies. By the definition of the variables δ_{ij}, we have

$$\delta_{ij} \in \{0,1\}, i = \overline{1,|I|}, j = \overline{1,|J|}. \tag{1}$$

To ensure that each resource is used at most once in the solution, we introduce the constraints

$$\sum_{i=1}^{|I|} \delta_{ij} \leqslant 1, j = \overline{1,|J|}. \tag{2}$$

We assume further, that the impurity content proportion in the volume of a requirement is equal to the arithmetic mean of the impurity content proportions in the resources that were used to fulfill it. This assumption, while simplified, is standard in the literature for bulk blending problems and is accepted by practitioners in the non-ferrous metals industry when mixing is homogeneous.

In order not to exceed the maximum allowable impurity content in the requirement volumes, it is necessary to introduce constraints on the maximum allowable silicon and iron impurity content. But first, let us consider a hypothetical situation in which the model turns out to be infeasible under the given constraints. Such a situation is quite realistic from a practical standpoint, and therefore the possibility of its occurrence cannot be completely ruled out.

To find a partial solution to the given problem (a maximum feasible subsystem of constraints), we introduce auxiliary variables $\hat{\delta}_i^{si} \in \{0,1\}$ for each $i = \overline{1,|I|}$. Let $\hat{\delta}_i^{si} = 1$ if for the i-th requirement it was not possible to find a solution satisfying the constraints on the maximum allowable silicon impurity content, and $\hat{\delta}_i^{si} = 0$ otherwise. Analogously for $\hat{\delta}_i^{fe} \in \{0,1\}$.

We introduce constraints on the maximum allowable proportions of silicon and iron content in the requirement volumes:

$$\sum_{j=1}^{|J|} \xi_j^{si} \delta_{ij} - K_i^{si} \hat{\delta}_i^{si} \leqslant \chi_i^{si}, i = \overline{1,|I|}, \tag{3}$$

$$\sum_{j=1}^{|J|} \xi_j^{fe} \delta_{ij} - K_i^{fe} \hat{\delta}_i^{fe} \leqslant \chi_i^{fe}, i = \overline{1,|I|}, \tag{4}$$

where $K_i^{si} > 0$ and $K_i^{fe} > 0$ are sufficiently large positive coefficients (greater than unity). The rationale behind it is to ensure that, in the case when the auxiliary variable for a given impurity component equals one, the constraint is guaranteed to be satisfied.

We introduce constraints on the fulfillment volume of the requirements. Since the volume constraints can only be satisfied with a certain precision, we introduce a parameter $\Delta > 0$, which specifies the maximum allowable deviation of the actual fulfillment volume of a requirement from its target weight. Also, as in the case of the constraints on the maximum allowable impurity content, it is necessary to introduce auxiliary variables $\hat{\delta}_i^m \in \{0,1\}$. We state $\hat{\delta}_i^m = 1$ if for the i-th requirement it was not possible to find a solution that satisfies the volume constraints, and $\hat{\delta}_i^m = 0$ otherwise. Thus, the constraints on the maximum allowable deviation of the actual fulfillment volume of a requirement from its target weight have the following form:

$$M_i - \Delta \leqslant \sum_{j=1}^{|J|} m_j \delta_{ij} + K_i^m \hat{\delta}_i^m \leqslant M_i + \Delta, i = \overline{1, |I|}, \tag{5}$$

where $K_i^m = M_i$ is a penalty coefficient. Its purpose is also to ensure that, in the case when the auxiliary variable $\hat{\delta}_i^m = 1$, the constraint is guaranteed to be satisfied.

Taking into account the introduced variables and constraints (1)–(5), we obtain the following minimization problem

$$\sum_{i=1}^{|I|}\sum_{j=1}^{|J|} \delta_{ij} + \sum_{i=1}^{|I|} K_i^{si}\hat{\delta}_i^{si} + \sum_{i=1}^{|I|} K_i^{fe}\hat{\delta}_i^{fe} + \sum_{i=1}^{|I|} K_i^m \hat{\delta}_i^m \longrightarrow \min_{\substack{\delta_{ij} \\ \hat{\delta}_i^{si}, \hat{\delta}_i^{fe}, \hat{\delta}_i^m \\ i=\overline{1,|I|}, j=\overline{1,|J|}}} \tag{6}$$

subject to

$$\begin{cases} \sum\limits_{i=1}^{|I|} \delta_{ij} \leqslant 1, j = \overline{1, |J|}, \\ \sum\limits_{j=1}^{|J|} \xi_j^{si}\delta_{ij} - K_i^{si}\hat{\delta}_i^{si} \leqslant \chi_i^{si}, i = \overline{1, |I|}, \\ \sum\limits_{j=1}^{|J|} \xi_j^{fe}\delta_{ij} - K_i^{fe}\hat{\delta}_i^{fe} \leqslant \chi_i^{fe}, i = \overline{1, |I|}, \\ M_i - \Delta \leqslant \sum\limits_{j=1}^{|J|} m_j\delta_{ij} + K_i^m\hat{\delta}_i^m \leqslant M_i + \Delta, i = \overline{1, |I|}, \\ \delta_{ij} \in \{0,1\}, i = \overline{1, |I|}, j = \overline{1, |J|}, \\ \hat{\delta}_i^{si} \in \{0,1\}, i = \overline{1, |I|}, \\ \hat{\delta}_i^{fe} \in \{0,1\}, i = \overline{1, |I|}, \\ \hat{\delta}_i^m \in \{0,1\}, \ i = \overline{1, |I|}. \end{cases} \tag{7}$$

As can be seen from the definition of objective (6), the variables δ_{ij} enter it with a coefficient of 1, while the auxiliary variables $\hat{\delta}_i^{si}$, $\hat{\delta}_i^{fe}$ and $\hat{\delta}_i^m$ enter it with large positive coefficients K_i^{si}, K_i^{fe} and K_i^m respectively. Such an approach is necessary so that the inclusion of the variables δ_{ij} in the solution is preferable, since the use of any of the auxiliary variables $\hat{\delta}_i^{si}$, $\hat{\delta}_i^{fe}$ and $\hat{\delta}_i^m$ (i.e., assigning it a value of 1) will lead to a significant increase in the value of the objective functional. If, in the solution, at least one of the auxiliary variables $\hat{\delta}_i^{si}$, $\hat{\delta}_i^{fe}$ and $\hat{\delta}_i^m$ is utilized, this will serve as an indicator that the given set of resources J is insufficient to fulfill the specified set of requirements I.

4 Comparison of MILP Solvers

To solve problem (6) with constraints (7), specialized software tools known as MILP solvers (Mixed Integer Linear Programming solvers) can be used. The choice of solver significantly impacts the practical applicability of the proposed approach, particularly when dealing with large-scale industrial problems. In this paper, two different approaches implemented in Python are utilized: the PuLP library employing the CBC solver (Coin-or Branch and Cut) and the SciPy library employing the HiGHS solver [1,9].

PuLP is a high-level library designed for formulating linear and mixed-integer linear programming problems. The main advantage of PuLP is its convenient and intuitive syntax, which allows for describing optimization models in terms of variables, objective functions, and constraints in a manner close to mathematical notation.

In contrast to PuLP, SciPy does not provide a declarative language for model description; however, it allows for formulating the problem in a standard matrix form, which ensures closer integration with numerical optimization methods. An important advantage of the HiGHS solver, in this context, is its focus on solving large-scale problems, which leads to its active use in various scientific and industrial applications.

Within the framework of the computational experiment, input data generators were implemented with the following characteristics:

- the set of requirements contains between 100 and 150 elements, each with a specified required volume and impurity content tolerances,
- the set of resources contains approximately 500 elements with small volumes and fixed impurity values,
- the values of K_i^{si} and K_i^{fe} were stated as $400M_i$ and $600M_i$ respectively,
- the value of K_i^m was stated as M_i,
- the value of Δ was stated as 500.

The results of the computational experiment are presented in Tables 2 and 3.

Table 2. Results of the resource allocation problem solution (baseline instance with parameters of $|I| \in [100, 150]$ and $|J| = 500$).

Parameter	PuLP + CBC	SciPy + HiGHS
Solution time, s	0.85	0.12
Solution status	Optimal	Optimal
Objective function value	488.0	488.0
Number of variables	60500	60500
Number of constraints	980	980
Variable type	Binary	Binary

Table 3. Detailed time characteristics ($|I| \in [100, 150]$, $|J| = 500$, 10 independent runs).

Run	Time PuLP+CBC, s	Time HiGHS, s	Speedup
1	0.82	0.11	7.5×
2	0.85	0.12	7.1×
3	0.79	0.10	7.9×
4	0.88	0.13	6.8×
5	0.84	0.11	7.6×
6	0.86	0.12	7.2×
7	0.81	0.11	7.4×
8	0.83	0.12	6.9×
9	0.87	0.13	6.7×
10	0.80	0.10	8.0×

The results of the computational experiment, presented in Tables 2 and 3, show that the HiGHS solver provides a significantly higher solution speed for large-scale problems. At the same time, the use of PuLP with the CBC solver offers greater model clarity and ease of development, which makes this approach convenient for educational and research purposes. This finding has important practical implications, as computational efficiency is often a limiting factor in industrial applications.

To study the effect of problem dimensionality, we varied the number of requirements I from 50 to 300 and resources J from 250 to 1500. The Table 4 below shows solution times for both solvers. As expected, HiGHS scales significantly better. For the largest instance ($|I| = 300, |J| = 1500, \sim 450000$ binary variables), CBC exceeded 600 s (time limit), while HiGHS solved in 18.4 s.

Table 4. Scalability of solvers with problem size (average time in seconds, 5 runs).

$\|I\|$	$\|J\|$	PuLP+CBC (s)	HiGHS (s)
50	250	0.21	0.05
100	500	0.85	0.12
150	750	2.43	0.31
200	1000	5.67	0.68
250	1250	12.1	1.52
300	1500	>600 (timeout)	18.4

A similar computational experiment was conducted on historical data from an actual metallurgical enterprise producing primary aluminum. The trained ML models (Linear Regression for Fe, Gradient Boosting for Si) were used to predict impurity values for 1000 test resources. We then solved the MILP problem using predicted values at first, and then using historical data. As a result of this experiment, a resource-to-requirements allocation was obtained that satisfies the specified technological constraints in 83% of cases. The corresponding figure based on historical data was 58%. Of the 83%, in 19% of cases the production plan was met exactly (the target impurity content indicators were achieved), and in 64% of cases it was exceeded (improvement over the stated target indicators was achieved). The analogous criteria for the historical data were 53.6% and 4.4%, respectively. Notably, even using Random Forest led to only a 1.2% additional objective degradation compared to Linear Regression (measured as increase in penalty terms or violation of constraints). It means, that the MILP model's penalty mechanism can compensate for small prediction errors. Thus, it can be concluded that the proposed algorithm enables consistently high-quality resource planning overall and represents a sufficiently effective approach from a practical standpoint. These results demonstrate the practical value of integrating ML-based forecasting with MILP optimization for real-world production planning.

5 Conclusion

This paper investigates the problem of production process planning at the stage of raw material resource allocation. The qualitative characteristics of the resources are known only approximately, which introduces an element of uncertainty into the problem statement. In this regard, a two-stage solution scheme is proposed. The proposed two-stage framework addresses a fundamental challenge in operations research: making optimal decisions under uncertainty. At the first stage, ML methods are applied to forecast raw material quality. An analysis of linear regression, random forest, and gradient boosting models is conducted. At the second stage, a MILP model is proposed for solving the main production planning problem. Within the framework of the MILP model, elements of the penalty function method are also employed to resolve the problem in an inconsistent formulation. To solve the resulting optimization problem, solvers from the open-source libraries PuLP and SciPy in Python are used. An analysis of their efficiency is carried out in the context of computational speed and the quality of the solutions obtained. Future investigations will address the limitations of this paper, including extending the impurity mixing model to account for non-linear interactions, incorporating more extensive benchmarking with other solvers (e.g., SCIP), performing a full robust optimization formulation instead of the penalty-based approach, as well as validating on additional industrial datasets from different production lines.

References

1. Achterberg, T.: SCIP: solving constraint integer programs. Mathe. Programm. Comput. **1**(1), 1–41 (2009)

2. Cheng, C., Yu, Q., Adulyasak, Y., Rousseau, L.M.: Distributionally robust facility location with uncertain facility capacity and customer demand. Omega 122(3), (2024). Article 102959
3. Clautiaux, F., Ljubić, I.: Last fifty years of integer linear programming: a focus on recent practical advances. Eur. J. Oper. Res. **324**(1), 701–731 (2024)
4. Dissanayake, I., Gunathilake, M.: A comprehensive survey of recent advances in facility location problems: models, solution methods, and applications. J. Sustain. Dev. Transp. Logistics **9**(2), 78–91 (2024)
5. Egorova, N.G.: Raspredelennye zadachi kalendarnogo planirovaniya proizvodstva: obzor literatury i algoritmy optimizacii rabot na zavode. In: Egorova, N.G., Sotskov, YU.N., Ekonomika, S. (eds.) modelirovanie, prognozirovanie: sbornik nauchnyh trudov/ Nauchno-issledovatel'skij ekonomicheskij institut Ministerstva ekonomiki Respubliki Belarus'; redkol.: M.K. Kravcov [i dr.]. Minsk **18**, 123–138 (2024). (in Russ.)
6. Géron, A.: Hands-On Machine Learning with Scikit-Learn, Keras, and Tensor Flow. 3rd ed., 856 p. O'Reilly Media (2022)
7. Golmohamadi, H.: Demand-side management in industrial sector: a review of heavy industries. Renew. Sustain. Energy Rev. **156** (2022). Article 111963
8. Haq, F.H.N., Chaerani, D., Triska, A.: Systematic literature review for robust mixed-integer linear programming models using benders decomposition in facility location problems. Jurnal Matematika Murni dan Aplikasi. **10**(1), 326–344 (2025)
9. Huangfu, Q., Hall, J.A.J.: Parallelizing the dual revised simplex method. Math. Program. Comput. **10**, 119–142 (2018)
10. Jaskólski, M., Bućko, P., Czapp, S.: The evolution of low- and medium-voltage distribution system development planning procedures and methods -a review. Energies **18**(13), 3461 (2025). https://doi.org/10.3390/en18133461
11. Ji, Y., Du, J., Wu, Y.: A mixed integer robust programming model for two-echelon inventory routing problem of perishable products. Physica A Stat. Mech. Appl. **548** (2020). Article 124481
12. Kaleta, M.: Robust co-optimization of medium- and short-term electrical energy and flexibility in electricity clusters. Energies **18**(3), 479 (2025). https://doi.org/10.3390/en18030479
13. Koch, T., Berthold, T., Pedersen, J., Vanaret, C.: Progress in mathematical programming solvers from 2001 to 2020. EURO J. Comput. Optim. **10** (2022). Article 100031
14. Kotsiantis, S.B.: Supervised machine learning: a review of classification techniques. Informatica **31**, 249–268 (2007)
15. Lima, D., Lezama, F., Soares, J., et al.: Modern distribution system expansion planning considering new market designs: review and future directions. Renew. Sustain. Energy Rev. **203** (2024). Article 114785
16. Rahmaniani, R., Crainic, T.G., Gendreau, M., Rei, W.: The Benders decomposition algorithm: a literature review. Eur. J. Oper. Res. **259**(3), 801–817 (2017)
17. Rasskazova, V.A.: A decomposition approach in the problem of distribution-type planning with priority constraints. Bull. South Ural State Univ. Ser. Math. Modell. Programm. Comput. Softw. **17**(3), 87–101 (2024)
18. Rasskazova, V.A., Skuridin, A.A.: The problem of allocation of production resources with a system of constraints. Modelirovanie i analiz dannykh [Modelling and Data Analysis] **13**(2), 142–150 (2023). (In Russ., abstr. in Engl.)
19. Sannigrahi, S., Ghatak, S.R., Acharjee, P.: Coordinated framework for active distribution system planning integrating active system management schemes. Int. J.

Numer. Model. Electron. Netw Devices Fields **2023**(6), 36 (2023). https://doi.org/10.1002/jnm.3113
20. Solat, S., Aminifar, F., Safdarian, A., Shayanfar, H.: An expansion planning model for strategic visioning of active distribution network in the presence of local electricity market. IET Gene. Transm. Distrib. **17**(24), 5410–5429 (2023)
21. Ustyugov, V.A.: On different methods for automated MILP solver configuration. In: 20-th International Asian School-Seminar on Optimization Problems of Complex Systems (OPCS), pp. 24–27 (2024)
22. Zavatteri, M., Raffaele, A., Ostuni, D., Rizzi, R.: An interdisciplinary experimental evaluation on the disjunctive temporal problem. Constraints **28**(1), 1–12 (2023)

A Matheuristic for Two-Level Resource-Constrained Assignment with Cost Balancing

Olga Samsonyuk(✉)

Matrosov Institute for System Dynamics and Control Theory, Irkutsk, Russia
olga.samsonyuk@icc.ru

Abstract. This paper investigates a class of assignment problems where tasks must be allocated to capacity-constrained agents partitioned into groups. Each group has an associated target cost, and the objective seeks to balance two competing goals: minimizing the total operational cost across the system while minimizing the maximum deviation of any group's cost from its target. These problems are motivated by applications in data center management, distributed computing, and workforce scheduling. The primary motivating application is virtual machine (VM) placement in data centers, where tasks correspond to VMs, agents correspond to physical machines (PMs), and groups correspond to cabinets. We develop a hybrid solution framework that integrates mathematical programming with problem-specific heuristics. The framework first employs a fast auxiliary model to identify a core set of agents, then applies a greedy procedure to fix a subset of assignment variables. An initial feasible solution is refined through a smoothing procedure that reallocates tasks to underutilized groups, thereby reducing the maximum cost deviation. Computational studies on benchmark VM placement instances demonstrate that the proposed method consistently yields superior solutions compared to the standalone SCIP solver under equivalent time constraints. The results show average improvements of more of 30% on individual instances, while requiring only 60–75 s of total computation time including heuristic preprocessing.

Keywords: Matheuristic · Assignment problems · Resource constraints · Cost balancing · Virtual machine placement · Mixed-integer programming

1 Introduction

Cloud resource management has received significant attention in recent years. The exponential growth of cloud services has led to large-scale virtualized infrastructures where computing resources must be carefully orchestrated to meet service quality objectives [12,14]. A fundamental challenge in this domain is the virtual machine placement problem: determining how to assign virtual machines

A. Kazakov et al. (Eds.): MOTOR 2026, LNCS 16687, pp. 424–436, 2026.
https://doi.org/10.1007/978-3-032-31390-4_28

(VMs) to physical servers (PMs) while respecting hardware capacities and operational policies [6,11]. The VM placement problem is known to be NP-hard [9]. Additional complexity arises in production environments, where physical machines (PMs) are organized into cabinets with fixed overhead costs, VM migrations incur expenses, certain workloads cannot co-reside due to security or performance isolation requirements, and migration limits are imposed to preserve system stability [15,16].

Beyond conventional cost minimization, data center operators face the additional requirement of balancing expenditures across cabinets. Uneven cost distribution can result in localized overloading, accelerated hardware degradation, and difficulties in meeting budget allocations for different facility zones. This gives rise to a optimization model where total system-wide costs must be minimized alongside the maximum deviation of any cabinet's cost from its target. Such problems belong to a broader family of two-level resource-constrained assignment problems with cost balancing, which also manifest in distributed computing, workforce planning, and logistics applications [7,10].

The inherent difficulty of these optimization problems arises from the intricate interplay of discrete decisions, multi-dimensional capacity limitations, placement conflicts, and the hierarchical coupling that links resource-level choices with group-level cost targets. Empirical evaluation reveals that general-purpose MILP solvers, when initialized without a feasible starting point, frequently struggle to identify good solutions within acceptable time frames, often terminating with significant optimality gaps [2,3]. This limitation motivates the development of specialized solution approaches that exploit problem structure.

To address this complexity, we propose a matheuristic, a methodology that combines exact mixed-integer programming (MIP) techniques with problem-specific heuristics. The term "matheuristic" was introduced to describe algorithms where mathematical programming techniques play a key role within a heuristic framework [4,5,8]. Unlike pure metaheuristics, matheuristics leverage the mathematical model of the problem (e.g., through relaxations or decomposition) to guide the search for high-quality solutions. This hybrid paradigm has been successfully applied to various challenging optimization tasks, including bin packing, routing, and assignment problems [4,10,13,14].

We develop a solution framework that integrates mathematical programming with problem-specific strategies. The methodology comprises three phases. First, an auxiliary model selects promising PMs based on knapsack-derived characteristics, identifying cabinets likely to host VMs efficiently. Second, a simplified model examines originally placed VMs to determine those that should optimally remain in place and thus avoid migration costs. Third, the solutions from both phases are combined to fix a subset of decision variables, after which the reduced problem is solved with a MILP solver under a time limit. The warm-starting approach provides an immediate primal bound, allowing the solver to concentrate on refining an already reasonable solution rather than exploring the search space from scratch. This technique has recently gained significant attention and is recognized as a key strategy for improving MILP performance in practice [3].

This paper is organized as follows. Section 2 presents the mathematical formulation of the problem and discusses its two-level structure, highlighting the role of cost balancing. Section 3 details the proposed matheuristic. Section 4 reports computational results.

2 Problem Statement

Let us define the following collections:

- $I = \{1, \ldots, n\}$ is a set of tasks (items);
- $J = \{1, \ldots, m\}$ is a set of agents (bins);
- $G = \{1, \ldots, \bar{g}\}$ is a set of groups (cabinets);
- $R = \{1, \ldots, \bar{r}\}$ is a set of resource types.

Additional data elements include:

- $I_j \subseteq I$ is the set of tasks that can be assigned to agent j;
- $I^j \subseteq I$ is a set of originally placed tasks for agent j;
- $J_i \subseteq J$ is the set of agents that can be assigned to task i;
- $J_g \subseteq J$ is the set of agents belonging to group g;
- $j_i \in J$ is an original location of task i;
- $V \subseteq I \times I$ is a set of pairs of tasks that cannot share an agent;
- $n_{\text{stay}} \geq 0$ is a minimum number of tasks that must remain with their original locations.

Each task $i \in I$ requires d_{ir} units of resource $r \in R$. Each agent $j \in J$ possesses capacity q_{jr} for resource type r. A cost c_{ij} is incurred if task i is assigned to agent j; this may represent migration, training, or communication overhead. Additionally, activating agent j incurs a fixed cost c_j.

Let us introduce binary variables:

$$y_j = \begin{cases} 1 & \text{if agent } j \text{ is used,} \\ 0 & \text{otherwise;} \end{cases} \qquad x_{ij} = \begin{cases} 1 & \text{if task } i \text{ is assigned to agent } j, \\ 0 & \text{otherwise.} \end{cases}$$

Each group g has a fixed operational cost C_g and a target expenditure $\overline{G}_g$. Denote by $Cost[g]$ the expenditures across group $g \in G$; i.e.,

$$Cost[g] := C_g + \sum_{j \in J_g} \left(c_j y_j + \sum_{i \in I_j} c_{ij} x_{ij} \right)$$

The control aim is to minimize the objective function:

$$J = \alpha \cdot \max_{g \in G} |Cost[g] - \overline{G}_g| + \beta \cdot \sum_{g \in G} Cost[g], \tag{1}$$

where $\alpha, \beta \geq 0$ are weighting coefficients.

The objective function (1) comprises two weighted components that capture the dual goals of the problem. The first term, weighted by α, minimizes the maximum absolute deviation between the actual cost $Cost[g]$ of group g and its target $\overline{G}_g$. This minimax formulation encourages balanced expenditures across all groups, preventing any single group from deviating excessively from its planned budget. The second term, weighted by β, minimizes the total cost summed over all groups, representing overall system efficiency. The nonnegative parameters α and β allow decision-makers to adjust the relative importance of cost balancing versus total cost minimization. Setting $\alpha = 0$ reduces the problem to pure total cost minimization, while $\beta = 0$ focuses exclusively on balancing group costs. Typical applications employ intermediate values to achieve a compromise between these two objectives.

Note that the objective function (1) is linearized in the MILP formulation by introducing a continuous auxiliary variable $P := \max_{g \in G} |Cost[g] - \overline{G}_g|$ representing the maximum absolute deviation, together with constraints

$$|Cost[g] - \overline{G}_g| \leq P \quad \forall g \in G,$$

and minimizing $\alpha P + \beta \sum_{g \in G} Cost[g]$.

Then, the optimization model is:

$$\text{Minimize} \quad \alpha P + \beta \sum_{g \in G} \left(C_g + \sum_{j \in J_g} \left(c_j y_j + \sum_{i \in I_j} c_{ij} x_{ij} \right) \right) \tag{2}$$

$$\text{subject to:} \quad \sum_{j \in J_i} x_{ij} = 1 \quad \forall i \in I, \tag{3}$$

$$\left| C_g + \sum_{j \in J_g} \left(c_j y_j + \sum_{i \in I_j} c_{ij} x_{ij} \right) - \overline{G}_g \right| \leq P \quad \forall g \in G, \tag{4}$$

$$\sum_{i \in I_j} d_{ir} x_{ij} \leq q_{jr} y_j \quad \forall j \in J,\ \forall r \in R, \tag{5}$$

$$x_{i_1 j} + x_{i_2 j} \leq 1 \quad \forall (i_1, i_2) \in V, \quad \forall j \in J_{i_1} \cap J_{i_2}, \tag{6}$$

$$\sum_{i \in I} x_{i j_i} \geq n_{\text{stay}}, \tag{7}$$

$$y_j \in \{0,1\},\ x_{ij} \in \{0,1\} \quad \forall i \in I_j,\ j \in J, \quad P \geq 0. \tag{8}$$

Equation (3) enforces that every task receives exactly one agent. Inequalities (4) define P as the maximum of group expenditures, embedding the balancing requirement. Constraints (5) represent resource limitations. Restrictions (6) prevent incompatible task pairs from co-locating. Constraint (7) requires that at least n_{stay} tasks remain in their originally assigned locations.

The problem (2)–(8) is inherently two-level: groups are selected at the upper level, while agents and task assignments are chosen at the lower level within each group. The upper-level objective minimizes the maximum absolute deviation of group costs from their targets (ensuring balanced expenditures), whereas the lower-level objective minimizes total operational cost. In essence, the two-level structure frames a trade-off between local efficiency (minimizing costs within a group) and global equity (ensuring all groups meet their targets). The presented matheuristic directly exploits this structure: the auxiliary model (Stage 1) pre-selects efficient agents at the lower level, while the smoothing procedure (Stage 4) rebalances costs at the upper level.

We aggregate both levels into a single objective (1) via linear convolution with weights α and β, a standard multiobjective technique. The interplay between α and β yields a Pareto frontier of trade-offs between balancing and total cost. A high α/β ratio prioritizes balancing at the possible expense of total cost; a low ratio emphasizes global efficiency. General approaches to approximate this frontier include the ε-constraint method [1], weighted sum with varying coefficients, or evolutionary algorithms [17]. In this paper, however, we focus on the practically relevant case $\alpha \gg \beta$ (both non-zero). Following the priorities in Table 1, we fix α and β and solve the resulting single-objective problem efficiently.

3 Proposed Matheuristic

3.1 Algorithm Description

Given the problem's computational difficulty, we design a hybrid approach that leverages both exact optimization and heuristic reasoning. The algorithm proceeds through five stages.

Stage 1: Agents Pre-selection via Auxiliary Model. We construct a simplified mathematical model (Model (A)) to identify a core set of agents (bins), hereinafter referred to as J^{good}, likely to appear in high-quality solutions.

For each agent j, we compute four metrics: F_j and M_j (based on all tasks), and F_j^{orig} and M_j^{orig} (based only on tasks originally placed on j). The latter metrics guide the decision of which originally placed tasks should remain fixed. These metrics are determined by a greedy fill procedure. The metric and parameter selection guidelines are described in Paragraph 3.2.

$$\textbf{Model } (A)\textbf{:} \text{ Minimize } \; J_A = h_1 P_A - h_2 \sum_{j=1}^{m} (M_j + M_j^{\text{orig}}) y_j, \tag{9}$$

$$\left| C_g + \sum_{j \in J_g} (c_j + F_j) y_j - \overline{G}_g \right| \leq P_A \quad \forall g \in G, \tag{10}$$

$$\sum_{i \in I} d_{ir} \leq \sum_{j \in J} q_{jr} y_j \qquad \forall r \in R, \tag{11}$$

$$\sum_{j \in J} y_j \leq \gamma \, |G|, \tag{12}$$

$$y_j \in \{0, 1\} \quad \forall j \in J, \;\; P_A \geq 0. \tag{13}$$

Here, the constraint (10) ensures balanced costs across groups (cabinets) using average cost estimates, (11) guarantees sufficient aggregate capacity, and (12) limits the total number of selected agents, where γ is a parameter controlling how many agents are chosen per group on average.

By definition, for every feasible solution to this model, $\sum_{j=1}^{m}(M_j + M_j^{\text{orig}}) > n$, where n is the number of tasks (items). On the other hand, when calculating P_A from constraint (10), we use the cost of only one average item per bin (F_j), so we use fewer than m "average" items for P_A. In the context of the problems, the value $\max_{g \in G} \sum_{j \in J_g} (c_j + F_j)$ is less than $\sum_{j=1}^{m}(M_j + M_j^{\text{orig}})$. To balance this fact, we set the weight coefficients to $h_1 = 2$ and $h_2 = 1$ in our computational experiments.

The constraint (12) includes a heuristic parameter γ. Its value is determined empirically, depending on the number of items ($n = |I|$), bins ($m = |J|$), and cabinets ($\bar{g} = |G|$) in the instances. If Model (A) returns infeasible, γ is incremented.

Remark 1. Model (A) is a heuristic surrogate. It replaces the original costs c_{ij} with average estimates F_j. The weights $h_1 = 2$ and $h_2 = 1$ were chosen empirically to mimic the effect of the original objective with $\alpha \gg \beta$. Specifically, minimizing $h_1 P_A$ roughly corresponds to reducing the maximum group deviation (like αP), while maximizing $\sum(M_j + M_j^{\text{orig}})y_j$ promotes packing more tasks, indirectly reducing total cost.

Although Model (A) is a mixed-integer programming problem and thus NP-hard in general, its size is significantly smaller than that of the original problem. We use the SCIP solver [2]. In all tested instances, SCIP solved it within a few seconds. The solution of Model (A) yields $J^{\text{good}} = \{j : y_j = 1\}$.

Stage 2: Variable Commitment. Based on J^{good}, we commit certain variables to reduce the problem size:

- Fix $y_j = 1$ for all $j \in J^{\text{good}}$.
- For each $j \in J^{\text{good}}$ and each originally placed task $i \in I^j$, fix $x_{ij} = 1$ if $c_{ij} \leq (1 + \zeta) \cdot F_j^{\text{orig}}$, where ζ is a tunable threshold.
- Ensure that the total number of such fixings does not exceed n_{stay} by discarding the most expensive commitments if necessary.

Note that $\zeta \in [-0.5, 0.5]$ depends on the problem size, with typical values of 0 for small instances and 0.2 to 0.5 for large ones.

Let the set of these indices $\{(i,j)\}$ be denoted as Fix_j^{orig} (indices of the fixed variables x_{ij}). We calculate the resource consumption for the fixed variables and the current cost of each cabinet, i.e.,

$$\Delta_g = C_g + \sum_{j \in J^{\text{good}} \cap J_g} (c_j + \sum_{(i,j) \in Fix_j^{\text{orig}}} c_{ij}) - \overline{G}_g.$$

We also calculate $\Delta_{\max} = \max_{g \in G} \Delta_g$.

If the number of fixed assignments exceeds n_{stay}, we discard the most expensive ones (i.e., those with the largest c_{ij}) until exactly n_{stay} tasks remain fixed. This ensures that the migration constraint (7) is not violated.

Stage 3 (optional): Baseline Optimization. With the fixed variables, we solve the original model (2)–(8) under a time limit. When executed, this stage produces an initial feasible solution that is passed to Stage 4. Otherwise, Stage 4 proceeds only with the variable commitments made in Stage 2. In our computational experiments in Sect. 4, Stage 3 was always omitted.

Stage 4: Expenditure Smoothing. This phase directly targets the reduction of P and is used for the case $n_{\text{orig}} < n$, where n_{orig} denotes the total number of originally located tasks.

Given $j \in J^{\text{good}}$, we fix some variables x_{ij} such that

$$\Delta_g + c_{ij} \leq \rho \cdot \Delta_{\max}.$$

Note that this equalizes the costs of cabinets. We use the coefficient $\rho = 1.5$ for all problems in Sect. 4. On the one hand, this allows us to increase the number of fixed variables for each cabinet; on the other hand, it avoids significantly increasing the current cabinet cost. The process repeats until no further assignments are possible.

At this stage of the algorithm, the conflict and resource constraints are under control, as well as the n_{stay} constraint. To avoid violating the n_{stay} constraint, we reserve resources for the items that are originally located but not fixed yet.

For each originally placed task that has not been fixed yet, we reserve the required resources on its original agent. This prevents the smoothing procedure from accidentally making it impossible to keep that task at its original location, which would violate the n_{stay} constraint.

Stage 5: Final Optimization. We introduce the linking constraints:

$$y_j \leq \sum_{i \in I_j} x_{ij} \quad \forall j, \qquad x_{ij} \leq y_j \quad \forall i, j,$$

and solve the reduced MILP (i.e., the original problem with the variables fixed in Stages 2 and 4) with the SCIP solver under a time limit of 60 s, returning the best solution found.

3.2 Metric and Parameter Selection Guidelines

The performance of the proposed matheuristic depends on several key parameters, which should be tuned according to the instance characteristics:

- The metrics F_j, M_j, F_j^{orig}, M_j^{orig}, $j = \overline{1, m}$, are described as follows:
  - F_j is an average migration cost of the most valuable tasks that can fit in agent j. To find F_j, consider bin j, and sort by increasing the values c_{ij} corresponding to j. For example, we obtained the set of index

$\{i_1, i_2, \ldots, i_{end}\}$. Then, $F_j = \frac{1}{s_j}\sum_{t=1}^{s_j} c_{i_t\,j}$, where s_j is such that

$$\sum_{t=1}^{s_j} d_{i_t\,r} \le q_{jr} \quad \forall r \in R \quad \text{and} \quad \sum_{t=1}^{s_j+1} d_{i_t\,r} > q_{jr} \quad \text{for at least one } r.$$

- M_j is the number of such tasks that can be assigned to agent j, i.e., $M_j := s_j$, $j = \overline{1,m}$.
- F_j^{orig} and M_j^{orig} calculate as F_j and M_j but considers only tasks originally placed on j. To find F_j^{orig}, consider bin j with originally located items, and sort by increasing the values $\{c_{ij} \mid i \in I^j\}$. Then, $F_j^{\text{orig}} = \frac{1}{s_j^{\text{orig}}}\sum_{t=1}^{s_j^{\text{orig}}} c_{i_t\,j}$, where s_j^{orig} is such that

$$\sum_{t=1}^{s_j^{\text{orig}}} d_{i_t\,r} \le q_{jr} \quad \forall r \in R \quad \text{and} \quad \sum_{t=1}^{s_j^{\text{orig}}+1} d_{i_t\,r} > q_{jr} \quad \text{for at least one } r,$$

$M_j^{\text{orig}} := s_j^{\text{orig}}$, $j = \overline{1,m}$.

- Parameter γ controls the number of bins selected per cabinet in Model (A). Based on empirical analysis, we recommend:

$$\gamma = \frac{1}{3} \cdot \frac{m}{\bar{g}} + \omega, \tag{14}$$

where $\omega \in [0, 1.5]$ adjusts for an instance density. For the tested instances, γ values ranged from 1.33 to 2.46 (see Table 1). The multiplier 1/3 allows us fixing about 30% of the average number of bins per cabinet. The second term ω enables fixing a larger number of bins in high-dimensional problems. The goal of the parameter γ is to identify a potentially high-quality number of bins that yields a very good feasible solution.
- Coefficient ζ is used in variable fixing to determine which originally placed items should remain. Recommended values:
  - Small instances ($n \le 500$): $\zeta \in [-0.5, 0]$ (conservative fixing)
  - Large instances ($n > 500$): $\zeta \in [0, 0.5]$ (more aggressive fixing)
- Weights h_1, h_2: In Model (A), we set $h_1 = 2$ and $h_2 = 1$ to balance the trade-off between minimizing the maximum deviation (via $h_1 P_A$) and maximizing the number of potentially fixed items (via $h_2 \sum(M_j + M_j^{\text{orig}})y_j$).
- Cost balancing threshold ρ: During the smoothing procedure, we use coefficient $\rho \in [1, 1.5]$ to ensure that new assignments do not increase the current cabinet costs beyond $\rho \cdot \Delta_{\max}$.

These parameters were determined through experimentation on the benchmark instances and were shown to provide robust performance across different problem sizes.

4 Computational Results

In this section, we interpret problem (2)–(8) in the context of virtual machine placement in data centers. Here, tasks correspond to virtual machines (VMs), agents to physical machines (PMs), and groups to cabinets. We consider instances P1–P10. The problem data are summarized in Tables 1 and 2.

Table 1. The problem data

Instance	VMs n	PMs m	Cab. $\bar{g}$	Loc. VMs	Res. $\bar{r}$	Confl. $\|V\|$	$\overline{AG}_{\text{plan}}$	n_{stay}	α	β	γ
P1	143	36	9	146	4	3	0	73	0.7	0.03333	1.33
P2	217	22	2	220	4	3	0	110	0.7	0.15	3.67
P3	948	135	15	951	4	3	0	476	0.7	0.02	4.0
P4	2114	133	38	1591	4	3	0	1059	0.7	0.00789	2.16
P5	2365	212	39	1728	4	3	0	1184	0.7	0.00769	2.37
P6	1167	141	46	840	4	3	191.5	585	0.7	0.0065	2.02
P7	2864	225	81	1931	4	3	198.1	1434	0.7	0.0037	1.93
P8	1160	76	25	792	4	3	188.0	582	0.7	0.012	2.01
P9	1587	112	35	1144	4	3	208.6	795	0.7	0.00857	2.07
P10	1856	162	37	1075	4	3	191.7	930	0.7	0.0081	2.46

Column descriptions in Table 1:

- Instance: Instance identifier.
- VMs (n): Total number of virtual machines in the instance.
- PMs (m): Total number of physical machines.
- Cab. ($\bar{g}$): Number of cabinets.
- Loc. VMs: Number of originally placed VMs.
- Res. ($\bar{r}$): Number of resource types.
- Confl. ($|V|$): Number of incompatible VM pairs that cannot be placed on the same PM.
- $\overline{AG}_{\text{plan}}$: Average planned (target) cost per cabinet; i.e., $\overline{AG}_{\text{plan}} := \frac{1}{\bar{g}} \sum_{g=1}^{\bar{g}} \overline{G}_g$. Zero values indicate that target costs were not specified for those instances.
- n_{stay}: Minimum number of VMs that must remain on their originally assigned PM (migration limit).
- α: Weight coefficient for the balancing component (maximum cabinet cost) in the objective function.
- β: Weight coefficient for the total cost component in the objective function.
- γ: A heuristic parameter defined by (14).

Table 3 reports the performance of SCIP in a cold start configuration; i.e., applied to the problems without our heuristic and without any initial feasible solution.

Table 2. Problem statistics

Instance	Number of Bin-variables	Number of constraints	Number of nonzero coef.
P1	3381	339	20489
P2	3098	338	18848
P3	84506	1663	508185
P4	189429	2920	1138540
P5	254086	3214	1526461
P6	109632	2001	768569
P7	430390	4174	3015092
P8	59702	1659	418968
P9	119129	2221	835235
P10	197840	2713	1386135

Table 3. SCIP cold start results with different time limits

Instance	Time = 60 s		Time = 3600 s		Time = 72000 s	
	Solution	GAP (%)	Solution	GAP (%)	Solution	GAP (%)
P1	210.0	3.47	208.5	0.15	208.5	0.14
P2	651.93	3.56	647.8	0.00	647.8	0.00
P3	1393.9	233.1	518.2	5.61	504.1	2.68
P4	–	∞	609.9	106.0	345.6	15.59
P5	–	∞	–	∞	401.5	11.52
P6	491.2	∞	491.1	242.6	193.6	34.02
P7	486.0	∞	486.0	364.7	486.0	241.82
P8	448.4	188.9	224.2	18.6	201.5	5.93
P9	432.2	∞	432.2	155.8	189.2	11.43
P10	676.6	∞	676.3	94.0	447.9	20.10

To demonstrate the effectiveness of our approach, we applied the heuristic algorithm to these problems and compared the results with the baseline. Let γ be calculated by the rule (14), where $\omega = \begin{cases} 0, & \text{for } n \leq 500, \\ 1, & \text{for } 500 < n \leq 1500, \\ 1.25, & \text{for } n > 1500. \end{cases}$
For instances P1–P10, the resulting γ values are reported in Table 1.

The computational results obtained by applying the proposed heuristic are summarized in Table 4.

In Table 4, in both cases, SCIP was executed with a time limit of 60 s or, for the cold start, until the first feasible solution was found – whichever occurred first. The "Cold start" columns report solutions obtained from a solver run with no initial feasible solution, forcing SCIP to discover feasible points entirely

Table 4. Computational results: cold start vs. warm start

Instance	Cold start		Warm start		Impr. (%)
	Solution	Time (sec.)	Solution	Total time (sec.)	
P1	210.0	60	209.3	60.1	0.33
P2	651.9	60	648.7	60.1	0.49
P3	1393.9	60	650.6	63.4	53.33
P4	609.9	90	465.3	66.2	23.71
P5	521.9	48829	499.7	68.2	100
P6	491.2	60	269.4	63.5	45.15
P7	486.0	74	346.0	74.6	28.81
P8	448.4	60	226.1	61.8	49.58
P9	432.2	60	310.5	63.8	28.16
P10	676.6	60	578.3	66.9	14.53

through its own search mechanisms. The "Warm start" columns show results when SCIP was initialized with a high-quality heuristic solution generated by our matheuristic. In this configuration, SCIP was also limited to 60 s of optimization time, while the additional time reported in the "Total time" column includes the seconds required by the heuristic to construct the initial solution. This warm start provides an immediate upper bound, enabling the solver to focus on improving the solution rather than exploring the search space from scratch. The substantial improvement rates (Impr, %) demonstrate the effectiveness of providing a warm start, even under identical time constraints.

Computational Environment. All experiments were conducted on a machine with AMD Ryzen 7 2700 Eight-Core Processor 3.20 GHz and 16 GB RAM. The matheuristic was implemented in Julia, and all MILP problems were solved using SCIP version 9.0 with default parameters. Time limits were set to 60 s for both cold start and warm start SCIP runs, with additional preprocessing time for the heuristic reported separately in the "Total time" column.

Analysis of Results. The computational results in Table 4 demonstrate the clear advantage of the warm start approach:

- Solution quality: Warm start solutions are consistently better than cold start solutions, with improvements ranging from 0.33% (P1) to 100% (P5). The average improvement across all instances exceeds 30%.
- Computational efficiency: Despite the additional preprocessing time (typically 0.1–14.6 s), warm start solutions are obtained within 60–75 s total, while cold start runs often require significantly more time to find comparable solutions.
- Extreme cases: For instance P5, the cold start required 48,829 s (over 13 h) to find a solution of 521.94, while warm start achieved a better solution (499.7) in just 68.2 s total.
- Consistency: The warm start approach successfully found feasible solutions for all instances, while cold start struggled with P5 (no solution within 3600 s).

These results validate the effectiveness of the proposed matheuristic in providing high-quality initial solutions that significantly accelerate the optimization process. The improvement shown in Table 4 becomes even more striking when compared to the cold start performance at 60 s (Table 3). For instance, P5 yields no feasible solution in 60 s under cold start, while warm start finds a solution of 499.7 in just 68 s total. For P3, cold start gives 1393.9, while warm start achieves 650.6 in 63.4 s. These comparisons confirm that the matheuristic provides both feasibility and quality where the standalone solver fails.

Acknowledgements. This work was supported by the state assignment within the framework of the research topic "Evolutionary and Dynamic Control Systems: Theory, Numerical Methods, and Applications" (project code FWEW-2026-0011, state registration No. 126021217177-7).

The author thanks the anonymous reviewers for their careful reading of the manuscript and for providing numerous constructive remarks that led to an improved final version.

References

1. Aminbakhsh, S., Sönmez, R., Atan, T.: ε-constraint procedures for Pareto front optimization of large size discrete time/cost trade-off problem. Eur. J. Oper. Res. **322**(3), 753–769 (2025). https://doi.org/10.1016/j.ejor.2024.11.032
2. Bolusani, S., Besançon, M., Bestuzheva, K., et al.: The SCIP Optimization Suite 9.0. ZIB-Report 24-01, Berlin (2024). https://doi.org/10.48550/arXiv.2402.17702
3. Bolusani, S., Besançon, M., Gleixner, A., et al.: The MIP Workshop 2023 Computational competition on reoptimization. Math. Prog. Comp. **16**, 255–266 (2024). https://doi.org/10.1007/s12532-024-00256-w
4. Boschetti, M., Letchford, A., Maniezzo, V.: Matheuristics: survey and synthesis. Int. Trans. Oper. Res. **30**, 2840–2866 (2023). https://doi.org/10.1111/itor.13301
5. Boschetti, M., Maniezzo, V.: Contemporary approaches in matheuristics an updated survey. Ann. Oper. Res. **343**, 663–700 (2024). https://doi.org/10.1007/s10479-024-06302-z
6. Challita, S., Paraiso, F., Merle, Ph.: A study of virtual machine placement optimization in data centers. In: 7th International Conference on Cloud Computing and Services Science (CLOSER 2017), Porto, Portugal (2017). https://doi.org/10.5220/0006236503430350
7. Ehrgott, M., Köksalan, M., Kadziński, M., Deb, K.: Fifty years of multi-objective optimization and decision-making: from mathematical programming to evolutionary computation. Eur. J. Oper. Res. **330**(1), 1–25 (2026). https://doi.org/10.1016/j.ejor.2025.06.012
8. Fischetti, M., Fischetti, M.: Matheuristics. In: Martí, R., Pardalos, P.M., Resende, M.G.C. (eds.) Handbook of Heuristics, pp. 121–153. Springer, Cham (2018). https://doi.org/10.1007/978-3-319-07124-4_14
9. Kaushal, P., Chander, S., Sinha, V.K.: Virtual machine placement in cloud computing: challenges, research Gaps, and future. Int. J. Wirel. Ad Hoc Commun. **3**(2), 64–71 (2021). https://doi.org/10.54216/IJWAC.030202

10. Lust, T., Teghem, J.: The multiobjective multidimensional knapsack problem: a survey and a new approach. Int. Trans. Oper. Res. **19**(4), 495–520 (2012). https://doi.org/10.1111/j.1475-3995.2011.00822.x
11. Pires, F.L., Barán, B.: A virtual machine placement taxonomy. In: 15th IEEE/ACM International Symposium on Cluster, Cloud and Grid Computing (CCGrid 2015), Shenzhen, Guangdong, China (2015). https://doi.org/10.1109/CCGrid.2015.15
12. Ray, S., Singhal, M.: Comprehensive survey on load balancing techniques in cloud computing environments. Int. J. Sci. Res. Sci. Technol. **11**(6), 853–860 (2024). https://doi.org/10.32628/IJSRST24114336
13. Schulte, J., Wetzel, D.: Two-phase matheuristic for assignment and truck loading problems. Eur. J. Oper. Res. **322**(1), 105–120 (2025). https://doi.org/10.1016/j.ejor.2024.10.020
14. Stillwell, M., Schanzenbach, D., Vivien, F., Casanova, H.: Resource allocation algorithms for virtualized service hosting platforms. J. Parallel Distrib. Comput. **70**(9), 962–974 (2010). https://doi.org/10.1016/j.jpdc.2010.05.006
15. Toutov, A., Toutova, N., Vorozhtsov, A., Andreev, I.: Optimizing the migration of virtual machines in cloud data centers. Int. J. Embedded Real-Time Commun. Syst. **13**(1), 1–19 (2022). https://doi.org/10.4018/IJERTCS.289200
16. Xu, F., Liu, F., Jin, H., Vasilakos, A.V.: Managing performance overhead of virtual machines in cloud computing: a survey, state of the art, and future directions. Proc. IEEE **102**(1), 11–31 (2014). https://doi.org/10.1109/JPROC.2013.2287711
17. Zheng, W., Doerr, B.: Approximation Guarantees for the Nondominated Sorting Genetic Algorithm II (NSGA-II). IEEE Trans. Evol. Comput. **29**(3), 891–905 (2025). https://doi.org/10.1109/TEVC.2024.3402996

Author Index

A. Kazakov et al. (Eds.): MOTOR 2026, LNCS 16687, pp. 437–438, 2026.
https://doi.org/10.1007/978-3-032-31390-4

MIX
Papier aus verantwortungsvollen Quellen
Paper from responsible sources
FSC® C105338

If you have any concerns about our products,
you can contact us on
ProductSafety@springernature.com

In case Publisher is established outside the EU,
the EU authorized representative is:
Springer Nature Customer Service Center GmbH
Europaplatz 3, 69115 Heidelberg, Germany

Printed by Libri Plureos GmbH
in Hamburg, Germany